XXII Symposium on Physics of Switching Arc 2017

Nove Mestro na Morave, Czech Republic
4 - 8 September 2017

ISBN: 978-1-5108-9536-2

Printed from e-media with permission by:

Curran Associates, Inc.
57 Morehouse Lane
Red Hook, NY 12571

Some format issues inherent in the e-media version may also appear in this print version.

Brno University of Technology
c/o Milada Bartlová - Department of Physics
Technical 2848/8, Zabovresky, 61600
Brno, Czech Republic

Phone: +420 541 143 431
Fax: +420 541 143 133

bartlova@feec.vutbr.cz

Additional copies of this publication are available from:

Curran Associates, Inc.
57 Morehouse Lane
Red Hook, NY 12571 USA
Phone: 845-758-0400
Fax: 845-758-2633
Email: curran@proceedings.com
Web: www.proceedings.com

TABLE OF CONTENTS

doi:10.14311/ppt.2017.3.203
Plasma Physics and Technology **4**(3):203–212, 2017

ADVANCED NONEQUILIBRIUM MODELLING OF DC TUNGSTEN-INERT GAS ARCS

M. Baeva*, D. Uhrlandt

Leibniz Institute for Plasma Science and Technology, Felix-Hausdorff-Strasse 2, 17489 Greifswald, Germany

* baeva@inp-greifswald.de

Abstract. The paper is concerned with the state-of-the-art nonequilibrium modelling of a DC tungsten-inert gas arc plasma. The advanced description involves the two-way interaction between the plasma and the electrodes. Results in atmospheric pressure argon demonstrating important features of the arc plasma are presented and discussed. First results in the presence of metal vapour released from the molten anode are presented. Outlook for further developments in nonequilibrium arc modelling are discussed.

Keywords: nonequilibrium, arc plasma, magnetohydrodynamic model, diffusion, plasma-electrode interaction.

1. Introduction

Significant progress has been continuously achieved in arc plasma modelling [1]. The unified description of the arc plasma and the electrodes has been an important step in the improvement of the boundary conditions of the plasma solution [2–4]. New models have been developed for atmospheric pressure DC arcs in argon with the simplified configuration of a tungsten-inert gas (TIG) arc. The latter includes a doped tungsten cathode with a conical tip and a flat anode made of copper or steel. The models usually consider axial symmetry and stationary conditions, and neglect metal evaporation. The state-of-the art models are aimed at improving the description in the regions with deviations from local thermodynamic equilibrium (LTE), i.e. the near-electrode regions and the arc fringes. The near-electrode regions are responsible for the arc attachment and the current transfer. The processes in the arc fringes have a strong influence on the arc shape. The models have to be capable of predicting many arc properties: distributions of electric current density and potential, temperatures of electrons and heavy particles, the arc voltage including contributions of the arc column and the near-electrode regions, and the heat loads on the electrodes.

Models solving the nonequilibrium near-electrode regions respectively on the cathode side [5] and on the anode side [6] and applying LTE conditions on the boundary between the equilibrium arc column and the nonequilibrium boundary layer have been developed. They are one-dimensional due to the big computational effort and do not cover the entire arc body because of the high degree of quasi-neutrality in the arc column. Mostly two- but also three-dimensional models of the near-cathode region have been published (see e.g. [7–9]), which apply the concept of a sub-division of the near-electrode regions into a space-charge sheath, a Knudsen layer, an ionization layer, and a layer of thermal nonequilibrium in the direction away from the electrode, and the model proposed by Benilov and Marotta [10]. Such models have been further adopted for a coupling to a model of the arc column. Approaches to arc plasma modelling, which take into account the properties of the near-electrode regions, have been developed so far. Their variety can be divided into three main groups. In the models of the first group, the LTE description of the arc column is combined with a nonequilibrium description of the near-cathode region including all sub-regions (see e.g. [11–14]). These models improve, in general, the arc plasma description since they take into account the contribution of the cathode boundary layer. However, the LTE description of the arc column neglects the electrical conductivity in the arc fringes and necessarily overestimates the current density, the plasma temperature, and the voltage drop in the arc column (see [15] and the references therein). The models of the second group employ ionization equilibrium and thermal nonequilibrium in the description of the arc plasma column. Therefore, the arc plasma column is characterized by two temperatures respectively associated with the electrons and the heavy particles of the plasma. The nonequilibrium boundary layer is described similarly to the models in the first group except that the layer of thermal relaxation is not necessarily required. The cathode boundary layer described according to [10] and coupled to a two-temperature arc plasma within the transfer function concept has been applied in [16]. Further developments have led to the model of nonlinear surface heating [7] which provided boundary conditions for the two-temperature arc plasma model [17], i.e. the cathode boundary layer is assumed independent of the plasma column. The models of the third group provide a fully nonequilibrium description of the arc plasma avoiding the assumptions of both thermal and ionization equilibria (see e.g. [18] and the most recent development [19]). If quasi-neutrality is assumed, the description of the arc plasma column is naturally extended to include the ionization layer. Hence, a description of the boundary layers is related only to the space-charge sheaths adjacent to the electrodes. The Knudsen layer, in which the transition from the ionization layer to the space-charge sheath occurs, is taken into account in the boundary conditions of the hydrodynamic description of the arc column. The models of the third group representing the most advanced modelling of TIG arcs up to date are considered in the present work.

2. General and advanced features of the arc model

An almost complete description of the modelling approach is given in a previous publication [19]. In what follows, a summary will be given outlining its main features and new developments. The modelling approach couples a magneto-hydrodynamic (MHD) simulation of the arc, a description of the arc-electrode interaction, and the heat and current balances of the electrodes. The MHD arc simulation is based on the Navier-Stokes equation system for conservation of mass, momentum, energy, and species. It is coupled with the Maxwell equations, Ohm law and the equation of state. The arc plasma is considered as quasi-neutral and contains electrons, atoms and singly charged ions of argon. Assumptions of thermal and ionization equilibrium are avoided. The MHD approach is valid not only in the arc column but also in the pre-sheath regions of the electrodes and in the arc fringes. Separate energy equations are solved for electrons and heavy particles, which account for a convective transport, the energy flux due to thermal conduction, transport of enthalpy due to diffusive fluxes and thermal diffusion, and gain and loss of energy due to Joule heating, elastic and inelastic collisions, and radiation. Assuming a Maxwellian energy distribution, the electrons are characterized by a temperature T_e while the the atoms and ions are characterized by a common temperature T. A general species equation is solved for the ions, which accounts for the transport processes of convection and diffusion as well as the production and loss due to ionization and recombination. The mass fluxes of the plasma species are expressed in terms of multicomponent diffusion based on the Stefan-Maxwell equations and the effective binary approach [20, 21]. These fluxes contain driving forces resulting from gradients in mass fractions, temperatures and pressure, and drift of charged particles in the electric field. Hence, the electric current density consists of a diffusive component and a drift component and represents the generalized formulation of the Ohm law. The atom density is determined with the equation of state and the Dalton law. The Lorenz force is considered in the momentum conservation equation to take into account the pressure gradient produced by the self-induced magnetic field. New developments of the nonequilibrium model reported in [19] are concerned with a more general representation of the terms caused by thermal diffusion and the transport parameters of the plasma. The latter are calculated in terms of the Chapman-Enskog theory with high order of approximation [22] and allow us an extended description in cases when metal vapour is released from the molten anode. Current and heat conduction are considered in the electrodes with temperature dependent thermal and electrical conductivity of the material. The quasi-neutral plasma and the electrodes are coupled through the tiny regions of the space-charge sheaths (with a thickness of the order of the Debye length) which are treated as dimensionless. The transition from the pre-sheath (included in the description of the arc column) to the space-charge sheath occurs in the Knudsen Layer (with a thickness of the order of the mean free-path of ion-neutral collisions) where the ions are accelerated up to the Bohm speed. The properties of this layer are accounted for in the boundary conditions of the hydrodynamic description [23]. The two-way interaction between the arc plasma and the electrodes is based on the balance of the electron energy and the total energy in the space-charge sheath as formulated in [23]. Contributions of ions and electrons from the plasma and electrons emitted from the metal electrode are taken into account for the continuity of the normal current density and the energy balance. The heat exchange with the electrodes includes the power deposited in the sheath and the cooling due to the transit of electrons from the metal into the plasma, the power released at the surface due to ion recombination, and black body radiation. The two-dimensional distribution of the normal current density and the voltage drop in space-charge sheath, and the arc voltage are obtained self-consistently for a given total arc current.

The modelling approach is applied to study TIG arcs in argon at atmospheric pressure. The schematic of the general arrangement is shown in figure 1 to the half of the arc geometry since axial symmetry is assumed. The cathode is a cylindrical rod made of ceriated tungsten (WC20) with a radius of 2 mm and various shapes of the tip. The anode is a plate made of steel or copper and is perpendicular to the arc axis. The inter-electrode distance d_{ca} is changed from 5 mm up to 120 mm to simulate cases of interest. Hence, the same physical model is used to study the arc column and the near-electrode regions for various arc lengths and operating conditions. The model is realized on the computational platform COMSOL Multyphysics [24] which is based on the Finite Element Method (FEM) with a computational domain of 5×10^4–9×10^4 mesh elements. Appropriate boundary conditions completing the model are presented in [19, 23]. All dependent variables have zero gradients on the axis of symmetry. The temperature of heavy particles approaches the surface temperature of the walls. The temperature of the gas inflow and nozzle/constrictor tube walls is set to 300 K and that at the base of the cathode is 500 K. At the bottom of the anode, the temperature

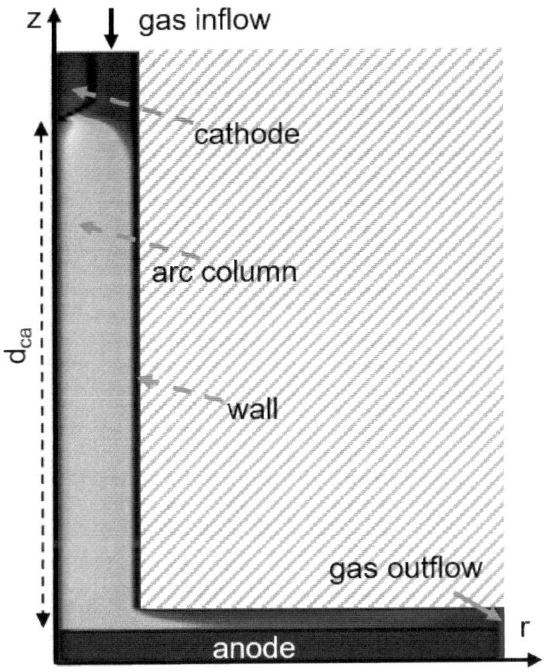

Figure 1. The general schematic view of the arc arrangement.

is set to 300 K to simulate a water cooling or a heat flux coefficient of 50 $\mathrm{Wm^{-2}K^{-1}}$ for an unforced air cooling is set. The outer boundaries are electrically and magnetically insulated except the cathode base and the anode bottom on which electric potential is specified (the anode is grounded).

3. Results and discussion

The results presented in this section are obtained under specific requirements depending on the focus of interest. Therefore, some deviation from the general conditions mentioned above can occur. These deviations are specified with the corresponding case of investigation - effects related to the cathode shape, the arc length, the processes in the anode region, and the presence of metal vapours in the arc plasma.

3.1. Effects of the cathode tip shape

The arc plasma parameters depend among others on the geometry of the cathode. Measurements of the plasma temperature concerned with the cathode shape have been reported by Haidar and Farmer [25], who have studied the plasma temperature (LTE conditions assumed) near the conical shaped cathode tip and the cathode surface temperature for cone angles from 12° to 150°. In these experiments, the dc arc was operated in argon at atmospheric pressure between a cathode made of thoriated tungsten and a copper plate (both electrodes were water-cooled). The measurements showed for all angles a maximum of the plasma temperature near the cathode tip. The plasma temperature was largest for a cone angle of 60°. The results were qualitatively explained by the interplay between the heat mechanisms of conduction, convection and radiation, and the thermionic cooling of the cathode. However, no modelling has been carried out to provide a detailed analysis.

We applied the advanced nonequilibrium model of TIG arcs for several shapes of the cathode tip, these are a cone with an angle of 30° and 60°, a hemisphere, and a flat truncation of the cylindrical rod. The conical tips are considered with a plateau radius of 0.2 mm and a fillet on the edge with a radius of curvature of 0.1 mm. The inter-electrode distance is $d_{\mathrm{ca}} = 5\,\mathrm{mm}$, the arc current is 200 A, and the flow rate is 15 slpm. The stick-off distance, i.e. the axial distance between the cathode tip and the end of the nozzle wall, is 4.5 mm.

Figures 2 and 3 show the radial profiles of the temperatures of electrons and heavy particles along a line-out at axial distance from the cathode tip of respectively 1 mm and 3 mm. For all shapes of the cathode tip, the arc plasma is in thermal equilibrium for radial positions up to approximately 2.5 mm (Fig. 2) and 4 mm (Fig. 3). Closer to the cathode tip, the common temperature is largest for a cone angle of 60° and smallest for the flat tip of the cathode with a maximum difference of about 10000 K. The variation of the plasma temperature becomes smaller as the axial distance from the cathode tip increases (Fig. 3). At an axial distance of 3 mm, the temperature variation with the cone angle is insignificant. This trend appears in the radial profiles of the magnitude of the current density (Fig. 4) so that closer to the cathode tip the current density is a bit higher for a cone angle of 60° in comparison to that for 30° and for the rounded tip compared to the flat one. Moving towards the anode, the variation in the radial profiles for the conical tips, on the one hand, and for the rounded and the flat tip, on the other hand, becomes very small.

The modelling results confirm the experimental findings by Haidar and Farmer [25] concerning the higher plasma temperature near the tip for a cone angle of 60° than for 30°. In order to analyse the results, we consider the conditions on the cathode surface, in the cathode bulk, and close the cathode tip in the plasma. Figure 5 shows the distributions of the temperature (lines with solid symbols) and the normal current density (lines with open symbols) on the cathode surface as a function of the distance measured from the centre of the cathode tip along the cathode surface. The distance of 0.2 mm corresponds to the radius of the plateau, which edge is rounded and forms the cathode tip. The conical part of the cathode differs in its length for both cone angles. Although the tip ends are of equal area, the normal current density and the surface temperature are slightly higher for a cone angle of 30°. The current density and Ohmic heating in the cathode bulk are presented respectively by means of arrow indicators and iso-lines with la-

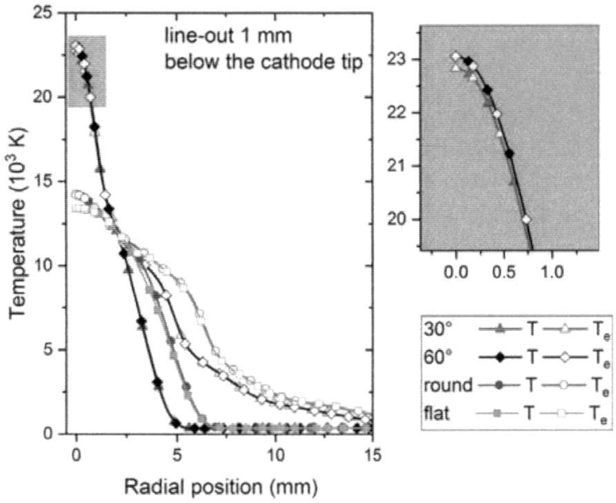

Figure 2. Radial profiles of the temperatures of electrons (lines with open symbols) and heavy particles (lines with solid symbols) for various shapes of the cathode tip along a line-out with an axial distance from the cathode tip of 1 mm.

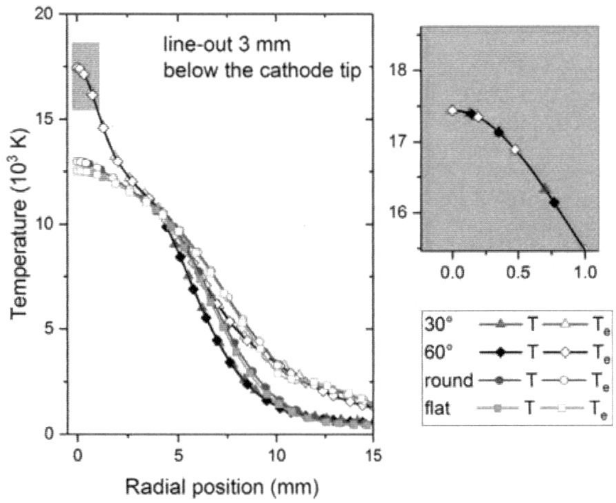

Figure 3. Radial profiles of the temperatures of electrons (lines with open symbols) and heavy particles (lines with solid symbols) for various shapes of the cathode tip along a line-out with an axial distance from the cathode tip of 3 mm.

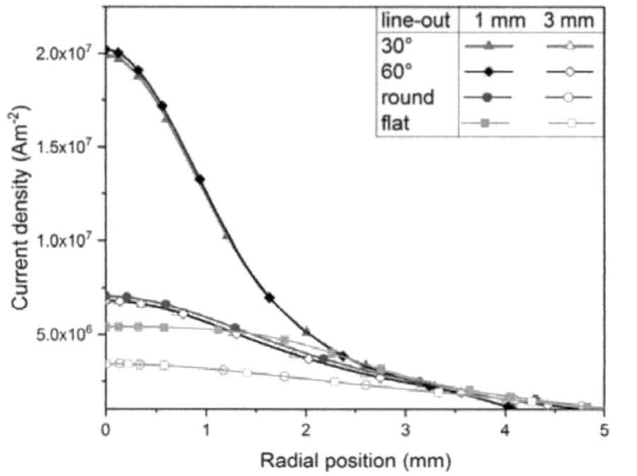

Figure 4. Radial profiles of the magnitude of the current density for various shapes of the cathode tip along a line-out with an axial distance from the cathode tip of 1 mm (lines with solid symbols) and 3 mm (lines with open symbols).

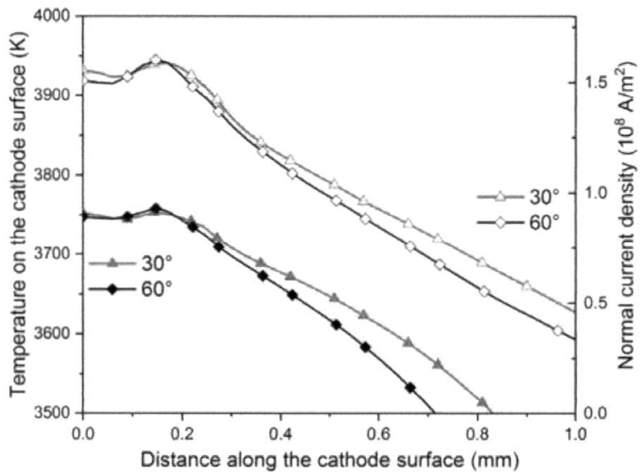

Figure 5. The temperature (solid symbols) and the normal current density (open symbols) on the cathode surface for a cone angle of 30° (triangles) and 60° (diamonds). The distance is measured from the centre of the cathode tip.

bels in units of $10^{10}\,\mathrm{W/m^3}$ in Figure 6. The current density and the ohmic heating are higher in the cathode bulk for a cone angle of 30° due to the smaller cross-section area. Well pronounced maxima near the inclined side of the cathode can be seen in both cases though at different axial positions. The higher current density and ohmic heating in the cathode bulk for the 30°-conical tip lead to the higher temperature of the cathode. However, in the plasma region close to the cathode tip, the conditions change(Figures 2 and 4). Figure 7 shows the terms of the conservation equation of electron energy (see e.g. Eq. (3) [28]) along a radial line-out with an axial distance from the cathode tip of 1 mm. The equation can be written into the form

$$\nabla \cdot \left(\frac{5}{2} n_e k_B T_e \mathbf{u}\right) + \nabla \cdot \mathbf{J}_\varepsilon + \frac{e}{m_e} \mathbf{J}_e \cdot \mathbf{E} + Q^{el}_{e-n} + Q^{in}_{e-h} + Q_{rad} = 0. \qquad (1)$$

In Equation (1), n_e, T_e, e and m_e denote respectively the number density, the temperature, the electric charge and the mass of electrons, $\mathbf{J}_\varepsilon$ is the density of the electron energy flux, $\mathbf{J}_e$ is the mass flux of electrons, and $\mathbf{E}$ is the electric field. k_B is the Botzmann constant. The first term in Equation (1) describing the change of electron energy change due to the enthalpy flux is denoted in Figure 7 as Q_{conv}. The second term represents the divergence of the density of the electron energy flux ($Q_{\mathrm{div} J_\varepsilon}$ in Figure 7). The third term is the ohmic heating of the electrons by the electric field (Q_{ohm}). The last three terms represent respectively

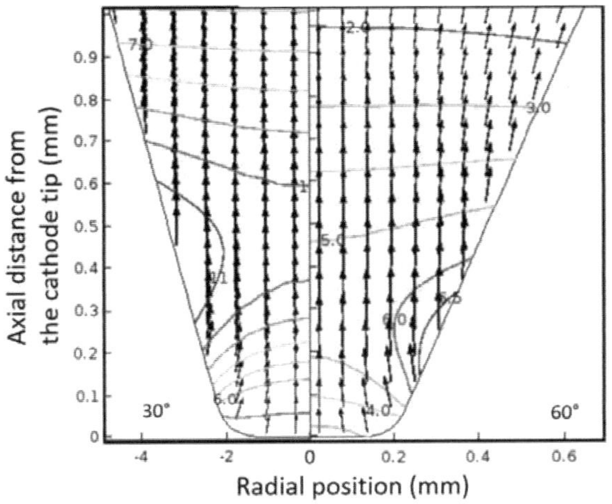

Figure 6. The current density (arrow indicators) and ohmic heating in units of 10^{10} W/m^3 (iso-lines) in the conical part of the cathode for cone angles of 30° and 60°.

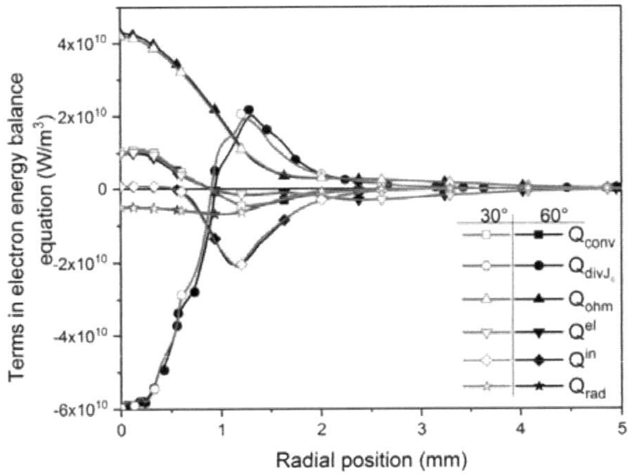

Figure 7. Terms of the electron energy balance for a cone angle of 30° (open symbols) and 60° (solid symbols) along a radial line-out in axial distance from the cathode tip 1 mm.

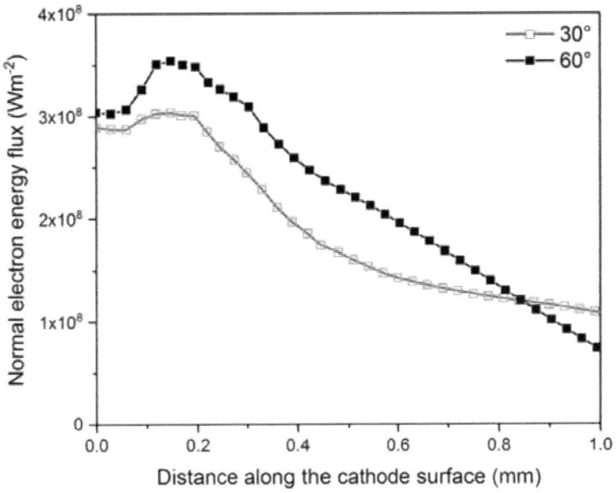

Figure 8. Normal component of the density of electron energy flux at the cathode for cone angles of 30° (open symbols) and 60° (solid symbols)

the energy change due to elastic (Q^{el}) and inelastic collisions (Q^{in}) between electrons and heavy particles, and radiation (Q_{rad}). In general, the results show very similar behaviour and values of the terms for both cone angles. Close to the axis, where small differences in the current density and temperature occur (Figures 2 and 4), the value of Q_{ohm} is slightly higher for the 60°-cathode. $Q_{\mathrm{div}J_\varepsilon}$ is the dominant term in Equation (1) and it ensures the energy balance by the transport of a local energy excess. The physical picture is very complex because of the and non-linear nature of the problem and the explanation of the temperature maximum for a cone angle of 60° seems to be much more sophisticated than the considerations suggested in [25]. A plausible explanation is provided by the two-directional plasma-electrode interaction. The detailed analysis of the heat load to the conical cathode of TIG arcs presented in [28] (see Figure 10 and Equation (29) therein) shows that a huge amount of the electric power is transferred from the space-charge sheath to the plasma electrons. This energy amount is less for a cone angle of 30° as the results in Figure 8 show.

3.2. Effects due to arc length variation

A variation of the arc length (the inter-electrode distance) touches on various important issues, e.g. industrial applications, plasma diagnostics, theoretical description and modelling of arc plasmas. Arc length variations occur in configurations with moving electrodes and in experiments concerned with the determination of the electric field strength among others. We consider in this part the impact of the arc length on the arc plasma.

Figure 9 presents the arc voltage as a function of the inter-electrode distance for an arc current of 100 A. The results are obtained with the nonequilibrium model and from experiments [26], in which the anode is a water cooled copper plate and the cathode has a conical tip of 60° and is made of ceriated tungsten (WC20). The slope of the nearly linear dependence of the arc voltage on the arc length at inter-electrode distances larger than 3 mm is used to estimate an average electric field of about 400 V/m. The non-linearity becomes stronger for distances below 3 mm. The extrapolation of this dependence to zero arc length provides the sum of the voltage drops in the space-charge sheaths, the cathode, and the anode. In this sum of about 8 V, the biggest contribution is assigned to the average voltage drop in the cathode space-charge sheath.

The two-dimensional distribution of the electron density for arc lengths of 2, 6 and 10 mm is shown in Figure 9. It indicates that a decrease of the arc length from 10 mm to 6 mm is related to a shortening of the plasma column in front of the anode. The region in

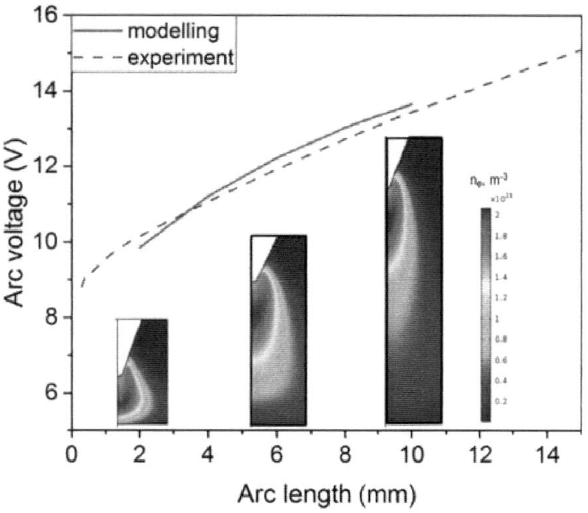

Figure 9. The arc voltage as a function of the arc length at arc current of 100 A. The two-dimensional distributions of the electron density for arc lengths of 2, 6, and 10 mm are shown to illustrate the arc structure.

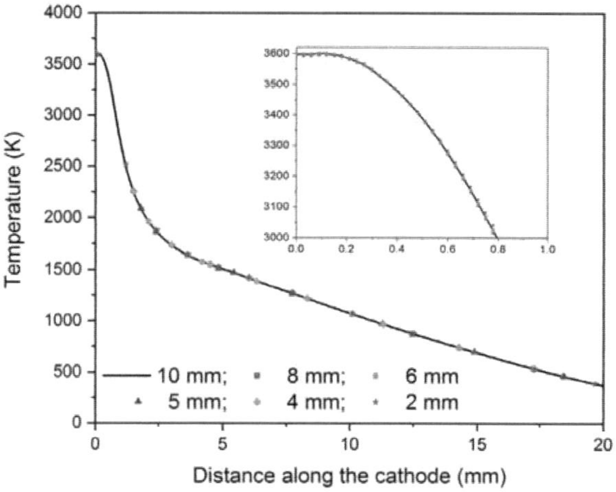

Figure 10. The temperature on the cathode surface as a function of the distance along the cathode. The distance is measured from the tip centre. The solid line presents the results for an arc length of 10 mm. Symbols present the results for shorter arc lengths.

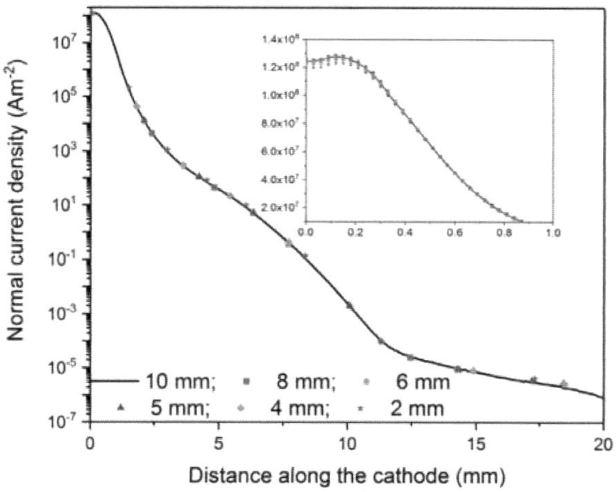

Figure 11. The normal current density on the cathode surface as a function of the distance along the cathode. The distance is measured from the tip centre. The notations are the same as in Fig. 10.

front of the cathode tip almost does not change. This effect results in the nearly linear dependence of the arc voltage on the arc length. The further decrease of the arc length, however, causes modifications in the arc structure. The hot arc core tends to occupy a great part of the inter-electrode gap and expands radially. As a result, the arc voltage decreases non-linearly with the arc length.

Figures 10, 11, and 12 present respectively the temperature on the cathode surface, the normal current density and the voltage drop in the cathode space-charge sheath for various arc lengths and an arc current of 100 A. The results obtained show a weak dependence on the arc length, i.e. the conditions in the cathode boundary layer are weakly affected by the rest of the arc. The temperature of the cathode, the normal current density, and the voltage drop decrease in the central part of the cathode tip for shorter arc lengths (see also the enlarged view of this area). The temperature and the normal current density slightly increase towards the end of the cathode for shorter arc lengths whereas the voltage drop in the space-charge sheath is consistently lower along the entire distance for shorter arc lengths. Moreover, the difference becomes much more pronounced. At distances beyond about 11.5 mm, the ion component of the normal current density exceeds the emission current component. Towards the end of the cathode, the ion current component is the dominant one. The voltage drop in the space-charge sheath is strong enough to suppress electrons diffusing from the plasma.

Models describing the cathode boundary layer, which includes the ionization layer and the space-charge sheath, are based on the assumption that the plasma-cathode interaction is independent of processes in the arc column [10, 27]. The weak dependence of the temperature and the normal current density on the arc length are likely to confirm this assumption. A detailed comparison of results for TIG arcs obtained by the advanced nonequilibrium model [19] and a model of the cathode boundary layer [7] shows a qualitative agreement in the behaviour of the temperature and the normal current density on the cathode surface [28]. Nevertheless, they differ since the plasma parameters obtained by means of the nonequilibrium model certainly differ from the integral values evaluated only in the cathode boundary layer.

3.3. Effects in the anode region

In this part, we discuss the influence of the conditions in the region between the tube end (1) and the anode surface on the arc attachment on the anode. The region has an axial length of 10 mm. Fig. 13 shows

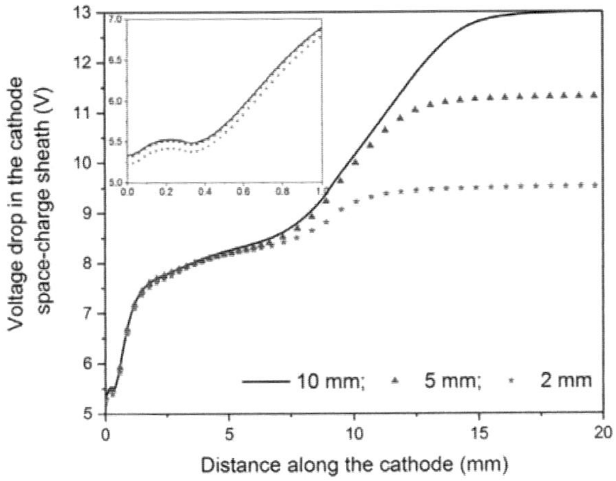

Figure 12. The voltage drop in the cathode space-charge sheath as a function of the distance along the cathode for arc lengths of 10 mm (solid line), 5 mm (triangles), and 2 mm (stars). The distance is measured from the tip centre.

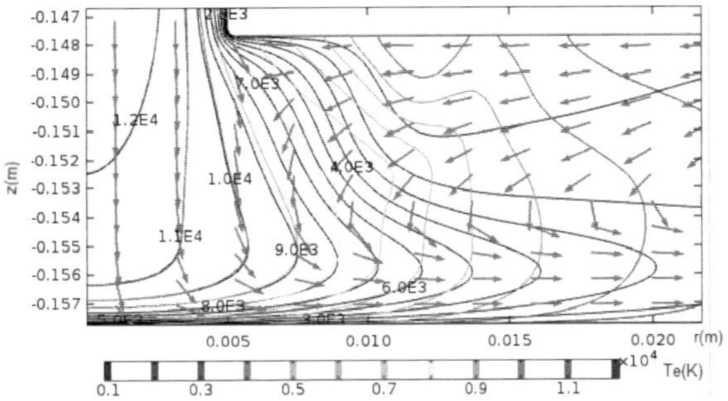

Figure 13. Iso-lines of temperatures in kelvin of heavy particles (black) and electrons (coloured) and normalized arrow indicators of the gas flow in the anode region of the arc at a gas flow rate of 4.5 slpm and an arc current of 200 A.

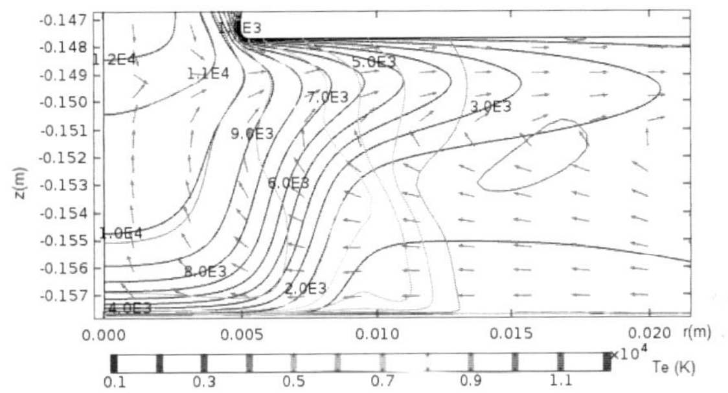

Figure 14. Iso-lines of temperatures in kelvin of heavy particles (black) and electrons (coloured) and normalized arrow indicators of the gas flow in the anode region of the arc at a gas flow rate of 1.3 slpm and an arc current of 200 A.

the two-dimensional distribution of the electron and heavy particle temperatures, and the flow direction obtained for an arc current of 200 A and a gas flow rate of 4.5 slpm. The hot arc region which is enclosed by the iso-line of 11000 K is in thermal equilibrium ($T = T_e$). A departure from thermal equilibrium occurs for temperatures below 10000 K. The hot gas flow impinging on the anode surface is directed towards the open boundary. A cold gas flow enters the upper patch of the open boundary and moves along the bottom tube edge towards the arc. The cathode jet dominates the arc and leads to the bell-shape formation which is typical for the diffuse arc attachment on the anode [29]. A completely different situation appears after reducing the gas flow rate to 1.3 slpm (Fig. 14). The hot arc region is significantly shorted, an anode jet is built and collides with the cathode jet approximately 1.5 mm below the tube edge. The hot gas flows along the bottom tube edge towards the open boundary, while the cold gas flows towards the arc. This causes a contraction of the bottom part of the arc so that the arc attachment on the anode becomes constricted. The results confirm the experimental findings in [30] and the modelling predictions in [31]. The role of macroscopic flows for the formation of a constricted arc attachment has been underlined in the overview of the processes and conditions in the anode region given in [29]. Not only the extension of the region of thermal nonequilibrium but also different behaviour related to the ionization nonequilibrium have been recently predicted [32]. In the diffuse mode, the nonequilibrium electron density exceeds the equilibrium value over the entire length of the anode region. In the constricted mode, this behaviour is observed till the stagnation position. Beyond this position, the nonequilibrium electron density is lower than the equilibrium one. Fig. 15 presents the axial flow velocity and the modulus of the axial electric field $|E_z|$ over the distance to the anode. The regions of respectively the cathode and the anode jets are indicated. At the position of stagnation, $|E_z|$ exhibits a local minimum in the constricted arc. In the vicinity of the anode, the electric field in both modes undergoes a field reversal (corresponding to a negative anode fall in the electric potential). However, the position of the field reversal is closer to the anode and the potential drop amounts approximately 1 V in the constricted mode. In the diffuse mode, the potential drop is about 3 V.

3.4. Effects due to the presence of metal vapour

Metal atoms, in general, have many excited levels below the ionization potential of the shielding gas and are therefore more likely to be excited and ionized. The radiative emission and the electric conductivity of the plasma is therefore expected to increase in the presence of metal vapours. It is important to take into account the presence of metal vapours for a better understanding of TIG welding arcs. The influence of the metal vapour on the arc characteristics has been studied over the course of many years in experimental and modelling works arriving at contradictory results. On the one hand, the model of a free-burning arc

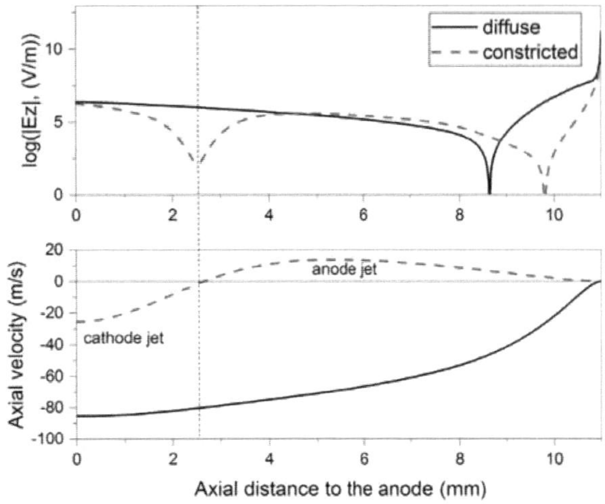

Figure 15. The flow velocity and the modulus of the electric field along the arc axis for the diffuse mode (solid line) and the constricted mode (dashed line) of arc attachment on the anode. Arc current 200 A.

by Gonzalez et al [33], which was based on the LTE assumption and assumed a temperature of 1000 K at the bottom surface of the anode, predicted a mass concentration of 9 % at the centre of the upper surface of the anode and 1 % at a distance of 10 mm from the anode surface for an arc current of 200 A and an inter-electrode distance of 10 mm. In this simulation, the temperature of the plasma column decreased significantly due to the increasing radiation losses. The effect led to a reduction of the heat flux to the anode and to a shrinking of the weld pool. On the other hand, the experimental studies in a free-burning arc with an inter-electrode distance of 5 mm and an arc current of 200 A by Farmer et al [34] showed a minor effect on the temperature in the presence of metal vapours released from the molten stainless-steel anode. Fe and Cr atoms were observed within a distance of 0.5 mm from the anode. More recent modelling results by Mougenot et al [35] at an arc current of 200 A showed a mass concentration of metal vapour of respectively 0.6 % at the anode centre and 0.1 % at a distance from the anode of 1 mm. Tanaka et al [36] studied an arc in argon at a current of 150 A and an inter-electrode distance of 5 mm and reported a maximum value of about 0.25 mol%. The models in [35, 36] are transient and developed respectively in 3D and 2D assuming LTE conditions. The main advance in these works was that flow in the molten weld pool was considered to provide a more realistic temperature at the upper anode surface. The temperature of the bottom anode surface was set at 300 K. A mass conservation equation is solved for the iron vapour and the viscosity approximation was used to express the diffusion coefficient.

The nonequilibrium model [19] has recently been extended to gas mixtures containing metal vapour. The general diffusion formulation includes atoms and singly charged ions of iron additionally to those of argon as the shielding gas. The concentration of iron vapour on the anode surface is used as a boundary condition. It is determined by the saturation vapour pressure (related to the temperature at the anode surface) and the ambient pressure [37]. The iron vapour pressure increases rapidly as the temperature at the anode surface approaches the boiling point of the material. The anode surface is assumed to be non-deformable and flat. A flow in the weld pool is not taken into account in the model at the current stage. The temperature at the bottom surface of the anode follows from the heat flux of unforced air cooling.

The two-dimensional distribution of the mass fraction of iron atoms and ions is shown respectively in Fig. 16 and Fig. 17 for an arc current of 150 A and an inter-electrode distance of 5 mm. The maximum mass fraction of Fe atoms and ions in the centre of the anode plate is about 0.6 %, which is close to the value reported in [35] at a higher arc current. The maximum temperature at the anode surface obtained with the nonequilibrium model is about 2050 K (Fig. 18). This value is lower than the value of 2200 K reported in [36] for an arc current of 150 A and that of 2310 K reported in [35] for an arc current of 200 A. The different results concerning the amount of metal vapours in [35, 36] can be probably explained by the different arc current values since both models consider the flow in the liquid weldpool and similar boundary conditions although the thermodynamic and transport properties applied in the models can differ. The main differences of the models [35, 36] to the nonequilibrium model is, however, the description of the arc plasma and the plasma-electrode interaction. Since no recirculations of the fluid flow can be seen in the region with iron atoms and ions (see the arrow indicators in Fig. 18), the peculiar structures shown in Figures 16 and 17 are the effects of species diffusion. The Fe atoms diffusing into the arc periphery are effectively ionized. At a distance of about 0.5 mm away from the anode surface, the mass fraction of Fe ions exceeds that of Fe atoms. The spatial distributions of Fe atoms and ions in the arc region point at off-axis maxima. The Fe atoms follow the fluid flow far away from the arc core towards the outlet, while the mass fraction Fe ions rapidly decreases. The appearance of an off-axis maximum in the distribution of Fe metal vapour was reported also in [35].

The results of the nonequilibrium model confirm the weak influence of iron metal vapours for a mass concentration up to 0.6 % and argon shielding gas reported in [35]. A stronger influence has been observed in helium arc plasma [36, 38] as the iron vapour has been found to increase the electrical conductivity at low temperatures. The decrease in the electron and heavy particle temperatures on the arc axis in the presence of iron vapours in comparison to the values in a pure argon plasma is almost negligible. Also, quite small is the increase in the voltage drop over the

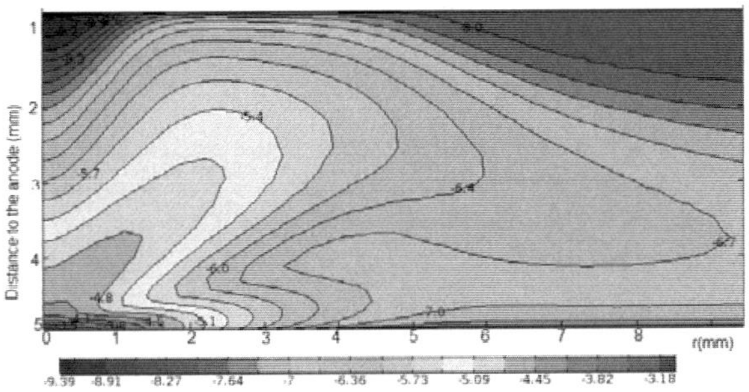

Figure 16. Mass fraction of iron atoms in the arc plasma at an arc current of 150 A and an inter-electrode distance of 5 mm. The presentation is in a logarithmic scale.

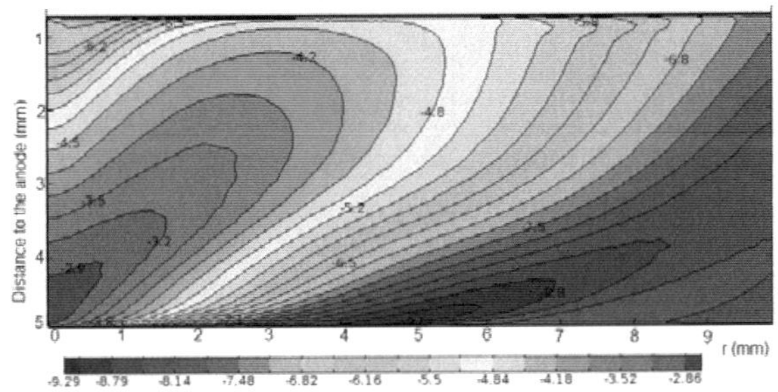

Figure 17. Mass fraction of iron ions in the arc plasma at an arc current of 150 A and an inter-electrode distance of 5 mm. The presentation is in logarithmic scale. The presentation is in a logarithmic scale.

arc column due to the increased radiative emission and the corresponding arc cooling.

Although the nonequilibrium model already significantly improves the description of the arc plasma and its interaction with the electrodes, further developments such as a proper treatment of the liquid metal phase, the deformation of the anode surface, and the motion of the arc relative to the anode are needed to extend its predictive capability to real welding applications. From a physical point of view, studies concerning the role of metastable states of the atoms of the shielding gas on the excitation of the metal ions will provide the answer whether the metal ions are in thermal equilibrium with the other heavy particles.

4. Conclusions

In this work, the state-of-the-art nonequilibrium modelling of DC tungsten-inert gas arcs is presented. One and the same physical model is applied to outline the effects of the cathode tip shape, the variation of the arc length, the formation of diffuse and constricted arc attachment modes on the anode, and the presence of metal vapour in atmospheric pressure argon arcs. The results are analyzed and discussed in relation to previously published experimental findings and modelling predictions.

It is shown that the arc plasma parameters strongly depend on the geometry of the cathode tip. The results obtained for a 60°-cathode and an arc current of 100 A show a weak dependence of the conditions near the cathode on the arc length in the range of inter-electrode distances 2–10 mm.

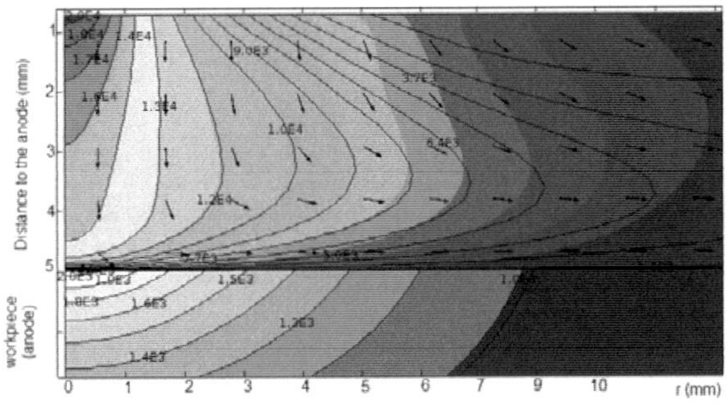

Figure 18. Temperature in kelvin of heavy particles (black lines) and electrons (filled coloured contours) in the arc region and in the workpiece. The flow field is presented by arrow indicators. Arc current 150 A, inter-electrode distance 5 mm.

The formation of a constricted arc attachment on the anode is discussed in relation to the macroscopic flow conditions and the thermal and ionization nonequilibria occuring in the anode region.

For an anode made of iron, a maximum mass fraction of 0.6 % of iron atom and ions is predicted in the centre of the anode plate for an arc current of 150 A. The presence of metal vapour in this concentration causes a small decrease in the electron and heavy particle temperatures on the arc axis and a small decrease in the voltage drop over the arc column in comparison to the values in a pure argon plasma.

Acknowledgements

This work was supported by the DFG (German Science Foundation) under Grant UH106/11-1.

References

[1] A. B. Murphy. A perspective on arc welding research: The importance of the arc, unresolved questions and future directions. *Plasma Chemistry and Plasma Processing*, 35:471–489, 2015. doi:10.1007/s11090-015-9620-2

[2] J. J. Lowke and A. B. Murphy. Plasma flows. In R. W. Johnson, editor, *The handbook of fluid dynamics*, chapter 15. CRC Press, Boca Raton, 1998.

[3] M. S. Benilov. Understanding and modelling plasma-electrode interaction in high-pressure arc discharges: a review. *J. Phys. D: Appl. Phys.*, 41(14):144001 (30pp), 2008. doi:10.1088/0022-3727/41/14/144001

[4] J. Mentel and J. Heberlein. The anode region of low current arcs in high intensity discharge lamps. *J. Phys. D: Appl. Phys.*, 43:023002, 2010. doi:10.1088/0022-3727/43/2/023002

[5] N. A. Almeida, M. S. Benilov, and G. V. Naidis. Unified modelling of near-cathode plasma layers in high-pressure arc discharges. *J. Phys. D: Appl. Phys.*, 41:245201, 2008. doi:10.1088/0022-3727/41/24/245201

[6] I. L. Semenov, I. V. Krivtsun, and U. Reisgen. Numerical study of the anode boundary layer in atmospheric pressure arc discharges. *J. Phys. D: Appl. Phys.*, 49:105204, 2016. doi:10.1088/0022-3727/49/10/105204

[7] M. S. Benilov and M. Cunha. Heating of refractory cathodes by high-pressure arc plasmas I. *J. Phys. D: Appl. Phys.*, 35:1736–1750, 2002. doi:10.1088/0022-3727/35/14/314

[8] M. S. Benilov, M. Carpaij, and M. D. Cunha. 3D modelling of heating of thermionic cathodes by high-pressure arc plasmas. *J. Phys. D: Appl. Phys.*, 39:2124–2134, 2006. doi:10.1088/0022-3727/39/10/024

[9] F. Cayla, P. Freton, and J.-J. Gonzalez. Arc/cathode interaction model. *IEEE Transac. Plasma Sci.*, 36(4):1944–1953, 2008. doi:10.1109/TPS.2008.927378

[10] M. S. Benilov and A. Marotta. A model of the cathode region of atmospheric pressure arcs. *J. Phys. D: Appl. Phys.*, 28:1869–1882, 1995. doi:10.1088/0022-3727/28/9/015

[11] J. Wendelstorf. Ab initio modelling of thermal plasma gas discharges (electric acrs). Ph.D. Thesis, Univ. Braunschweig, Germany, 2000.

[12] J. J. Gonzalez et al. Two-dimensional self-consistent modelling of the arc/cathode interaction. *J. Phys. D: Appl. Phys.*, 42:145204, 2009. doi:10.1088/0022-3727/42/14/145204

[13] S. Gorchakov et al. Nonequilibrium arc model for the description of arc-electrode interaction. *Proceedings 27th ICEC*, 22-26 June Dresden, Germany, VDE Verlag, 2014.

[14] J. Shirvan and I. Choquet. Gtaw process - a review of cathode plasma coupling modelling. *Welding in the World*, 60:821835, 2016.

[15] M. Baeva. Thermal and chemical nonequilibrium effects in free-burning arcs. *Plasma Chemistry and Plasma Processing*, 36:151–167, 2016. doi:10.1007/s11090-015-9650-9

[16] J. Wendelstorf. Two-temperature, two-dimensional modelling of cathode-plasma interaction in electric arcs. *Proceedings 24th ICPIG*, 11-16 July Warsaw, Poland, v. 2 p. 227, 1999.

[17] H.-P. Li and M. S. Benilov. Effect of a near-cathode sheath on heat transfer in high-pressure arc plasmas. *J. Phys. D: Appl. Phys.*, 40:2010–2017, 2007. doi:10.1088/0022-3727/40/7/024

[18] M. Baeva et al. Two-temperature chemically non-equilibrium modelling of transferred arcs. *Plasma Sources Sci. Technol.*, 21:055027, 2012. doi:10.1088/0963-0252/21/5/055027

[19] M. Baeva et al. Novel non-equilibrium modelling of a DC electric arc in argon. *J. Phys. D: Appl. Phys.*, 49:245205, 2016. doi:10.1088/0022-3727/49/24/245205

[20] V. M. Zhdanov. *Transport phenomena in multicomponent plasma.* London: Taylor and Francis, 2002.

[21] J. D. Ramshaw. Self-consistent effective binary diffusion in multicomponent gas mixtures. *J. Non-Equilib. Thermodyn.*, 15:295–300, 1990. doi:10.1515/jnet.1990.15.3.295

[22] V. Rat et al. Treatment of non-equilibrium phenomena in thermal plasma flows. *J. Phys. D: Appl. Phys.*, 41:183001, 2008. doi:10.1088/0022-3727/41/18/183001

[23] M. S. Benilov et al. Account of near-cathode sheath in numerical models of high-pressure arc discharges. *J. Phys. D: Appl. Phys.*, 49:215201, 2016. doi:10.1088/0022-3727/49/21/215201

[24] COMSOL Multiphysics v.5.2a (Stockholm: COMSOL) http://www.comsol.com

[25] J. Haidar and A. J. D. Farmer. Large effects of cathode shape on plasma temperature in high-current free-burning arcs. *J. Phys. D: Appl. Phys.*, 27:555–560, 1994. doi:10.1088/0022-3727/27/3/019

[26] M. Baeva, E. Siewert, and D. Uhrlandt. Electric field and voltage of tig arcs from non-equilibrium modeling and experiment. *Proceedings Gas Discharges and Their Applications*, 11-16 September Nagoya, Japan, vol. 1, pp. 73-76, 2016.

[27] M. S. Benilov. Theory of nonlinear surface heating. *Phys. Scripta*, T84:22–46, 2000. doi:10.1238/Physica.Topical.084a00022

[28] M. Baeva. Non-equilibrium modeling of tungsten-inert gas arcs. *Plasma Chem. Plasma Process.*, 37:341–370, 2017. doi:10.1007/s11090-017-9785-y

[29] J. Heberlein, J. Mentel, and E. Pfender. The anode region of electric arcs: a survey. *J. Phys. D: Appl. Phys.*, 43:023001, 2010. doi:10.1088/0022-3727/43/2/023001

[30] R. V. Hartmann and J. V. Heberlein. Quantitative investigations on arc-anode attachments in transferred arcs. *J. Phys. D: Appl. Phys.*, 34:2972–2978, 2001. doi:10.1088/0022-3727/34/19/307

[31] T. Amakawa et al. Anode-boundary-layer behaviour in a transferred, high-intensity arc. *J. Phys. D: Appl. Phys.*, 31:2826–2834, 1998. doi:10.1088/0022-3727/31/20/017

[32] M. Baeva. A survey of chemical nonequilibrium in argon arc plasma. *Plasma Chem. Plasma Process.*, 37:513–530, 2017. doi:10.1007/s11090-016-9778-2

[33] J.-J. Gonzalez et al. Mathematical modeling of a free-burning arc in the presence of metal vapor. *J. Appl. Phys.*, 74(5):3065–3070, 1993. doi:10.1063/1.354624

[34] A. J. D. Farmer, G. N. Haddad, and L. E. Cram. Temperature determinations in a free-burning arc: III. measurements with molten anodes. *J. Phys. D: Appl. Phys.*, 19:1723–1730, 1986. doi:10.1088/0022-3727/19/9/016

[35] J. Mougenot et al. Plasma-weld pool interaction in tungsten-inert gas configuration. *J. Phys. D: Appl. Phys.*, 46:135206, 2013. doi:10.1088/0022-3727/46/13/135206

[36] M. Tanaka et al. Time-dependent calculations of molten pool formation and thermal plasma with metal vapour in gas tungsten arc welding. *J. Phys. D: Appl. Phys.*, 43:434009, 2010. doi:10.1088/0022-3727/43/43/434009

[37] G. Lago et al. A numerical modelling of an electric arc and its interaction with the anode: Part i. the two-dimensional model. *J. Phys. D: Appl. Phys.*, 37:883–897, 2004. doi:10.1088/0022-3727/37/6/013

[38] A. B. Murphy et al. Modelling of thermal plasmas for arc welding: the role of the shielding gas properties and of metal vapour. *J. Phys. D: Appl. Phys.*, 42:194006, 2009. doi:10.1088/0022-3727/42/19/194006

doi:10.14311/ppt.2017.3.213
Plasma Physics and Technology **4**(3):213–224, 2017

MODELLING AND SIMULATION OF SF_6 HIGH-VOLTAGE CIRCUIT-BREAKERS - AN OVERVIEW ON BASICS AND APPLICATION OF CFD ARC SIMULATION TOOLS

F. Reichert[a,*], A. Petchanka[a], P. Freton[b], J.J. Gonzalez[b]

[a] *Siemens AG, Nonnendammallee 104, 13629 Berlin, Germany*
[b] *University Paul Sabatier, LAPLACE, 118 route de Narbonne, CNRS-UPS, 31062 Toulouse, France*
[*] frank.reichert@siemens.com

Abstract. The paper gives an overview on the basics of CFD arc simulation tools with respect to the simulation of the fluid mechanical processes in the interrupter unit of SF_6 high-voltage circuit-breakers at no-load and short-circuit switching-off processes. On the example of SF_6 self-blast circuit-breakers the complete process from the analysis of the switching-off process to the creation of a modular simulation model consisting of several sub models is illustrated. Details to the modelling in the particular sub modules and to the implementation are given. The capability of a CFD arc simulation tool based on the program package ANSYS/FLUENT is demonstrated on the basis of selected simulation results. Furthermore case examples for the application of the presented CFD arc simulation tool in the development process of high-voltage circuit-breakers are given.

Keywords: high-voltage circuit-breaker, CFD arc simulation tool, modelling, application.

1. Introduction

High-voltage circuit-breakers (HVCB) are exposed to several loads during operation. Figure 1 shows selected loads and the resulting stresses connected with high-voltage circuit-breakers [1]. The load by a continuous current leads due to the Ohmic heating of the current paths to a temperature rise and thus to a thermal stress. A short-circuit current leads to thermal stresses too and to mechanical stresses due to the high electromagnetic forces. Applying a voltage leads to a dielectric stress. Mechanical stresses can be also generated by loads at the terminals or seismic activities.

Loads		Stresses	Expected results from simulation
Continuous current		Thermal	
Short-time current		Thermal, Mechanical	
Voltage		Dielectrical	
Mechanics		Mechanical	
Switching	Switching-off of short-circuit currents (Terminal fault, Out-of-phase, Short-line fault)	Dielectrical, Mechanical, Thermal	Gas behaviour (pressure build-up, gas temperature and density), Switching capacity
	Switching of capacitive currents	Dielectrical	⇧
	Switching of small inductive currents	Dielectrical	CFD arc simulation tool

Figure 1. Loads, stresses and expected simulation results connected with high-voltage circuit-breakers.

According to the requirements at its installation location high-voltage circuit-breakers have to handle different switching cases which lead to a mixture of dielectrical, mechanical and thermal stresses. In this context a CFD (Computational Fluid Dynamics) arc simulation tool should be able to simulate the gas flow processes inside the interrupter unit of high-voltage circuit-breakers during the switching-off process of short-circuit currents, see Figure 1. The CFD arc simulation tool should give information on the gas behaviour as the pressure build-up inside the interrupter unit and e.g. the distribution of gas temperature and density. Furthermore the CFD arc simulation tool should give information on the switching capacity for the different switching cases. Such information can be used in the development process of high-voltage circuit-breakers.

In this publication the basics of the modelling and simulation with respect to SF_6 self-blast circuit-breakers are given, which enable such a functionality of the CFD arc simulation tool. Based on the analysis of the switching-off process in Section 2.1 the overall physical system is identified which is based on the combination and interaction of several sub systems. In Section 2.2 the main aspects of the modelling in the separated sub systems are explained. Section 2.3 gives specific details to the implementation of CFD arc simulation tools. Considering the given information on modelling and implementation the set-up of a CFD arc simulation tool implemented in ANSYS/FLUENT is introduced in Section 2.4. In order to demonstrate the capability of this tool Section 2.5 presents selected simulation results. On the basis of three case examples it is shown in Section 3 how the introduced CFD arc simulation tool can be applied in the development process of high-voltage circuit-breakers.

Finally, the conclusions and an outlook are given in Section 4.

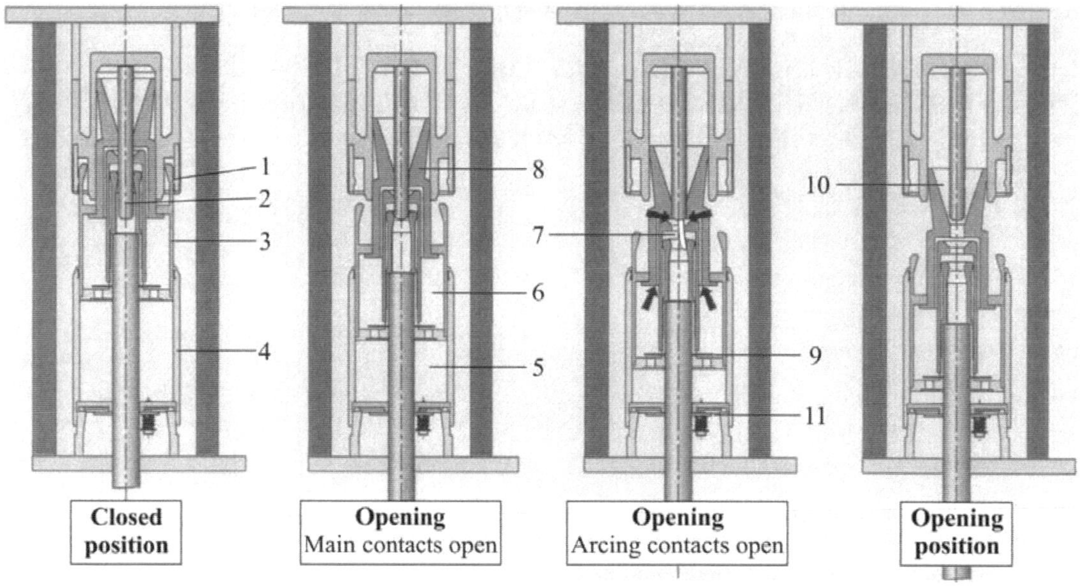

Figure 2. Illustration of variation of interrupter unit geometry during switching-off process.

2. Basics of modelling and simulation

2.1. Structuring of physical system of a SF_6 self-blast circuit-breaker

In case of SF_6 self-blast circuit-breakers the gas pressure needed for the flow of the arc extinguishing medium is created by the thermal energy of the arc itself. This is partially realized by a complex variation of the interrupter unit geometry during the switching-off process [2], see Figure 2. In the closed position the current flows over the main contacts (1). During the switching-off process the drive operates and the movable parts of the interrupter unit are moved downwards. By the opening of the main contacts the current commutates to the arcing contacts (2). Subsequently, the switching arc is ignited by the separation of the arcing contacts. Simultaneously, the cylinder (3) moves into the socket (4) which leads to the compression of the arc quenching gas inside of the compression volume (5). The gas flows through the heating volume (6) into the arcing volume (7) and quenches the arc. This behaviour of the gas flow is valid at the switching-off process of low-current arcs and during no-load operations. At the switching-off process of fault currents with significant current values the conditions change. In that case the strong Ohmic heating by the arc and the ablation of the insulating nozzle (made by polytetrafluoroethylene, PTFE) (8) lead to a high pressure build-up in the arcing volume forcing a hot gas flow into the heating volume. Thus the pressure increases inside the heating volume very strongly and valve 1 (9) closes supporting the pressure build-up furthermore. When the arcing contact enables the opening of the upper nozzle outlet (10) the gas from the heating volume flows back into the arcing volume and the switching arc is quenched in the current zero. To enable such a switching-off process a pressure equalization is realized by a second valve (11).

Following this analysis of the switching-off process the physical system of a SF_6 self-blast circuit-breaker can be identified, see Figure 3. As it can be seen from Figure 3 the physical system is based on the combination and interaction of the following sub systems

- Drive and switching mechanics
- Flow geometry
- Gas flow
- Switching arc
- Recovery after current zero

In the high-current phase the system input is represented by the current and the system output by the arc voltage. In the post arc phase the transient recovery voltage (TRV) acts as the system input and the recovery of the switching gap as the system output. The physical system can be coupled with an external sub system named External circuit, see Figure 3. In this case the output of the physical system is the ohmic resistance of the switching gap which is considered in the electrical network of the external circuit.

For each sub system the corresponding sub model can be developed. Details to the modelling are given in the following section.

2.2. Modelling for the various sub models

2.2.1. Drive and switching mechanics

The consideration of the operation of drive and switching mechanics depends on the features of the CFD program package in use. The illustrations in this paper concern the utilization of ANSYS/FLUENT. In this case the solver requires a velocity value in each time step to control the flow geometry variation. The

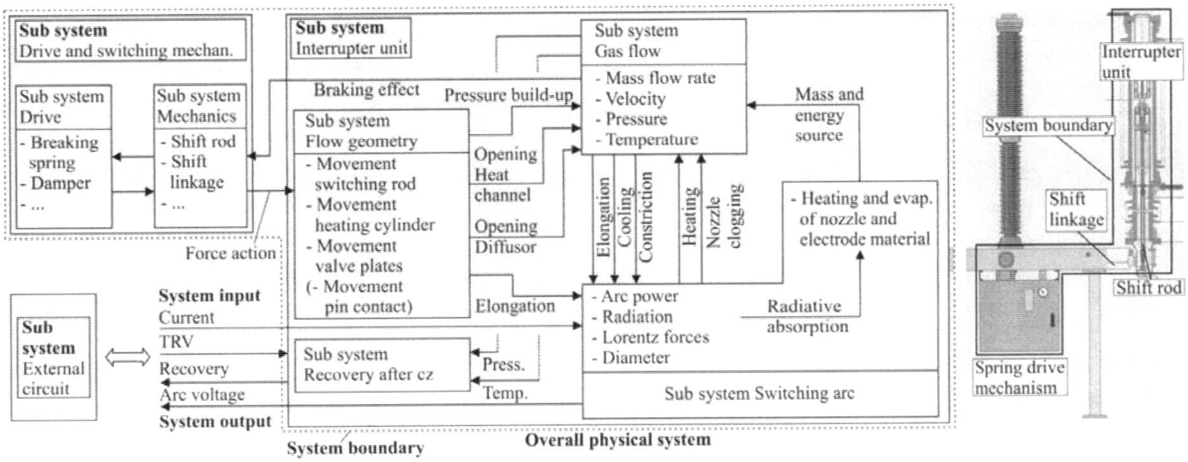

Figure 3. Physical system of a SF_6 self-blast circuit-breaker.

respective velocity value can be calculated by two different models

- □ Interpolation model
- □ Dynamic model

Interpolation model

In case of interpolation model a velocity course evaluated from measurements is provided to the CFD program at the beginning of the simulation. During the simulation an interpolation algorithm depending on simulation time delivers the velocity value in each time step. This is the most common approach in the literature. The important drawback of this approach is that the modelling does not take into account the feedback effect of the pressure build-up inside the interrupter unit on the operation of drive and switching mechanics.

Dynamic model

In case of dynamic model the pressure force determined from the flow simulation can be considered in the equation of motion which should reflect the operation of drive and switching mechanics. In [3], for example, the movement of movable parts is described using the following equation of motion

$$m\frac{\mathrm{d}^2x}{\mathrm{d}t^2} = F - k\cdot x - F_\mathrm{P} - f\left(x, \frac{\mathrm{d}x}{\mathrm{d}t}\right) \qquad (1)$$

In eq. (1), m is the weight of movable parts and driving part, x the displacement, F the driving force, k the spring constant, F_P the pressure force and f the damping force at displacement x and velocity $\mathrm{d}x/\mathrm{d}t$. With eq. (1) it is possible to investigate the influence of pressure build-up inside the interrupter unit on the stroke of movable parts.

The 1D approach according to eq. (1) cannot be used to describe the operation of drive and switching mechanics in a more complex system as depicted in Figure 4. The dynamic behaviour of such system can be described analytically using the Lagrangian mechanics. With the following assumptions

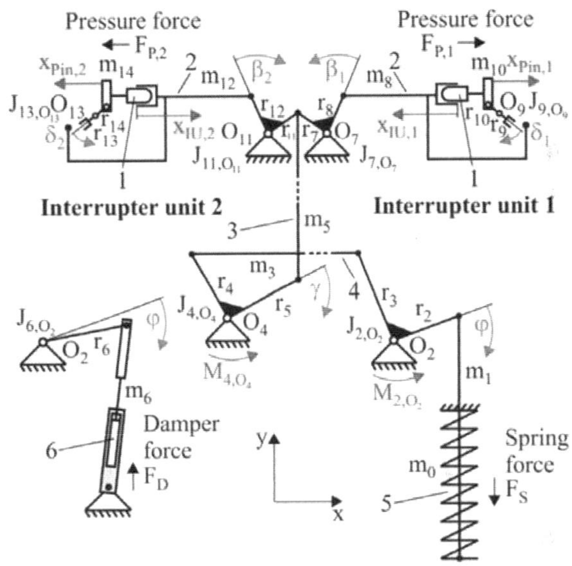

Figure 4. Scheme of switching mechanics for a circuit-breaker with two interrupter units (1: pin contact, 2: moving contact, 3: switching rod, 4: driving rod, 5: breaking spring, 6: hydraulic damper).

- □ Pure translational movement of individual parts
- □ No consideration of bouncing of individual parts
- □ No consideration of changing in the kinematics of individual parts

the system can be modelled by a rigid body model. Despite the complexity of the system the degree of freedom is one and the mathematical model consists of only one differential equation for the angle φ. All other angles are linearly depending on φ. Due to the existence of conservative forces as well as non-conservative forces, the Lagrange equation is as follows

$$\frac{\mathrm{d}}{\mathrm{d}t}\left(\frac{\partial L}{\partial \dot{\varphi}}\right) - \left(\frac{\partial L}{\partial \varphi}\right) - Q = 0, \qquad (2)$$

whereas L is the Lagrange function according to

$$L = T - U. \tag{3}$$

In eq. (2), the term Q represents the generalised forces of non-conservative nature (e.g. the pressure force, the breaking torques and the force component from the hydraulic damper). In eq. (3), the term T summarises the parts concerning the kinetic energy as follows

$$T = m_0 \frac{(\dot{y})^2}{2} + \cdots = m_0 \frac{(r_2 \dot{\varphi})^2}{2} + \cdots \tag{4}$$

and U concerning the potential energy (by the gravitational force on the switching rod and by the spring force) as follows

$$U = \frac{c_S}{2} (L_S - y)^2 + \cdots = \frac{c_S}{2} (L_S - r_2 \varphi)^2 + \cdots. \tag{5}$$

The substitution of evaluated derivatives for the Lagrange function and the term for Q in eq. (2) leads to the equation of motion for the system:

$$\left(m_0 r_2^2 + \cdots\right) \ddot{\varphi} - c_S r_2 \left(L_S - r_2 \varphi\right) + \cdots = 0. \tag{6}$$

The evaluation of eq. (6) in each time step delivers the angular velocity for the angle φ. Taking into account the mentioned linear relationships between φ and the other angles the interrupter unit velocity and the pin contact velocity are estimated in each time step. With these velocity values the variation of discretisation of the simulation model from time step to time step is updated in a continuous way. Note that via the term for the pressure force F_P (see Figure 4) the flow simulation and the simulation of the mechanical system are linked. Further details to the presented approach can be found in [4, 5].

2.2.2. Flow geometry

As mentioned in Section 2.1 the interrupter unit geometry changes notably during the switching-off process. The CFD arc simulation tool should be able to handle such a complex variation of the flow geometry, including the mesh modification, during flow simulation. In this context the elimination or generation of cells near to moving boundaries (Layering mesh method), the sliding of adjacent parts of the mesh relative to each other (Sliding mesh method) and the remeshing in separated regions (Remeshing mesh method) should be at least possible. Figure 5 illustrates a typical variation of solution domain and mesh between closed and opening position which was realized using these features with ANSYS/FLUENT. The illustration in Figure 5 shows e.g. the separation of pin and tube contact, the decreasing of the compression volume and the movement of the valve plate (valve 1) between heating and compression volume.

The movement of the valve plate should be controlled depending on the acting pressure force and in case of the pressure release valve additionally depending on a spring force on the plate. The respective

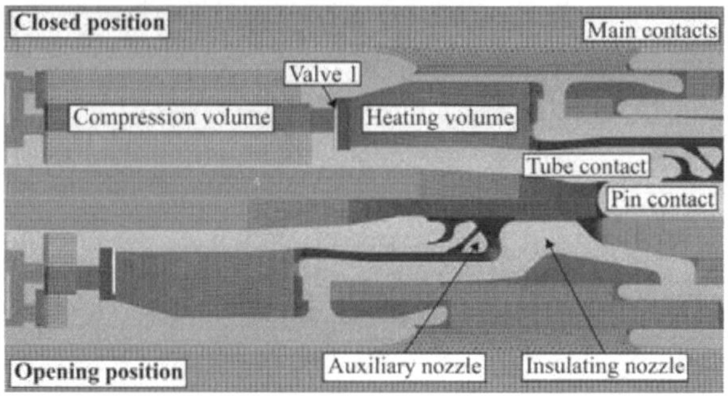

Figure 5. Illustration of mesh modification during switching-off process.

plate velocity v_{pl} can be calculated by the following equation of motion

$$v_{\mathrm{pl}}^{t+\Delta t} = v_{\mathrm{pl}}^{t} + \frac{\Delta t}{m_{\mathrm{pl}}} F_{\mathrm{pl}} \tag{7}$$

whereas m_{pl} is the plate mass. The term F_{pl} represents the acting pressure force and as mentioned in case of the pressure release valve the sum of acting pressure force and spring force.

2.2.3. Gas flow

The gas flow in the interrupter unit of high-voltage circuit-breakers has some unique characteristics. Independent from the circuit-breaker type, puffer or self-blast breaker, a changing pressure drop between inlet and outlet of the nozzle configuration is generated by piston-like geometries and by the switching arc during the switching-off process. The implications of this for the gas flow are as follows

- Starting flow process with low velocities (Mach number $M < 1$)
- Generation of compression shocks with changing localization
- Propagation of expansion and shock waves
- Flow degeneration at the end of switching-off process

The CFD arc simulation tool must be capable to reflect such a flow behavior. The respective functionality can be verified, for example, by the simulation of the stationary flow in a Laval nozzle and by the simulation of the transient flow in a shock tube. In [5] this verification process is illustrated for the CFD program package ANSYS/FLUENT.

2.2.4. Switching arc

In [6] the most important physical phenomena that have to be considered in the simulation of gas circuit-breakers are summarized. Based on this summarization the relevant phenomena for the modelling and simulation of ablation controlled arcs in the interrupter unit of high-voltage circuit-breakers can be identified as follows:

- Compressible flow including friction and turbulence

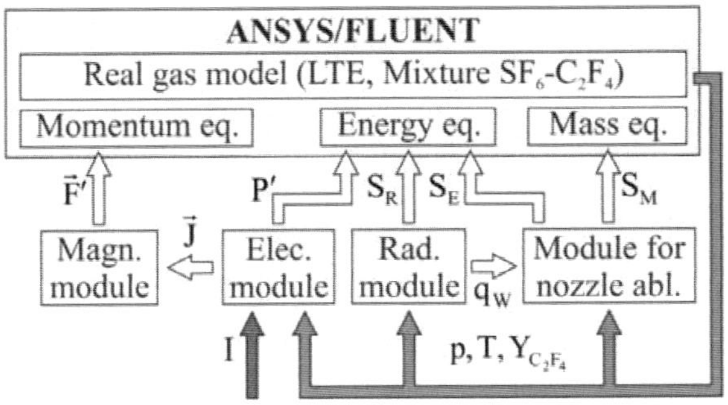

Figure 6. Link of conservation equations and sub models via source terms, vectorial and scalar quantities.

- □ Radiation transfer
- □ Ablation of nozzle wall material (C_2F_4)
- □ Electrode evaporation (WCu)
- □ Multi species transport (SF_6, C_2F_4, W, Cu)
- □ Electromagnetism

In the literature a distinction is made between an integral and a differential approach for the modelling of ablation controlled arcs [7].

In the intergal approach the switching arc is assumed as cylindrical plasma column with uniform temperature surrounded by an isothermal and non-conducting layer of ablated nozzle-vapour. This so-called two-zone model requires the input of several integral parameters derived from experimental data, see e.g. [8] or [9], or from CFD simulations, see e.g. [10]. The model delivers as the main predictions the arc axis temperature, the central pressure, the arc voltage, the arc radius and the ablating mass flux and can be used to describe the mass and energy sources in a CFD arc simulation, see e.g. [11].

In the more general differential approach the switching arc is described by a set of differential equations. In doing so the switching arc is integrated in the flow simulation using several source terms in the conservation equations which are evaluated by the solution of the respective differential equations. As an example Figure 6 shows a possibility for the implementation of an ablation controlled arc taking into account the ablation of nozzle wall material.

The module for nozzle ablation reproduces the heating and vaporization of nozzle material and delivers both a source term S_M in the mass equation and a source term S_E in the energy equation depending essentially on the ratio of wall radiative flux and effective specific ablation enthalpy of PTFE. A detailed description of the modelling in this module can be found in [12]. The module was further improved in [13] to take into account the nozzle wall deformation due to ablation.

The radiative transfer inside the switching arc and into its environment is reflected by the source term S_R in the energy equation. This source term is calculated in the radiation module. The most commonly used radiation models are the semi-empirical model based on the net emission coefficient (NEC) [14], the two standard methods for solving the radiative transfer equation (RTE), the model based on the spherical harmonics method (P1 model) [15] and the model based on the discrete ordinate methode (DO model) [16] or a combination of P1 and DO model [17]. Furthermore the radiation module provides the wall radiative flux q_W which is necessary for the calculation of nozzle ablation.

The source term P' in the energy equation for the consideration of the Joule effect is calculated in the electric module. In the most cases this source term is evaluated solving the potential scalar equation with appropriate boundary conditions, see e.g. [6] or [18].

The self-induced magnetic field due to the current flow is calculated in the magnetic module using the current density vector $\vec{J}$ provided from the electric module. Several approaches exist to calculate the self-induced magnetic field such as the Maxwell-Ampere formulation [19], the Biot & Savart formulation (BF) [20], the vector potential approach (VP) [21] or a combination of BF and VP [18]. From the self-induced magnetic field the Lorentz force vector $\vec{F'}$ can be deduced whose components represent the respective source terms in the momentum equations.

To solve the plasma flow the thermodynamic and transport properties of the arc plasma must be provided to the flow solver of ANSYS/FLUENT. This is realized by a real gas model for the mixture SF_6 - C_2F_4 based on the assumption of local thermodynamic equilibrium (LTE). The scalar quantities pressure p, temperature T and mass fraction $Y_{C_2F_4}$ calculated by ANSYS/FLUENT represent the input variables to the module for nozzle ablation, the radiation module and the electric module. A further input variable to the electric module is the current I.

The plasma flow is strongly governed by turbulence effects. The Prandtl mixing length model and the realizable $k-\varepsilon$ model are the most frequently used turbulence models in HVCB geometries. The Prandtl mixing length model is validated for the description of turbulence in the arc core region, see e.g. [22]. In [23], it is shown that the realizable $k-\varepsilon$ model clearly describes the flow conditions inside the heating volume compared with experimental results. The two models were compared in [24] and it was concluded that the realizable $k-\varepsilon$ model represents a reliable option to describe the turbulent flow inside the entire HVCB geometry.

2.2.5. Recovery after current zero

Concerning the modelling for this sub model it is necessary to distinguish between the behaviour at terminal fault (TF) and the behaviour at short-line fault (SLF).

<u>Behavior at terminal fault</u>

At a TF interruption the switching gap fails by

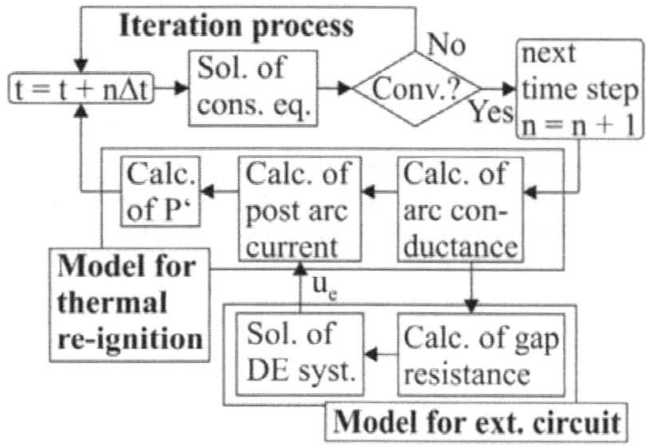

Figure 7. Flow chart for the simulation in the post arc phase at SLF interruption.

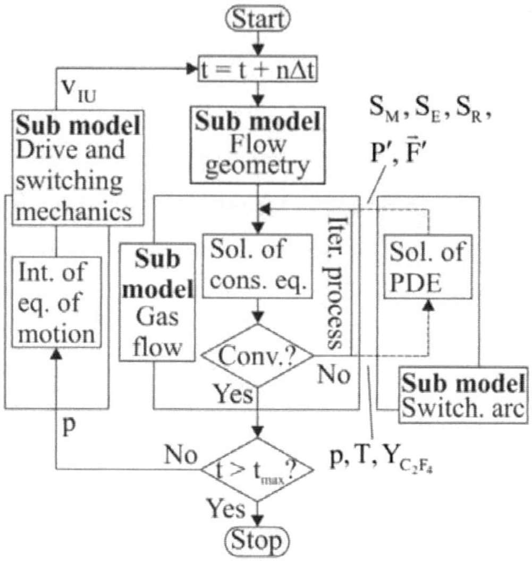

Figure 8. Flow chart for the simulation in the high current phase.

dielectric breakdown. Owing to its simplicity, the most commonly used approach for the study of the dielectric breakdown probability is the utilization of the so-called dielectric criterion k according to eq. (8) (see e.g. [25])

$$k = \frac{E_{\mathrm{geom}}}{E_{\mathrm{cr}}}. \quad (8)$$

Values for k greater than 1 characterize a high dielectric breakdown probability. For the estimation of k, an electrostatic field simulation is carried out at the same time as the flow simulation. The geometric electric field E_{geom} is evaluated from the simulated distribution of the electrostatic potential depending on the TRV which is applied between pin and tube contact after current zero. Using the results from the hot gas flow simulation after current zero, the critical electric field E_{cr} is calculated. For this purpose a pre-calculated data bank for the critical electric field depending on pressure, temperature and mass fraction must be provided to the flow simulation [26]. A more complex approach can be found in [27]. Here, a prediction method is presented which is built on spatial integration of critical and geometric electric field along several integration paths starting from the pin tip.

Behavior at short-line fault

At a SLF interruption the switching gap fails by thermal re-ignition due to the very fast rise of the TRV. Several approaches for the evaluation of the SLF interruption capability exist in the literature. For instance in [28] the trend of the post arc current calculated with different constant rises of recovery voltage is taken to evaluate the SLF interruption capability. In [29] a SLF interruption performance index is used which combines the pressure in the heating volume and the density above the arc zone.

Another possibility for the evaluation of the SLF interruption capability is given by an approach based on the coupling of a model for the thermal re-ignition and a model describing the transient response of an external circuit (e.g. high-voltage circuit) [30]. According to Figure 7 the solution of the equation system for the circuit in the model for the external circuit delivers the TRV u_{e} which is used in the model for the thermal re-ignition to calculate the post arc current and finally the source term in the energy equation P'. The simulated trend of the TRV can be used to evaluate the SLF interruption capability. Thus it is possible to visualize the simulated switching capacity of the regarded circuit-breaker in diagrams similar to test protocols.

2.3. Implementation of CFD arc simulation tools

Essentially, there are two main possibilities for the implementation of CFD arc simulation tools

- □ Framework with one CFD solver (Finite volume (FV) formulation)
- □ Framework with different solvers (FV formulation and finite element (FE) formulation)

Framework with one CFD solver

Figure 8 illustrates this framework. As it can be seen from Figure 8 the sub model Drive and switching mechanics is called at the end of the iteration process at each time step. Taking into account the simulated pressure distribution p the equation of motion is intergrated. This integration delivers the velocity of the movable parts of the interrupter unit v_{IU} which is used in the sub model Flow geometry to update the solution domain discretisation at the beginning of the new time step. After the discretisation manipulation the iteration process for the solution of the conservation equations and the partial differential equations describing the switching arc is started. As the data exchange between the sub model Gas flow and the sub model Switching arc is carried out at each iteration step a strong coupling between both sub models is achieved.

Framework with different solvers

As examples for this framework the simulation system using the coupling server MpCCI and the program package STAR-CCM+ should be named here.

In the first approach the data exchange and interpolation between the two customized commercial codes ANSYS/EMAG (FE solver) and ANSYS/FLUENT (FV solver) is realized by the coupling server MpCCI [31]. The ANSYS/EMAG solver is used to solve the electric and magnetic equations. The conservation equations as well as the radiation equations are solved by means of the ANSYS/FLUENT solver. Additonal sub models for e.g. electrode erosion and wall ablation are implemented using the programming interface of ANSYS/FLUENT.

According to [32] STAR-CCM+ innately provides a data mapper functionality for coupling of FE and FV solver. The gas flow and the electrical current are simulated on a polyhedral grid using the FV solver. The magnetic field is simulated on a standard shape finite element grid by means of the FE solver. To couple both solvers e.g. the source current density $\vec{J}$ is mapped onto the FE grid and the magnetic flux density $\vec{B}$ onto the FV grid.

2.4. Set-up of a CFD arc simulation tool based on ANSYS/FLUENT

The simulation results presented in this contribution have been achieved with the following set-up for the CFD arc simulation tool:

- ☐ ANSYS/FLUENT version 15
- ☐ Valve model
- ☐ Drive model
- ☐ 2 Species SF_6 - C_2F_4 real gas mixture model
- ☐ Realizable $k - \varepsilon$ model
- ☐ Non-grey DO model (7 bands)
- ☐ Switching arc model
- ☐ SLF or TF interruption model
- ☐ No consideration of electrode evaporation

2.5. Selected simulation results

In order to demonstrate the functionality of the presented CFD arc simulation tool selected simulation results are visualized in the following sub sections.

2.5.1. Simulation of the switching-off process in a model chamber

For the purpose of tool validation a model chamber has been assembled. The arc quenching area of the model chamber is presented in Figure 9. The slit in the insulating nozzle enables the access for optical emission spectroscopy (OES) measurements from which radial temperature distributions have been evaluated [33]. Besides the spectroscopic measurements, the stroke of moving contact, the arc voltage and the pressure build-up in the arc and heating volume have been measured.

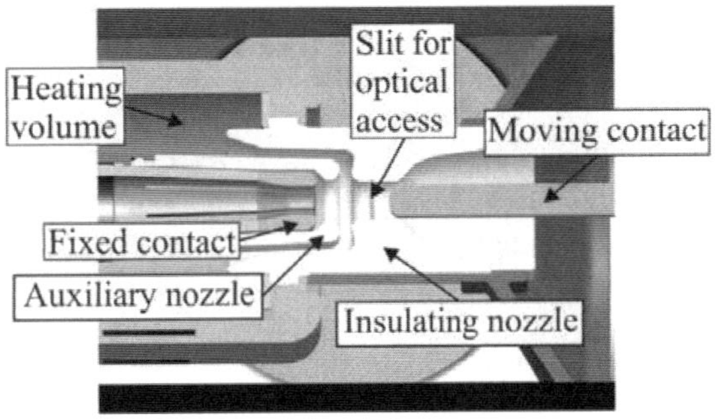

Figure 9. Arc quenching area of model chamber.

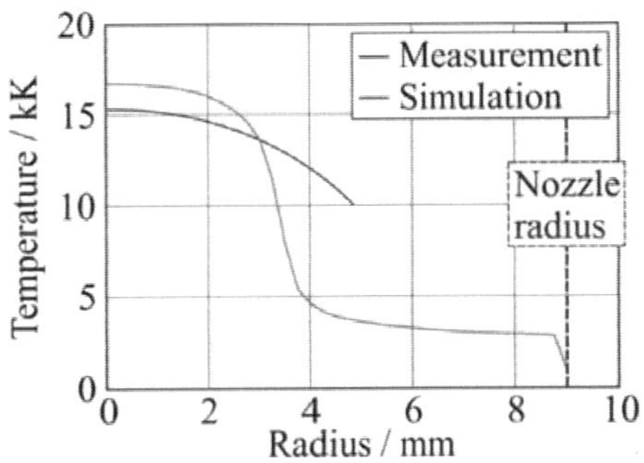

Figure 10. Comparison of radial temperature profiles.

By means of the CFD arc simulation tool the switching-off process in the model chamber has been simulated for an interrupting current with a root mean square (rms) value of $I_{\text{rms}} = 10\,\text{kA}$ and an arcing time of $t_{\text{arc}} = 10\,\text{ms}$. Details on the set-up of the CFD arc simulation tool can be found in [34].

In Figure 10 the simulated temperature profile is compared with the temperature profile evaluated from the OES measurements. The temperature profiles are compared at the time instant corresponding to an electrical current of $I_{\text{instant}} = 1.7\,\text{kA}$. As it can be seen from Figure 10 there is a good agreement between the simulated and the measured temperature profile validating the CFD arc simulation tool. The presented simulated temperature profile characterizes that there is no wall stabilization of the arc at the regarded time instant. Indeed, the temperature profile width is about 3 mm in contrast to the insulating nozzle radius of 9 mm. Assuming the square root dependence of arc core radius from electrical current, the arc wall stabilization should occur at currents higher than 15 kA.

2.5.2. Influence of nozzle degradation on pressure build-up

It is well known that insulating and auxiliary nozzle degrade during successive interruptions influencing arc power and pressure build-up inside the interrupter unit. To show the capability of the presented CFD arc simulation tool with respect to the reflection of

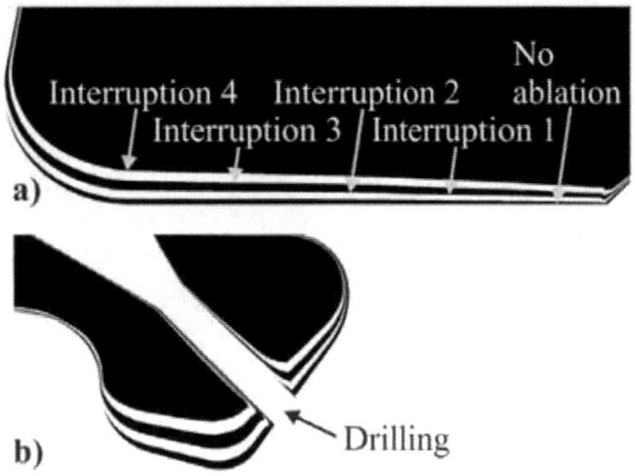

Figure 11. Degradation of the throat of the insulating nozzle (a) and of the nose part of the auxiliary nozzle (b).

this aspect a test series consisting of a set of four interruptions at $I_{\text{rms}} = 25, 30, 35$ and $40\,\text{kA}$ have been simulated [13]. After the simulation of the respective on-load switching-off process a no-load switching-on process has been simulated to reach the closed position of main and arcing contacts. As a result of this simulation sequence Figure 11 shows the accumulative wall degradation of insulating and auxiliary nozzle. The remarkable change in the shape of the nozzles can be clearly seen. The throat of the insulating nozzle takes on a conical profile instead of a cylindrical one and its cross-section increases. The part of the auxiliary nozzle near the moving contact becomes smaller. The drilling in the auxiliary nozzle changes its shape and becomes larger too. These changes in the geometry of the PTFE nozzles can also be observed in experiments.

The consideration of the accumulated PTFE nozzle deformation is essential for the simulation of the gas flow inside the interrupter unit. The difference between the pressure build-up trends simulated under the assumption of original and degraded nozzle geometries becomes much more remarkable with the increase in the number of interruptions in the test series and with the increase in the current. For the 4th interruption at the current $I_{\text{rms}} = 40\,\text{kA}$ the comparison of the pressure build-up in the heating volume is presented in Figure 12. The pressure build-up obtained with the degraded nozzle geometry is in very good agreement with the measurements. The simulation with the original nozzle geometry overestimates the pressure build-up very strongly. This behaviour can be explained as follows:

- □ The flow cross-section inside the PTFE nozzles increases remarkably due to the accumulated PTFE nozzle degradation, as demonstrated in Figure 11. This leads to a faster escape of hot gas from the arcing volume decreasing the pressure here. The smaller amount of hot gas flowing to the heating volume results in a smaller pressure build-up. In the case of the simulation with original nozzle geoemtry this phenomenon takes no place resulting in a too high pressure build-up.

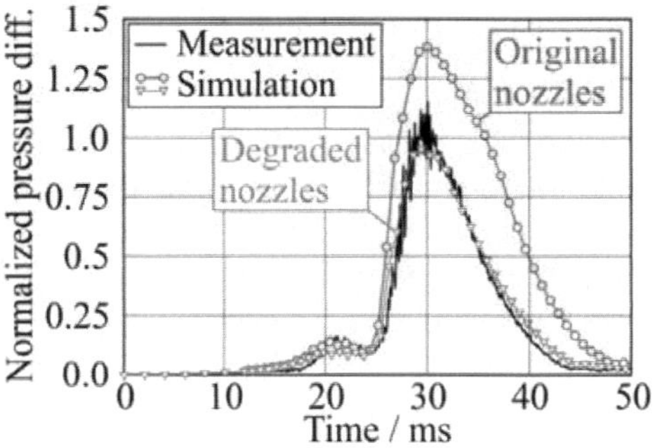

Figure 12. Pressure build-up in the heating volume depending on nozzle contours.

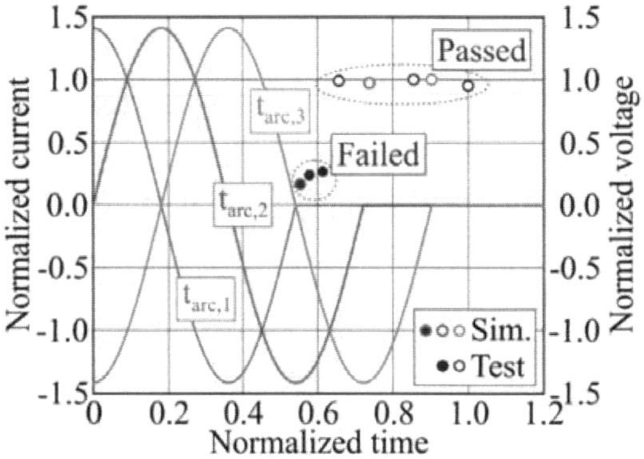

Figure 13. Current trends and peak values of TRV.

- □ The inner diameter of the insulating nozzle increases due to ablation. The amount of the ablated PTFE is smaller in the case of degraded nozzle geometry in comparison with the case of original nozzle geometry. This leads to the increase in the diameter of the plasma column. Thus the arc power is reduced resulting in a decreased pressure build-up.

2.5.3. Evaluation of switching capacity at SLF interruption

With the approach presented in Section 2.2.5 the SLF interruption capacity of a real SF_6 self-blast circuit-breaker has been investigated for a complete test series consisting of the switching-off process at short arcing time ($t_{\text{arc},1}$), medium arcing time ($t_{\text{arc},2}$) and long arcing time ($t_{\text{arc},3}$) [30]. Figure 13 shows the respective current trends which were multiplied for the simulation by a current rms value of $I_{\text{rms}} = 36\,\text{kA}$. The time line of Figure 13 is normalized by the maximum arcing time and the voltage values by the maximum peak value of the TRV in the test series. Additionally the measured and simulated peak values of the TRV are depicted in comparison. As it can be seen the simulation model predicts the switching capacity very similar to the test series.

From Figure 13 it can clearly be seen that the conditions with respect to a successful switching-off process are clearly worse at $t_{\text{arc},1}$ than at $t_{\text{arc},2}$. The

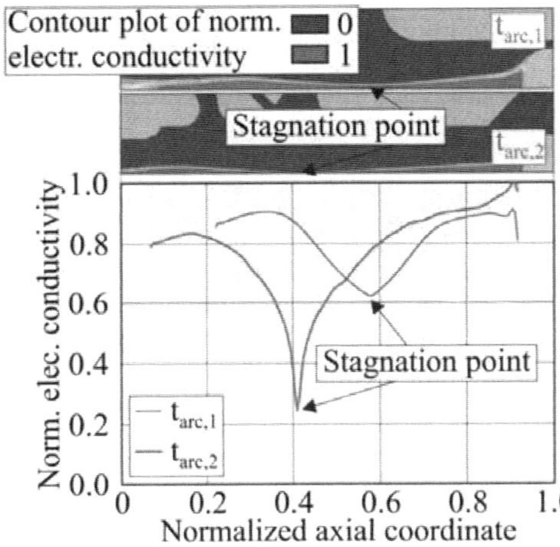

Figure 14. Axial distribution of electrical conductivity at current zero.

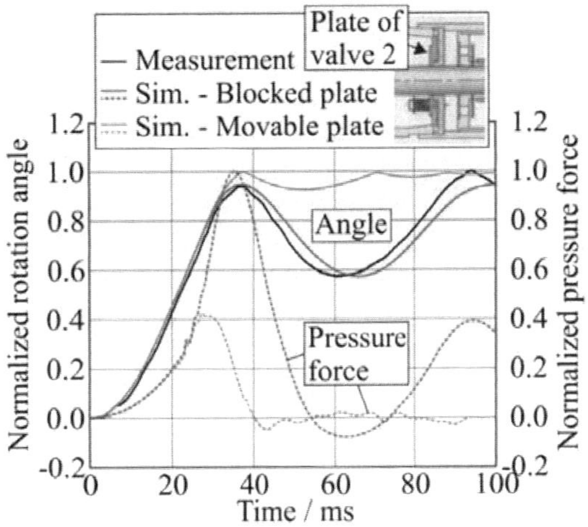

Figure 15. Course of rotation angle and pressure force.

pressure build-up in the heating volume during the high-current phase is clearly lower at $t_{arc,1}$ which leads to a reduced gas flow from the heating volume to the arcing volume at current zero. Thus there are lower values of pressure and higher values of temperature in the arcing volume at $t_{arc,1}$ at current zero leading (especially in the stagnation point) to higher values of the electrical conductivity in the arcing volume and to a greater diameter of the remaining arc channel, see Figure 14. The higher values of the electrical conductivity and the greater diameter of the remaining arc channel in case of $t_{arc,1}$ lead to a lower resistance of the contact gap in and just after the current zero when the contact gap is stressed by the TRV. By this reason the TRV causes a post arc current through the remaining arc channel high enough to heat up the leftover plasma and finally the thermal re-ignition of the contact gap.

3. Application of CFD arc simulation tools

CFD arc simulation tools are able to support the development process of high-voltage circuit-breakers. In this connection their application can yield benefits for the following selected aspects:

- □ Generation of a better understanding of physical processes in the interrupter unit during the switching-off process
- □ Improvement of existing interrupter unit designs
- □ Development of new interrupter unit designs
- □ Savings in costs for assembling and performance testing of prototypes
- □ Savings in development time

To illustrate some of these aspects three case examples are highlighted in the following sections.

3.1. Detection of a mechanical failure in the interrupter unit

By the utilization of the CFD arc simulation tool it is possible to locate mechanical failures in the interrupter unit. Figure 15 shows an experimental rotation angle (black curve, measured during SLF interruption at O_4 in Figure 4) which characterizes an unintended backward movement of the moving contact during the switching-off process. The cause for the depicted strong backward movement of the moving contact could be detected by several CFD arc simulations in form of a blocked plate of valve 2 (see top right in Figure 15). By the blocked valve plate a very strong pressure increase in the compression volume is generated resulting in an excessive pressure force which decelerates the drive and causes the strong backward movement (see red curves in Figure 15). In the CFD arc simulation with a moveable plate of valve 2 the pressure force reaches only about 40% resulting in a very slight backward movement (see blue curves in Figure 15). Using these results the design near valve 2 was improved and the problem of the strong backward movement of the moving contact did no longer exist in ensuing tests.

3.2. Reduction of the transient pressure rise at a rupture disc

Selected circuit-breaker types feature rupture discs for pressure equalization in case of fault arcs. In this connection the rupture disc must not operate during a regular switching-off process. That means that the transient pressure rise at the rupture disc caused by a switching-off process of short-circuit currents must not exceed the opening pressure of the rupture disc. The CFD arc simulation tool can be used in the process of rupture disc qualification. In a first simulation at an asymmetrical TF interruption the transient pressure rise at the rupture disc has been simulated for a first design (design 1). From the black curve in Figure 16 it can be seen that the maximum value

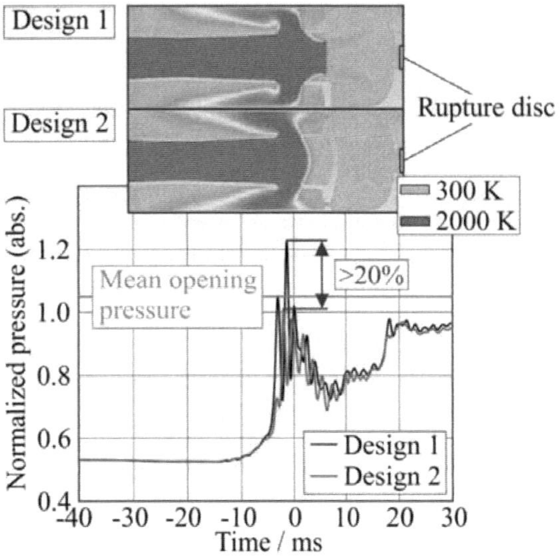

Figure 16. Pressure build-up at the rupture disc and contour plot of temperature at current zero ($t = 0$).

of the simulated pressure at the rupture disc clearly exceeds the opening pressure identifying the necessity for a design improvement. By the modification of the interrupter unit geometry in a second design (design 2) the maximum value of the simulated pressure at the rupture disc could be reduced by about 20% (see the red curve in Figure 16). The comparison of the contour plots of temperature in Figure 16 shows that in case of design 2 the hot gas cannot flow directly into the cover. Thus the inflow of hot gas into the cover is reduced and with it the pressure rise at the rupture disc.

3.3. Estimation of minimal arcing time at TF interruption

With the approach for the study of the dielectric breakdown probability presented in Section 2.2.5 a first estimation of the minimal arcing time at TF interruption in a virtual prototype of a high-voltage circuit-breaker can be performed using the CFD arc simulation tool. Figure 17 shows for three different current trends with increasing arcing time the distribution of dielectric criterion in the arcing volume at the time when the TRV reaches its maximum value. The completely red region of the dielectric criterion ($k \geq 1$) between pin tip and tube contact in case of the smallest arcing time $t_{\mathrm{arc},1}$ characterises a high dielectric breakdown probability. The comparison of the dielectric criterion distributions illustrates that the region with low values for the dielectric criterion increases with rising arcing time. At the arcing time $t_{\mathrm{arc},3}$ almost the complete arcing volume shows low values for the dielectric criterion characterising a low dielectric breakdown probability. These simulation results lead to a minimal arcing time for the investigated virtual prototype between $t_{\mathrm{arc},2}$ and $t_{\mathrm{arc},3}$.

As it can be seen from Figure 17 the pressure build-

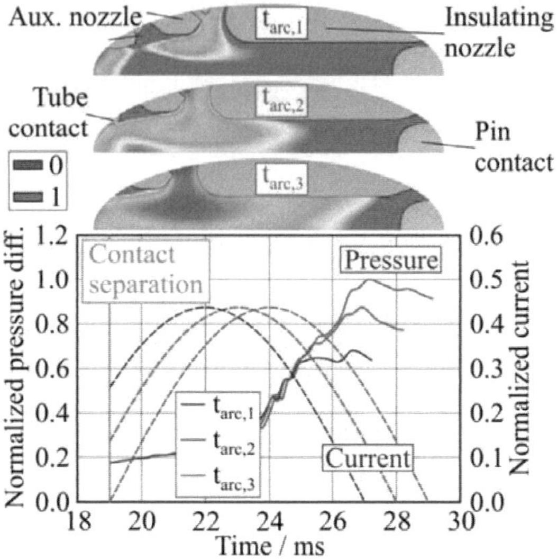

Figure 17. Distribution of dielectric criterion in the arcing volume at maximum value of TRV and corresponding trends of current and pressure in the heating volume.

up in the heating volume increases also with rising arcing time. The high pressure build-up in the heating volume at $t_{\mathrm{arc},3}$ forces a very strong gas flow back from the heating volume towards arcing volume at current zero. Thus, the region between the pin and the tube contact is almost completely filled with cold gas leading to high values of the critical electric field E_{cr} and to low values of the dielectric criterion k.

4. Conclusions and outlook

The present paper focuses on the basics of CFD arc simulation tools and its application with respect to the simulation of the fluid mechanical processes in the interrupter unit of SF_6 high-voltage circuit-breakers. Following the analysis of the switching-off process in a SF_6 self-blast circuit-breaker the physical system is structured which is based on the combination and interaction of several sub systems. Next, the modelling in the separated sub systems and aspects to the implementation of CFD arc simulation tools are discussed in detail. The set-up of a CFD arc simulation tool based on the program package ANSYS/FLUENT is introduced. The presented set-up does not consider electrode evaporation. It is demonstrated how this CFD arc simulation tool can be used to describe the switching-off process in a model chamber, the influence of the nozzle degradation on the pressure build-up inside the interrupter unit and the SLF interruption capacity of a real SF_6 self-blast circuit-breaker. Finally, case examples for the application of the presented CFD arc simulation tool in the development process of high-voltage circuit-breakers are illustrated.

Starting from the presented level of modelling

the capability of CFD arc simulation tools could be further improved with respect to the high-current phase e.g. by introducing a model for the electrode erosion. Essentially, this would involve the consideration of solid electrodes, the description of arc roots and metal vapour generation and the enhancement of the real gas mixture model by the new metal species. A further important improvement could be the consideration of the deviation from LTE especially in the modelling for the post arc phase. However, it can be expected that these improvements will increase both the complexity of the CFD arc simulation tool and the numerical effort resulting in longer computation times.

References

[1] M. Kriegel et al. Simulation and calculation as verification tools. *Electra*, 234:13–18, 2007.

[2] J. Sedlacek et al. Optimization of high-voltage self-blast interrupters by gas flow and electric field computations. *IEEE Transactions on Power Delivery*, 18(4):1228–1234, 2003. doi:10.1109/TPWRD.2003.817536.

[3] N. Osawa et al. Investigation of the optimum design of thermal puffer type gas circuit breaker with secondary chamber. In *15th International Conference on Gas Discharges and their Applications*, pages 13–16, Toulouse, France, 2004.

[4] F. Reichert et al. Modellierung und Simulation eines Federspeicher-Antriebssystems. *Konstruktion*, 64(3):67–73, 2012.

[5] F. Reichert. *Numerische Simulation strömungsmechanischer Vorgänge in SF_6-Hochspannungsleistungsschaltern.* Postdoctoral thesis. University press Ilmenau, Germany, 2014.

[6] J. Ostrowski et al. Computational magnetohydrodynamics in the simulation of gas circuit breakers. *International Journal of Computational Science and Engineering*, 9(5/6):433–444, 2014. doi:10.1504/IJCSE.2014.064528.

[7] C. Lüders. *Vergleich von Strahlungs- und Turbulenzmodellen zur Modellierung von Lichtbögen in SF_6-Selbstblasleistungsschaltern.* Thesis. RWTH Aachen, Germany, 2005.

[8] L. Niemeyer. Evaporation dominated high current arcs in narrow channels. *IEEE Transactions on Power Apparatus and Systems*, PAS-97(3):950–958, 1978. doi:10.1109/TPAS.1978.354568.

[9] P. Kovitya et al. Theoretical predictions of ablation-stabilised arcs confined in cylindrical tubes. *J. Phys. D: Appl. Phys.*, 17(6):1197–1212, 1984. doi:10.1088/0022-3727/17/6/016.

[10] M. Seeger et al. An integral arc model for ablation controlled arcs based on CFD simulations. *J. Phys. D: Appl. Phys.*, 39(10):2180–2191, 2006. doi:10.1088/0022-3727/39/10/029.

[11] M. Claessens et al. Simulation of gas flow phenomena in high-voltage self-blast circuit breakers at heavy fault current interruption. *IEEE Transactions on Plasma Science*, 25(5):1001–1007, 1997. doi:10.1109/27.649618.

[12] J.J. Gonzalez et al. PTFE vapor contribution to pressure changes in high-voltage circuit breakers. *IEEE Transactions on Plasma Science*, 43(8):2703–2714, 2015. doi:10.1109/TPS.2015.2450536.

[13] A. Petchanka et al. Modelling of the deformation of PTFE-nozzles in a high voltage circuit breaker due to multiple interruptions. *J. Phys. D: Appl. Phys.*, 49(13):135201, 2016. doi:10.1088/0022-3727/49/13/135201.

[14] Y.J. Kim et al. SF_6 arc plasma simulation and breakdown performance prediction using computational fluid dynamics and arc modelingoriginal. *Thin Solid Films*, 521:206–211, 2012. doi:10.1016/j.tsf.2011.11.076.

[15] M.J. Ha et al. Simulation of hot gas flow in a high voltage circuit breaker with P1 radiation model. In *20th Symposium on Physics of Switching Arc*, pages 167–171, Brno, Czech Republic, 2013.

[16] A.A. Iordanidis et al. Self-consistent radiation-based simulation of electric arcs: II. Application to gas circuit breakers. *J. Phys. D: Appl. Phys.*, 41(13):135206, 2008. doi:10.1088/0022-3727/41/13/135206.

[17] F. Reichert et al. Modelling and simulation of radiative energy transfer in high-voltage circuit breakers. *J. Phys. D: Appl. Phys.*, 45(37):375201, 2012. doi:10.1088/0022-3727/45/37/375201.

[18] P. Freton et al. Magnetic field approaches in dc thermal plasma modelling. *J. Phys. D: Appl. Phys.*, 44(34):345202, 2011. doi:10.1088/0022-3727/44/34/345202.

[19] J.J. Lowke et al. A simplified unified theory of arcs and their electrodes. *J. Phys. D: Appl. Phys.*, 30(14):2033–2042, 1997. doi:10.1088/0022-3727/30/14/011.

[20] L.Z. Schlitz et al. High-resolution transmission electron microscopy of some $Ti_{n+1}AX_n$ compounds (n=1, 2; A=Al or Si; X=C or N). *J. Appl. Phys.*, 85(5):2540–2546, 1999. doi:10.1063/1.371089.

[21] A.B. Murphy. A self-consistent three-dimensional model of the arc, electrode and weld pool in gas-metal arc welding. *J. Phys. D: Appl. Phys.*, 44(19):194009, 2011. doi:10.1088/0022-3727/44/19/194009.

[22] J.D. Yan et al. A comparative study of turbulence models for SF_6 arcs in a supersonic nozzle. *J. Phys. D: Appl. Phys.*, 32(12):1401–1406, 1999. doi:10.1088/0022-3727/32/12/317.

[23] R. Bini et al. Arc-induced turbulent mixing in an SF_6 circuit breaker model. *J. Phys. D: Appl. Phys.*, 44(2):25203–25212, 2011. doi:10.1088/0022-3727/44/2/025203.

[24] J.J. Gonzalez et al. Turbulence and magnetic field calculations in high-voltage circuit breakers. *IEEE Transactions on Plasma Science*, 40(3):936–945, 2012. doi:10.1109/TPS.2011.2180404.

[25] X. Li et al. Breakdown electric field calculation of hot SF_6 and its application to high voltage circuit breakers. In *58th IEEE Holm Conference on Electrical Contacts*, pages 43–48, Portland, OR, US, 2012.

[26] M. Yousfi et al. Breakdown electric fields in dissociated hot gas mixtures of sulfur hexafluoride including teflon: Calculations with experimental validations and utilization in fluid dynamics arc simulations. *J. Appl. Phys.*, 121(10):103302, 2017. doi:10.1063/1.4977864.

[27] S. Aded Hussein et al. Detailed investigation of breakdown prediction models for high voltage circuit breakers. *Plasma Physics and Technology*, 2(2):108–111, 2015.

[28] J.C. Lee et al. Numerical study on switching arcs and turbulence models inside a SF_6 self-blast interrupter. In *20th Symposium on Physics of Switching Arc*, pages 232–235, Brno, Czech Republic, 2013.

[29] H.K. Kim et al. Analysis of SLF interruption performance of self-blast circuit breaker by means of CFD calculation. *J. Electr. Eng. Technol.*, 9(1):254–258, 2014. doi:10.5370/JEET.2014.9.1.254.

[30] F. Reichert et al. Studies on the thermal re-ignition in SF_6 high-voltage circuit-breakers by means of coupled simulation. *Plasma Physics and Technology*, 2(2):195–198, 2015.

[31] C. Rümpler et al. Arc modeling challenges. *Plasma Physics and Technology*, 2(3):261–270, 2015.

[32] S. Hartridge et al. Aspects of plasma simulations in STAR-CCM+. NAFEMS World Congress, Stockholm, Sweden, 2016.

[33] R. Methling et al. Spectroscopic study of arc temperature profiles of a switching-off process in a model chamber. *Plasma Physics and Technology*, 2(2):163–166, 2015.

[34] A. Petchanka et al. CFD arc simulation of a switching-off process in a model chamber. *Plasma Physics and Technology*, 2(1):63–66, 2015.

Using the FRAM Method to Analyse Human/System Interactions on Submarines

Ari Antonovsky, Katrina Hosszu, & Karina Jorritsma

School of Psychological Science, University of Western Australia

THE UNIVERSITY OF WESTERN AUSTRALIA

CENTRE FOR SAFETY

This project was conducted under Defence Science and Technology Research Agreement Number: 6963

Today's Presentation

- Human Factors in a complex STS: the Future Submarine
- Applying the FRAM in this project
- Thematic analysis of interview data
- Development of a FRAM database
- Summary: Potential applications to solving Human Factors problems

2

Future Submarine Human/Systems Integration

Aims

- "...to provide advice to the future submarine program in respect to the 'work systems' design, that are likely to maximise the performance and resilience of 'lean' future submarine crews"

➲ **Map current processes**

- Develop process maps of the current systems, communications patterns & interactions between different functions
  - *Propulsion Control system, Surface Running, Damage Control & Weapons Handling*

➲ **Provide recommendations to improve processes**

- identify work design risks and possible responses to inform the future submarine design

3

A Submarine as a Complex Socio-Technical System

➲ **Similar human factors to an aircraft**

- Large metal tube transporting many people
- Complex, highly-automated systems (e.g. Fly-by-wire)
- Safety needs to be assured for an inherently unsafe proposition
- Greatest risks are due to interactive complexity and fast coupling

➲ **Some basic differences**

- Numerous people on board to keep systems functioning
- Submarines can be repaired in transit
- Fewer examples compared to numbers of commercial aircraft

4

Functional Maps:
Functional Resonance Analysis Method (FRAM)

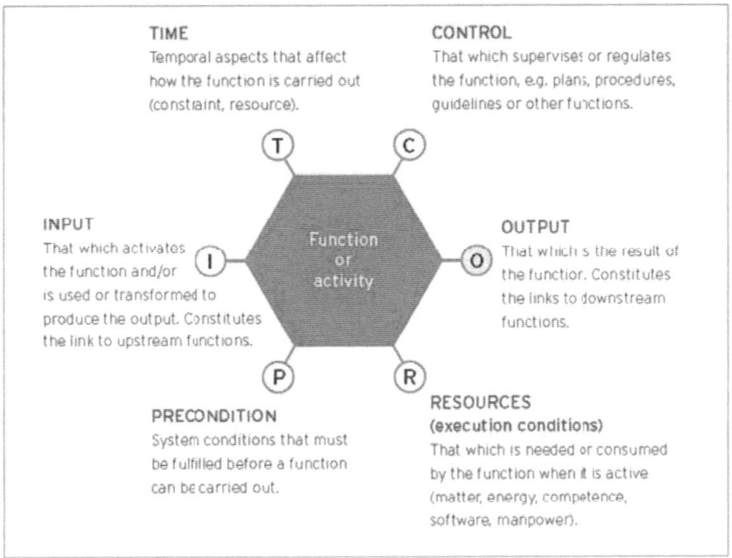

E. Hollnagel, FRAM: the Functional Resonance Analysis Method (2012)

6

Analysing the Human Factors in a S-T System

- **Two roles of a Human Factors analysis:**
  - **#1.** Understand the processes-esp. human /system interactions
  - **#2.** Develop improvements to optimise work design, human/system interfaces and equipment usability
- **Functional Resonance Analysis Method is:**
  - one of the better ways of representing #1
  - able to facilitate #2 if supplemented by qualitative data
- **Semi-structured interviews are a way of eliciting information**
  - Provides more depth than surveys

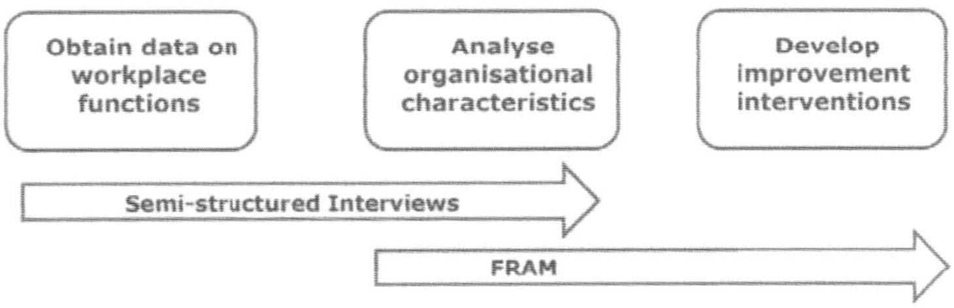

7

Project Methodology

- **Interviews based on FRAM structure to:**
  - Determine inputs & outputs, pre-conditions, resources, controls, & time
  - Construct FRAM map of work design processes

- **Thematic analysis of interview statements**
  - *Using IPA or Grounded Theory:*
  - Parse statements into discrete ideas
  - Identify the concepts expressed by participants
  - Assess the relevance of specific themes

- **Further investigation of data**
  - Analyse frequencies of themes
  - Develop database to collect:
    - *additional qualitative information*
    - *quantitative data*

8

FRAM Example: Damage Control Process

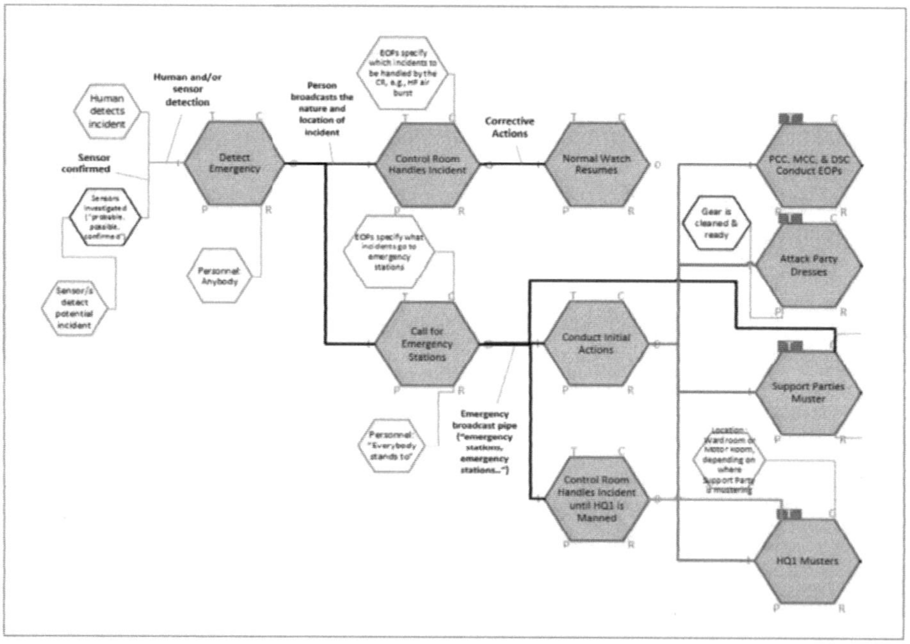

9

Thematic Analysis of Interview Data

Step:	Record distinct statements	Identify repeating ideas that are conceptually similar	Identify areas of commonality within the framework	Identify broad groupings of themes that generate a new concept
	Text	**Assign sub-themes**	**Group sub-themes into themes**	**Combine themes into a Super-Ordinate Theme**

↓

	A	B	D	G	H	J	M
1	ID	System	Text	Sub-Theme_1	Sub Theme_2	Theme	Super-Ord Theme
218	DC32	Damage Control	People assess them on getting dressed. These guys are tweaking things, if they change the way they get into suit to save time (e.g., save 5 seconds overall).	Personnel are assessed how quickly they can get dressed	Personnel make tweaks to quicken the process of gettinginto the suit	**Emergency training essential**	Training

10

Findings from Interview Themes

Automation

- Opportunities for automation
- Risks of automation
- Resistance to automation
- Automation reduces workload and/or frees up personnel
- Increasing trust in automation
- Impact of automation on human error
- Automation will increase maintenance load

Crewing & Workload

- Crew optimisation
- General workload
- Cognitive workload
- Work design
- Fatigue issues
- Limitations of a lean crew

Design & Reliability

- Need to improve design
- Systems worth preserving
- System redundancy is critical
- Equipment & system reliability
- Technical issue

Control Systems

- Sensors not sufficiently reliable
- Need for more monitoring and trending
- Alert management
- Need for more sensors
- More control in manual mode

Training

- Need to develop & maintain expertise
- Insufficient sea training opportunities
- Training must keep up with continual change
- Trade-offs in training
- Importance of maintaining consistency in training

Maintenance

- Supply & storage of spares
- Issues of the operator-maintainer model
- Equipment maintenance access
- Better maintenance strategy is needed

Communications & Procedures

- Opportunities to improve communication efficiency
- Procedures to ensure work quality

11

Thematic Analysis: Theme Frequencies

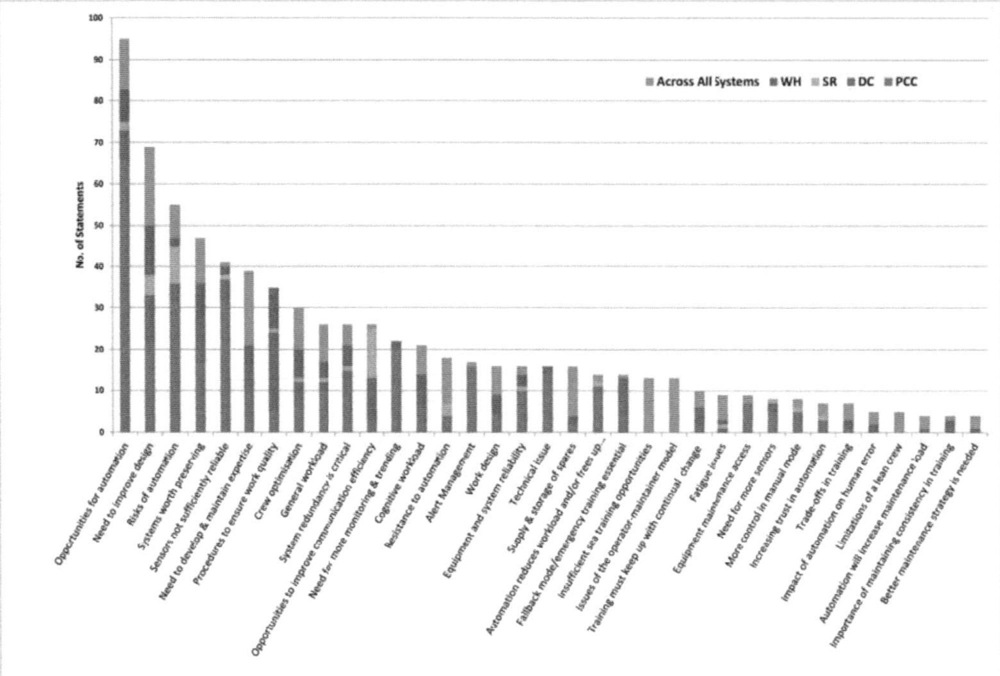

12

7 Super-ordinate Themes (SOTs) Developed from 34 Themes

Aggregated SOTs and their frequencies

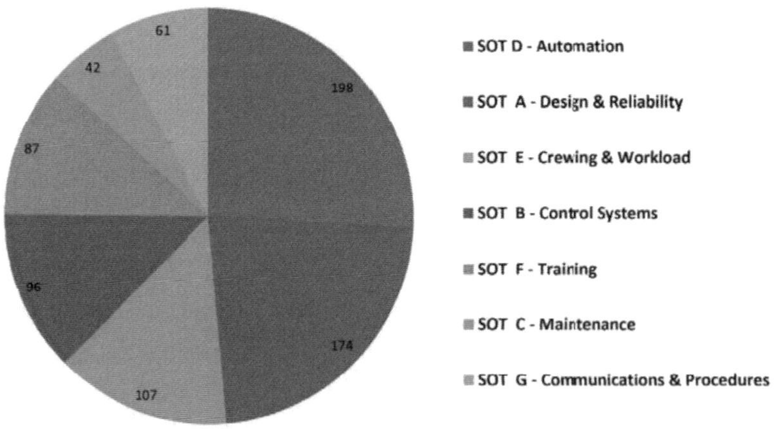

13

Create Database Record for FRAM Data

FRAM Functions = Tasks / Activities

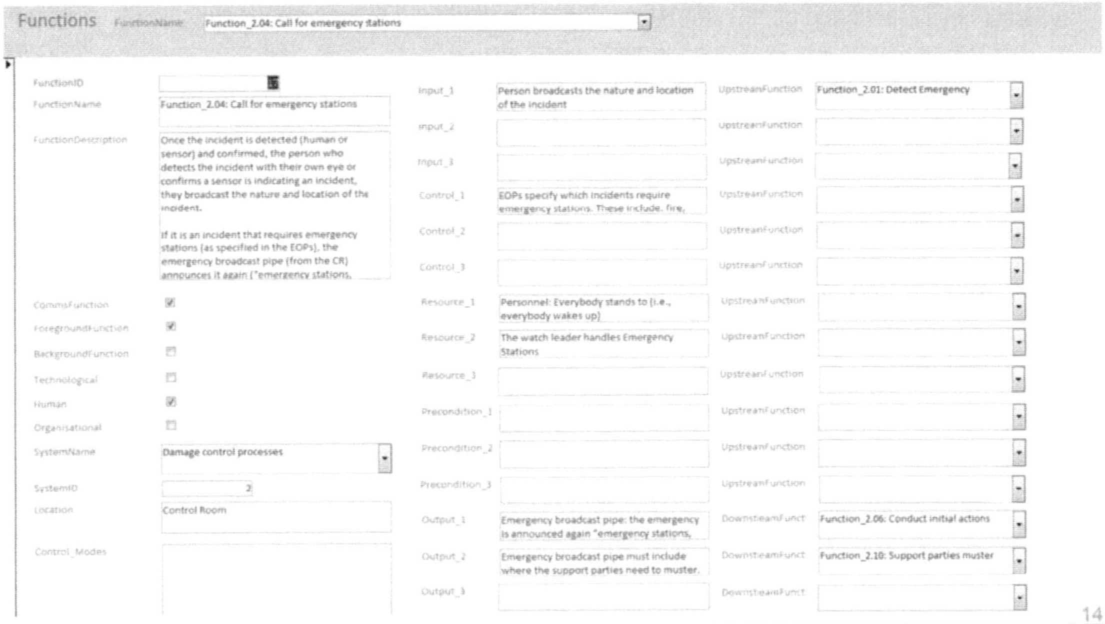

14

MS Access Database linked to FRAM Map

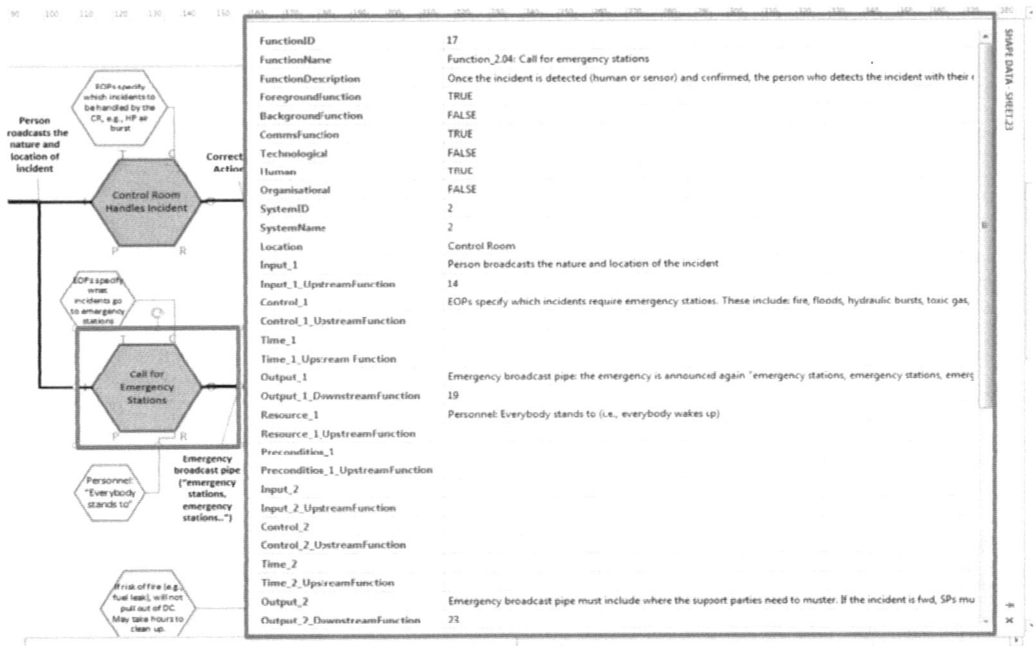

15

Summary: Using the FRAM to Solve HF Problems

- **Human Factors is concerned with understanding processes to:**
  - Characterise the activities involved
  - Identify risks and opportunities for improvement
- **The FRAM method offered an effective starting point to:**
  - Map basic workflow characteristics (resources, controls, time, pre-conditions)
  - Obtain data on how well these processes function
  - Store qualitative & quantitative data via MS Access database
- **Thematic Analysis of interview data**
  - Identify ideas & themes of concern to participants
  - Quantify the frequency of recurring themes
  - Develop work design interventions for socio-technical systems under investigation
- **Potential Applications**
  - Improve the interfaces & usability of automated systems
  - Resolve work design problems e.g. with workflow or proceduralised processes
  - Compare work design processes before & after interventions

16

doi:10.14311/ppt.2017.3.225
Plasma Physics and Technology **4**(3):225–233, 2017

NON-EQUILIBRIUM STUDIES IN SWITCHING ARC PLASMAS IN JAPAN

Y. Tanaka[a,*], K. Tomita[b], Y. Inada[c], A.Kumada[d], K. Hidaka[d], T. Fujino[e], K. Suzuki[f], T. Shinkai[g]

[a] *Kanazawa University, Kakuma, Kanazawa, Ishikawa 920-1192, Japan*
[b] *Kushu University, Kasuga, Fukuoka 816-8580, Japan*
[c] *Saitama University, 542 Shimo-Okubo, Sakura, Saitama 338-8570 Japan*
[d] *The University of Tokyo,7-3-1 Hongo, Bunkyo, Tokyo 113-8654, Japan*
[e] *University of Tsukuba, 1-1-1 Tennodai, Tsukuba, Ibaraki 305-8577, Japan*
[f] *Tokyo Denki University, 5 Senju Aasahi, Adachi, Tokyo 120-8551, Japan*
[g] *Tokyo University of Technology, 1404-1 Katakura, Hachioji, Tokyo 192-0982, Japan*
[*] tanaka@ec.t.kanazawa-u.ac.jp

Abstract. This paper briefly introduce research work examples of non-equilibrium studies in switching arcs. In understanding arc behavior, one often assumes local thermodynamic equilibrium (LTE) condition in the arc plasma. However, actual arc plasmas are not completely and not always in LTE state because of strong temperature change temporally and spatially, and high electric field application etc. Recently, we have a collaboration work in numerical simulations and experimental approaches for decaying arcs without LTE assumption. First, our numerical model is presented for decaying arcs without chemical equilibrium assumption. Secondly, two experimental methods are introduced for measuring electron density in decaying arcs without LTE assumption: Laser Thomson Scattering method and the Schack-Hartmann method. Finally, comparison results is shown between the LTE simulation, the chemically non-equilibrium simulation, and the above experimental measurements.

Keywords: Local thermodynamic equilibrium, chemically non-equilibrium, thermally non-equilibrium, electron density measurement.

1. Introduction

Thermal plasmas/arc plasmas are widely used for various applications such as plasma spray coating, rapid surface modification, fabrication of functional films, plasma arc cutting, nanopowder synthesis, and carbon nanotube synthesis etc. Switching arcs are also important applications of arc plasmas in an electric power transmission system. In understanding the behavior of thermal plasmas/arc plasmas, local thermodynamic equilibrium (LTE) models have been widely adopted. This LTE model assumes the same temperatures for kinetic energies of electrons and heavy particles. The LTE model also assumes chemically equilibrium condition, which corresponds to extremely high reaction rates for all reactions including ionization, electronic, vibrational and rotational excitations. Therefore, the LTE assumption is equivalent to one-temperature chemically equilibrium (1T-CE) model. Such the LTE model is greatly useful to predict the mass and energy transport in thermal plasmas or arcs for a conventional use, and it successfully interprets behaivor of the thermal/arc plasmas. However, more detailed understanding of the distributions of chemical species in arcs and thermal plasmas has been required for advanced thermal plasma materials processes, as well as switching arcs.

One phenomenon which causes chemically non-equilibrium effects is high spatial gradient in number density at medium-temperature or low-temperature region of arcs and thermal plasmas. In medium or low temperature regions, which are always present surrounding the high-temperature region of the arcs, the LTE assumption cannot always be established. It is necessary to account for chemically non-equilibrium effects in this case [1, 2]. Another important phenomenon for deviation from chemical equilibrium condition is a rapid change in time, i.e., a transient phenomenon in arcs and thermal plasmas. For example, switching arcs or modulated induction thermal plasmas [3, 4] are fundamentally under transient conditions. To study the transient state in arcs and thermal plasmas, attention should be paid to chemically non-equilibrium effects. On the other hand, a deviation from thermal equilibrium condition may be due to high electric field application. In switching arcs, the transient recovery voltage (TRV) is applied to a residual arc between the electrodes after current zero. Such a high voltage application may increase the electron temperature, leading to two-temperature state in residual arcs. To elucidate the aforementioned phenomena in detail, non-equilibrium effects must be considered.

This paper briefly introduces an example of non-equilibrium modeling in arc plasmas [5, 6]. In addition, some recent experiments will be introduced for decaying arcs with a detailed control of applied voltage and injected current using some power semiconductors [7]. In the experiments, two measurement methods were adopted to determine the electron density without LTE assumption: the laser Thomson scattering method [8] and the Shack-Hartmann method [9]. Finally, we compare results between the LTE simulation, the chemically non-equilibrium simulation, and the above experimental measurements [6].

2. Non-equilibrium effects for switching arcs

For simulation of switching arcs, the LTE or one-temperature chemical equilibrium (1T-CE) models has been adopted. However, the following non-equilibrium effects may be taken into account for its further understanding:

(a) Chemically non-equilibrium effects: In arc plasmas, there are lots of reactions including molecular dissociation, association, ionization, recombination, attachment and detachment. Any reaction requires a certain finite time to reach its chemical equilibrium because the rate coefficients of reactions are also finite. On the other hand, switching arcs are essentially under transient state, and they have steep gradients in densities of chemical species. Chemically equilibrium (CE) in the switching arcs is therefore achieved only if the relaxation time for any reaction is much shorter than the characteristic time of diffusion on the particle composition as well as the transient time of the states. In other words, a rapid time change and steep gradient in switching arcs may involve chemically non-equilibrium (non-CE) state.

(b) Thermally non-equilibrium effects: In high pressure arcs, the electron temperature T_e becomes close to the heavy particle temperature T_h if the electron sufficiently transfers its kinetic energy to the heavy particle through elastic collisions compared to the energy gained from the external electric field. Establishement of thermal equilibrium state needs much less electron energy gain by the electric field than the thermal energy of electrons. Otherwise a thermal non-equilibrium condition $T_\mathrm{e} >> T_\mathrm{h}$ can be obtained. In switching arcs, the off-axis region has relatively lower temperature and lower electron density, which results in lower collision frequency between electrons and heavy particles. At the same time, the electric field is applied to the off-axis region as well as on-axis region in the arc. This situation may offer thermally non-equilibrium state. Furthermore, to the switching arcs, a high transient recovery voltage (TRV) is applied after current zero. This high voltage application may lead residual arcs to ones in thermally non-equilibrium state.

(c) Non-Boltzmann distribution for a population of excited particles: The population of excited particles does not follow the Boltzmann distribution if the population is more influenced by processes such as the radiative de-excitation than by thermal excitation and de-excitation. This concept is important particularly for investigating a physical state in thermal plasmas from spectroscopic observation results.

(d) Non-Maxwellian energy distribution function: The electron energy distribution function (EEDF) cannot be Maxwellian if the electron–electron energy relaxation time is much longer than the electron acceleration time by an external electric field, or if the vibrational and rotational excitation cross section is comparable with the momentum transfer cross section by the electron impact. In such cases, the Boltzmann equation should be solved considering such cross sections with an applied electric field. This situation appears in prediction of arc re-ignition phenomena in a circuit breaker.

The present paper only focuses (a) chemically non-equilibrium (non-CE) effects.

3. Numerical approaches: Modeling on chemically non-equilibrium effects

3.1. Mass conservation equation

Among various non-equilibrium effects, this section treats chemically non-equilibrium (non-CE) effects because the non-CE effects affect the behavior of arc plasmas. In non-CE state, it is difficult to adopt an LTE assumption with the mass action law such as the Guldberg–Waage and Saha equations to obtain the particle composition distribution and its transient evolution. In these cases, the following mass conservation equation for each of species j should be solved considering convection and diffusion effects and reaction rates to determine the time evolution in the particle composition distribution:

$$\frac{\partial(\rho Y_j)}{\partial t} + \nabla\cdot(\rho \boldsymbol{u} Y_j - \rho D_j \nabla Y_j) = S_j \qquad (1)$$

$$S_j = m_j \sum_{\ell=1}^{L} (\beta^\mathrm{r}_{j\ell} - \beta^\mathrm{f}_{j\ell}) \left(\alpha^\mathrm{f}_\ell \prod_{i=1}^{N} n_i^{\beta^\mathrm{f}_{i\ell}} - \alpha^\mathrm{r}_\ell \prod_{i=1}^{N} n_i^{\beta^\mathrm{r}_{i\ell}} \right) \qquad (2)$$

where Y_j signifies the mass fraction of species j, ρ stands for the mass density of the bulk plasma, D_j denotes the effective diffusion coefficient of j, α^f_ℓ and α^r_ℓ respectively stand for the rate coefficients of forward and backward reaction ℓ; also, n_i is the density of species i, $\beta^\mathrm{f}_{i\ell}$ and $\beta^\mathrm{r}_{i\ell}$ are the stoichiometric coefficient in reaction ℓ for forward and backward directions for species i, and N and L respectively denote the number of species and reactions considered. Thermodynamic and transport properties of plasmas should be calculated at each position at each time using the calculated temperature, pressure and particle composition combined with the collision integrals between the species.

3.2. Examples of chemically non-equilibrium modeling for SF_6 and N_2 arcs

Some research efforts have been undertaken to treat such chemically non-equilibrium effects in high-P and high-T plasmas for a long time [10]–[14]. Some non-equilibrium effects of SF_6 arc plasmas were studied theoretically or numerically from the 1990s [15]–[31]. Among these studies, chemically non-equilibrium modelling on SF_6 arcs had been conducted by Gleizes et al. [16]–[20] and the present authors [5],[23]–[31] .

Here, two examples of our recent non-CE modeling is introduced for SF_6 and N_2 arcs [5, 6].

3.2.1. Assumptions

The developed model assumes the following things in the decaying arc plasma for simplicity: (i) The arc plasma has the axisymmetric structure. (ii) The gas flow is laminar flow, thus the turbulent effect is neglected. (iii) The optically thin condition is assumed for radiation. (iv) All the temperatures including electron temperature, heavy particle temperature and excitation temperatures are identical. (v) Electrons, ions and neutral species have the same flow velocity. (vi) Electrode surface phenomena such as electron emission from the electrodes and surface recombination of ions were not considered for simplicity. (vii) Evaporation of the electrode and the nozzle was neglected. (viii) For SF_6 arcs, 19 species SF_6, SF_5, SF_4, SF_3, SF_2, SF, S_2, F_2, S, F, SF^+, S_2^+, F_2^+, SF^-, S^+, F^+, S^-, F^- and e^- were taken into account as constituents in SF_6 arc plasmas. On the other hand, N_2 arcs were assumed consisting of 5 species N_2, N_2^+, N, N^+ and e^-. (ix) Quasi-neutrality for electronic charge is established.

3.2.2. Governing equations

The following equations were solved to simulate an SF_6 decaying arc plasmas on the basis of the assumptions in the previous section as follows:

Mass:

$$\frac{D\rho}{Dt} = -\rho(\nabla \cdot \boldsymbol{u}), \tag{3}$$

Momentum:

$$\rho\frac{D\boldsymbol{u}}{Dt} = -\nabla p + \nabla \cdot \boldsymbol{\tau}, \tag{4}$$

$$\boldsymbol{\tau} = \tau_{ij} = 2\eta\left[e_{ij} - \frac{1}{3}\delta_{ij}(\nabla \cdot \boldsymbol{u})\right] \tag{5}$$

Energy:

$$\rho C_{\mathrm{v}}\frac{DT}{Dt} = -p(\nabla \cdot \boldsymbol{u}) + \nabla \cdot (\lambda_{\mathrm{tr}}\nabla T) + Q_{\mathrm{heat}} \tag{6}$$

$$Q_{\mathrm{heat}} = \sum_{j=1}^{N} \nabla \cdot (\rho D_j h_j \nabla Y_j) + \sum_{\ell}^{L} \Delta Q_\ell - P_{\mathrm{rad}} \tag{7}$$

$$h_j = \frac{1}{m_j}\left(\frac{5}{2}kT + kT^2\frac{\partial \ln Z_j}{\partial T}\right) \tag{8}$$

$$C_{\mathrm{v}} = \sum_{j=1}^{N}\frac{Y_j}{m_j}\left[\frac{3}{2}k + \frac{\partial}{\partial T}\left(kT^2\frac{\partial \ln Z_j}{\partial T}\right)\right] \tag{9}$$

Mass of species j:

$$\rho\frac{DY_j}{Dt} = \nabla \cdot (\rho D_j \nabla Y_j) + S_j, \tag{10}$$

$$S_j = m_j\sum_{\ell}^{L}(\beta^{\mathrm{r}}_{j\ell} - \beta^{\mathrm{f}}_{j\ell})\left(k^{\mathrm{f}}_\ell\prod_{i=1}^{N} n_i^{\beta^{\mathrm{f}}_{i\ell}} - k^{\mathrm{r}}_\ell\prod_{i=1}^{N} n_i^{\beta^{\mathrm{r}}_{i\ell}}\right) \tag{11}$$

The equation of state:

$$\rho = \frac{p}{\kappa\sum_j^N \frac{Y_j}{m_j}T} \tag{12}$$

where ρ: mass density, t: time, $\boldsymbol{u}$: gas flow vector, p: pressure, $\boldsymbol{\tau}$: stress tensor, η: viscosity, e_{ij}: unit tensor, δ_{ij}: Kronecker delta, T: temperature, C_{v}: effective specific heat at constant volume, λ_{tr}: translational thermal conductivity, D_j: effective diffusion coefficient, h_j: enthalpy of species j, Y_j: mass fraction of species j, P_{rad}: radiation power, ΔQ_ℓ: heating power from reaction heat of reaction ℓ, m_j: mass of species j, $\beta_{j\ell}$: stoichiometric coefficient of species j for reaction ℓ, $k_\ell^{\mathrm{f,r}}$: reaction rate coefficient for reaction ℓ, n_j: density of species j, κ: Boltzmann constant, $\frac{D}{Dt} = \frac{\partial}{\partial t} + \boldsymbol{u} \cdot \nabla$ is the Lagrangian derivative. Solving (10) calculates the particle composition in the arc plasma considering reaction rate, convection and diffusion effects.

For N_2 arc plasmas, a similar set of governing equations was solved but in forms of Fluent [6].

3.2.3. Properties and rate coefficients for reactions

The thermofluid calculation for arc plasmas needs the thermodynamic and transport properties such as mass density ρ, specific heat C_{v}, enthalpy h_j, translational thermal conductivity λ_{tr}, viscosity η, P_{rad}. These properties were calculated using the particle composition and the temperature computed and the collision integrals at each position at each calculation step. The calculation of the transport properties were carried out by the first order approximation of Chapman-Enskog method.

For SF_6 arcs, 64 forward reactions including the electron impact ionization, the electron attachment, the electron impact dissociation, and other chemical reactions were taken into account as well as their backward reactions [5]. These temperature-dependent coefficients were obtained in literatures [16]–[20]. In the present work, ion-ion recombination reactions such as $SF^+ + F^- + F \rightarrow SF + F + F$, $SF^+ + F^- + F \rightarrow F_2 + S + F$, and $SF^+ + F^- + F \rightarrow F_2 + SF$ were also taken into account to study the effects of these reactions on the composition in residual arcs, especially ion remaining between the electrodes. For N_2 arcs, totally 22 reactions were taken into account including 11 forward reactions and their backward reactions. The rate coefficients for these reactions were obtained in literatures [6].

Figure 1 depicts the calculation space. This space was divided into 153×170 grids for the r-z two dimensional space. This configuration is the same to that in

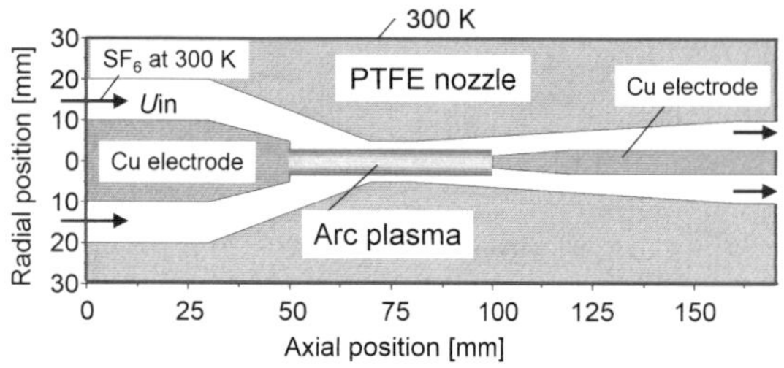

Figure 1. Calculation domain.

our experimental device. First, the 100 A dc current was applied to the arc in the steady state. Then, the transient state calculation was performed for a current stepped down from 100 A to 0 A, i.e. for free recovery condition for a fundamental study.

3.3. Examples of calculation results

3.3.1. SF_6 arcs: Transient distribution of F^- with and without ion-ion recombination reactions

Figure 2 indicates time evolutions in F^- density distribution under free recovery condition. We here treat transient behavior of F^- because F^- production is related to the electron density and may influence electrical conduction just around current zero [5]. In addition, we study non-CE effects on ion remaining between the electrodes. The upper panels in this figure thus show the results with consideration of ion-ion recombination reactions such as $SF^+ + F^- + F \rightarrow SF + F + F$, $SF^+ + F^- + F \rightarrow F_2 + S + F$, and $SF^+ + F^- + F \rightarrow F_2 + SF$, whereas the lowers are the results without consideration of ion-ion recombination reactions. By comparing the upper panel and the lower panel, we can investigate influence of ion-ion recombination reactions. At $t = 0\,\mu s$, F^- has higher density about $10^{20}\,m^{-3}$ around the boundary between the arc and cold SF_6 gas flow. At $t = 5\,\mu s$, F^- density increases rapidly with time around the fringe of the arc plasma. In particular, the arc fringe has higher F^- density more than $10^{21}\,m^{-3}$ around the nozzle throat inlet, i.e. positions $55 < mm\ z < 75\,mm$. This rapid increase in F^- density arises mainly from two-body dissociative attachment reactions with diatomic molecules $F_2 + e \rightarrow F^- + F$ and $SF + e \rightarrow F^- + S$ just after the current down. After that, F^- density is still high around $10^{21}\,m^{-3}$ around the nozzle throat inlet until $t = 10\,\mu s$. Even at $t = 50\,\mu s$, F^- becomes distributed with $10^{21}\,m^{-3}$ in whole arc region.

Consideration of the ion-ion recombination reactions influences transient behavior of F^- density distribution. At $t = 5\,\mu s$, F^- remains around nozzle throat inlet but in inner region. The density of F^- becomes lower at the fringe of the arc plasma. This is attributed to the contribution of the ion-ion recombination reactions considered. At $t = 10$–$50\,\mu s$ F^- density is predicted to be lower with the ion-ion recombination. However, it is noteworthy that F^- still remains with densities of $10^{20}\,m^{-3}$. This implies that F^- is one of the key charged species for electrical conduction in an SF_6 residual arc.

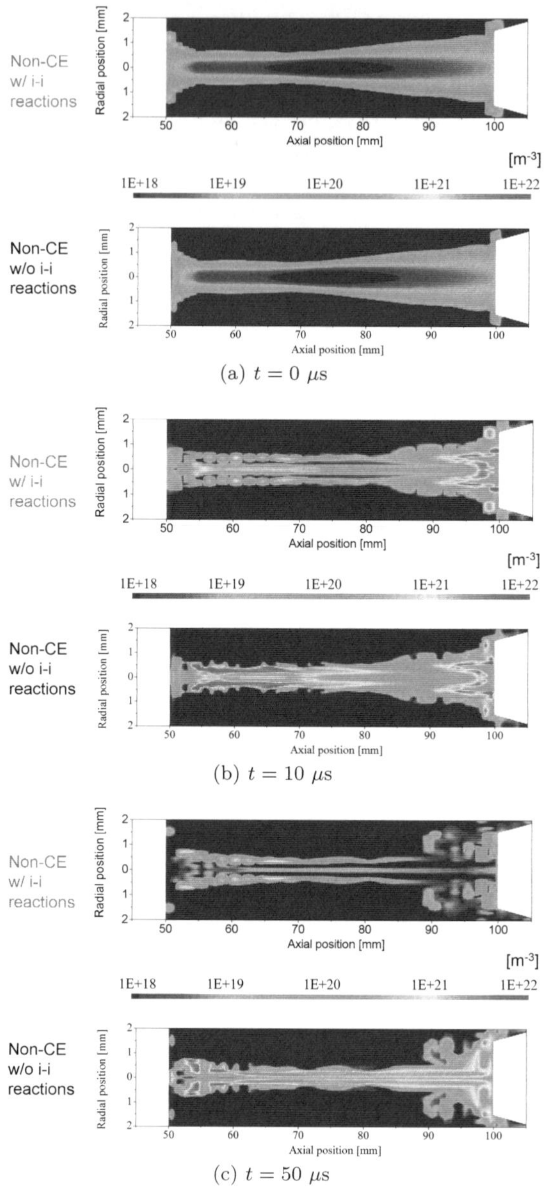

Figure 2. F^- density distribution in an SF_6 arc at $10\,\mu s$ after free recovery condition initiation. The upper panel: with ion-ion recombination; the lower panel: without ion-ion recombination.[5]

3.3.2. N_2 arcs: Temperature and composition change by chemically non-equilibrium model

We calculated N_2 arc behavior in the same arc device using a similar non-CE model [6]. Figure 3 presents the time variations in the temperature at axial position of nozzle throat at the center axis under free recovery condition. This figure includes two results by the LTE model and the non-CE model. During $t = 0$–$15\,\mu s$, the temperature decay calculated by the non-CE model is similar to that by the LTE model.

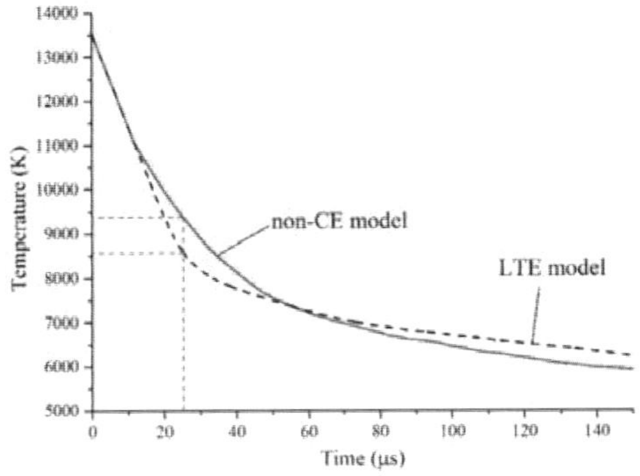

Figure 3. Time evolution in the temperature at nozzle throat in N_2 arcs [6].

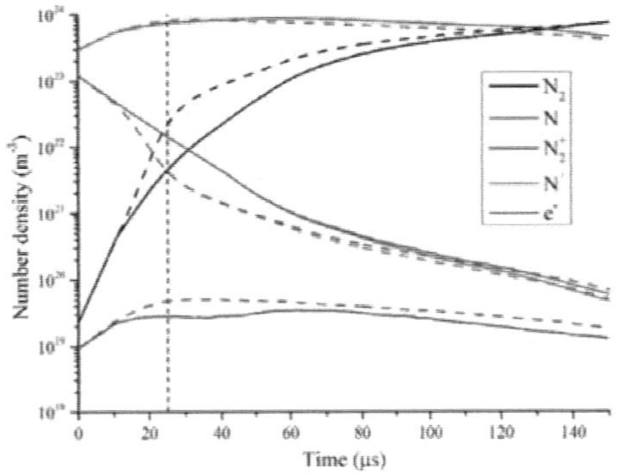

Figure 4. Time evolution of particle composition at nozzle throat in N_2 arcs [6].

The reason is that at the high temperature region more than 11000 K, the reaction rates are enough high and all the reactions tend to reach their equilibrium. At $t = 15$–25 µs, the LTE model predicts a more rapid temperature decay than the non-CE model. After $t = 50$ µs, the temperature in the non-CE model decreases with a higher decrease rate. After $t = 55$ µs, the temperature in the non-CE model is lower than that by the LTE model. The above difference in the temperature decreases by the LTE model and the non-CE model is attributed to reaction rates [6].

Figure 4 shows the time evolution in particle composition at nozzle throat in N_2 arcs under free recovery condition. Solid curves are the results by the non-CE model, whereas the dashed curves are the results by the LTE model. From $t = 0$ to 10 µs, two kinds of results hardly have deviation from each other. This is because the temperature is still higher than 10000 K, and thus the reaction rates for dominant reactions are high. However, at $t = 10$–25 µs, the electron number density, as well as the N^+ number density, decreases more slowly in the non-CE model than those in the LTE model. This higher electron density is due to the larger amount of exothermic recombination reactions. At $t > 25$ µs, N_2 density predicted by the non-CE model recovers more slowly than that by the LTE model. This arises from the delay in association reaction $N + N + M \rightarrow N_2 + M$.

4. Experimental approaches: Electron density measurement without LTE assumption

We have performed the electron density measurement in an arc under free recovery condition in a arc test device without LTE assumption. This measurement result can be compared with the numerical simulation result to study arc behavior including non-equilibrium effects.

4.1. Gas-blast arc device

Figure 5 shows a photo and the cross section diagram of the nozzle used in our experiments [7]. The nozzle

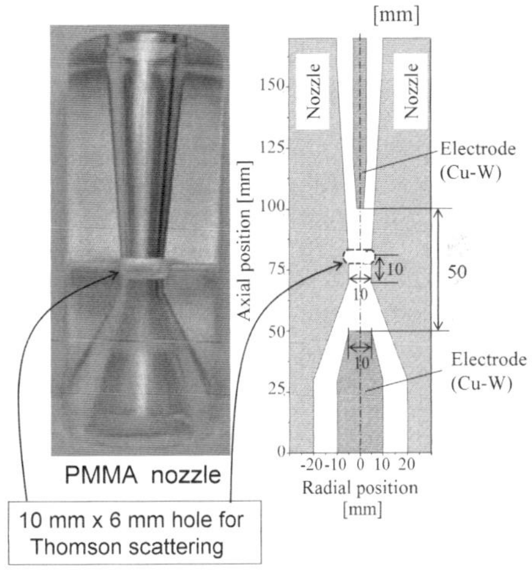

Figure 5. Configuration of the electrodes and the nozzle.

and the electrodes have a cylindrical axis-symmetric structure. The tip of the electrodes were made of copper-tungsten alloy. The lower electrode is fixed, while the upper electrode can move. The distance between the electrodes is 50 mm in open position. Between these electrodes, the arc plasma is formed. The nozzle is made of polymethyl methacrylate (PMMA), and then the visible light from the arc plasma can be observed through the nozzle. The nozzle throat width and length are 10 mm and 10 mm respectively. At the nozzle throat outlet, there is a observation hole with 10 mm × 6 mm for measuring the electron density in arc plasmas by laser Thomson scattering method. Gas flow is supplied from the downer side along the lower nozzle to the arc plasma axially.

Figure 6 depicts the experimental setup including the arc device and the electrical circuit. The nozzle and the electrodes in Fig. 5 are located inside a chamber made of stainless steel. Gas flow is supplied from a high-pressure gas tank to the lower side of the arc device. The upper electrode can open for a 50 mm stroke in about 80 ms by high-pressure air. The arc device was connected with a dc inverter current source. This

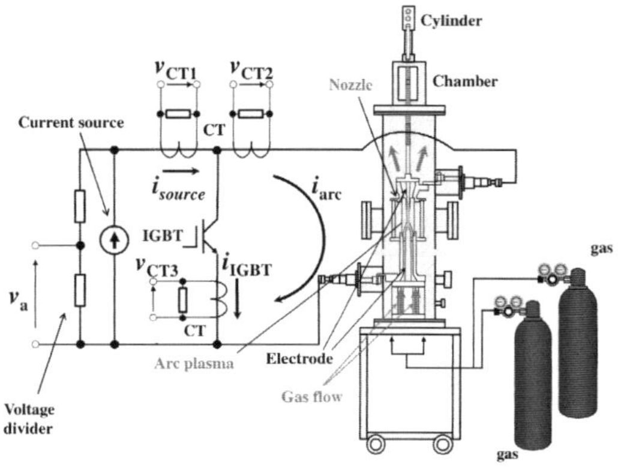

Figure 6. Experimental setup for decaying arc discharges.

inverter current source is originally used for plasma cutting arcs, and it is rated at 450 V/ 300 A. In parallel to the arc device, an insulated gate bipolar transistor (IGBT) semiconductor was connected. Switching on the IGBT commutates the source current from the arc device to the IGBT element to make the arc under free recovery condition. In the experiment, we measured the source current i_{source}, the arc current i_{arc}, the IGBT current i_{IGBT} using the current transducers (CT). The arc voltage was measured with a CR potential divider. At different specified timing after current commutation, the electron density at the nozzle throat was measured using a laser Thomson scattering method [8].

4.1.1. Laser Thomson scattering method

Laser Thomson scattering (LTS) is well used for precise measurements of electron density in arc plasmas. When a laser light is injected to an arc plasma, free electrons in the arc plasma are excited by electromagnetic field of the laser light, resulting in the light scattering. This is called Thomson scattering. The LTS method yields local electron density and electron temperature in plasmas with high spatial and temporal resolutions without LTE assumption. In case of the arc plasma which has a short Debye length, a incident laser light with a wavelength larger than the Debye length is often used from the electron term to determine the electron density and electron temperature. We have measured the electron density in an arc under free recovery condition for our arc device depicted in Fig. 5 without LTE assumption by LTS method [8]. In this study, the second harmonic light of an Nd:YAG laser at a wavelength of 532 nm was used for LTS. The 90° scattered light from the laser path was detected.

Figure 7 indicates an example of LTS spectrum from the arc in Ar/20% SF_6 at a flow rate of 100 L/min [8]. We used an 2D-ICCD camera as a detector, the LTS spectra along the laser path, corresponding the radial direction of the arc, were simultaneously obtained as

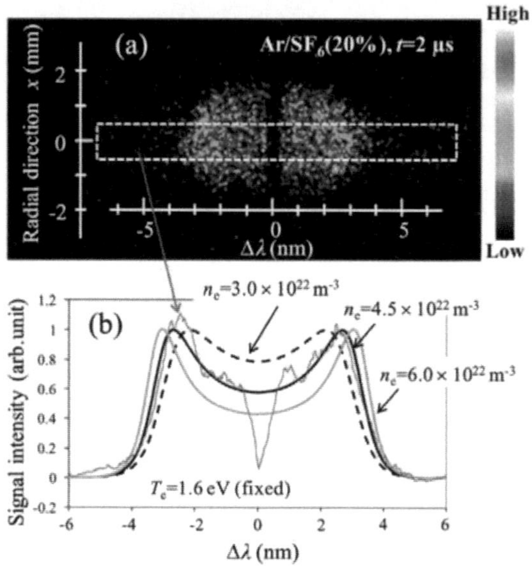

Figure 7. Laser Thomson scattering spectrum measured for Ar/SF_6 arcs and theoretical fitting curves [8].

indicated in Fig. 7 (a). Figure 7 (b) shows the measured LTS spectrum at the center of the arc, and the theoretical curves of LTS for different electron densities. In this case, the electron density and the electron temperature were estimated to be $4.5\times10^{22}\,\mathrm{m}^{-3}$ and 1.6 eV, respectively, by fitting the theoretical curve to the measured spectrum.

Figure 8 presents the time evolutions in the measured electron density at the center position in nozzle throat region in Ar/SF_6 arcs under free recovery condition. The electric current was changed from 50 A to 0 A at $t = 0\,\mu\mathrm{s}$. The gas flow rate of Ar/SF_6 is 100 L/min. As seen, the electron density reaches to $8\times10^{22}\,\mathrm{m}^{-3}$ at 50 A at $t = 0\,\mu\mathrm{s}$. After current down to 0 A at $t > 0\,\mu\mathrm{s}$, the electron density decays almost exponentially. In particular, the 100%SF_6 arc has a most rapid decay in the electron density with a time constant of 10 μs. In this way, the electron density could be determined for decaying arcs at different gas mixture and different gas kinds without LTE assumption. The comparison of the results by such LTS method with the theoretically calculated result in time evolution in the electron density will be introduced in later section.

4.1.2. Shack-Hartman method

Recently, two-dimensional distribution of electron density in a plasma is measured by Shack-Hartmann type laser wavefront sensor [9]. Figure 9 shows the basic concept of a Shack-Hartmann sensor for measuring electron density distribution in an arc plasma. A Shack-Hartmann sensor is composed of an image sensor, e.g. an ICCD camera, and a sheet of micro-lens arrays. Micro-lens arrays convert the localized wavefront gradients of a laser beam into the shifts

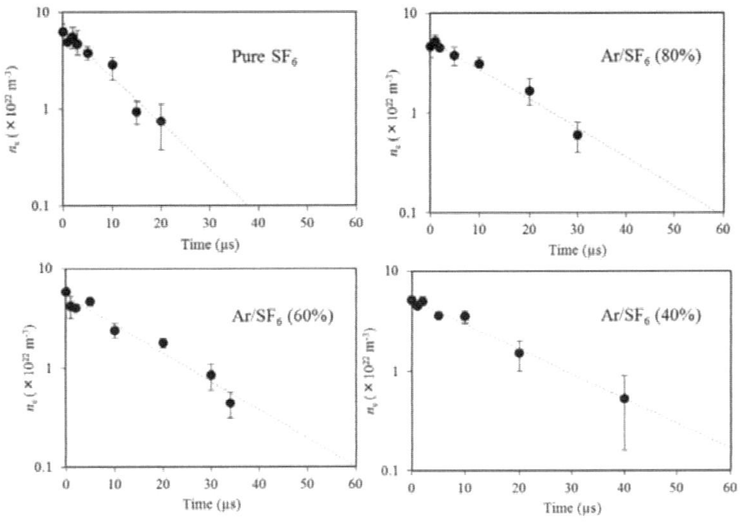

Figure 8. Temporal evolution in measured electron density in Ar/SF_6 (40, 60, and 80% SF_6) arcs current commutation under free recovery condition [8].

of the focal spot positions. A laser beam with the flat wavefront is irradiated to the arc plasmas. The laser wavefront profile through the plasma is locally distorted by different refractive indexes from the local electron density in the plasma. This distorted wavefront makes the shifted focal spot positions by the micro-lens array. This shifted spot positions were measured by the imaging sensor. This shifted displacement depends on the gradient in electron density. Thus, the electron density profile can be obtained without LTE assumption.

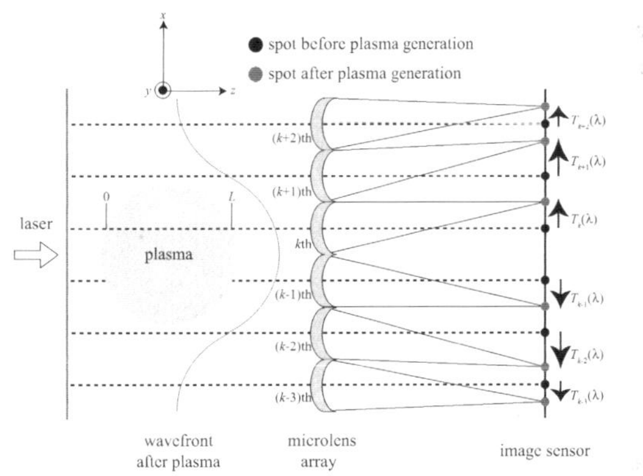

Figure 9. A schematic of a Shack-Hartmann type laser wavefront sensor for measuring electron densities in a plasma [9].

Figure 10 indicates the two-dimensional electron density distribution at $t = -10, 10, 30$ and 50 μs after current commutation from 50 A to 0 A to produce free recovery condition. The two-dimensional region of 4.5 mm × 4.5 mm in the nozzle throat was taken as a region for electron density measurement. The CO_2 gas is supplied from the lower region at a flow rate of 100 L/min. At $t = -10$ μs, the dc arc with an electric current of 50 A is established between the electrodes. As seen in this figure, at $t = -10$ μs, a two-dimensional high electron density area can be obtained with magnitudes above 10^{23} m^{-3} from upstream (lower) to downstream (upper). After a current down to 0 A at $t > 0$ μs, the electron density decreases with time under free recovery condition, and the high electron density region becomes narrower. At $t = 30$ μs, the electron density decreases to 3×10^{22} m^{-3}. At $t = 50$ μs, the upstream portion has the decreased electron density less than 10^{21} m^{-3} Such a two-dimensional electron density distribution can be obtained by Shack-Hartmann method.

4.2. Comparison of measurement results by Laser Thomson scattering with results calculated by chemically non-equilibrium model

The electron density measured by laser Thomson scattering (LTS) method was compared with those calculated by the non-CE model and the LTE model [6]. Figure 11 shows the comparison for a N_2 decaying arc under free recovery condition from 50 A to 0 A. As seen in this figure, at $t = 0$–10 μs, the measurement results fairly agree with both those by the non-CE model and the LTE model. At $t > 10$ μs, the LTE model predicted a rapid decay in the electron density. However, at $t > 10$ μs, the electron density measured by the LTS method is higher than that by the LTE model. On the other hand, the non-CE model predicted the electron density fairly agreeing with that

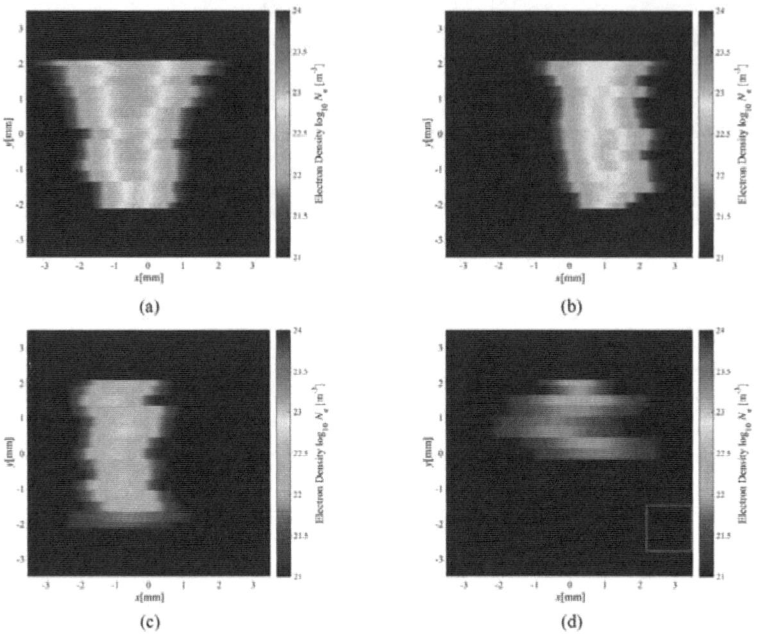

Figure 10. Two-dimensional electron density distribution in a CO_2 decaying arc plasma at t = (a) 0.1, (b) 10, (c) 30 and (d) 50 μs [9].

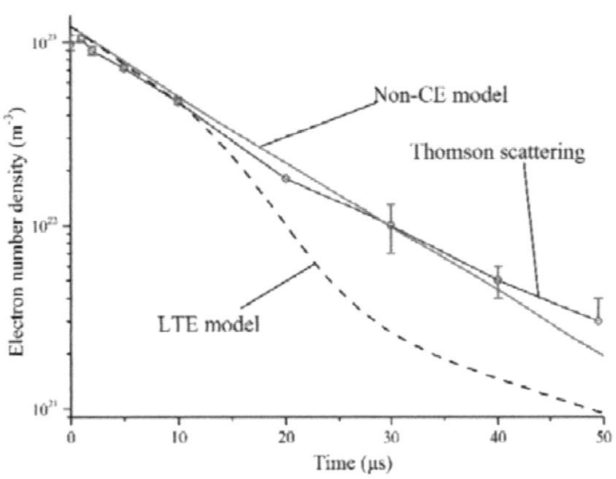

Figure 11. Comparison of electron density measurement results by laser Thomson scattering with calculation results of electron density predicted by chemically non-equilibrium model [6].

by the LTS method. The difference in electron density by the non-CE model from that by the LTE model is due to consideration of reaction rate of $N_2^+ + e \rightarrow N + N$. This implies a validity of the non-CE model, and an importance of considering non-CE effect in decaying arcs.

5. Conclusions

This paper briefly summarizes one recent approach for switching arcs in Japan: a modeling work of arc plasmas considering non-equilibrium effects in high-pressure high-temperature plasmas, and experimental works for electron density measurement without LTE assumption. Results indicates an importance of chemically non-equilibrium effects in a switching arc.

References

[1] A. Gleizes et al. Thermal plasma modelling. *J. Phys. D: Appl. Phys.*, 38(9):R153, 2005. doi:10.1088/0022-3727/38/9/R01

[2] V. Rat et al. Treatment of non-equilibrium phenomena in thermal plasma flows. *J. Phys. D: Appl. Phys.*, 41(18):183001, 2008. doi:10.1088/0022-3727/41/18/183001

[3] Y. Tanaka, T. Muroya, K. Hayashi, and Y. Uesugi. Control of nitrogen atomic density and enthalpy flow into reaction chamber in Ar-N_2 pulse-modulated induction thermal plasmas. *IEEE Trans. Plasma Sci.*, 35:197-203, 2007. doi:10.1109/TPS.2007.892709

[4] Y. Tanaka, K. Hayashi, T. Nakamura, and Y. Uesugi. Influence of ontime on increased number density of excited nitrogen atom in pulse modulated induction thermal plasmas. *J. Phys. D: Appl. Phys.*, 41(18):185203, 2008. doi:10.1088/0022-3727/41/18/185203

[5] Y. Tanaka, K. Suzuki, and T. Iijima. Role of ion-ion recombinations in decaying SF_6 arc plasmas. *Proc.21th Int. Conf. Gas Discharges and their Applications GD2016*, A15, pp.57-60, 2016.

[6] H. Sun, Y. Tanaka, K. Tomita, Y. Wu, M. Rong, Y. Uesugi, and T. Ishijima. Computational non-chemically equilibrium model on the current zero simulation in a model N_2 circuit breaker under the free recovery condition. *J. Phys. D: Appl. Phys.*, 49(5):055204, 2015. doi:10.1088/0022-3727/49/5/055204

[7] Y.Tanaka, T.Nakano, H.Shimizu, K.Tomita, and K.Suzuki. Fundamental investigation technique on gas-blast arc behaviors in decaying and re-ignition processes using power semiconductors. *Trans. IEEJ PE*, 135(11):661-668, 2015. doi:10.1541/ieejpes.135.661

[8] K. Tomita, D. Gojima, T. Shimizu, K. Uchino, T. Nakano, Y. Tanaka, K. Suzuki, T. Iijima, and T. Shinkai. Thomson scattering diagnostics of SF_6 gas-blasted arcs confined by a nozzle under free recovery conditions. *J. Phys. D: Appl. Phys.*, 48(26):265201, 2015. doi:10.1088/0022-3727/48/26/265201

[9] Y. Inada, A. Kumada, H. Ikeda, K. Hidaka, T. Nakano, K. Murai, Y. Tanaka, and T. Shinkai. Comparative study on extinction process of gas-blasted air and CO_2 arc discharge using twodimensional electron density imaging sensor. *J. Phys. D: Appl. Phys.*, 50:175202, 2017. doi:10.1088/1361-6463/aa652b

[10] G.Y. Zhao, J. Mostaghimi, and M.I. Boulos. The induction plasma chemical reactor: Part II. Kinetic Model. *Plasma Chem. Plasma Process.*, 10(1):151–166, 1990. doi:10.1007/BF01460453

[11] J. B. Belhaouari, J.J. Gonzalez, and A. Gleizes. Simulation of a decaying SF_6 arc plasma: hydrodynamic and kinetic coupling study. *J. Phys. D: Appl. Phys.*, 31(10):1219–1232, 1998. doi:10.1088/0022-3727/31/10/014

[12] A. B. Murphy and T. McAllister. Modeling of the physics and chemistry of thermal plasma waste destruction. *Phys. Plasmas*, 8(5):2565–2571, 2001. doi:10.1063/1.1345884

[13] Y. Tanaka, T. Michishita, and Y. Uesugi. Hydrodynamic chemical non-equilibrium model of a pulsed arc discharge in dry air at atmospheric pressure. *Plasma Sources Sci.& Technol.*, 14(1):134–151, 2005. doi:10.1088/0963-0252/14/1/016

[14] S.A. Al-Mamun, Y. Tanaka, and Y. Uesugi. Chemically non-equilibrium modelling of Ar inductively coupled thermal plasma with C-H-O systems. *J. Plasma & Fusion Res. Ser.*, 8:1243–1247, 2009.

[15] A. Gleizes, F. Mbolidi, and A.A.M. Habib. Kinetic model of a decaying SF_6 plasma over the temperature range 12000 K to 3000 K. *Plasma Sources Sci. Technol.*, 2(3):173-179, 1993. doi:10.1088/0963-0252/2/3/007

[16] R. Girard, J.J. Gonzalez, and A. Gleizes. Modelling of a two-temperature SF_6 arc plasma during extinction. *J. Phys. D: Appl. Phys.*, 32(11):1229-1238, 1999. doi:10.1088/0022-3727/32/11/308

[17] A. Gleizes, B. Chervy, and J.J. Gonzalez. Calculation of a two-temperature plasma composition: bases and application to SF_6. *J. Phys. D: Appl. Phys.*, 32(16):2060-2067, 1999. doi:10.1088/0022-3727/32/16/315

[18] R. Girard, J.B. Belhaouari, J.J. Gonzalez, and A. Gleizes. A two-temperature kinetic model of SF_6 plasma. *J. Phys. D: Appl. Phys.*, 32(22):2890-2901, 1999. doi:10.1088/0022-3727/32/22/311

[19] I. Coll, A.M. Casanovas, L. Vial, A. Gleizes, and J. Casanovas. Chemical kinetics modelling of a decaying SF_6 arc plasma in the presence of a solid organic insulator, copper, oxygen and water. *J. Phys. D: Appl. Phys.*, 33(3):221-229, 2000. doi:10.1088/0022-3727/33/3/308

[20] J.J. Gonzalez, R. Girard, and A. Gleizes. Decay and post-arc phases of a SF_6 arc plasma: a thermal and chemical non-equilibrium model. *J. Phys. D: Appl. Phys.*, 33(21):2759-2768, 2000. doi:10.1088/0022-3727/33/21/314

[21] Y. Tanaka, Y. Yokomizu, M. Ishikawa, and T. Matsumura. Electrical conductivity and enthalpy of two-temperature SF_6 plasma at a pressure of 0.1 MPa. *Proc. Int. Conf. on Electr. Eng. '97*, 726-729, 1997.

[22] T. Matsumura, Y. Yokomizu, T. Matsubara, and Y. Tanaka. Electrical conductivity and enthalpy of SF_6 plasma in two-temperature state. *Proc. 12th Int. Conf. on Gas Discharges & Their Appl.*, I, 94-97, 1997.

[23] Y. Tanaka, Y. Yokomizu, T. Matsubara, and T. Matsumura. Particle composition of two-temperature SF_6 plasma in pressure range from 0.1 to 1 MPa. *Proc. 12th Int. Conf. on Gas Discharges & Their Appl.*, II, 566-569, 1997.

[24] T. Matsubara, Y. Tanaka, Y. Yokomizu, and T. Matsumura. Deviation from local thermal equilibrium in SF_6 post-arc channel exposed to transient recovery voltage. *1998 IEEE Int. Conf. on Plasma Sci.*, 150, 1998. doi:10.1109/PLASMA.1998.677556

[25] Y. Tanaka, T. Matsubara, Y. Yokomizu, T. Matsumura, and T. Sakuta. Calculation of SF_6 plasma composition in two-temperature steady state using reaction kinetics. *Proc. Int. Conf. on Electr. Eng. ICEE'98*, 583-586, 1998.

[26] Y. Tanaka and T. Sakuta. Numerical approach for analysing transient behaviour of SF_6 induction thermal plasma using reaction kinetics. *Proc. Int. Symp. on Plasma Chem.*, 245-250, 1999.

[27] Y. Tanaka and K. Suzuki. *Proc. Int. Conf. Elect. Power Equip. ICEPE-ST 2011*, 2011.

[28] Y. Tanaka, K.Suzuki, T.Iijima, and T.Shinkai. Development of a thermally and chemically non-equilibrium model for decaying SF_6 arc plasmas. *2nd Int. Conf. Electric Power Equip. ICEPE-ST 2013*, 2013. doi:10.1109/ICEPE-ST.2013.6804353

[29] Y. Tanaka and K.Suzuki. Numerical Simulation on Two-Temperature Chemically Non-Equilibrium States in Decaying SF_6 Arcs after Application of Recovery Voltage. *20th Int. Conf. Gas Discharges and Their Appl. GD2014*, A34, pp.179-182, 2014.

[30] Y. Tanaka and K. Suzuki. Non-Equilibrium Effects on SF_6 Arc Plasmas in Decaying Phases. *Int. Conf. Electric Power Equipment ICEPE-ST 2015*, GO-181, 2015. doi:10.1109/ICEPE-ST.2015.7368379

[31] Y. Tanaka and K.Suzuki. A Fundamental Study on Chemically Non-Equilibrium State in Decaying SF_6 Arc Plasmas. *Int.Conf.Fluid Dynamics (ICFD2016)*, 2016.

InterSafe
Protecting People

Challenging Slips and Falls

1 December 2017

Brendan McDougall

InterSafe

Why focus on slips and falls at the same level?

- 15% of hospital admissions due to injury and poisoning
- 14.6% of all serious injury (>5 days Lost Time)
- 12.2% of >60 days Lost Time Injuries - Qld Mining Industry

Protecting People

InterSafe — Contributing Factors

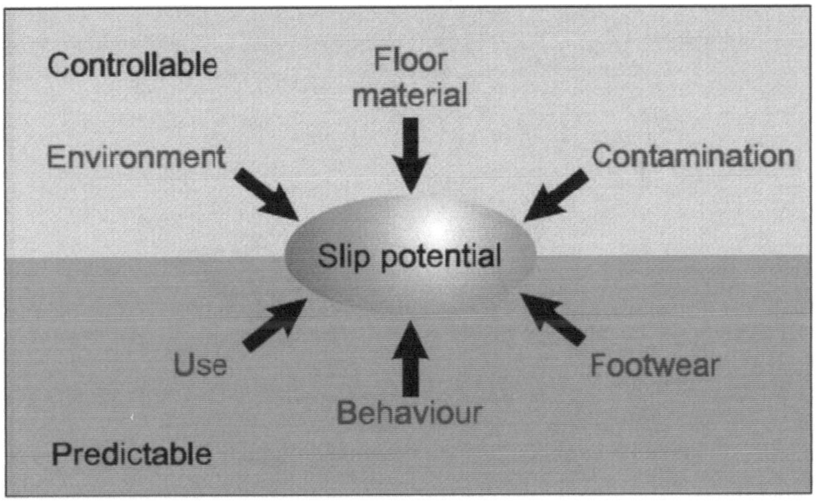

Protecting People

InterSafe — Source of information

- AS/NZS 4586 (2004/2013)
- HB 197 (1999)
- HB 198 (2014)
- National Construction Code

Protecting People

Wet Slip Resistance Test Methodology

InterSafe

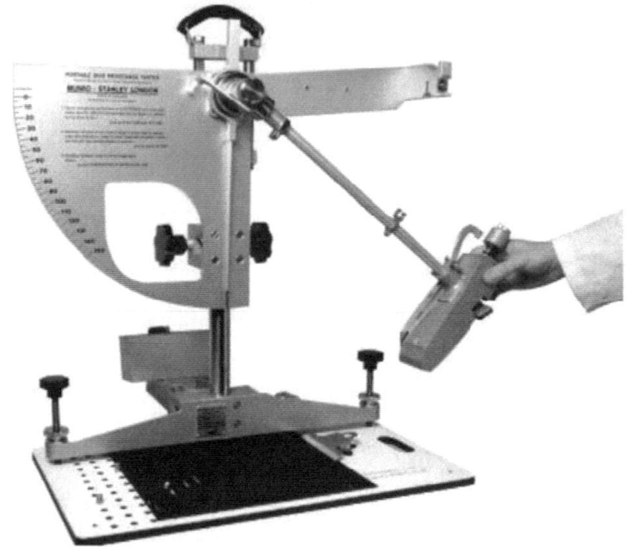

Protecting People

Slip Resistance Contribution to risk of pedestrian slipping

InterSafe

Class	Pendulum mean (BPN)*		Contribution of the floor surface to the risk of slipping when wet
	Four S rubber	TRRL rubber	
V	> 54	> 44	**Very low**
W	45-54	40-44	**Low**
X	35-44	-	**Moderate**
Y	25-34	-	**High**
Z	**< 25**	**-**	**Very high**

Protecting People

InterSafe

HB 197
Specific Recommendations

Location	Pendulum
External colonnade, walkways and pedestrian crossings	W
External ramps	V
Entry foyers hotel, office, public buildings – wet	X
Entry foyers hotel, office, public buildings – dry	Z
Shopping centre excluding food court	Z
Shopping centre – food court	X
Internal ramps, slopes (greater than 2 degrees) - dry	X
Lift lobbies above external entry level	Z
Other separate shops inside shopping centres	Z
Other shops with external entrances – entry area	X
Fast food outlets, buffet food servery areas	X
Hospitals and aged care facilities – dry areas	Z
Hospital and aged care facilities – ensuites	X
Supermarket aisles except fresh food areas	Z
Shop and supermarket fresh fruit and vegetable areas	X
Communal changing rooms	X
Swimming pool surrounds and communal shower rooms	W
Swimming pool ramps and stairs leading into water	V
Toilet facilities in offices, hotels, shopping centres	X
Undercover concourse areas of sports stadium	X
Accessible internal stair nosings (dry) – handrails present	X
Accessible internal stair nosings (wet) – handrails present	W
External stair nosings	W

Protecting People

InterSafe

UK Slip Resistance Group Guidelines

Pendulum Value (SRV) (Four-S Rubber)	Potential for Slip*
0 – 24	High
25 – 35	Moderate
36 – 64	Low

Protecting People

InterSafe Basis for the risk ratings

Friction utilised during normal walking determined from studies of pedestrians walking over force plates

Risk	Straight walking	Turning: Left Foot	Turning: Right foot
1 in 1,000,000	0.36	0.40	0.36
1 in 100,000	0.34	0.38	0.34
1 in 10,000	0.29	0.34	0.33
1 in 200	0.27	0.31	0.32
1 in 20	0.24	0.27	0.29

Protecting People

InterSafe Can such research be extended to predict risk of pedestrians slipping?

- Whether the mathematical probability that required coefficient of friction will exceed the available coefficient of friction provides an accurate insight into actual slip and fall incidents and how the measure of risk should be interpreted to compare the potential for injury with other risks which exit is workplaces and public places.

Protecting People

InterSafe Analysis

100 randomly selected slip and fall incidents investigated by InterSafe where the person sustained some degree of permanent damage

- Limited to 18-65 yr olds
- Limited to level walking surfaces
- Surface tested in a similar condition to that present at the time of the incident

Protecting People

Observations

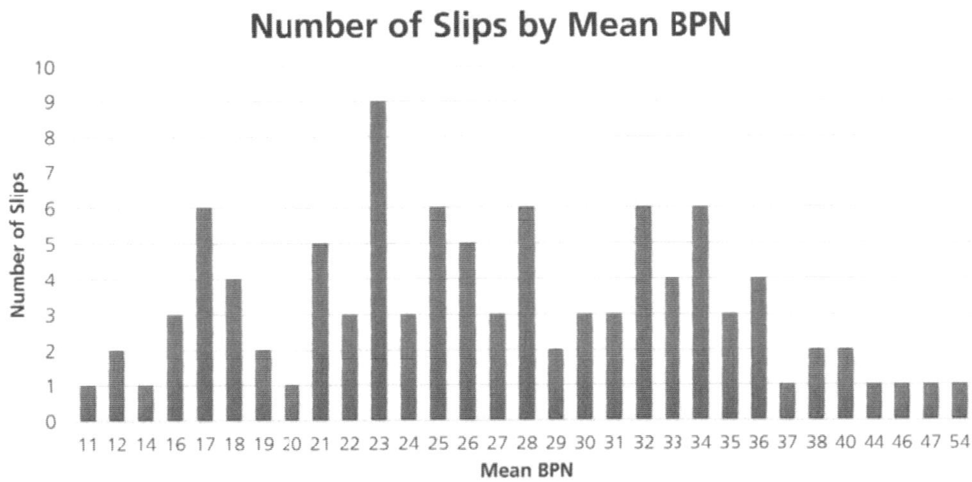

Protecting People

InterSafe Findings

InterSafe

- 40% occurred on surfaces with < 25BPN
- 44% occurred on surfaces with a slip resistance in the 25 – 34 range
- 13% occurred on surfaces with a slip resistance in the 35 – 44 range
- 3% occurred on surfaces with a slip resistance of > 45 range.

Protecting People

Conclusions

InterSafe

- A minimum slip resistance of 40BPN (or even 45) for a walking surface which could become wet is required
- The data challenges the UK Slip Resistance Group Guidelines' recommended BPN of 35 for low risk
- The data challenges HB 197's recommended Class X slip resistance for most internal surfaces which become wet (BPN as low as 35)

Protecting People

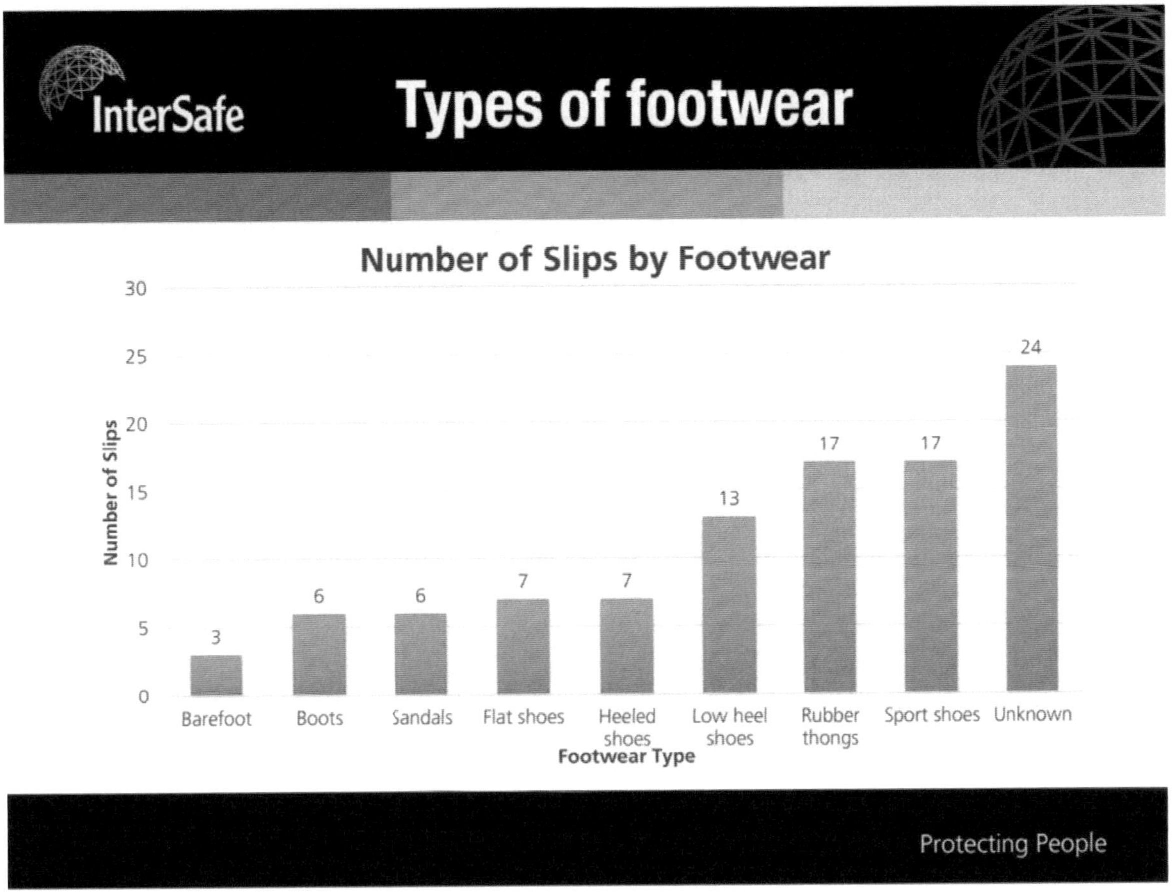
InterSafe
Types of footwear
Number of Slips by Footwear
Number of Slips
30
25
20
15
10
5
0
3
6
6
7
7
13
17
17
24
Barefoot
Boots
Sandals
Flat shoes
Heeled shoes
Low heel shoes
Rubber thongs
Sport shoes
Unknown
Footwear Type
Protecting People

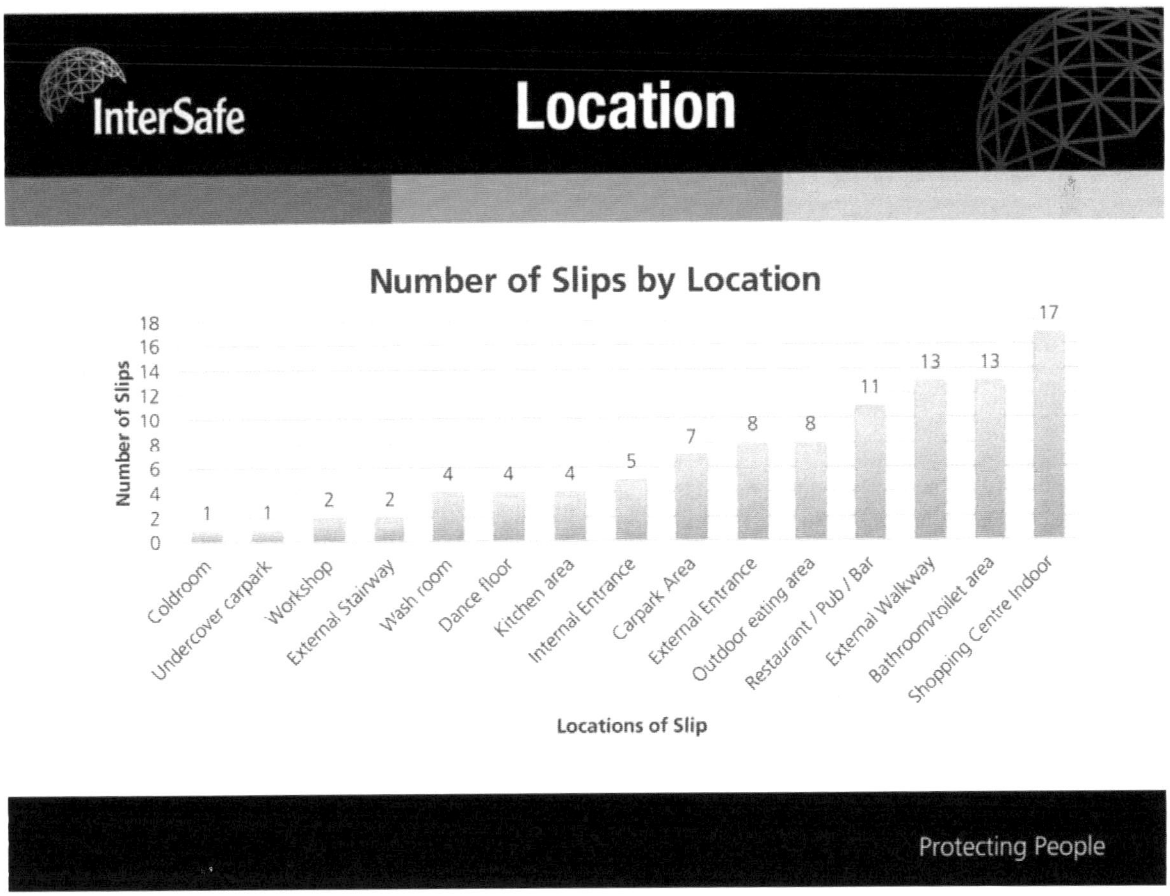
InterSafe
Location
Number of Slips by Location
Number of Slips
18
16
14
12
10
8
6
4
2
0
1
1
2
2
4
4
4
5
7
8
8
11
13
13
17
Coldroom
Undercover carpark
Workshop
External Stairway
Wash room
Dance floor
Kitchen area
Internal Entrance
Carpark Area
External Entrance
Outdoor eating area
Restaurant / Pub / Bar
External Walkway
Bathroom/toilet area
Shopping Centre Indoor
Locations of Slip
Protecting People

InterSafe

AS 4586 – 2013 / HB 198

Classification	Wet pendulum SRV	
	Slider 96 Rubber	Slider 55 Rubber
P5	>54	>44
P4	45 – 54	40 – 44
P3	35 – 44	34 – 39
P3	25 – 34	20 – 34
P1	12 – 24	<20
P0	<12	

Protecting People

InterSafe

HB 198 Specific Recommendations

Location	Wet pendulum test	Oil-wet inclining platform test
Hotels, Offices, Public Buildings, Schools and Kindergartens		
Entries and access areas including hotels, offices, public buildings, schools, kindergartens, common areas of public buildings, internal lift lobbies		
Wet area	P3	R10
Transitional area	P2	R9
Dry area	P1 (see Note 3)	R9
Toilet facilities in offices, hotels and shopping centres	P3	R10
Hotel apartment bathrooms, en suites and toilets	P2	A
Hotel apartment kitchens and laundries	P2	R9
Supermarkets and Shopping Centres		
Fast food outlets, buffet food servery areas, food courts and fast food dining areas in shopping centres	P3	R10
Shop and supermarket fresh fruit and vegetable areas	P3	R10
Shop entry areas with external entrances	P3	R10
Supermarket aisles (except fresh food areas)	P1 (see Note 3)	R9
Other separate shops inside shopping centres—wet	P3	R10
Other separate shops inside shopping centres—dry	P1 (see Note 3)	R9
Loading Docks, Commercial Kitchens, Cold Stores, Serving Areas		
Loading docks under cover and commercial kitchens	P5	R12
Serving areas behind bars in public hotels and clubs, cold stores and freezers	P4	R11

ting People

InterSafe

Recommendation

InterSafe's analysis of 100 randomly selected actual slip and fall incidents strongly supports the need for level walking surfaces which could become wet to have a slip resistance of at least 40 BPN (or SRV) **maintained** throughout the life of that surface.

Protecting People

doi:10.14311/ppt.2017.3.234
Plasma Physics and Technology **4**(3):234–240, 2017

MECHANISMS RESPONSIBLE FOR ARC COOLING IN DIFFERENT GASES IN TURBULENT NOZZLE FLOW

Y. Guo[a], H. Zhang[a], Y. Yao[a], Q. Zhang[b], J.D. Yan[b,*]

[a] *Pinggao Group Co. Ltd., Pingdingshan City, Henan, China, 467001*
[b] *Department of Electrical Engineering and Electronics, the University of Liverpool, Liverpool L69 3GJ, the U.K.*
[*] yaneee@liverpool.ac.uk

Abstract. A high voltage gas blast circuit breaker relies on the high speed gas flow in a nozzle to remove the energy due to Ohmic heating at high current and to provide strong arc cooling during the current zero period to interrupt a fault current. The physical mechanisms that are responsible for the hugely different arc cooling capabilities of two gases (SF_6 and air) are studied in the present work and important gas material properties controlling the cooling strength identified.

Keywords: SF_6 replacement, SF_6 alternative gases, switching arc.

1. Introduction

SF_6 has long been exclusively used in gas blast circuit breakers at voltage levels above 245 kV because of its excellent dielectric strength and current interruption capability. It is however a strong greenhouse gas with a Global Warming Potential of 23,500 [1]. There has been increasing worldwide effort in the last 10 years to search for alternatives gases that can replace SF_6 for high current switching. Most of the work carried out so far has however focused on the dielectric performance of potential gases such as CF_3I, $C_5F_{10}O$ and C_4F_7N and their mixtures with CO_2 [2–4], operating temperature of gas mixture [5], gas decomposition [6] and toxicity [6]. There is a limited amount of experimental work on the interruption capability of the potential alternative gases [7–9], but little work towards a quantitative understanding of the mechanisms responsible for the hugely different interruption capabilities of different gases.

The present work is aimed towards a quantitative explanation of the relevant importance of different energy exchange mechanisms participating in the arc cooling process and the identification of the causes that control their relevant cooling strength. The arc model will be first introduced with a discussion on the choice of the turbulent models. This is followed by a verification of the model using existing experimental results for which test conditions are known. The temperature distribution of the arc column and the energy exchange fluxes due to thermal conduction (including turbulent enhanced heat exchange), convection and radiation will be analysed to identify the mechanisms through which different gases produce different arc cooling effect. It is expected that the findings will be directly relevant to the composition or selection of SF_6 alternative gases by relating the interruption capability of a gas to its material properties.

2. Arc model

2.1. Governing equation

Local thermodynamic equilibrium (LTE) is a commonly accepted assumption for the plasma state in switching arcs. Gas flow inside and around the arc column which is confined in a nozzle is turbulent in nature and can be described by the time averaged Navier-Stokes equations modified to take into account the effects of Ohmic heating, radiation transfer and electromagnetic field. By assuming axisymmetry for the switching arc, the conservation equations are given below in cylindrical coordinates:

$$\frac{\partial(\rho\phi)}{\partial t} + \frac{1}{r}\frac{\partial[r\rho v\phi - r\Gamma_\phi \frac{\partial\phi}{\partial r}]}{\partial r} + \frac{\partial[\rho w\phi - \Gamma_\phi \frac{\partial\phi}{\partial z}]}{\partial z} = S_\phi \tag{1}$$

where ϕ is the dependent variable and ρ the gas density. v and w are respectively the radial and axial velocity components. The source terms (S_ϕ) and the diffusion coefficients (Γ_ϕ) are listed in Table 1 where all notations have their conventional meaning. The subscript l denotes the laminar part of the exchange coefficient and t the turbulent part. Viscous heating due to molecular and turbulent stresses is given in the source term for the enthalpy equation (Table 1).

The equation of state and the thermodynamic properties and transport coefficients including electrical conductivity are determined by the gas temperature and pressure only under LTE and usually given in the form of data tables. These data are taken from [10] for SF_6, and [11, 12] for air.

For low current nozzle arc, the radial component of electrical field is negligible in comparison with the axial component and the radial variation of the axial component is much smaller than its magnitude. Therefore, the axial electrical field is considered to be constant over the arc cross-section, which can be calculated by the simplified Ohmic law

Equation	ϕ	Γ_ϕ	S_ϕ
Continuity	1	0	0
Z-momentum	w	$\mu_l + \mu_t$	$-\frac{\partial p}{\partial z}$
R-momentum	v	$\mu_l + \mu_t$	$-\frac{\partial p}{\partial r} - (\mu_l + \mu_t)\frac{v}{r^2}$
Enthalpy	h	$\frac{k_l+k_t}{C_p}$	$\frac{dp}{dt} + \sigma E^2 - q + (\mu_l + \mu_t)\{2[(\frac{\partial v}{\partial r})^2 + \frac{v^2}{r^2} + (\frac{\partial w}{\partial z})^2] + (\frac{\partial v}{\partial z} + \frac{\partial w}{\partial z})^2\}$

Table 1. Terms in governing equations (1).

$$i = E\int_0^\infty \sigma\, 2\pi\, r\, \mathrm{d}r \qquad (2)$$

where i is the instantaneous current and σ the electrical conductivity.

For an axisymmetric arc with monotonically decreasing radial temperature profile, radiation transport can be calculated with the approximate model of Zhang et al. [13] which calculates the volumetric radiative energy loss in the arc core (from axis up to R_{83} which is the radius corresponding to 83 % of the axis temperature) based on the concept of net emission coefficient (NEC) and radiation absorption (from R_{83} to R_{4K} which is the radius corresponding to 4000 K) in the surrounding gas layer. The NEC values as a function of pressure and temperature under LTE is from [14] for SF_6 and [15, 16] for air and nitrogen. The NEC is defined for an isothermal cylindrical column of infinite length. In switching arc applications, the arc column is never isothermal. Therefore the use of the NEC is only approximate and the definition of the arc radius will affect the accuracy of the calculation of the emitted power from the arc core. By comparing with the measured arc temperature, it was found that the NEC data based on an emission radius defined as $0.5(R_{83} + R_{4K})$ needs to be multiplied by a factor of 2.5 to achieve good agreement. This approximate model has been proven sufficiently accurate in the modelling of nozzle arcs. The percentage of the radiation flux from the arc core that is absorbed at the arc edge is a parameter in the approximate model. It is 80 % for SF_6 and 60 % for air based on previous studies.

2.2. Turbulence models

There are numerous turbulence models, however there is no general theoretical guidance regarding the choice of turbulence models for arcs in supersonic flow. Prandtl mixing length model has achieved considerable success in predicting turbulent arc behavior. The standard K-epsilon model with the default values for the five parameters and two of its variants (the renormalization group, commonly known as the RNG model and Chen-Kim model) have been used for the modelling of turbulent arc flow in circuit breakers with contradictory claims regarding their successes. The Prandtl mixing length model relates the turbulence length scale to the width of the jet which marks the boundary of the high velocity core. It is calculated by

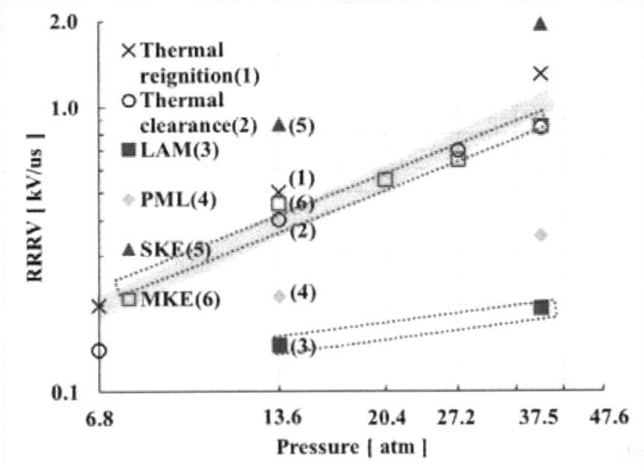

Figure 1. Predicted critical rate of rise of recovery voltage (RRRV) of air as a function of upstream stagnation pressure with di/dt = 13.5 A/µs. Simulation conditions are identical to those used in the experiment [9].

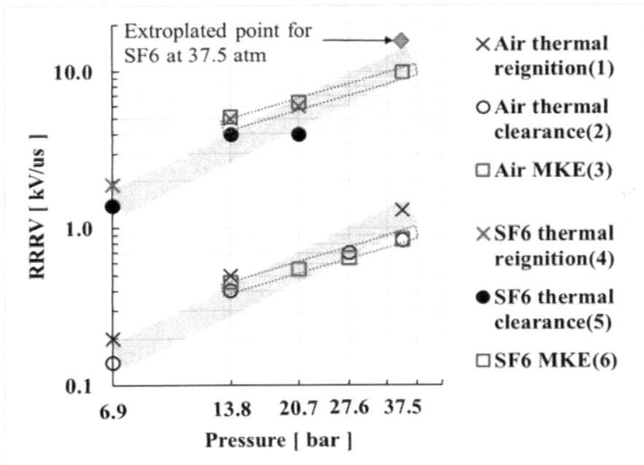

Figure 2. Predicted RRRV for SF_6 and air as a function of upstream stagnation pressure with di/dt = 13.5 A/µs. Experimental results are from [9].

$$\lambda_c = c r_\delta = c\sqrt{\int_0^\infty (1 - \frac{T_\infty}{T}) 2r\, \mathrm{d}r} \qquad (3)$$

where T_∞ is the temperature near the nozzle wall where the radial temperature gradient is negligible. c is a turbulence parameter the value of which is found by the best fit between model prediction and experimental results. The eddy viscosity is related to the turbulence length scale and the mean velocity

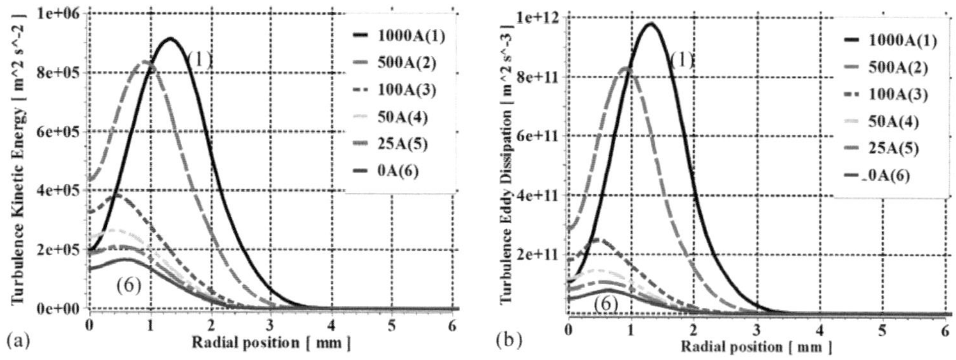

Figure 3. Radial distribution of turbulence kinetic energy and its dissipation rate in air arc at the axial location of 17 mm downstream the nozzle throat [9].

gradient by

$$\mu_t = \rho\lambda_c^2\left(\left|\frac{\partial w}{\partial r}\right| + \left|\frac{\partial v}{\partial z}\right|\right) \quad (4)$$

The existence of turbulence eddies in the main flow enhances the energy exchange process when a temperature gradient exists. In analogue to thermal conduction, the turbulent counterpart to the laminar thermal conductivity is related to the eddy viscosity through a unit Prandtl number by

$$Pr_t = \frac{\mu_t}{(k_t/C_p)} = 1 \quad (5)$$

Thus we are able to quantitatively account for the effect of turbulent cooling by the use of a turbulent thermal conductivity k_t. The standard K-epsilon model (SKE) and its variants consider the conversion of the main flow kinetic energy into the chaotic turbulence kinetic energy, k, as well as the destroy of turbulence eddies through a turbulence kinetic energy dissipation rate, ε:

$$\frac{\partial(\rho k)}{\partial t} + \nabla\cdot(\rho\mathbf{V}k - \frac{\rho\nu_t}{\sigma_k}\nabla k) = \rho(P_k - \varepsilon) \quad (6)$$

$$\frac{\partial(\rho\varepsilon)}{\partial t} + \nabla\cdot(\rho\vec{V}\varepsilon - \frac{\rho\nu_t}{\sigma_\varepsilon}\nabla\varepsilon) = \rho\frac{\varepsilon}{k}(C_{1e}P_k - C_{2e}\varepsilon) \quad (7)$$

where P_k represents the generation of turbulence kinetic energy due to the existence of mean flow velocity gradient, which is given by

$$P_k = \nu_t[2(\frac{\partial w}{\partial z})^2 + 2(\frac{\partial v}{\partial r})^2 + 2(\frac{v}{r})^2 + (\frac{\partial w}{\partial r} + \frac{\partial v}{\partial z})^2] \quad (8)$$

The turbulence length and velocity scales are respectively defined as $\lambda_c \propto k^{1.5/\varepsilon}$ and $V_c \propto k^{0.5}$.

The eddy viscosity is expressed as

$$\mu_t = \rho C_\mu \frac{k^2}{\varepsilon} \quad (9)$$

There are altogether five model constants in the k-Epsilon model with the default values of $\sigma_k = 1.0$,

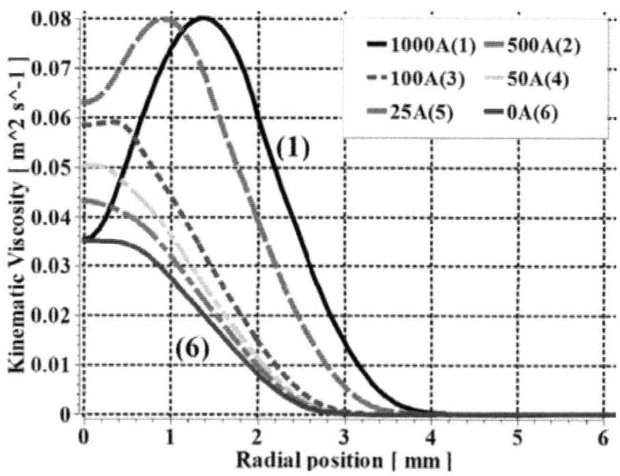

Figure 4. Radial distribution of turbulent kinematic viscosity in air arc at the axial location of 17 mm downstream the nozzle throat [9].

$\sigma_\varepsilon = 1.3$, $C_{1e} = 1.44$, $C_{2e} = 1.92$ and $C_\mu = 0.09$. By calibrating this model and examining its validity against experimental results, it has been found that acceptable agreement can be achieved by adjusting C_{1e} from 1.44 to 1.62. For comparison, the Chen-Kim K-epsilon model and the RNG K-epsilon model were also used in the calibration process [17]. Results shown in Figure 1 show that the prediction made by laminar flow assumption is simply too low. The Prandtl mixing length model (PML) also produces interruption capability that is significantly below the measurement while the standard K-epsilon model (SKE) gives much higher prediction. However the modified K-epsilon model (MKE) gives acceptable agreement for both DC at different current [18] as well as transient arcs at different upstream pressure [9]. We thus have confidence in the MKE model to represent the turbulence effect in the arcing process and the results using the MKE model will be studied to identify the dominant mechanisms responsible for the cooling effect of different gases.

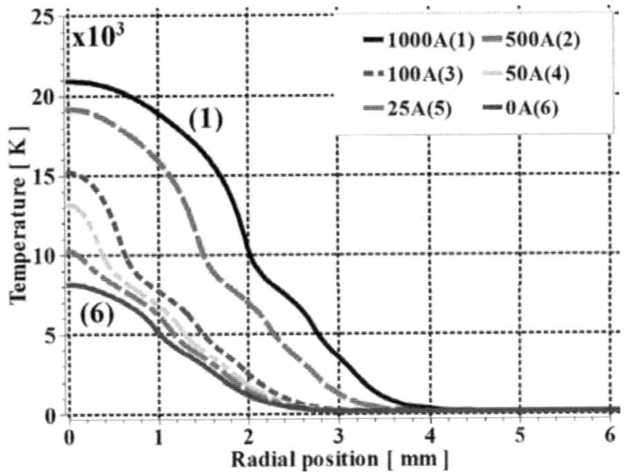

Figure 5. Radial distribution of air arc temperature at the axial location of 17 mm downstream the nozzle throat [9].

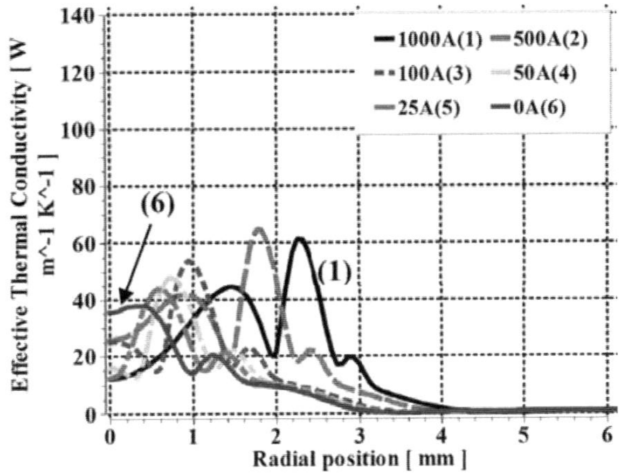

Figure 6. Radial distribution of the effective turbulent thermal conductivity in air arc at the axial location of 17 mm downstream the nozzle throat [9].

3. Comparative analysis of the energy exchange mechanisms in different gases

3.1. Difference in interruption capability of SF_6 and air

It is well know that the current interruption capability of SF_6 is much higher than that of the air, as experimentally proved by Frind and Rich [9] in a supersonic nozzle. Figure 2 shows the relative largeness of the interruption capability in terms of RRRV. Different from the dielectric strength which is a well-defined material property that only depends on the state of the gas, the current interruption capability of a gas not only depends on the type of gas, but also depends on the flow field, which explains the difference in interruption capabilities obtained in different experiments. For example, the interruptible RRRV ratio of SF_6 to air in a supersonic nozzles with a fixed upstream pressure of 37.5 bar and a $\mathrm{d}i/\mathrm{d}t$ immediately before current zero of 13.5 A/µs is 1 : 0.1 [9] whereas the interruptible $\mathrm{d}i/\mathrm{d}t$ (immediately before current zero) ratio obtained from a model circuit breaker is 1 : 0.28 [8].

The difference in interruption capability between SF_6 and air is also predicted by our arc model (Figure 2) where good agreement with measurement is observed. Results in Figure 1 also shows that despite the interruption capability of air is significantly lower than SF_6, turbulence is still important because without including turbulence the predicted RRRV is 30% or even lower than the measured values when the upstream pressure is higher than 13.6 bar.

3.2. The role of turbulence

The presence of turbulence eddies in the flow promotes momentum and energy exchange by increasing the effective viscosity and thermal conductivity of the gas. Since the turbulence kinetic energy generation term (Equation (8)) depends on the velocity gradient, it is expected that the kinematic viscosity will be largest at the arc edge where the velocity profile is the steepest. Figure 3 shows that the radii at which the maximum value of the turbulence kinetic energy and its dissipation rate occur are the same and decrease when the current linearly ramps down towards current zero. At 1 kA and 500 A, the radius of the arc core is larger than 1 mm. It is apparent that diffusion fails to spread the turbulence towards the centre of the arc column when convection in the axial direction is strong and the radial gradient of the axial velocity becomes smaller towards the arc centre. As a result turbulent kinematic viscosity reaches its maximum at the arc edge (Figure 4).

When the current reduces towards its zero point, the size of the arc core becomes smaller (Figure 5) and the maximum kinematic viscosity is the largest at the arc centre (Figure 4). It must however be noted that turbulence enhanced energy transfer in terms of the turbulent thermal conductivity as given in Equation (5) is the product of density, specific heat at constant pressure and the turbulent kinematic viscosity. Since the specific heat represents the energy density per unit mass, it directly affects the net energy exchange flux when there exists a temperature gradient. Thus the effective turbulent thermal conductivity has a more complex radial distribution, as shown in Figure 6. It is no longer monotonic and has two peaks. This is the result of the multiple peaks in the specific heat as a function of temperature. The product of density and specific heat (hereafter referred to as ρC_p for convenience) of three gases is shown in Figure 7 where there are two peaks above 4000 K (air at this temperature no longer conducts electricity).

Since the arc column is surrounded by cold gas, the temperature of the gas has to change from a high value at the arc centre to the cold gas temperature. The existence of radial temperature gradient enables the turbulent thermal conductivity to have an important role in shaping the radial temperature profile despite

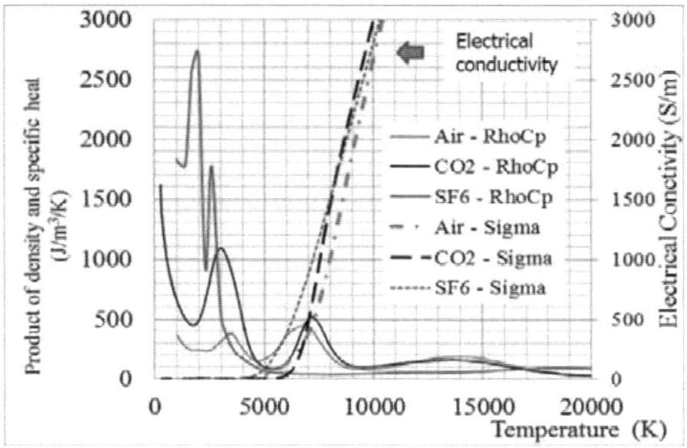

Figure 7. Product of density and specific heat of three gases as a function of temperature at 1 bar.

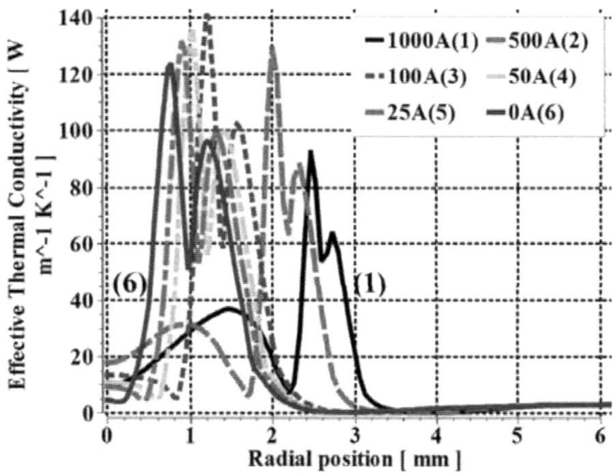

Figure 8. Radial distribution of the effective turbulent thermal conductivity in SF_6 arc at the axial location of 17 mm downstream the nozzle throat [9].

convection and radiation have also influence on it. Results in Figure 5 clearly show that the non-monotonic radial distribution of the effective turbulent thermal conductivity leads to the inflection points as labelled. From Figure 5, there will be two inflection points in the radial temperature profile as long as the arc centre temperature is higher than 10,000 K. The immediate consequence of the existence of the inflection points is that the arc column (electricity conduction region) becomes larger in size.

For comparison, SF_6 has consistently low ρC_p in the temperature range above 4,000 K when it starts to become electrically conductive (rapidly increasing electrical conductivity). The very high ρC_p below 4,000 K means highly efficient energy removal in the cooler surrounding gas so below 4,000 K the radial temperature gradient would be small. The low ρC_p above 4,000 K means the temperature gradient has to be large to maintain a radial energy flux that the surrounding cooler gas can absorb. The distribution of the effective turbulent thermal conductivity for SF_6 under identical arcing conditions is given in Figure 8 and the radial temperature in Figure 9. The only inflection point in the arc column for SF_6 is that

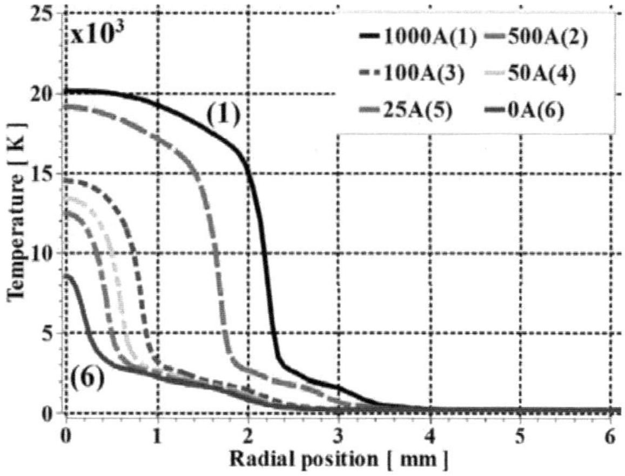

Figure 9. Radial distribution of SF_6 arc temperature at the axial location of 17 mm downstream the nozzle throat [9].

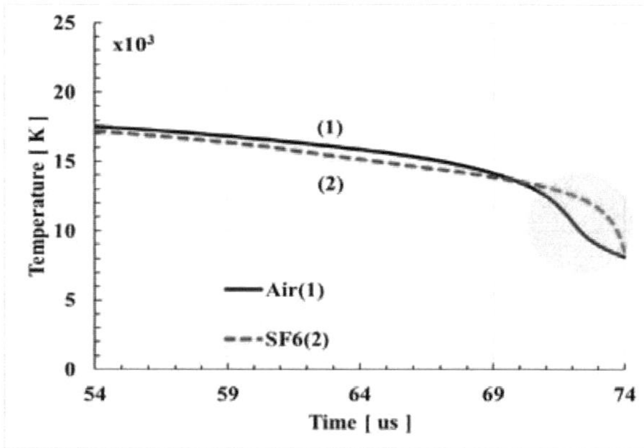

Figure 10. Variation of the axis temperature at the axial location of 17 mm downstream the nozzle throat [9]]. The current at 54 µs is 270 A, nearly reaching 0 A at 74 µs.

near the conducting temperature of SF_6 (4,000 K), i.e. close to the cooler surrounding gas. This means that because the ρC_p peaks for SF_6 lies below the conducting temperature while that of the air lies above the conducting temperature, the arc column of air arc is therefore broadened.

3.3. Energy exchange mechanisms leading to different current interruption capability

Arc cooling depends on the energy removal rate from the conducting column, or the arc column. At high current where Ohmic heating is strong, energy removal heavily relies on radiation and convection. However when the arcing current rapidly decreases towards its zero point, the arc column rapidly shrinks and turbulence enhanced thermal conduction becomes important or even dominant. Since the energy transfer mechanisms are closely coupled through the conservation equations, it is impossible to obtain analytic solution to the conservation equations. An approximate order of magnitude analysis shows that the character-

Gas	Current (A)	Radial thermal conduction	Radial convection	Radiation	Total radial cooling (%)	Axial cooling (%)
Air	500	13.6	-22.3	13.6	5.0	86.7
	50	20.1	11.6	1.3	33	57.4
	25	22.2	20.8	0.4	43.4	50
SF_6	500	33.9	0	9.7	43.5	47.7
	50	45	27.3	4.8	77	20.5
	25	45.2	35.1	6.2	86.5	12.6

Table 2. Percentage weighting of different energy exchange mechanisms for the whole arc column in SF_6 and air. The sum of Ohmic heating and reduction rate of the energy storage in the arc column is taken as 100%.

istic time for cooling by different mechanisms points to the relationship of $\tau_{r,\mathrm{tur}} \sim r_a$ for radial turbulent cooling, $\tau_{r,\mathrm{con}} \sim r_a/v_b$ for radial convective cooling where v_b is a positive radial velocity at the conduction boundary of the arc column, and $\tau_{z,\mathrm{con}}$ does not depends on r_a for axial convective cooling (r_a is the conducting column radius). This means energy removal across the radial boundary of the arc column becomes stronger when the arc radius decreases while the axial convective cooling is not sensitive to the change in arc radius.

A broadened arc column such as in air will lead to larger radial characteristic cooling time, thus lower RRRV values in comparison with SF_6 under identical flow conditions. Results in Table 2 clearly show that at 500 A, radial convection does not contribute to the cooling process instead it brings energy into the arc column. Turbulent enhanced radial thermal conduction already takes away 34 % of the total energy loss in SF_6 arc at 500 A while in air arc it is less than 14 %. This is directly a consequence of the broadening of the arc column. Near current zero (25 A), the total radial cooling effect accounts for 86 % of the total cooling in SF_6 while for air it is only 43 %. The difference is expected to be even larger when the current further reduces. Results in Figure 10 affirm our findings where the axis temperature in the SF_6 arc starts to reduce much more rapidly than the air arc when the current approaches zero due to much stronger turbulent cooling effect of SF_6.

4. Conclusion

A detailed study into the causes of SF_6's excellent current interruption capability in comparison with air has been carried out. It is shown that the huge difference in the interruption capability of SF_6 and air, when the arc is quenched in a supersonic nozzle, originates from the difference in their material properties, or the product of density and specific heat at constant pressure as a function of temperature. More specifically, it is the ρC_p peaks of air at temperatures above the conducting temperature (4,000 K) that broadens the arc column, consequently reduces the effectiveness of turbulent cooling. This is in contrast to SF_6 whose large ρC_p peak is below the conducting temperature. The consistently low ρC_p value of SF_6 above the conducting temperature leads to a sharp edge of the arc column and a smaller arc radius, enabling efficient turbulent cooling. Therefore, for the purpose of selecting or chemically composing SF_6 alternative gas or gas mixtures, one of the criteria will be that the ρC_p values above their conducting temperature should be consistently low and that below the conducting temperature should be high.

References

[1] Fifth assessment report (AR5) of the intergovernmental panel on climate change (IPCC), 2013.

[2] H. Katagiri, H. Kasuya, H. Mizoguchi, and S. Yanabu. Investigation of the performance of CF_3I gas as a possible substitute for SF_6. *IEEE Transactions on Dielectrics and Electrical Insulation*, 15(5):1424–1429, 2008. doi:10.1109/TDEI.2008.4656252.

[3] H.E. Nechmi, A. Beroual, A. Girodet, and P. Vinson. Fluoronitriles/CO_2 gas mixture as promising substitute to SF_6 for insulation in high voltage applications. *IEEE Transactions on Dielectrics and Electrical Insulation*, 23(5):2587–2593, 2016. doi:10.1109/TDEI.2016.7736816.

[4] P. Simka and N. Ranjan. Dielectric strength of C5 perfluoroketone. In *19th International Symposium on High Voltage Engineering, Pilsen, Czech Republic*, 2015.

[5] J.D. Mantilla, N. Gariboldi, S. Grob, and M. Claessens. Investigation of the insulation performance of a new gas mixture with extremely low GWP. In *Electrical Insulation Conference (EIC), 2014*, pages 469–473, 2014. doi:10.1109/EIC.2014.6869432.

[6] A. Beroual and A.M. Haddad. Recent advances in the quest for a new insulation gas with a low impact on the environment to replace sulfur hexafluoride (SF_6) gas in high-voltage power network applications. *Energies*, 10(8):1216, 2017. doi:10.3390/en10081216.

[7] G. Frind. *EPRI Report*, 284:5, 1977.

[8] P.C. Stoller, M. Seeger, A.A. Iordanidis, and G.V. Naidis. CO_2 as an arc interruption medium in gas circuit breakers. *IEEE Transactions on Plasma Science*, 41(8):2359–2369, 2013. doi:10.1109/TPS.2013.2259183.

[9] G. Frind and J.A. Rich. Recovery speed of axial flow gas blast interrupter: Dependence on pressure and di/dt for air and SF_6. *IEEE Transactions on Power Apparatus and Systems*, 5:1675–1684, 1974. doi:10.1109/TPAS.1974.293900.

[10] L.S. Frost and R.W. Liebermann. Composition and transport properties of SF_6 and their use in a simplified enthalpy flow arc model. *Proceedings of the IEEE*, 59(4):474–485, 1971. doi:10.1109/PROC.1971.8206.

[11] J.M. Yos. *AVCO Technical Release*, 28, 1967.

[12] P.J. Shayler and M.T.C. Fang. The transport and thermodynamic properties of a copper-nitrogen mixture. *Journal of Physics D: Applied Physics*, 10(12):1659–1670, 1977. doi:10.1088/0022-3727/10/12/015.

[13] J.F. Zhang, M.T.C. Fang, and D.B. Newland. Theoretical investigation of a 2 ka dc nitrogen arc in a supersonic nozzle. *Journal of Physics D: Applied Physics*, 20(3):368–379, 1987. doi:10.1088/0022-3727/20/3/020.

[14] R.W. Liebermann and J.J. Lowke. Radiation emission coefficients for sulfur hexafluoride arc plasmas. *Journal of quantitative spectroscopy and radiative transfer*, 16(3):253–264, 1976. doi:10.1016/0022-4073(76)90067-4.

[15] J.J. Lowke. Predictions of arc temperature profiles using approximate emission coefficients for radiation losses. *Journal of Quantitative Spectroscopy and Radiative Transfer*, 14(2):111–122, 1974. doi:10.1016/0022-4073(74)90004-1.

[16] P.J. Shayler and M.T.C. Fang. Radiation transport in wall-stabilised nitrogen arcs. *Journal of Physics D: Applied Physics*, 11(12):1743–1756, 1978. doi:10.1088/0022-3727/11/12/013.

[17] Q. Zhang, J.D. Yan, and M.T.C. Fang. The modelling of an SF_6 arc in a supersonic nozzle: I. cold flow features and dc arc characteristics. *Journal of Physics D: Applied Physics*, 47(21):215201, 2014. doi:10.1088/0022-3727/47/21/215201.

[18] M.T.C. Fang, S. Ramakrishnan, and H.K. Messerle. Scaling laws for gas-blast circuit-breaker arcs during the high current phase. *IEEE Transactions on Plasma Science*, 8(4):357–362, 1980. doi:10.1109/TPS.1980.4317340.

doi:10.14311/ppt.2017.3.241
Plasma Physics and Technology **4**(3):241–244, 2017

SYNTHETIC TESTING OF LOAD CURRENT INTERRUPTION IN MEDIUM VOLTAGE LOAD BREAK SWITCHES

M. Bendig[a,*], M.Ksoll[a], A. Kalter[b], M. Schaak[b], K. Ermeler[c], A. Schnettler[a]

[a] *Institute for High Voltage Technology, RWTH Aachen University, Schinkelstr. 2, 52056 Aachen, Germany*
[b] *Siemens AG, Energy Management Division, Carl-Benz-Str. 22, 60386 Frankfurt am Main, Germany*
[c] *Siemens AG, Energy Management Division, Nonnendammallee 104, 13629 Berlin, Germany*
[*] bendig@ifht.rwth-aachen.de

Abstract. To conduct the mainly active load current test duty according to IEC 62271-103, a directly powered test circuit and therefore a medium voltage connection or a power generator is needed. A newly developed synthetic test circuit allows to replicate the current as well as the full transient recovery voltage (TRV) and the power frequent recovery voltage (RV) of the direct test circuit up to its crest value. It is dimensioned for voltage classes up to 52 kV and can be adapted to test currents between 630 A and 1250 A. The test circuit allows a detailed investigation of load break switches without costly high power sources and loads. Many parameters, like current and voltage steepness at current zero, as well as different parameters defining the TRV steepness can be varied individually.

Keywords: Load switching, synthetic testing, circuit simulation, load break switch.

1. Introduction

In todays medium voltage grids load break switches (LBS) are often used. In the voltage range from 1 kV to 52 kV they serve as an economical alternative to relatively expensive circuit breaker/disconnector combinations.

To improve present LBS as well as to develop future switchgear, an intensive experimental testing is necessary. For high voltage circuit breakers a synthetic test circuit is known, giving a high flexibility in current and voltage waveforms [1]. The type testing of LBS is standardized by IEC 62271-103 as the mainly active load current test duty TD_{load} with a directly powered test circuit [2]. To perform these tests a medium voltage connection or a high power generator is needed.

Previous investigations show different approaches in designing an easier, more flexible test by either modifying the direct test to only meet the IEC standard in the first few hundred microseconds after current zero (CZ) [3] or breaking the test down to a two-part synthetic test with one test for the first few hundred microseconds, the so-called transient recovery voltage (TRV), and one test for the latter part of the recovery voltage, the power frequent recovery voltage (RV) [4].

In this paper a newly developed synthetic test circuit for medium voltage LBS testing is presented. In a first step, the current and voltage forms in the IEC direct test are analyzed and the requirements for the synthetic test circuit are formulated. Afterwards, the design of the new test circuit is described in detail. The dimensioning is shown and results of circuit simulations carried out with MATLAB Simulink are compared to the test circuit requirements.

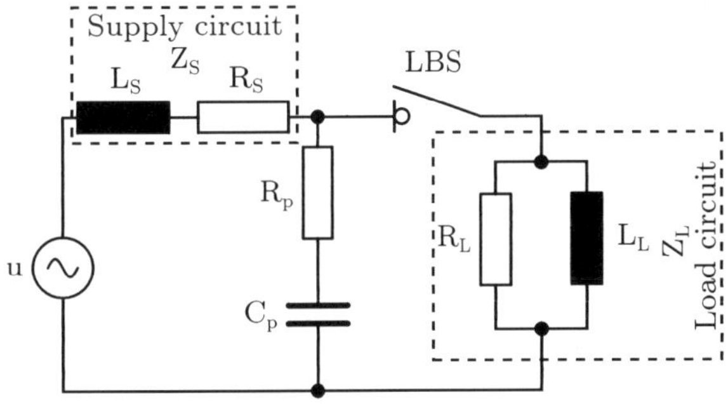

Figure 1. Circuit diagram of the single phase direct test circuit according to IEC 62271-103

2. Test circuit requirements

The direct test circuit specified in IEC 62271-103 is shown in Figure 1. It can be divided into a supply and a load circuit with the impedances Z_S and Z_L respectively. The voltage u is defined as $U_{Test} = k_{pp} \times U_r/\sqrt{3}$ with a first-pole-to-clear factor of $k_{pp} = 1.5$. The switching current is determined by the total impedance $Z_S + Z_L$ of the circuit. The ratio of the impedances is defined as $|Z_S| = (0.15 \pm 0.03)|Z_L|$. The power factor of the two impedances should be in the range of $\cos\varphi_S \leq 0.2$ and $0.65 \leq \cos\varphi_L \leq 0.75$, respectively. The parallel capacity as well as the resistor are chosen with regard to the rise time and peak value of the prospective TRV in case of a short-circuited load. The corresponding values are given in [2].

The direct test circuit is dimensioned for the voltages U_r = 7.2 kV, 12 kV, 24 kV, 36 kV and 52 kV. Simulations are carried out in MATLAB Simulink to ana-

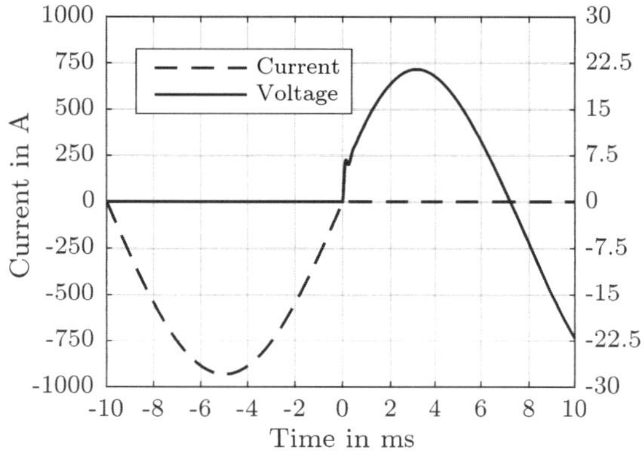

Figure 2. Current an voltage of simulated 24 kV direct test

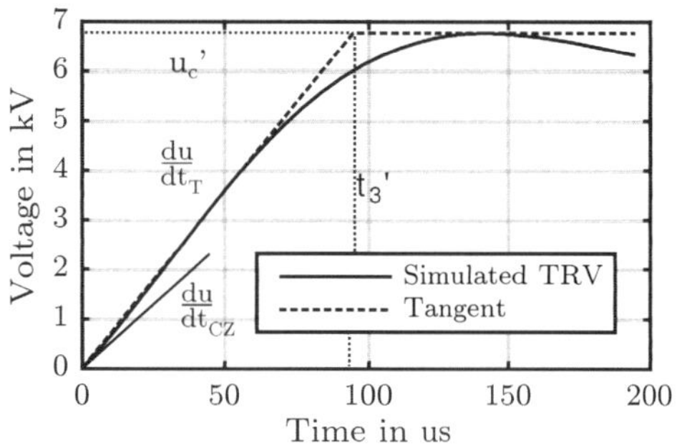

Figure 3. First 200 µs of the simulated TRV of the direct test circuit

lyze the shape of the resulting recovery voltage. The shape of the current as well as the recovery voltage for a test with a rated voltage of $U_r = 24$ kV and $I = 630$ A are shown in Figure 2. The supply circuit has a power factor of $\cos\varphi = R_S/Z_S = 0.2$ and the load circuit has a power factor of $\cos\varphi = Z_L/R_L = 0.75$. The resulting recovery voltage is the difference of the voltage on the supply side, which is a fast rise to the grid voltage, and the voltage on the load side, an exponentially decaying DC-voltage. This leads to a two-step recovery voltage with a relatively steep first part until a first maximum plateau, the TRV, and a slower rising second part until the crest value, the RV.

The first 200 µs after CZ are shown in Figure 3. The shape of the TRV is analyzed as in [3] considering the maximum of the first peak u_c', the voltage steepness at CZ du/dt_{CZ} as well as the steepness of an aligned tangent du/dt_T and the rise time t_3'. The tangent and rise time are defined as is [2]. The resulting values are shown in Table 1. Due to the slight $(1 - \cos)$-form of the TRV, the voltage steepness at CZ is lower than the steepness of the tangent for all rated voltages.

The RV part is analyzed regarding the crest value of the voltage u_{cr} and the time at which the crest value occurs t_{cr}. The resulting values are shown in Table 2. The time is constant for all voltages as it is defined by the sine-wave of the grid voltage and not longer by the elements in the circuit. The crest values

U_r[kV]	u_c'[kV]	t_3'[µs]	$\frac{du}{dt}_{CZ}[\frac{V}{\mu s}]$	$\frac{du}{dt}_{T}[\frac{V}{\mu s}]$
7.2	1.85	54.3	29.2	34.3
12	3.13	63.8	43	49.5
24	6.8	94.6	62.24	71.6
36	10.65	118	77.9	90.92
52	16.6	155	93.2	107.1

Table 1. Characteristics of the first part of the TRV in a direct test

U_r[kV]	u_{cr}[kV]	t_{cr}[ms]
7.2	6.45	3.15
12	10.75	3.15
24	21.51	3.15
36	32.26	3.15
52	46.61	3.15

Table 2. Characteristics of the second part of the TRV in a direct test

are only 73 % in amplitude and occur later than the unaffected voltage peak $\hat{U} = \sqrt{2}U_r/\sqrt{3} \times k_{pp}$. Both effects can be explained by the decaying DC voltage in the load circuit. Measurement results could show a higher and earlier amplitude in comparison to the simulations, due to a lower DC voltage on the load site.

Besides these definitions for direct testing, other requirements for the synthetic test are derived from [1] as the following:

1. The arcing power should be the same as in the direct test
2. There should be no time interval between current and voltage stress
3. The TRV of the synthetic circuit should be the same as the standardized TRV but not less in the first 1/8 of a 50 Hz period
4. There should be no damping of the TRV by the post-arc current
5. A switch failure should be clearly detectable.

3. Design of a synthetic test circuit

The novel synthetic test circuit for load break switches is a combination of three individual circuits. One circuit provides the high current and two circuits are shaping the recovery voltage. To avoid a gap between current and voltage stress, the current-superposition-method is chosen. Shortly before CZ a smaller current with a higher frequency is superimposed over the high current. While the high current is then interrupted by an auxiliary breaker, the superimposed current – with a slightly later CZ – is interrupted by the device under test. The final circuit diagram of the synthetic test circuit is shown in Figure 4. For reasons of simplicity, only a single phase test circuit is designed.

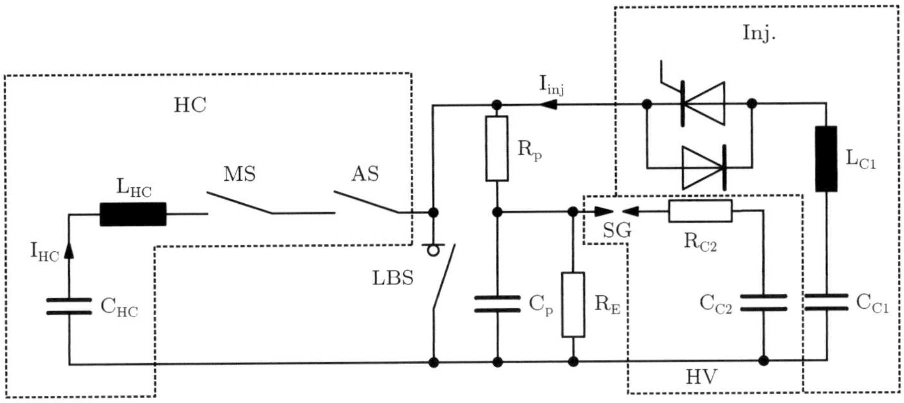

Figure 4. Circuit diagram of the synthetic test circuit

The high current (HC) circuit is a simple LC resonant circuit containing a making switch (MS) and an auxiliary switch (AS). The frequency of this circuit is determined by the values of C_{HC} and L_{HC} and should be set to nearly 50 Hz. Additionally the elements should be chosen in a way, that the charging voltage for a certain test current is significantly above the arcing voltage of the test device to avoid damping of the current.

As the current I_{HC} approaches CZ the thyristor in the injection circuit (Inj.) is fired, injecting the superimposed current I_{inj} into the test object. The value of the inductance L_{C1} is determined by the steepness of the injection current at current zero:

$$\frac{di_{\mathrm{Inj}}}{dt}\Big|_{t=t_{\mathrm{CZ}}} \stackrel{!}{=} \frac{di_{\mathrm{HC}}}{dt}\Big|_{t=t_{\mathrm{CZ}}}, \tag{1}$$

$$\frac{di_{\mathrm{Inj}}}{dt}\Big|_{t=t_{\mathrm{CZ}}} = \frac{U_{C_{\mathrm{C1}}}|_{t=t_0}}{L_{\mathrm{C1}}} \tag{2}$$

After the current is successfully interrupted, the capacitance C_{C1} is charged negatively and the first part of the TRV is applied to the test object through C_{C1}, L_{C1} and the parallel elements R_{p} and C_{p}. The thyristor is now changing back to blocking mode and the charging current is carried by the free wheeling diode until the first maximum of the TRV is crossed. The voltage steepness at CZ is only defined by the value of R_{p}:

$$\frac{du}{dt}\Big|_{t=t_{\mathrm{CZ}}} = R_{\mathrm{p}} \times \frac{di_{\mathrm{Inj}}}{dt}\Big|_{t=t_{\mathrm{CZ}}} \tag{3}$$

The voltage shape until the first maximum is dependent on all the components and, as C_{C1}, L_{C1} and R_{p} are set, can be influenced by changing the value of C_{p}.

As soon as the TRV crossed its first peak, the diode is blocking and the spark gap (SG) of the high voltage circuit (HV) is triggered. The capacitance C_{C2} is negatively charged and therefore the absolute value of the recovery voltage rises to its crest value. The rise time is defined by the resistor R_{C2}. The resistor R_{E} is a discharge resistor and does not affect the highly transient events in the first few milliseconds after CZ.

U_r [kV]	L_{C1} [mH]	R_{p} [Ω]	U_{C1} [kV]	U_{C2} [kV]
7.2	5.18	107	1.45	7
12	8.93	155	2.5	11.5
24	19.29	225	5.4	23
36	30.36	278	8.5	34.4
52	46.43	332	13	50

Table 3. Component values and charging voltages for the synthetic test circuit

3.1. Dimensioning

The dimensioning of the circuit is obtained partly analytical using the above equations and partly using the simulation software MATLAB Simulink. For the high current circuit a charging voltage of $U_{\mathrm{c}} = 1$ kV for a test current of $I = 630$ A is chosen to be significantly above the arcing voltage of the test object. This leads to an inductance of $L_{\mathrm{HC}} = 3.57$ mH and a capacitance of $C_{\mathrm{HC}} = 2.8$ mF. The values of the three capacitances C_{p}, C_{C1} and C_{C2} are constant for all rated voltages with $C_{\mathrm{p}} = 160$ nF, $C_{\mathrm{C2}} = 3.88$ µF and $C_{\mathrm{C1}} = 1$ µF respectively.

As the rise time for the RV is constant as well (s. Table 2), the resistor R_{C2} is consequently determined for all voltage levels with $R_{\mathrm{C2}} = 6.3$ kΩ. Due to the different voltage steepnesses for the different levels the values of L_{C1} and R_{p} as well as the charging voltages for the capacitances C_{C1} and C_{C2} have to be chosen individually. The calculated values are shown in Table 3.

3.2. Simulative analysis

The voltage waveform of the resulting TRV of the synthetic test circuit is analyzed regarding the same characteristics as in section 2. The resulting values for the different voltage levels are shown in Table 4. As in Table 1 the tangent steepness is higher than the voltage steepness at CZ due to the (1 − cos)-form. The values show a high correspondence to the results of the direct test, with a smaller rise time t_3 and therefore a higher tangent steepness du/dt_{T} for all rated voltages.

U_r[kV]	u'_c[kV]	t'_3[µs]	$\frac{du}{dt}_{CZ}[\frac{V}{\mu s}]$	$\frac{du}{dt}_{T}[\frac{V}{\mu s}]$
7.2	1.87	50.97	30.02	37.13
12	3.18	63.32	43.46	50.43
24	6.88	93.25	63.02	74.14
36	10.86	117.8	77.96	92.59
52	16.71	148.1	93	113.3

Table 4. Characteristics of the first part of the TRV in a synthetic test

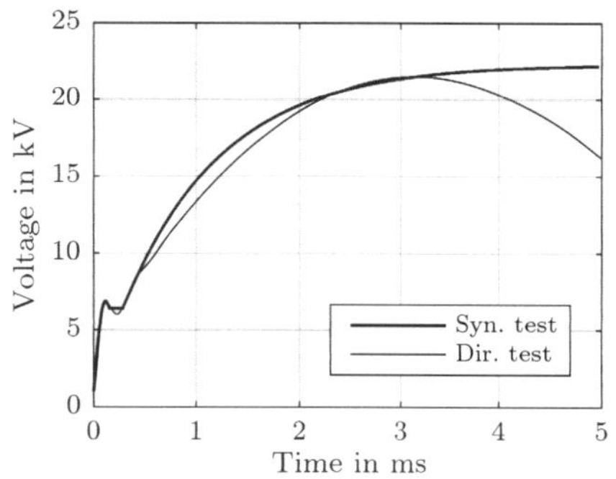

Figure 5. Comparison of the TRV of the synthetic test circuit and the direct test

Maximum u'_c and voltage steepness at CZ du/dt_{CZ} are nearly identical to the direct test.

The RV meets the exact requirements shown in Table 2 for all voltage levels but shows a higher steepness in the beginning due to its exponential character.

4. Discussion

Figure 5 shows the comparison of the recovery voltage of the synthetic test circuit and the standardized direct test circuit. The first 150 µs are shown in Figure 6. For the first 50 µs the voltages of the two circuits show no significant difference. Afterwards the TRV of the synthetic test is slightly above the the TRV of the direct test, which leads to the higher tangent steepness and the lower rise time. In the RV part the voltage of the synthetic circuit is rising faster due to its exponential character. Both leads to test conditions which are equal or harder than in the direct test and therefore fulfills the requirements defined in section 2. After the crest value, the voltage of the synthetic test circuit is nearly constant while the voltage of the direct test circuit decreases and approximates the grid sine-wave. However, during this time-period, the critical dielectric phase for the LBS is assumed to have ended.

A limitation of the synthetic test is, as for synthetic testing of circuit breakers, that the load break switch must interrupt the current in the first half wave, due to the polarity of U_{C2}. It is possible to redesign the test in a way, that current interruption in the second half wave is tested. But in that case the switch must not clear in the first half wave. This can be archived

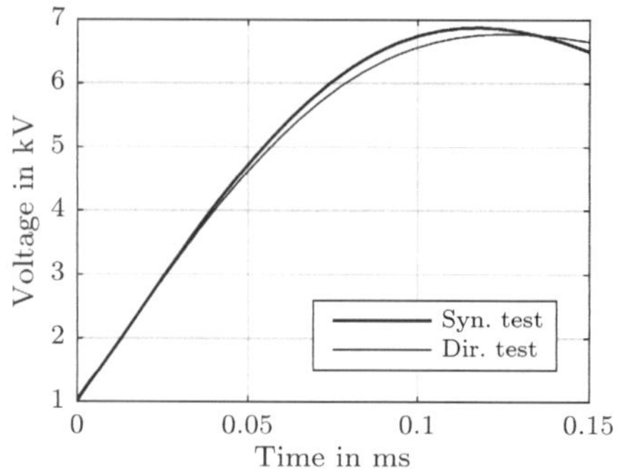

Figure 6. First 150 μs of the TRVs of the synthetic and the direct test

by an additional re-ignition circuit. Furthermore, a natural damping occurs in the test circuit. Therefore, if multiple current zero crossings are tested, the current's amplitude as well as its steepness at current zero decrease with every current zero crossing.

5. Conclusion

The proposed synthetic test circuit for load current interruption tests of medium voltage load break switches offers a highly flexible and cheap alternative to the standardized direct test. Simulations in MATLAB Simulink show a high correspondence of the different TRV characteristics with the direct test. The shape of the recovery voltage can be easily changed and adapted, as i.e. the two parts – TRV and RV – can be varied individually. The synthetic test circuit has the limitation that the current interruption must take place in a predefined half wave.

Acknowledgements

The authors of this paper would like to thank Hannes Geertz for his support in the development of this test circuit.

References

[1] IEC 62271-101:2012 High-voltage switchgear and controlgear - Part 101: Synthetic testing. Standard, International Electrotechnical Comission, 06 2012.

[2] IEC 62271-103:2011 High-voltage switchgear and controlgear - Part 103: Switches for rated voltages above 1 kV up to and including 52 kV. Standard, International Electrotechnical Comission, 06 2011.

[3] E. Jonsson and M. Runde. Medium voltage laboratory for load break switch development. In *International Conference on Power System Transients*, 07 2013.

[4] A. Islam, D. Birtwhistle, T. K. Saha, and B. Diverall. Two-part synthetic test procedures for the testing of medium-voltage load break switches. In *IEEE Transactions on Power Delivery*, volume 31, pages 1645–1654, 08 2016. doi:10.1109/TPWRD.2015.2511188.

MACQUARIE University

Human factors issues in digital health and their effects on patient safety

A/Prof. Farah Magrabi

HFESA
Wollongong
28 November 2017

Outline

MACQUARIE University

1. Healthcare delivery challenges
2. IT and patient safety: human factors issues are a major risk
3. Types of human factors issues

MACQUARIE University

1. Current safety challenges in healthcare

Patient safety is a major public health crisis

MACQUARIE University

Hospitals

- 10% of admissions associated with patient harm
- 1 in 5 lead to permanent disability or death
- 50% were preventable

General practice

- 10% of patients experience an adverse drug event
- 1 million Australians every year

Thomas 2000; Miller 2006; Roughead 2006

Medical error, third leading cause of death in the USA

MACQUARIE University

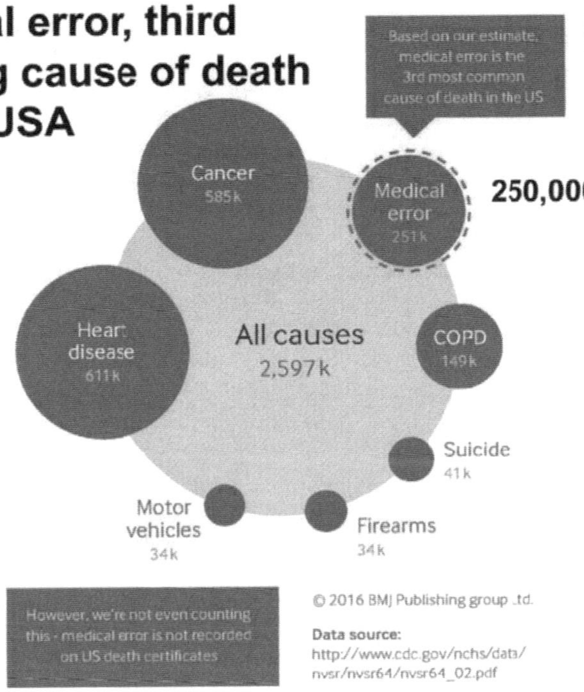

250,000 deaths

© 2016 BMJ Publishing group Ltd.

Data source: http://www.cdc.gov/nchs/data/nvsr/nvsr64/nvsr64_02.pdf

5

Care delivery is highly variable and inappropriate

MACQUARIE University

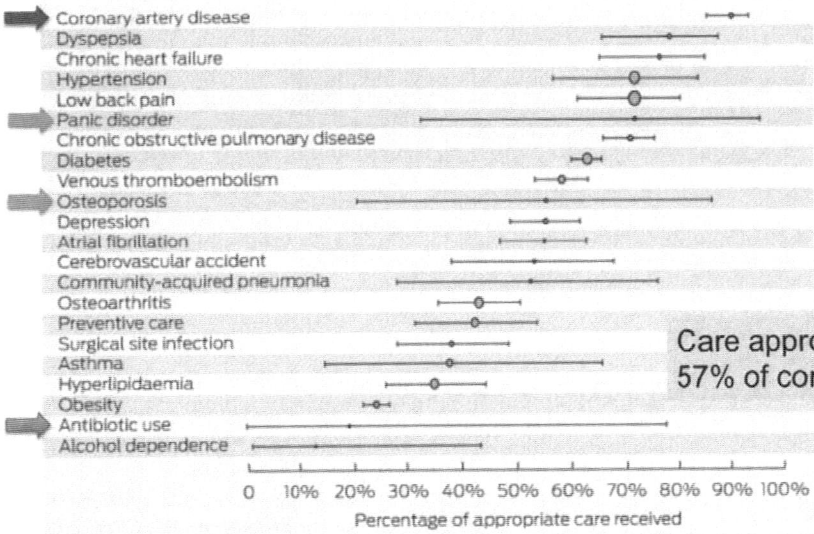

Care appropriate in 57% of consultations

CareTrack Australia

AUSTRALIAN INSTITUTE OF HEALTH INNOVATION
FACULTY OF MEDICINE AND HEALTH SCIENCES

Runciman et al. MJA 2012

Digital health is essential to system reform

decision-making   messaging   record-keeping

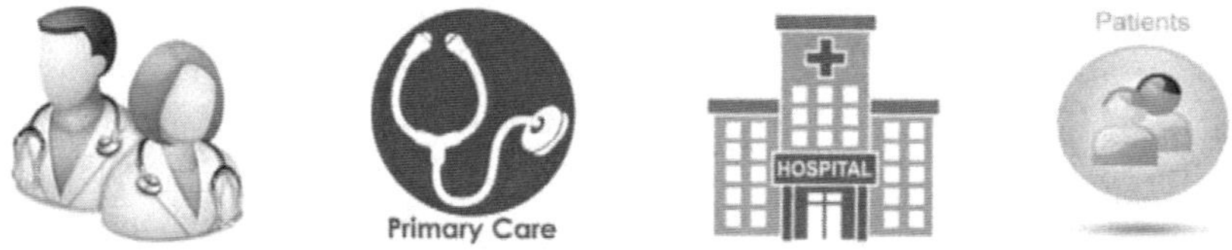

AUSTRALIAN INSTITUTE OF HEALTH INNOVATION
FACULTY OF MEDICINE AND HEALTH SCIENCES

7

IT systems support critical health processes

Pathology testing

test ordered — specimen collected — results available — results read

Medical devices are networked

8

Consumers are more engaged in their health

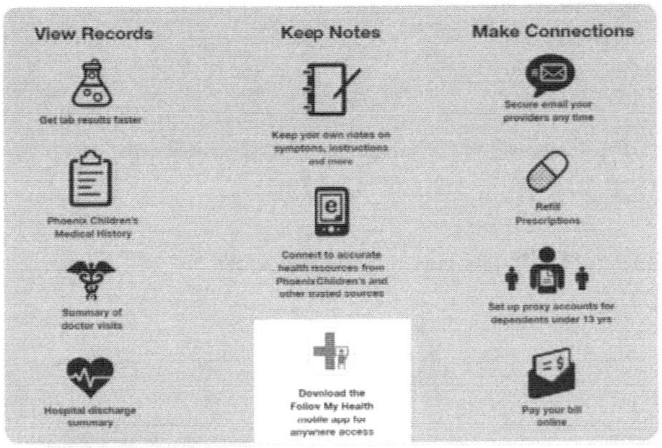

phoenixchildrens.org

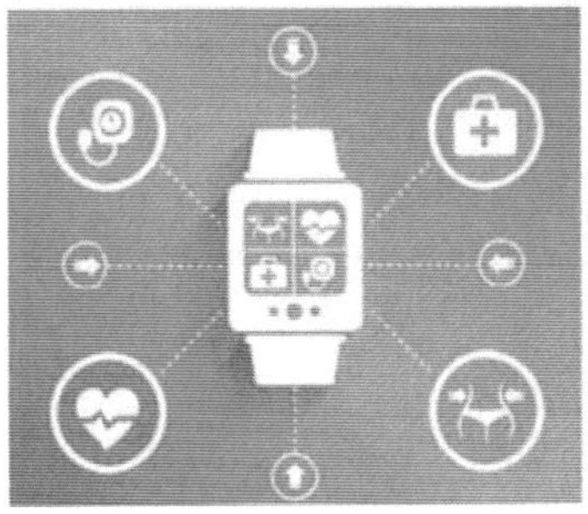

9

MACQUARIE University

2. Evidence about patient safety risks

Alongside benefits digital health can pose risks to patient safety

MACQUARIE University

- Digital health is the use of IT in care delivery.
- Problems with IT can **disrupt care delivery** and **introduce new clinical errors** that can harm patients. (US IOM 2012)
- Safety risks are a side effect or **unintended consequence** of IT. (Ash, Berg & Coiera, 2004)

Reports of IT-related harms are growing

MACQUARIE University

2010 | 2011 | 2012 | 2013 | 2014 | 2015 | 2016

SA Health

FDA

ECRI Institute

Dutch CMR

Patient Safety Authority

NHS Wales GIG Cymru

VA Health Care

NHS England

The Joint Commission

TechWatch

Journal of the American Medical Informatics Association, 24(2), 2017, 246–260
doi: 10.1093/jamia/ocw154
Review

Special Focus on Safety

Problems with health information technology and their effects on care delivery and patient outcomes: a systematic review

Mi Ok Kim, Enrico Coiera, and Farah Magrabi

Centre for Health Informatics, Australian Institute of Health Innovation, Macquarie University, Sydney, Australia

Corresponding author: Dr Mi Ok Kim, Postdoctoral Research Fellow, Centre for Health Informatics, Australian Institute of Health Innovation, Macquarie University, Sydney, NSW 2109 Australia. E-mail: mi.kim@mq.edu.au; Tel: +61 2 9850 2454; Fax: +61 2 9850 2499

Abstract

Objective: To systematically review studies reporting problems with information technology (IT) in health care and their effects on care delivery and patient outcomes.
Materials and methods: We searched bibliographic databases including Scopus, PubMed, and Science Citation Index Expanded from January 2004 to December 2015 for studies reporting problems with IT and their effects. A framework called the information value chain, which connects technology use to final outcome, was used to assess how IT problems affect user interaction, information receipt, decision-making, care processes, and patient outcomes. The review was conducted in accordance with the Preferred Reporting Items for Systematic Reviews and Meta-Analyses (PRISMA) statement.
Results: Of the 34 studies identified, the majority ($n = 14$, 41%) were analyses of incidents reported from 6 countries. There were 7 descriptive studies, 9 ethnographic studies, and 4 case reports. The types of IT problems

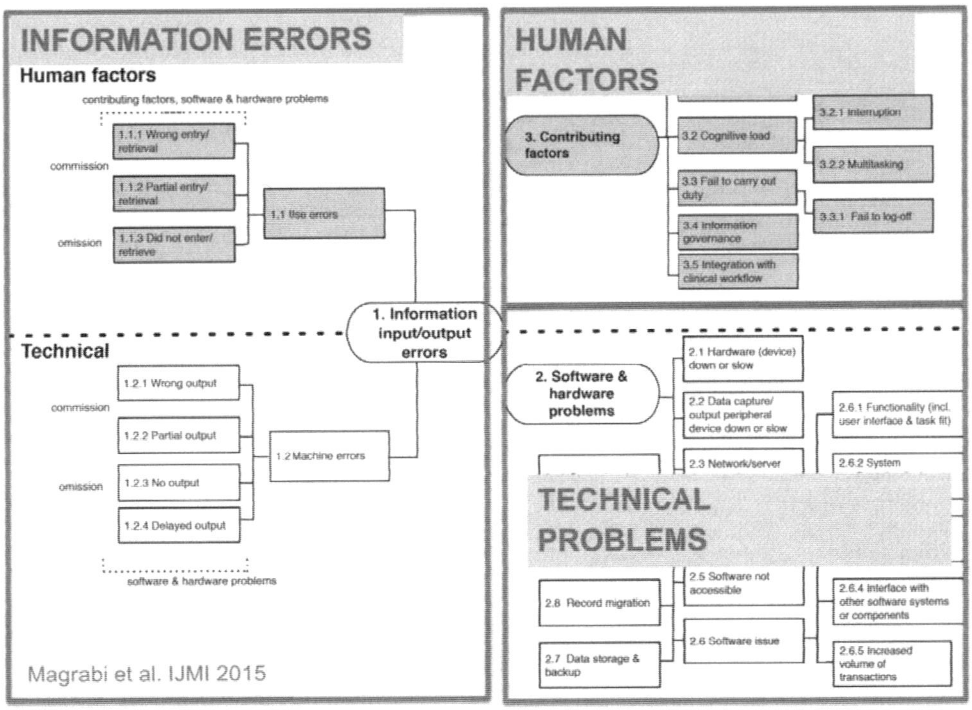

Human factors problems were proportionally higher in patient harm events

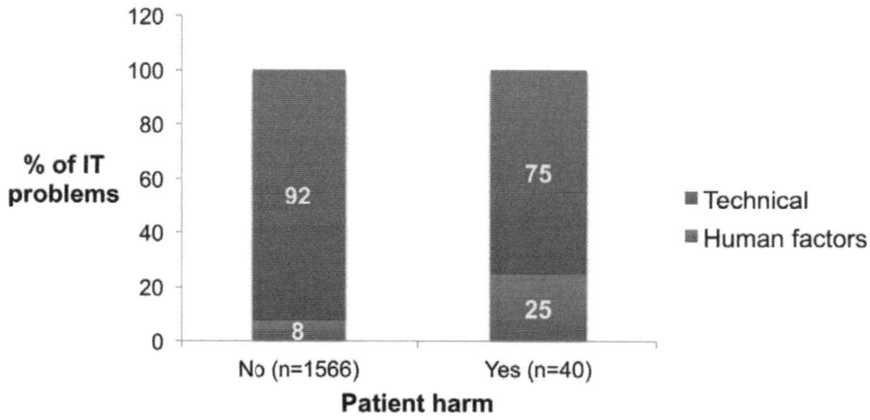

4 times as likely to result in patient harm than technical problems

- 25% vs. 8% (Chi sq =13, df =1, $p<0.001$)
- Odds ratio 4 (2 to 8)

Magrabi et al. IJMI 2015

MACQUARIE University

3. Types of human factors issues

Knowledge & skills of users

MACQUARIE University

Use error: A patient who was seen with **another patient's records** in general practice was prescribed that patient's medication and died later the same day from taking it. No further details were available.

Use error: A doctor intended to prescribe 4 mg trandolapril for an elderly male patient, but **mistakenly prescribed** Amaryl 4 mg (glimepiride). On taking the medication the patient went into a hypoglycemic coma and had seizures. He was resuscitated in an ICU and admitted to hospital for a week.

System limitations*: a clinician prescribed the wrong medication, by **wrongly assuming** that the system would have alerted them if a mistake had been made.

Cognitive resources devoted to system use

MACQUARIE University

Slip of concentration: Avanza (mirtazepine) was prescribed instead of Avandia (rosiglitazone) due to a **slip in concentration**. A pharmacist detected the error because the patient did not suffer from diabetes and contacted the doctor to issue a new prescription.

Multi-tasking, multiple patient files open: A doctor mistakenly prescribed a medication for the wrong patient when **two patient files were opened up simultaneously** on the computer screen. The doctor noticed the error and corrected it.

Interruption: A doctor wrote a prescription for the wrong patient when **interrupted by a phone call** during a consultation. At the end of the call the doctor returned to the wrong patient record. The error was detected by a pharmacist and returned to the doctor.

Organizational policies and procedures

MACQUARIE University

Policy for training and system use: A radiologist who missed a training session had been reporting reporting old films and using the new film as a comparison for 6 months.

Access: System access was erroneously given to all users rather than 14 users who had been trained.

Information governance: An **HIV test** ordered during hospital stay was **not followed-up** after discharge. When the patient was re-admitted, the admitting doctors were unable to access the HIV test result because the **test request was hidden** from them. The patient developed and died from pneumocystis pneumonia.

Summary

MACQUARIE University

- Impact of IT incidents on patient safety is small but noticeable
- Magnitude of risk is not known- ***tip of the iceberg?***
- Human factors issues require close attention
  - knowledge & skills of users
  - safe use including multi-tasking and interruption handling
  - policies and procedures

Thank you

farah.magrabi@mq.edu.au

MACQUARIE University

References

MACQUARIE University

1. Coiera E, Guide to Health Informatics. Boca Raton, FL, USA: CRC Press, Taylor & Francis Group; 2015.
2. Kim MO, Coiera E, Magrabi F. Problems with health information technology and their effects on care delivery and patient outcomes: a systematic review. J Am Med Inform Assoc. 2017;24(2):246-50.
3. Magrabi F, Baker M, Sinha I, Ong MS, Harrison S, Kidd MR, et al. Clinical safety of England's national programme for IT: A retrospective analysis of all reported safety events 2005 to 2011. Int J Med Inform. 2015;84(3):198-206.
4. Magrabi F, Liaw T, Arachi D, Runciman WB, Coiera E, Kidd MR. Identifying patient safety problems associated with Information Technology in general practice: an analysis of incident reports. BMJ Qual Saf. 2015; Nov 5.
5. Runciman W, Hibbert P, Thomson R, Van Der Schaaf T, Sherman H, Lewalle P. Towards an International Classification for Patient Safety: key concepts and terms. Int J Qual Health Care. 2009;21(1):18-26.
6. Magrabi F et al. Death despite known drug allergy. In: Johnson JK, Haskell HW, Barach PR, editors. Case Studies In Patient Safety: Foundations for Core Competencies. Burlington MA: Jones & Bartlett Learning; 2015. p. 247-60.

MACQUARIE University

Patient work in consumer health informatics: A scoping review

HUONG LY TONG, LILIANA LARANJO, PAIGE MARTIN, ANNIE YS LAU

Centre for Health Informatics, Australian Institute of Health Innovation, Faculty of Medicine and Health Sciences, Macquarie University

Human Factors and Ergonomics Society of Australia (HFESA) Conference 2017, 27th November

Background

MACQUARIE University

CONSUMER HEALTH INFORMATICS (CHI)

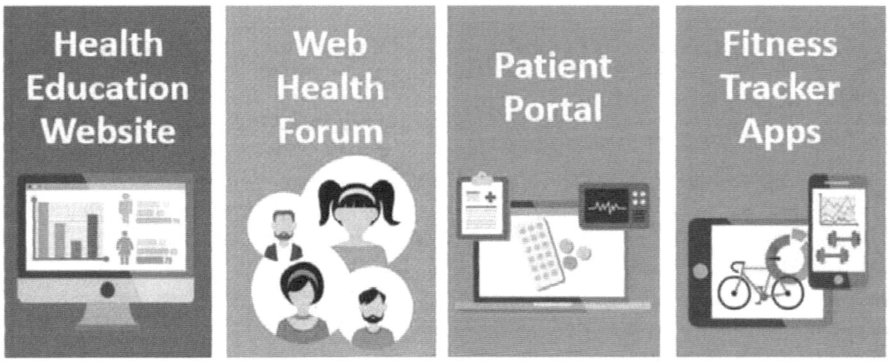

"The development and evaluation of **technology** to support **patients' health management**"[1]

[1] Flaherty, D. et al., 2015. What is consumer health informatics? A systematic review of published definitions. *Informatics for Health and Social Care*, *40*(2), pp.91-112.

Background

PATIENT WORK DEFINITION

"The exertion of effort and time by patients or their family members to accomplish a health goal" [2]

[2] Strauss, A. L. 1993. *Continual permutations of action,* New York, Aldine de Gruyter.

Background

PATIENT WORK CLASSIFICATION

- Health work
- Everyday life work
- Illness-adjusted work [3]

[3]Valdez, R. S. et al. 2015. Transforming consumer health informatics through a patient work framework: connecting patients to context. *Journal of the American Medical Informatics Association,* 22, 2-10.

Research question

MACQUARIE University

What is known from the existing literature about the characteristics of CHI interventions that support patient work?

Specifically:

(1) What activities of patient work are supported by CHI interventions? For what health conditions?
(2) Technology mode? Functions?
(3) Main outcomes?

Methodology

MACQUARIE University

SCOPING REVIEW VS SYSTEMATIC REVIEW

Characteristics	Scoping review	Systematic review
Research question	Broadly defined	Highly focused
Study selection	All study types	Defined study types (normally RCTs)
Data extraction	Chart data according to key themes	Synthesise findings and assess the effectiveness of interventions

[4] Arksey, H. & O'Malley, L. 2005. Scoping studies: towards a methodological framework. *International journal of social research methodology*, 8, 19-32.

Methodology

SEARCH STRATEGY

MACQUARIE University

- Databases: Medline, Embase, CINAHL, ACM Digital, Web of Science, Scopus
- Key terms: patient work, patient context and consumer health informatics
- Search reference lists of included articles
- Grey literature: Google Scholar

Methodology

ELIGIBILITY CRITERIA

MACQUARIE University

Included if the studies

(1) Focused on patients or caregivers

(2) Involved the use of CHI

(3) Examined how health tasks fit into patients' settings

Methodology

SCREENING PROCESS

MACQUARIE University

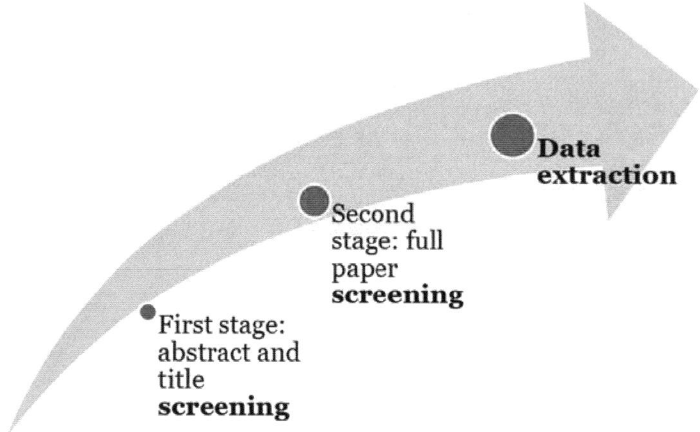

Methodology

DATA EXTRACTION AND SYNTHESIS STRATEGY

MACQUARIE University

(1) General characteristics

- Study author, year, location
- Health domain
- Study type

(2) Participants

- Number
- Age
- Sex

(3) Interventions

- Health activities
- Technology mode
- Functions
- Context

(4) Main findings

Results

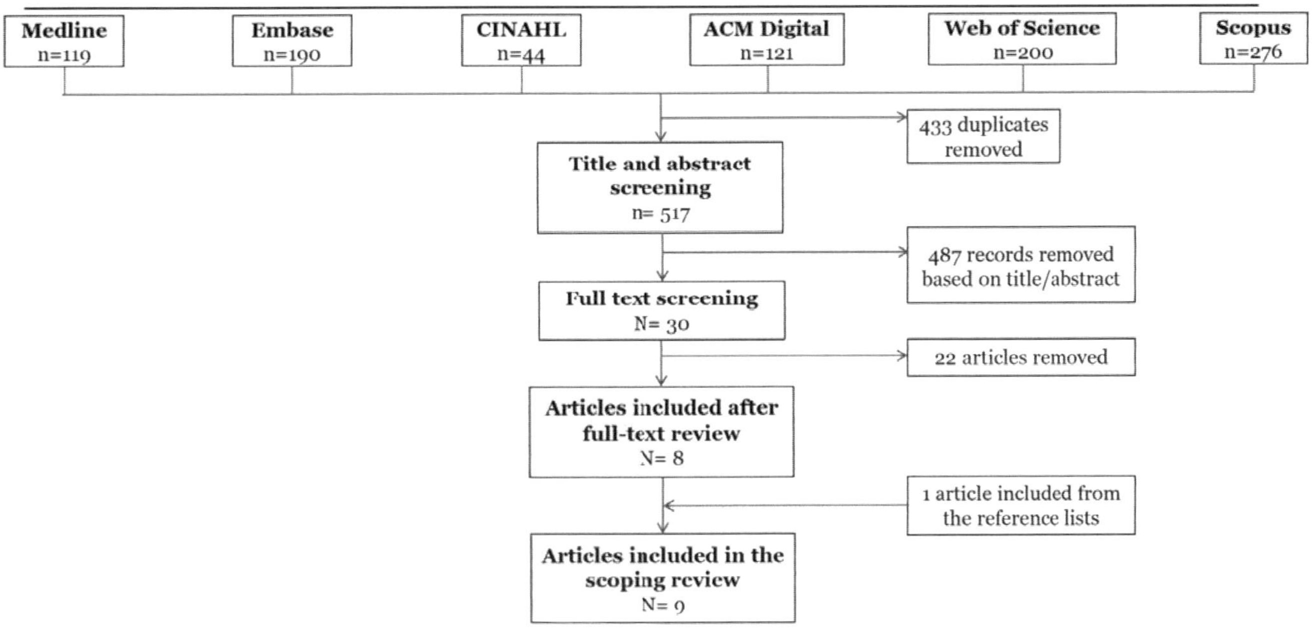

Results

BRIEF DESCRIPTION OF INCLUDED STUDIES

Location

- 7 studies from US
- 1 from Sweden
- 1 from Thailand

Health domains

- 4 studies: chronic diseases
- 2 studies: aged care and home nursing care

Methodology

- 3 qualitative studies: interviews and surveys
- 3 system design studies
- 3 reviews

Results

MACQUARIE University

Participant characteristics

- Total = 1,992 participants
- 3 studies focused on patients, 2 on caregivers
- Mostly in the older age group (aged 65+)

Technology mode

- Web-based platform was the most popular.
- Sensors were often used in conjunction with other interventions.

Discussion

MAIN RESULTS

MACQUARIE University

- The first scoping review which examines CHI applications that support patient work
- The literature is still in its infancy
- No agreed upon conceptual understanding of patient work

Discussion

THEME 1: CLASSIFICATION OF PATIENT WORK ELEMENTS

MACQUARIE University

Patient work framework

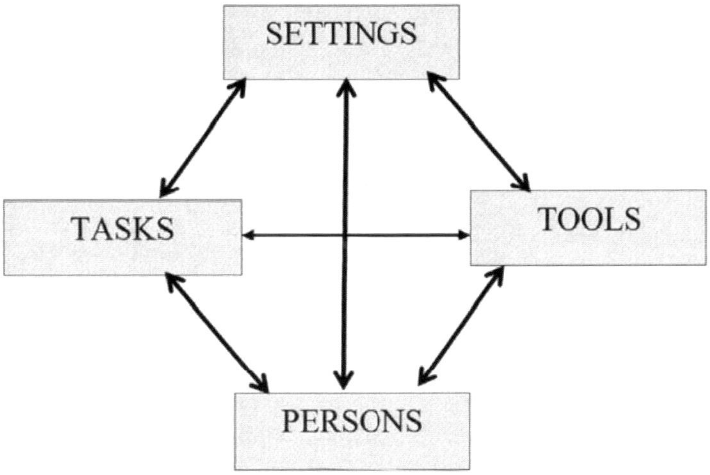

Discussion

THEME 2: IMPLICATIONS ON METHODOLOGICAL

MACQUARIE University

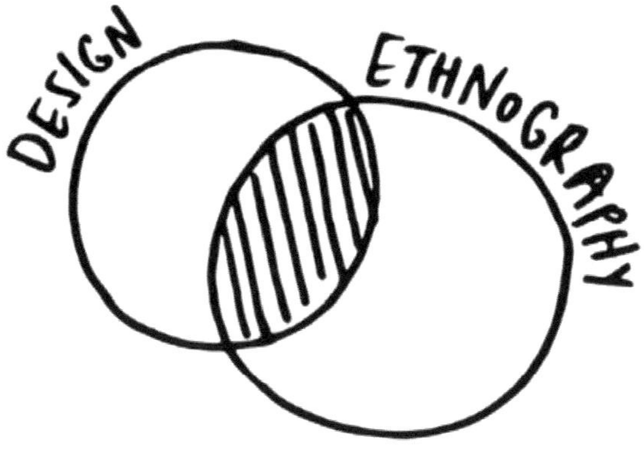

Strengths and limitations

Strengths	Limitations
Extensive search across multiple databases Screening form was piloted before actual screening Kappa score of 0.54	Could not evaluate interventions' effectiveness Exclude non-English papers

In summary…

- A patient work lens can be used to examine health activities in larger context.
- More ethnographic studies are needed to examine what activities consumers actually do for their health.
- User-centred design is essential.

MACQUARIE University

Thank you!

Contact: Ms Huong Ly Tong

huong-ly.tong@students.mq.edu.au

OFFICE | FACULTY | DEPARTMENT 19

HFESA

Journal Club – a sustainable professional development model for the HFESA

Jennifer Long

Jennifer Long Visual Ergonomics, Katoomba, NSW

School of Optometry and Vision Science, UNSW, Sydney

Christine Aickin

Workability, Sydney, NSW

HFESA

What is journal club?

HFESA

Sydney, September 2013:

The first HFESA Journal Club meeting

HFESA

By August 2015:

The HFESA had hosted more than 20 Journal Club meetings

HFESA

Aims of this study

1. What do members find valuable with Journal Club?

2. How can we improve meeting delivery and participation?

Human Factors & Ergonomics Society of Australia Inc.

HFESA

Process

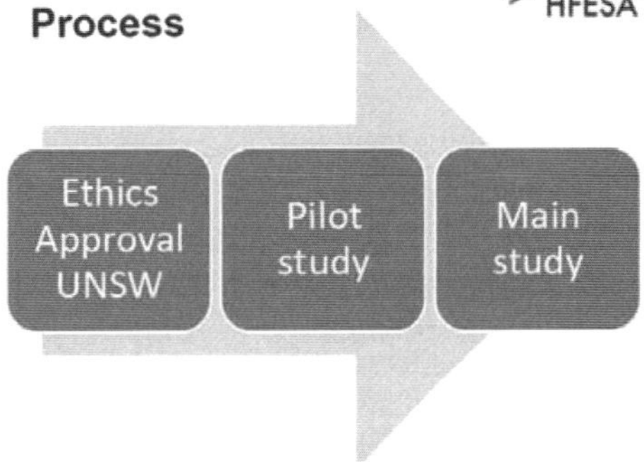

Human Factors & Ergonomics Society of Australia Inc.

HFESA

February 2016 – March 2016

Online survey of HFESA members

Email to members
Email to Branch chairs
Email to SIG chairs
Invitation in HFESA Connect

Human Factors & Ergonomics Society of Australia Inc.

HFESA

47 participants (response rate = 8.6%)

- 46 based in Australia
- All Australian States except NT

Human Factors & Ergonomics Society of Australia Inc.

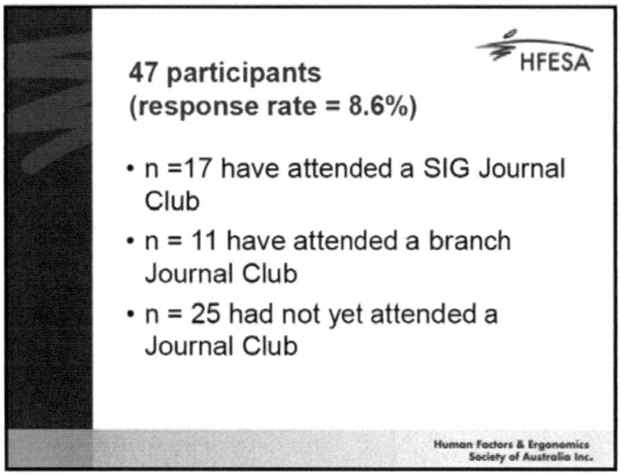
HFESA
47 participants
(response rate = 8.6%)
• n =17 have attended a SIG Journal Club
• n = 11 have attended a branch Journal Club
• n = 25 had not yet attended a Journal Club
Human Factors & Ergonomics Society of Australia Inc.

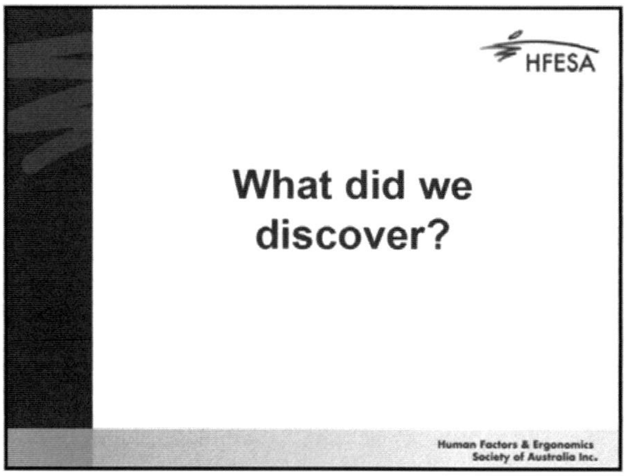
HFESA
What did we discover?
Human Factors & Ergonomics Society of Australia Inc.

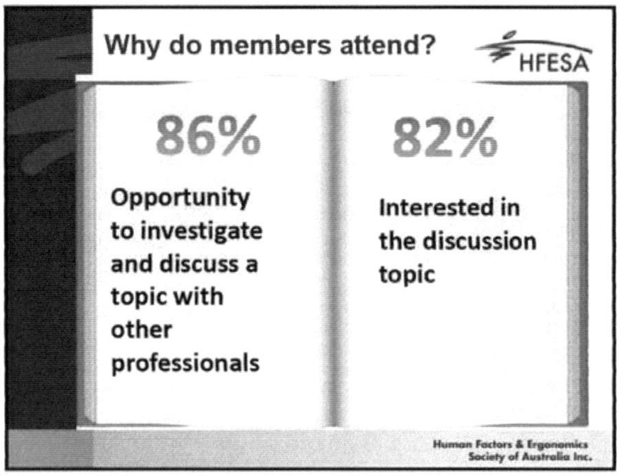
Why do members attend?
HFESA
86%
Opportunity to investigate and discuss a topic with other professionals
82%
Interested in the discussion topic
Human Factors & Ergonomics Society of Australia Inc.

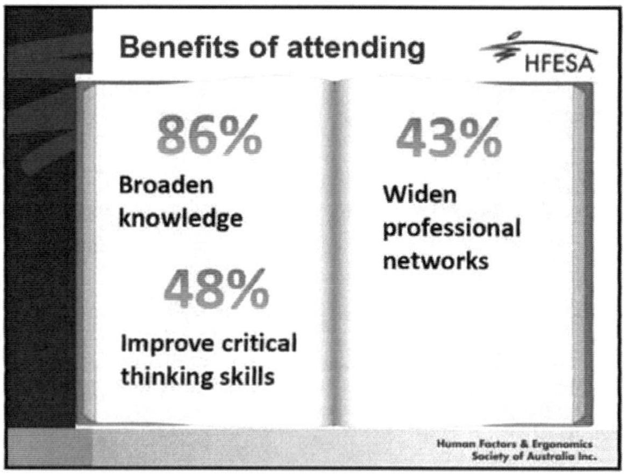
Benefits of attending
HFESA
86%
Broaden knowledge
48%
Improve critical thinking skills
43%
Widen professional networks
Human Factors & Ergonomics Society of Australia Inc.

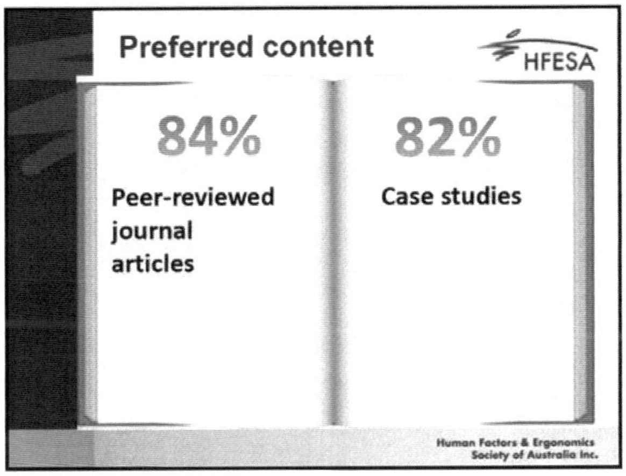
Preferred content
HFESA
84%
Peer-reviewed journal articles
82%
Case studies
Human Factors & Ergonomics Society of Australia Inc.

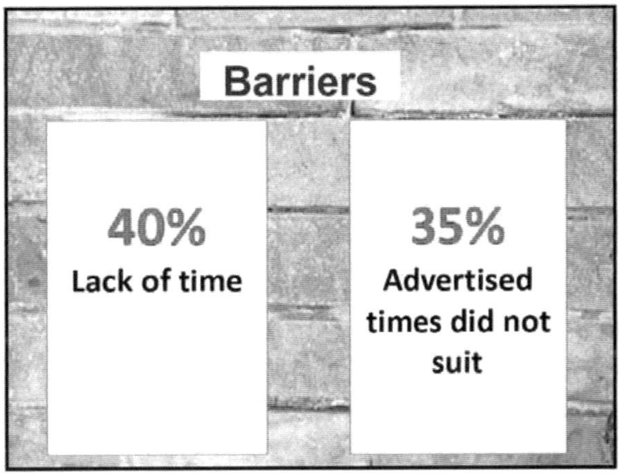
Barriers
40%
Lack of time
35%
Advertised times did not suit

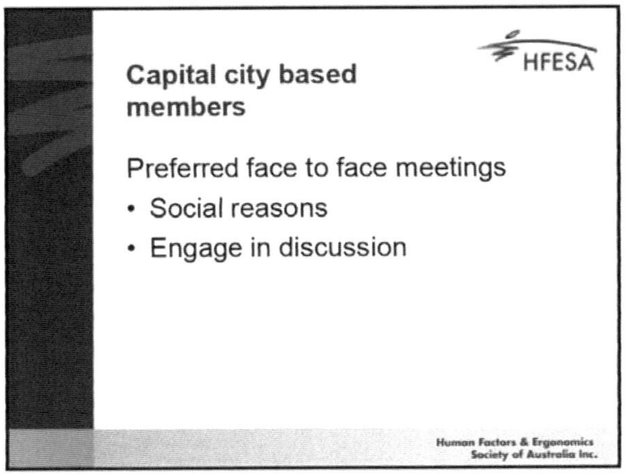
Capital city based members
HFESA
Preferred face to face meetings
• Social reasons
• Engage in discussion
Human Factors & Ergonomics Society of Australia Inc.

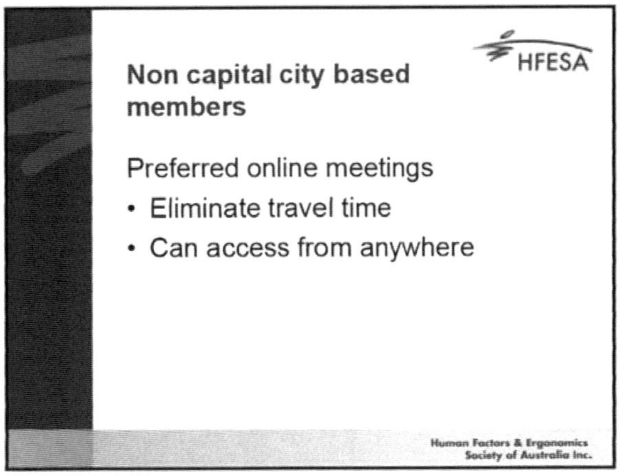
Non capital city based members
HFESA
Preferred online meetings
• Eliminate travel time
• Can access from anywhere
Human Factors & Ergonomics Society of Australia Inc.

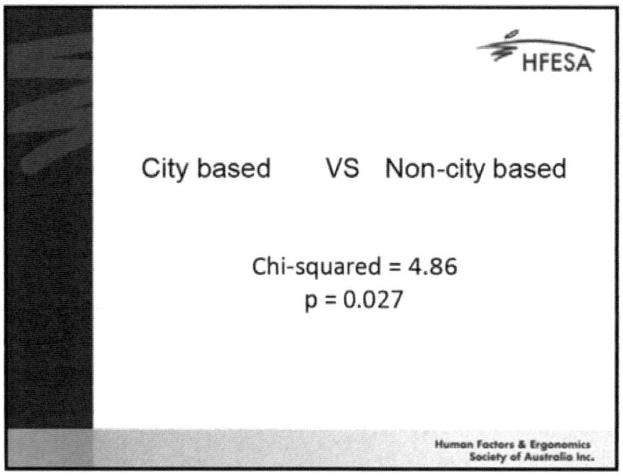
HFESA
City based VS Non-city based
Chi-squared = 4.86
p = 0.027
Human Factors & Ergonomics Society of Australia Inc.

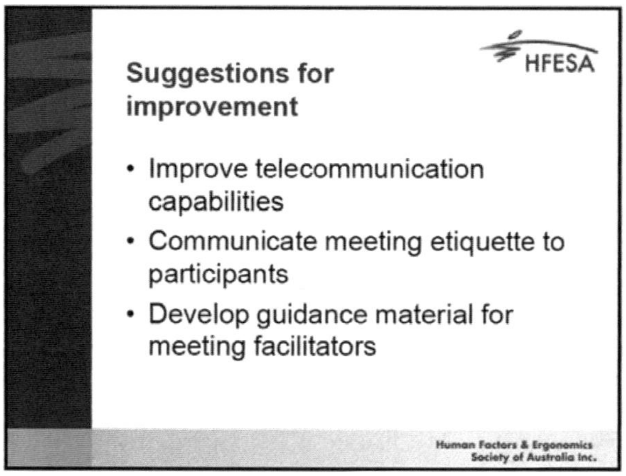
HFESA
Suggestions for improvement
• Improve telecommunication capabilities
• Communicate meeting etiquette to participants
• Develop guidance material for meeting facilitators
Human Factors & Ergonomics Society of Australia Inc.

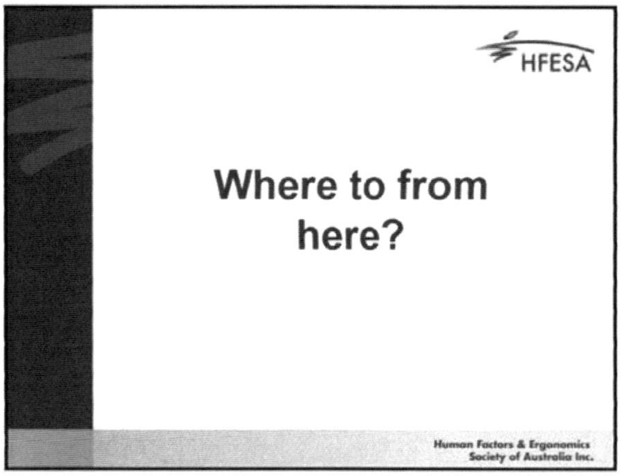
HFESA
Where to from here?
Human Factors & Ergonomics Society of Australia Inc.

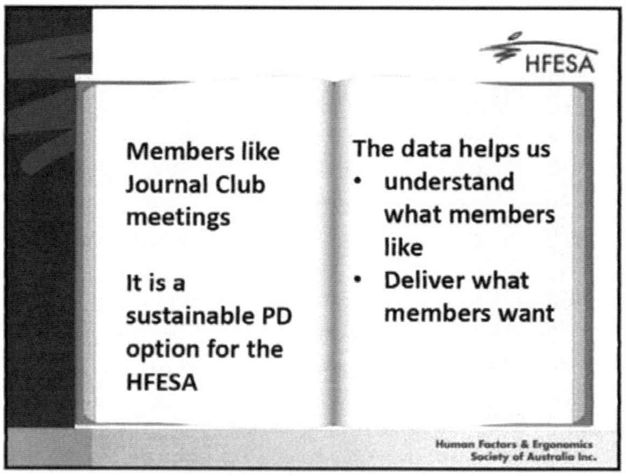
HFESA
Members like Journal Club meetings
It is a sustainable PD option for the HFESA
The data helps us
• understand what members like
• Deliver what members want
Human Factors & Ergonomics Society of Australia Inc.

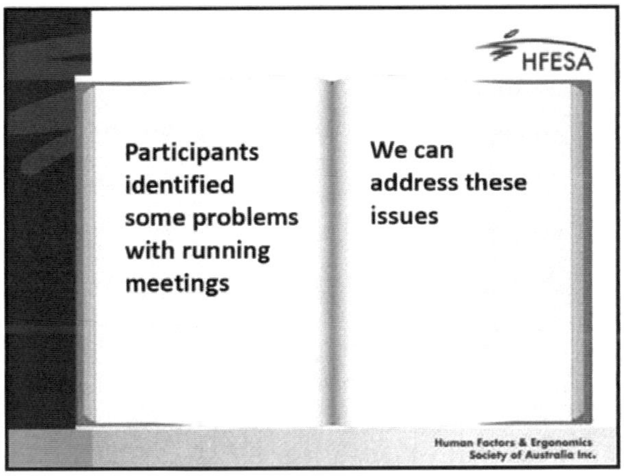
HFESA
Participants identified some problems with running meetings
We can address these issues
Human Factors & Ergonomics Society of Australia Inc.

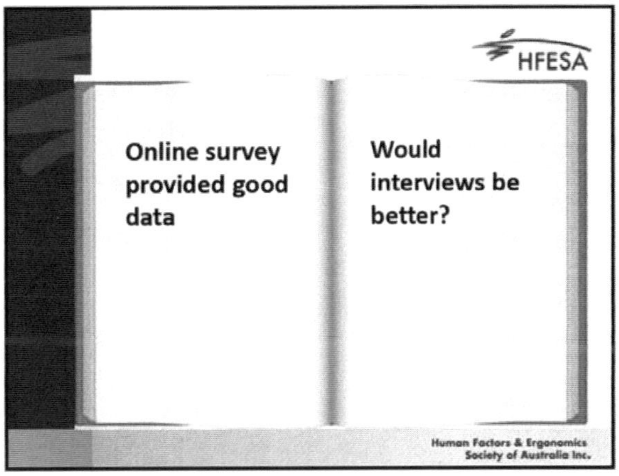
HFESA
Online survey provided good data
Would interviews be better?
Human Factors & Ergonomics Society of Australia Inc.

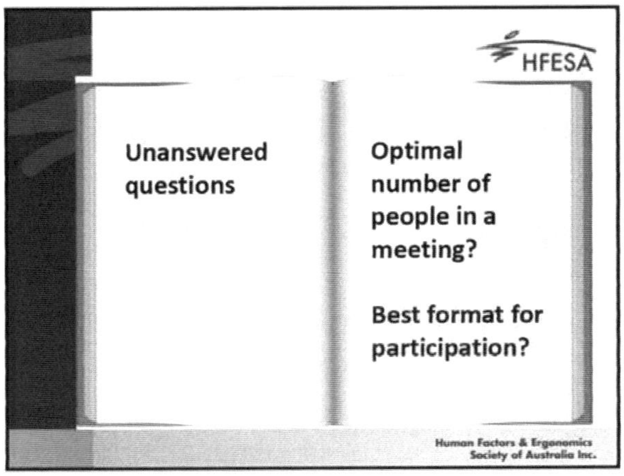
HFESA
Unanswered questions
Optimal number of people in a meeting?
Best format for participation?
Human Factors & Ergonomics Society of Australia Inc.

HFESA
Resources for the HFESA
Human Factors & Ergonomics Society of Australia Inc.

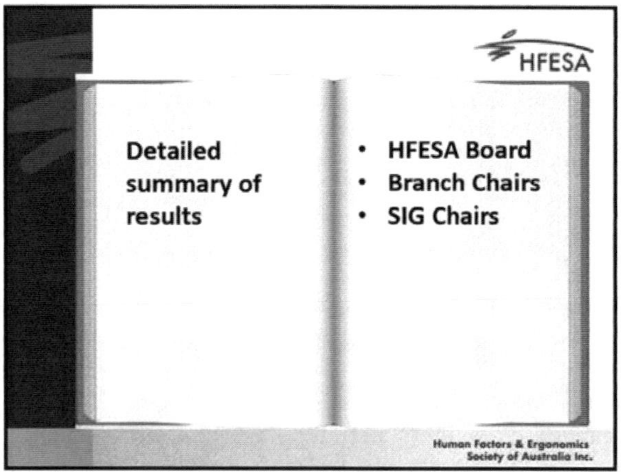
HFESA
Detailed summary of results
• HFESA Board
• Branch Chairs
• SIG Chairs
Human Factors & Ergonomics Society of Australia Inc.

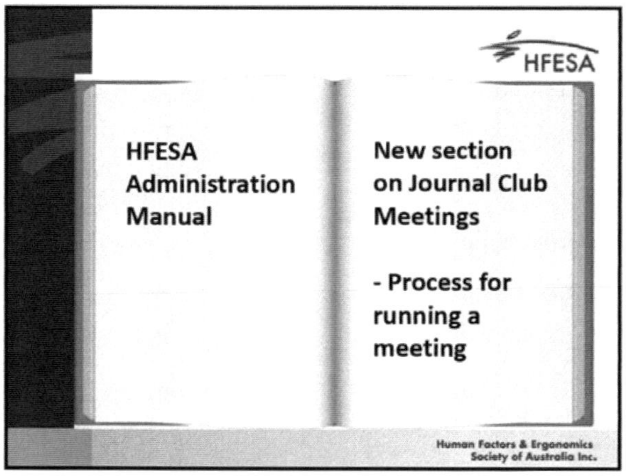
HFESA
HFESA Administration Manual
New section on Journal Club Meetings
- Process for running a meeting
Human Factors & Ergonomics Society of Australia Inc.

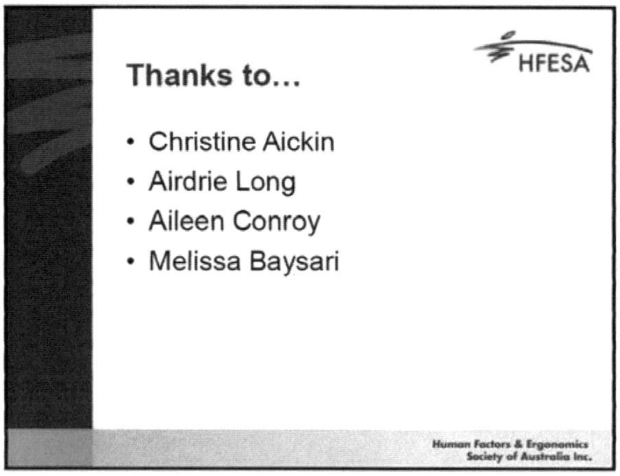
Thanks to...
HFESA
• Christine Aickin
• Airdrie Long
• Aileen Conroy
• Melissa Baysari
Human Factors & Ergonomics Society of Australia Inc.

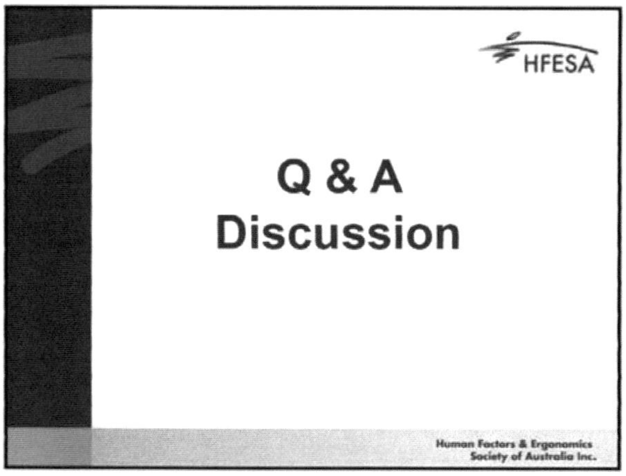
HFESA
Q & A
Discussion
Human Factors & Ergonomics Society of Australia Inc.

doi:10.14311/ppt.2017.3.249
Plasma Physics and Technology **4**(3):249–252, 2017

EMISSION SPECTROSCOPY DURING HIGH-CURRENT ANODE MODES IN VACUUM ARC

A. Khakpour*, R. Methling, St. Franke, S. Gortschakow, D. Uhrlandt

Leibniz-Institut for Plasma Science and Technology, Felix-Hausdorff-Str. 2, 17489 Greifswald, Germany

* alireza.khakpour@inp-greifswald.de

Abstract. A vacuum interrupter reaches its interruption limit once high-current anode phenomena occur. High-current anode modes lead to an increase of the anode surface temperature and an increased generation of metal vapor, which may result in a weakening of the dielectric recovery strength after current zero. In this work, different discharge modes in a vacuum arc for AC 50 Hz including diffuse, footpoint, anode spot type 1 and type 2, and anode plume are investigated. Electrodes made of CuCr7525 with diameter of 10 mm are used. The final gap length is about 20 mm. Time and space resolved optical emission spectroscopy is used to examine the temporal and spatial distribution of atomic and ionic copper lines. The distribution of atomic and ionic lines parallel and perpendicular to the anode surface is investigated. Radiator density is also determined for Cu I, Cu II, and Cu III near the anode surface.

Keywords: vacuum arc, high-current anode phenomena, emission spectroscopy.

1. Introduction

The interest in developing vacuum circuit breakers for high-voltage applications is still growing, whereas the high-current anode phenomena are important due to their impact on the contact erosion and interruption failure [1, 2]. Different high-current anode modes including footpoint, anode spot type 1 and type 2, and intense mode in which the anode plays an active role are investigated in [3]. The transition from the low-current mode to high-current modes is associated with considerable changes in the optical, thermal, and electrical properties near to the anode, the cathode, and in the inter-electrode gap [4, 5].

Different discharge modes have been investigated by means of the evaluation of high-speed camera images and electrical characteristics [5]. The temporal and axial distribution of different atomic and ionic lines during transition between different discharge modes are investigated in [4]. Two types of anode spot modes are investigated in [5] which differ in the voltage course and the radiation intensity. The anode spot type 2 appears at higher gap length and even lower current comparing to anode spot type 1. An abrupt change in the arc voltage and intensity of Cu I–III lines is observed during transition to anode spot type 2. However, the previous work is focused on the axial distribution of line intensity of different copper species. Nevertheless, more information regarding different discharge modes in vacuum arc and recent research can be find in [6].

In this work the temporal distribution of Cu I, Cu II, and Cu III lines is examined near the anode for AC 50 Hz during transition between different high-current anode modes. Temporal and spatial distribution of radiating density is presented for different high-current anode modes.

2. Experimental setup

The schematic representation of the experimental setup is discussed in [4]. The power source consists of capacitors, inductances, charging and control units and enables the possibility to realize different current waveforms. In the present work a sinusoidal alternating current at 50 Hz is used. To generate an AC 50 Hz waveform, five LC elements are applied in parallel. The capacitance and inductance of each branch is 512 µF and 4700 µH, respectively. This part is in series with inductance of 2730 µH. Different current levels can be achieved by adjusting the charging voltage of the capacitors.

The ultrahigh vacuum chamber with basic pressure of about 2×10^{-8} mbar is equipped with mounting adapters for different electrodes. In this work, cylindrical electrodes of CuCr7525 with a diameter of 10 mm are used. A mechanical-pneumatic actuator system is designed for the separation of electrodes. The upper electrode is fixed and connected to the power supply, whereas the lower electrode can be moved down. Constant opening velocities are obtained that can be chosen between 1 m/s to 4 m/s depending on the starting pressure in the actuator. The maximum gap length is about 20 mm. The delay between start of current and electrode separation can be adjusted freely with a total jitter below 100 µs [4]. So, the arc voltage is appears at the moment of the opening the contacts.

The registration of the arc dynamics and observation of anode modes is realized by means of a high-speed video camera (Y4 series from Integrated Design Tools) with a recording speed of 20 000 fps. The spectroscopic setup consists of an imaging spectrograph with 0.5 m focal length (SP2500 from Princeton Instruments) combined with high-speed camera (IDT Y4) with a recording speed of 10 000 frames per second. Usually, that technique is named as video spectroscopy. In our case, it enables to study the temporal evolution of lines of different atoms and ions including their spatial profiles during one pulse with 100 μm resolution. Therefore, it is appropriate also for cases of limited reproducibility. The area at a distance of 1.0 mm to the anode is imaged to the entrance slit of the spectrograph using an entrance slit width of 50 μm and grating with 150 lines per mm.

For registration of electrical signals a system consisting of a Pearson current monitor (Model 1423) is used with a resistance of 1 mΩ and with low and high frequency 3 dB cut-off of 1 Hz and 1.2 MHz, respectively. To measure the voltage, a voltage probe (Tektronix 6015A) with a bandwidth of 75 MHz is used.

3. Experimental results

Experiments have been performed applying AC waveforms and peak currents of up to 4.5 kA. Contact separation speed has been set to 1.2 m/s and the final gap distance was 20 mm. The successive transition from the diffuse to the different high-current anode modes is illustrated in the following by current and voltage waveforms together with high-speed camera images. The time behavior of Cu atomic and ionic lines is investigated in cases parallel and perpendicular to the anode surface.

3.1. Electrical characteristics and high-speed camera images

Fig. 1 presents the formation of different discharge modes of the vacuum arc including diffuse, footpoint mode, anode spot type 1, anode spot type 2, and anode plume for AC 50 Hz. The corresponding current and voltage traces are presented in Fig. 2. The opening time is about 0.40 ms and the maximum current about 4.4 kA in this example.

At the beginning of the discharge, a diffuse mode can be observed (Fig. 1a), after that the footpoint mode appears until 3.6 ms when anode spot type 1 occurs (Fig. 1b). Anode spot type 1 continues up to 5.4 ms. Notice, with the transition to anode spot type 1 at about 3.6 ms, the average of the arc voltage jumps up by about 10 % (Fig. 2). Then anode spot type 2 appears corresponding to brighter illumination areas at both anode and cathode compared to anode spot type 1 (cf. Fig. 1c to f). As presented in Fig. 2 the transition to anode spot type 2 is associated with the abrupt change in the arc voltage of about 35 %. After extinction of anode spot type 2 (Fig. 1g) the anode plume appears on the anode (Fig. 1f) and it continues up to the extinction of the arc.

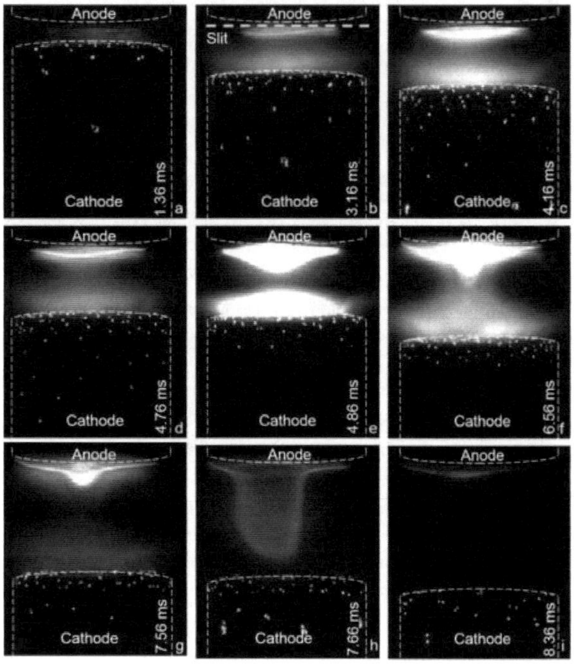

Figure 1. The formation of different discharge modes: diffuse (a), footpoint (b), anode spot type 1 (c and d), anode spot type 2 (e, f. and g), anode plume (h), and diffuse (i).

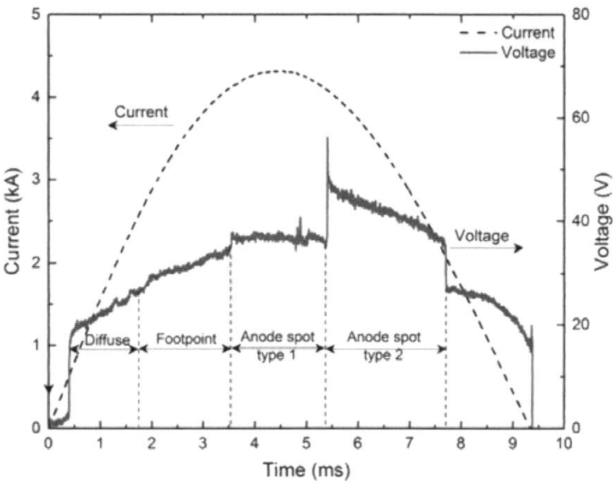

Figure 2. The current and voltage waveforms during transition to different anode spot modes. The time instant of the formation of different high-current anode modes are presented.

3.2. Spectroscopy parallel to the anode surface

To get information about distribution of atomic and ionic line radiation during the anode spot modes the spectrograph has been aligned parallel to the anode surface at a distance of 1.0 mm from the anode surface (cf. dotted line in Fig. 1b). The spectral range 440–580 nm is chosen that contains well-resolved spectral lines of different ionization states of Cu and Cr. The local emission coefficients of the Cu I lines can be estimated based on Abel inversion of parts of the line

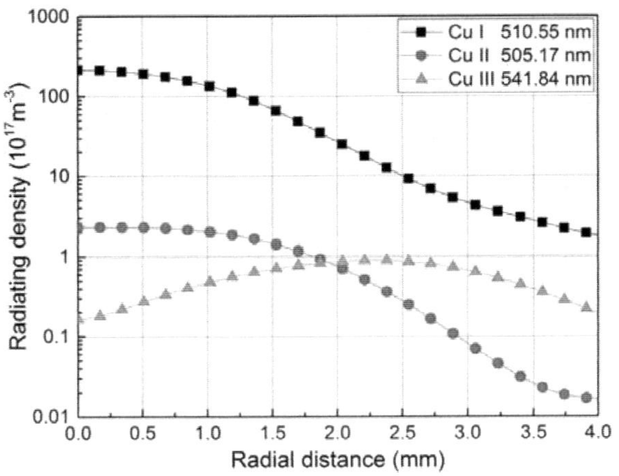

Figure 3. The density of radiating species of Cu I [510.55 nm], Cu II [505.17 nm] and Cu III [541.84 nm] during anode spot type 1 at 4.5 ms and at a distance of 1.0 mm from anode surface in dependence on the distance from the anode spot center.

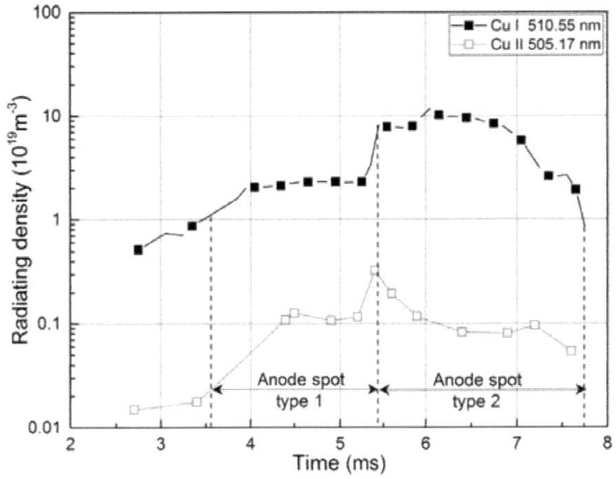

Figure 4. The temporal distribution of the radiating density for Cu I [510.55 nm] and Cu II [505.17 nm]in the anode spot center and at a distance of 1.0 mm from anode surface.

integrated Cu I radiation after absolute calibration of the measured intensities. The local emission coefficient ε of a transition from an upper level u to a lower level l is related to the local radiating density n_{u}, i.e. the density of excited Cu atoms of the corresponding upper level u can be calculated according to

$$\varepsilon = \frac{1}{4\pi}\frac{hc}{\lambda}n_{\mathrm{u}}A_{\mathrm{lu}} \qquad (1)$$

where λ is the transition line wavelength, A_{ul} is the transition probability, h is the Planck constant, and c is the speed of light. Results for the upper level excited state densities corresponding to the lines Cu I [510.55 nm], Cu II [505.17 nm] and Cu III [541.84 nm] during anode spot type 1 (at 4.5 ms after contact opening) are presented in Fig. 3.

The radiating density of Cu I at 510.55 nm at the center of anode spot type 1 is about $2 \times 10^{19}\,\mathrm{m}^{-3}$, whereas the density of radiating species of Cu II at 493.16 nm and Cu III at 541.84 nm are about $2 \times 10^{17}\,\mathrm{m}^{-3}$ and $2 \times 10^{16}\,\mathrm{m}^{-3}$, respectively. Notice, outside the anode spot at about 2 mm the radiating density of Cu III is higher than Cu II which can be also confirmed by different intensity distribution of anode spot.

The temporal distribution of the radiating density of Cu I [510.55 nm] at the center of anode spot and a distance of 1.0 mm from anode surface is presented in Fig. 4. During footpoint mode the radiating density is in the range of $5 - 7 \times 10^{18}\,\mathrm{m}^{-3}$. During anode spot type 1 the radiating density is about $1 - 2 \times 10^{19}\,\mathrm{m}^{-3}$. After transition to the anode spot type 2 at 5.4 ms, the radiating density is about $9 \times 10^{19}\,\mathrm{m}^{-3}$ and it starts to decrease at 6.5 ms. After extinction of the anode spot type 2 at 7.7 ms, the radiating density decreases strongly. Notice that the radiating density for the time instant before 2.5 ms is not presented due to the weak emission lines.

Notice, that conclusions from the radiating densities to ground state densities of atoms and ions are not possible without knowledge of additional physical quantities because of the non-equilibrium nature of the vacuum arc plasma.

3.3. Spectroscopy perpendicular to the anode surface

The axial intensity distribution of atomic and double charge ion lines are selected here for comparison. The spectrograph has been aligned for an observation along a line perpendicular to the anode surface at the center of the electrodes. The axial distributions during the transition from the diffuse mode to high-current anode modes including footpoint mode, anode spot type 1 and 2 are presented in Fig. 5. The approximate positions of anode and cathode are also presented using the high-speed camera images. The intensity of each species is normalized with respect to the peak intensity of anode spot type 2.

The relative intensity of the Cu I line near to the anode becomes larger with each anode mode. Cu I line intensity shows a pronounced decrease towards the cathode. Notice, the distribution of Cu I line in case of anode spot type 2 is broader compared to other discharge modes which is in a good agreement with the radiation intensity of anode spot type 2 presented in Fig. 1. In different discharge modes, the intensity of the Cu I line changes slightly near the cathode.

The relative intensities of the Cu II line during footpoint mode and anode spot modes are similar to those of the Cu I line near to the anode and the cathode. Note, that a graphic illustration is omitted here for the sake of brevity. However, the profile near to the anode is much broader compared to the Cu I line. Notice, in case of diffuse mode the intensity of the Cu II line near to the anode is very low.

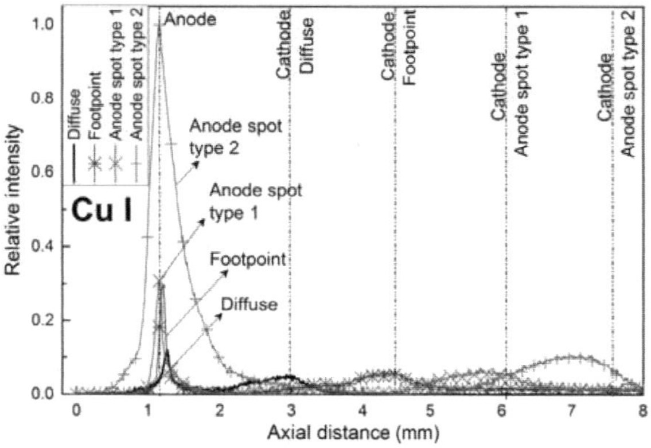

Figure 5. Axial intensity distributions of the line Cu I [510.55 nm] during transition from diffuse mode to different high-current anode modes. The dashed lines correspond to the corresponding electrode positions.

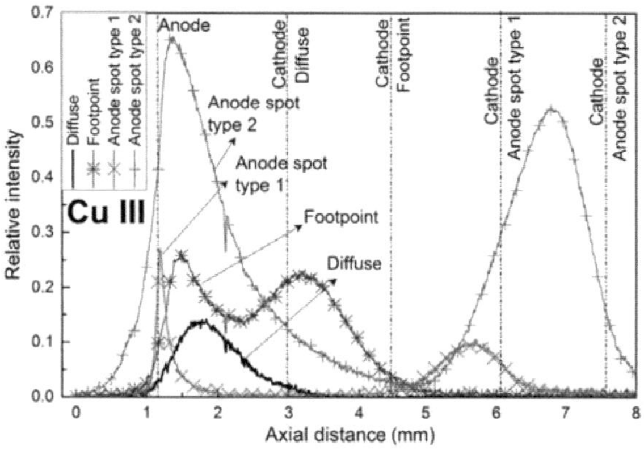

Figure 6. Axial intensity distribution of the line Cu III [509.42 nm] during transition from diffuse mode to different high-current anode modes. The dashed lines correspond to the corresponding electrode positions.

The relative intensity of the Cu III line shows a different behavior than Cu I with a much broader spatial profile along the discharge axis. There is a clear change in the distribution of the Cu III line during transition from diffuse mode to footpoint mode and anode spot modes near the anode, the cathode and in the inter-electrode gap. It seems that the intensity of the Cu III line has also two maxima near to the anode and the cathode but more distant from the surfaces compared to Cu I. The two maxima cannot be resolved in the diffuse mode and are still overlapping in the footpoint mode. The line intensity of Cu III in case of footpoint mode and anode spot type 1 mode is almost the same near the anode.

However, the intensity near the cathode is higher in case of footpoint mode. During anode spot type 2 the intensity of the Cu III line increases near to the anode and cathode and has a much broader profile there comparing to anode spot type 1.

The high intensity of the Cu III line near the anode during anode spot of both types give strong hint that double charged Cu ions near the anode originate from the plasma emitted from the anode.

The absolute error can be estimated to 20-30 % for both intensity and density of radiating species. Relative errors influencing the shape of the profile can be estimated to few percent of the overall intensity. Considering the logarithmic scale in Fig. 4, the limited dynamic range of the high-speed camera (10 bit) has to be considered. Although it can be neglected for higher densities and is acceptable between about 4 and 7.5 ms, the detected signals are rather low compared to the noise level before and after this time. The densities before 4 ms can have uncertainties of more than 100 %. Nevertheless the information about the rise of density and the order of magnitude is valuable.

4. Summary and outlook

In this work, the radial and axial distributions of atomic and ionic lines during transitions from diffuse mode to different high-current modes are investigated. During transition from anode spot type 1 to anode spot type 2 an abrupt change appears in the arc voltage and in both intensities near the anode and cathode.

The determined radial distributions of radiating densities of the Cu I and Cu II lines show that the radiating density of the atomic line is higher than that of the ionic line during anode spot modes. In addition, the atomic radiating density during anode spot type 2 is 5 times higher than during anode spot type 1. The density of the Cu III line outside the high intensity shell of the anode spot is higher than that of Cu II.

Determining the ground state density and temperature during different high current anode modes is scheduled as a future work considering the non-equilibrium vacuum arc plasma.

References

[1] E Schade and E Dullni. Recovery of breakdown strength of a vacuum interrupter after extinction of high current. *IEEE Trans. Dielectr. Elect. Insul.*, 9(2):207–215, 2002. doi:10.1109/94.993737.

[2] E. Dullni and E. Schade. Investigation of high-current interruption of vacuum circuit breakers. *IEEE Trans. Dielectr. Elect. Insul.*, 28(4):607–620, 1993. doi:10.1109/14.231543.

[3] H. C. Miller. A review of anode phenomena in vacuum arcs. *Contrib. Plasma Phys.*, 29(3):223–249, 1989. doi:10.1109/TPS.1985.4316413.

[4] A. Khakpour et al. Video spectroscopy of vacuum arcs during transition between different high-current anode modes. *IEEE Trans. Plasma Sci.*, 44(10):2462–2469, 2016. doi:10.1109/TPS.2016.2602384.

[5] A. Khakpour et al. Optical and electrical investigation of transition from anode spot type 1 to anode spot type 2. *IEEE Trans. Plasma Sci.*, 45(8):2126–2134, 2017. doi:10.1109/TPS.2017.2690572.

[6] P. G. Slade. *The Vacuum Interrupter: Theory, Design, and Application.* Boca Raton, FL, USA: CRC Press, 2008.

doi:10.14311/ppt.2017.3.253
Plasma Physics and Technology **4**(3):253–256, 2017

RADIATION TRANSFER IN ARC PLASMAS

N. Bogatyreva[a,*], M. Bartlová[a], V. Aubrecht[b], P. Kloc[b]

[a] *Department of Physics, Faculty of Electrical Engineering and Communication, Brno University of Technology, Technická 3058/10, 616 00 Brno, Czech Republic*

[b] *Department of Power Electrical and Electronic Engineering, Faculty of Electrical Engineering and Communication, Brno University of Technology, Technická 3058/12, 616 00 Brno, Czech Republic*

[*] bogatyreva@feec.vutbr.cz

Abstract. In this paper, attention has been given to the absorption properties of the arc plasma at the different pressure conditions. Calculations of the absorption coefficients for a thermal plasma have been performed as a function of the temperature and the frequency. Methods for prediction of the average absorption coefficients were described and compared in detail.

Keywords: air plasma, radiation transfer, absorption coefficient.

1. Introduction

Currently arc discharges are used increasingly in various fields of modern science and technology. Certain plasma parameters make its employment very promising for creation of new technologies. Applications of the plasma are quite diverse: from equipment for the modification and processing of polymer materials and plasma cutting machines till using plasma in high voltage circuit breakers and as well as household applications, etc.

In most cases optimization of technological conditions is the result of experiment during development of advanced equipment employing plasma. However, the experimental conditions are rather extreme (high temperature, pressure and gas velocity). Also, carrying out experiments is associated with substantial material expenditures. Therefore, it is very important to develop mathematical approximations and process modeling of parameters in the arc plasma. Mathematical modeling is also quite complex and it requires certain computational resources, so it is carried out in several stages. Absorption coefficients are primary parameters in the simulation. Based on that it is possible to define other parameters, which are necessary for analyzing of energy transfer in arc plasma.

2. Energy balance of arc plasma

Electric discharge plasmas transmit relatively great deal of power. The ratios while transmission are described by the energy balance of the plasma. In the theory of the energy balance of the plasma only the element of the plasma volume is taken into consideration and heat convection is neglected.

The energy balance of the arc plasma in stationary state can be described by Ehlenbaas–Heller equation [1]

$$\sigma_E E^2 = -\mathrm{div}\,[\lambda \cdot \mathrm{grad}(T)] + \mathrm{div}(\mathbf{W}_\mathrm{R}), \qquad (1)$$

where $\sigma_E E^2$ denotes the input electric power (σ_E is the electric conductivity, E is the electric field), $\lambda \cdot \mathrm{grad}(T)$ denotes the energy flux due to heat conduction (λ is the heat conductivity, T is the plasma temperature, $\mathrm{div}(\mathbf{W}_\mathrm{R})$ denotes the losses of energy by radiation ($\mathbf{W}_\mathrm{R}$ is the radiation flux).

Therefore, for modeling of arc plasma including radiation it is necessary to determine the radiation flux and its divergence. The solution of the equation of radiation transfer makes possible to estimate these parameters.

The complete stationary equation of radiation transfer for an absorbing and emitting medium is

$$\mathbf{\Omega} \cdot \mathrm{grad}(I_\nu) = \kappa_\nu [B(\nu) - I_\nu], \qquad (2)$$

where $\mathbf{\Omega}$ is the unit direction vector, I_ν is the radiation intensity, κ_ν is the absorption coefficient, $B(\nu)$ is Planck's spectral radiation intensity for equilibrium radiation, ν is the frequency [2].

The equation of transfer is obviously very complicated. The spectral intensity, which is the dependent variable in this equation, depends in general upon independent variables $(\mathbf{r}, \nu, \mathbf{\Omega})$. One must approximate the equation of transfer, either analytically or numerically, in order to obtain a solution. Due to the non-linearity of equations describing the radiation field and strong dependence of input parameters on the radiation frequency, various approximate methods are used. One of them is the method of spherical harmonics – P_N-approximation [3–6].

The great advantage of the method of spherical harmonics is the conversion of the equation of transfer to relatively simple partial differential equations. However, the low-order approximations are usually only accurate in optically thick media and for higher-order approximations the mathematical complexity increases rapidly. It is known from neutron transport theory that approximations of odd orders are more accurate than even ones. Due to its simplicity, mainly the lowest order P_N solution corresponding to $N = 1$

(P_1-approximation) is usually used (so called diffusion approximation). The diffusion approximation describes in good accuracy the radiation field in many problems of radiation hydrodynamics.

In P_1-approximation we suppose that the angular dependence of the specific intensity can be represented by the first two terms in a spherical harmonic expansion

$$I_\nu(\mathbf{r},\nu,\mathbf{\Omega}) = \varphi_1(\mathbf{r},\nu) + 3\varphi_2(\mathbf{r},\nu)\cdot\mathbf{\Omega}, \tag{3}$$

where $\mathbf{r}$ is the position vector fixing the location of a point in space, φ_1 and φ_2 correspond to the density of the radiation field multiplied by velocity of light c, and to the radiation flux [7].

The spectral density of the radiation field is

$$U_\nu(\mathbf{r},\nu) = \frac{1}{c}\int_0^{4\pi} I_\nu(\mathbf{r},\nu,\mathbf{\Omega})\mathrm{d}\Omega = \frac{4\pi}{c}\varphi_1(\mathbf{r},\nu). \tag{4}$$

Likewise, for radiation flux we obtain

$$\mathbf{W}_\nu = (\mathbf{r},\nu)\int_0^{4\pi} I_\nu(\mathbf{r},\nu,\mathbf{\Omega})\cdot\mathbf{\Omega}\mathrm{d}\Omega = 4\pi\varphi_2(\mathbf{r},\nu). \tag{5}$$

By integration of the equation of radiation transfer Eq. (2) over all solid angles we obtain the diffusion equation for the spectral density of radiation field

$$\begin{aligned}-\mathrm{div}\left\{\frac{c}{3\kappa_\nu}\mathrm{grad}\left[U_\nu(\mathbf{r},\nu)\right]\right\} + \kappa_\nu c U_\nu(\mathbf{r},\nu)\\ = \kappa_\nu 4\pi B(\nu),\end{aligned} \tag{6}$$

and for the radiation flux

$$\mathbf{W}_\nu(\mathbf{r},\nu) = -\frac{c}{3\kappa_\nu}\mathrm{grad}\left[U_\nu(\mathbf{r},\nu)\right]. \tag{7}$$

The diffusion approximation is valid under assumption that the spectral radiation intensity is almost isotropic.

3. Multigroup method

One of the methods for handling the frequency variable in the equation of transfer is the multigroup method [8], which leads to its discretization. One assigns a given photon to one of G frequency groups, and all photons within a given group are treated the same from the point of view absorption properties of the medium, the absorption coefficient for given frequency group k is supposed to be constant with certain average value

$$\kappa_\nu(\mathbf{r},\nu,T) = \kappa_k(\mathbf{r},T), \nu_k \le \nu \le \nu_{k+1}, k = 1,\ldots,G. \tag{8}$$

The equation of transfer for the given frequency group can be treated as equation for grey medium:

$$\mathbf{\Omega}\cdot\mathrm{grad}\left[I_\nu(\mathbf{r},\mathbf{\Omega})\right] = \overline{\kappa_k}\left[\int_{\nu_k}^{\nu_{k+1}} B(\nu)\mathrm{d}\nu - I_k(\mathbf{r},\mathbf{\Omega})\right], \quad 1 \le k \le G, \tag{9}$$

where $\overline{\kappa_k}$ is the mean absorption coefficient defines as

$$\overline{\kappa_k} = \frac{\int_{\nu_{k-1}}^{\nu_k} \kappa_\nu(\nu)\left[B(\nu) - I_\nu(\mathbf{r},\nu,\mathbf{\Omega})\right]\mathrm{d}\nu}{\int_{\nu_{k-1}}^{\nu_k}\left[B(\nu) - I_\nu(\mathbf{r},\nu,\mathbf{\Omega})\right]\mathrm{d}\nu}. \tag{10}$$

For P_1-approximation the multigroup diffusion equation has the form

$$\begin{aligned}-\mathrm{div}\left\{\frac{c}{3\overline{\kappa_k}}\mathrm{grad}\left[U_k(\mathbf{r})\right]\right\} + \overline{\kappa_k} c U_k(\mathbf{r})\\ = \overline{\kappa_k}4\pi\int_{\nu_{k-1}}^{\nu_k} B(\nu)\mathrm{d}\nu.\end{aligned} \tag{11}$$

4. Mean absorption coefficients

For the multigroup method to be useful, one must be able to compute or estimate the mean values of absorption coefficients Eq. (10). An exact calculation of these group constants involves knowledge of the spectral intensity $I(\mathbf{r},\nu,\mathbf{\Omega})$ which is unknown. The underlying assumption in the multigroup method is that the group constants are relatively insensitive to the weighting functions $I(\mathbf{r},\nu,\mathbf{\Omega})$ used in computing the averages over frequency. The accuracy of this method depends also on the interval splitting. As the group width approaches zero the group constants become independent upon the estimate made for $I(\mathbf{r},\nu,\mathbf{\Omega})$.

The absorption coefficients are generally complex and widely varying functions of frequency (as can be seen at Fig. 1. The use of different weighting functions $I(\mathbf{r},\nu,\mathbf{\Omega})$ can lead to the quite different results. Generally, $\overline{\kappa_k}$ is taken as either Rosseland or Planck mean.

The Rosseland mean, also called mean free path of radiation is appropriate when the system approaches equilibrium (almost all radiation is reabsorbed). It has the form

$$\overline{\kappa_\mathrm{R}^{-1}} = \frac{\int_{\nu_{k-1}}^{\nu_k}\kappa_\nu^{-1}\frac{\mathrm{d}B(\nu,T)}{\mathrm{d}T}\mathrm{d}\nu}{\int_{\nu_{k-1}}^{\nu_k}\frac{\mathrm{d}B(\nu,T)}{\mathrm{d}T}\mathrm{d}\nu}. \tag{12}$$

The Planck mean is appropriate in the case of optically thin and emission dominated system. It can be expressed as

$$\overline{\kappa_\mathrm{P}} = \frac{\int_{\nu_{k-1}}^{\nu_k}\kappa_\nu B(\nu,T)\mathrm{d}\nu}{\int_{\nu_{k-1}}^{\nu_k} B(\nu,T)\mathrm{d}\nu}. \tag{13}$$

5. Absorption properties of plasma

Radiation in arc plasma depends, besides others physical quantities, on concentrations of all chemical species occurring in the plasma. Equilibrium compositions of

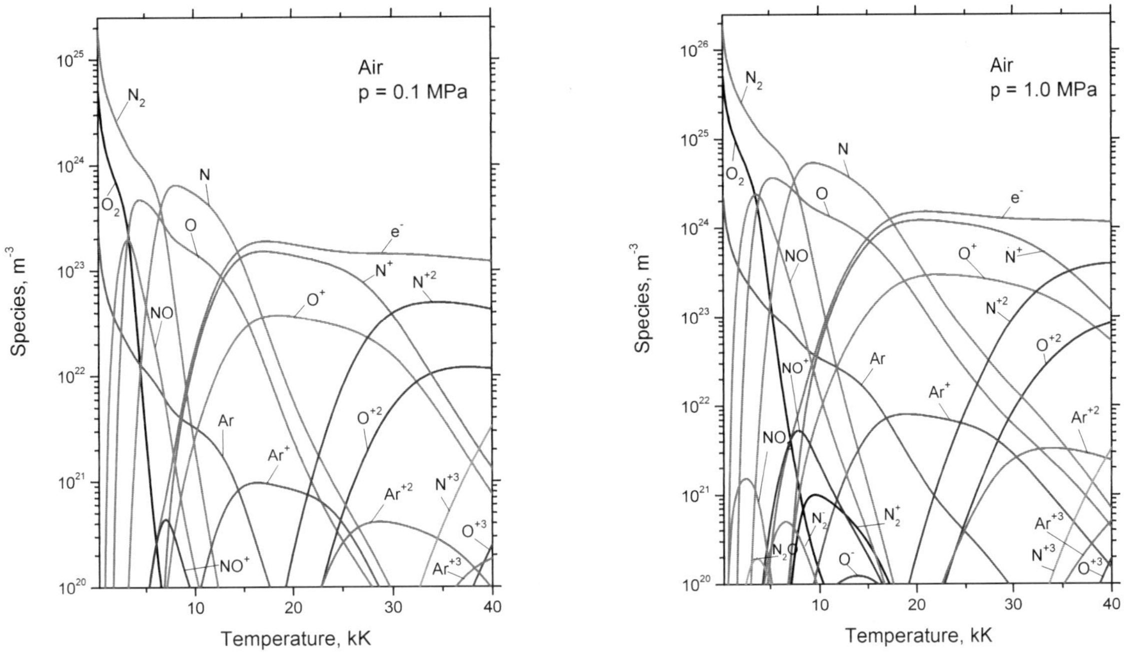

Figure 1. Concentration of different species in the thermal plasma as a function of temperature.

different types of plasma can be computed using some methods. At our Faculty, a computer code was developed for the calculation of absorption coefficients in thermal plasmas and a database of input parameters for many types of plasma was created. Here, these parameters were calculated with greater accuracy. One of the options considered is clean air without CO_2. In this case, we take into account availability in plasma some elements O_2, N_2, Ar, e^- and also ions of O, N, Ar. Figure 1 shows the concentrations of different species in the air plasma for two pressures 0.1 MPa and 1.0 MPa. It is obvious, that the concentration depends on the plasma pressure. Therefore values of these parameters influence the magnitude of the absorption coefficients.

The absorption coefficients of air plasma were calculated for eleven different pressures (0.01 MPa, 0.1 MPa, 0.2 MPa, 0.4 MPa, 0.6 MPa, 0.8 MPa, 1 MPa, 1.5 MPa, 2 MPa, 3 MPa and 4 MPa). The plasma temperature was considered in the range from 300 K to 35 000 K. The frequency interval for calculations was $(0.01 \text{ to } 10) \cdot 10^{15}$ Hz with increment of 5×10^5 Hz. It allows the accuracy increasing of the calculated data.

An example of absorption coefficient for the pressure of 0.1 MPa and temperatures 5000 K and 15 000 K is shown in Fig. 2. This spectrum is a complex of discrete and continuous spectra. High complexity of radiation absorption coefficients (with respect to the frequency and temperature) can be seen.

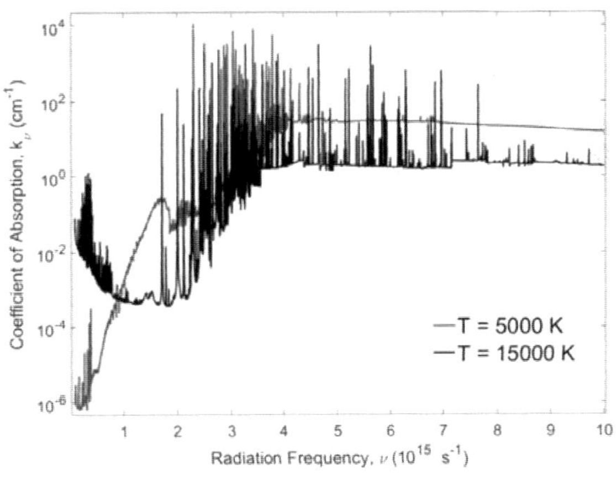

Figure 2. Absorption coefficient of air thermal plasma as a function of frequency.

After evaluation of obtained spectra for various pressures, expected regularities were revealed for values of absorption coefficient. The absorption coefficients at the pressure of 0.1 MPa and 1.0 MPa were taken for comparison. The corresponding graphs for temperatures of 5000 K and 15 000 K are shown in Fig. 3.

Given by the appearance of these spectra, computer code was created to analyzing and evaluation of the absorption coefficients dependence on frequency. This computer code makes possible to calculate mean

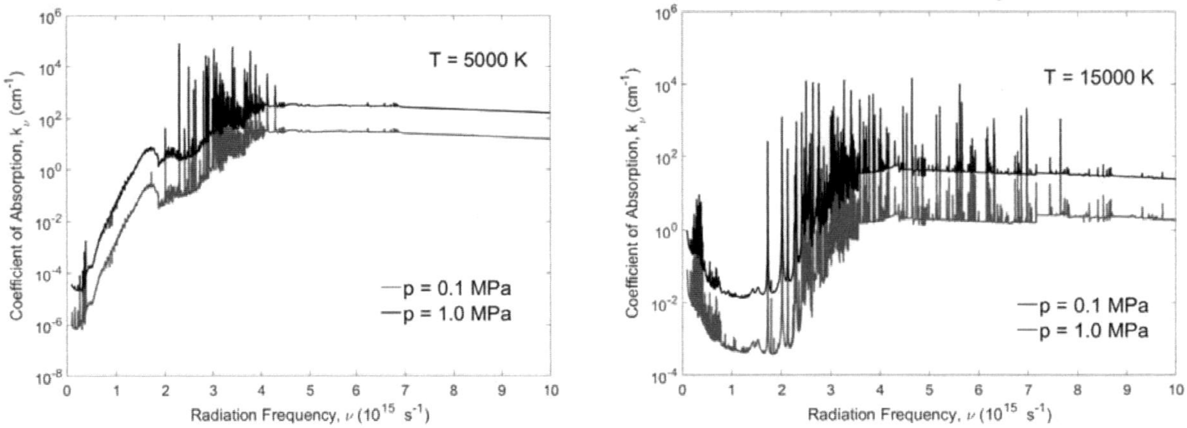

Figure 3. Comparison of absorption coefficients in air thermal plasma for two pressures.

absorption coefficients using methods described previously. Therefore, this software allows to divide the frequency interval into a certain number of groups.

As a result of these procedures, we obtain the mean absorption coefficients in relation to the temperature. Fig. 4 shows the example of results calculated for two frequency groups (the whole frequency interval was divided into 15 groups).

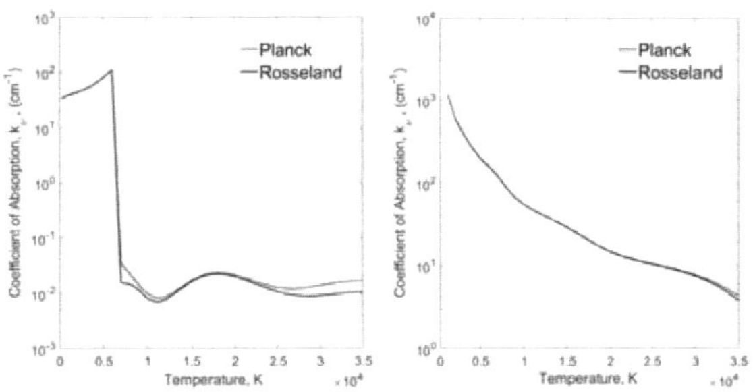

Figure 4. Mean absorption coefficients in air plasma for two frequency groups $(0.76-1.42)\cdot 10^{15}$ *Hz and* $(8.68-9.34)\cdot 10^{15}$ *Hz.*

6. Conclusions

Analyzing different ways of dividing into groups and different number of groups, we can conclude that these approximation methods give similar results in two thirds of our frequency interval. In estimating obtained data and graphs for an eleven different pressures we can conclude that the best results yields for Rosseland mean in frequency groups with small values of absorption coefficient and Planck's mean with high values of absorption coefficient. This occurs due to the properties of emission and re-absorption. From the comparison follows that Planck means generally overestimate the emission of radiation, and Rosseland means underestimate it.

In accordance with these conclusions we will use at each interval relevant method of obtainment the mean absorption coefficients for further calculation and determination of other parameters, which are necessary for resolve the equation of radiation transfer. Thus, using approximation we can significantly reduce computational time expenditures, while remaining a sufficiently high accuracy of estimations.

Acknowledgements

This research work has been carried out in the Centre for Research and Utilization of Renewable Energy (CVVOZE). Authors gratefully acknowledge financial support from the Ministry of Education, Youth and Sports of the Czech Republic under NPU I programme (project No. LO1210) and from the Czech Science foundation under project No. GA 15-14829S.

References

[1] B. Gross and O. Havelka. *Elektrické přístroje II.* VUT, Brno, 1980.

[2] Boulos. Maher I. *Thermal plasmas: fundamentals and applications.* 1. Plenum Press, New York, 1994.

[3] V. Aubrecht and M. Bartlová. Radiation transfer in air thermal plasmas. *In Proc. of XVIth Symposium on Physics of Switching Arc*, pages 9–12, 2005.

[4] J.H. Jeans. The equations of radiative transfer of energy. *Monthly Notices Royal Astronomical Society*, 78:28–36, 1917. `doi:10.1093/mnras/78.1.28`.

[5] V.G. Sevastyanenko. Radiation transfer in a real spectrum. integration over frequency. *J. Eng. Phys.*, 36:138–148, 1979. `doi:10.1007/BF00865111`.

[6] R.E. Marshak. Note on the spherical harmonics method as applied to the milne problem for a sphere. *Phys. Rev*, 71:443–446, 1947. `doi:10.1103/PhysRev.71.443`.

[7] B. N. Chetverushkin. *Matematicheskoe modelirovanie zadach dinamiki izluchajuschego gaza.* Nauka, Moskva, 1985.

[8] G.C. Pomraning. *The equations of radiation hydrodynamics.* Dover Publications, New York, 2005.

doi:10.14311/ppt.2017.3.257
Plasma Physics and Technology **4**(3):257–260, 2017

DIAGNOSTICS OF VARIOUS PHENOMENA IN LV DEVICES UNDER REAL SWITCHING CONDITIONS

J. Valenta*, M. Samohejl, M. Fendrych, P. Kloc, L. Dostál

Faculty of Electrical Engineering and Communication, Brno University of Technology, Technická 12, 616 00 Brno, Czech Republic

* valentaj@feec.vutbr.cz

Abstract. The article deals with issues to be tackled when performing experiments with low voltage devices under real switching conditions and subsequently discusses various phenomena in an experimental device. The first part describes optimum setting of diagnostic equipment - mainly for optical diagnostic methods. The second part describes some phenomena encountered during switching process under real switching conditions – arc roots movement (cathode and anode spots). These phenomena are not only important for experimental study itself but also form necessary input data for numerical models and their validation.

Keywords: electric arc, optical diagnostics, high speed video, arc root movement.

1. Introduction

When performing experiments in low voltage devices to study various phenomena under real switching devices, one has to get over effects affecting considerably quality of information gained. One of the most widespread method of arc study is based on optical diagnostics with the usage of high-speed camera. Here, the extreme light intensity accompanying arc burning at short-circuit currents and its variation plays a crucial role. Usually special filters like neutral density filters and bandpass filters are combined to get optimum result. The question is how to choose combination of camera setting (lenses, shutter time, iris) and filters close to the optimum from the very beginning for various phenomena to be studied and large scale of testing parameters (currents ranging from A to several tens of kA). The paper provides partly an answer to this question at the first part.

Arc extinguishing process in low voltage devices is usually connected with magnetic field influence on arc moving towards quenching chamber. The arc movement is also significantly influenced by arc root behavior - both on contacts and possibly various types of arc runners. This phenomenon is quite a complex one and very difficult to be accurately modeled in numerical models [1, 2]. To verify the models, a lot of experiments (statistical values) have to be carried out. To shorten the time necessary to evaluate a lot of data, a special code in Matlab software was prepared.

To take into account this phenomenon in the numerical model, a lot of experimental data are needed. Some of the experimental results are presented in this paper.

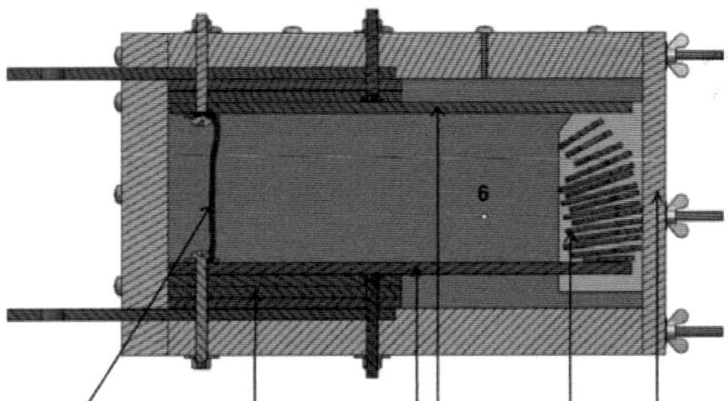

Figure 1. CAD model of the device used for experiments.

2. Experiment setting

2.1. Experimental model

Fig. 1 shows the model used for both optimal diagnostic equipment setting and arc root movement study. The model enables to perform experiments with currents up to 10 kA, measure pressure at three points (6), change exhaust array (5) (number and diameter of the holes on the right side of the model) behind the chamber, change space between arc runners (3) by adding auxiliary bars (2). The arc is ignited on the left side of the model by a small wire (1) and moves towards the chamber (splitter plates) (4) on the right side by the action of magnetic forces created by the loop of the current path. For optical diagnostics, front side and possibly left side is formed by PC transparent material. One has to consider that this type of material absorbs UV part of arc spectra!

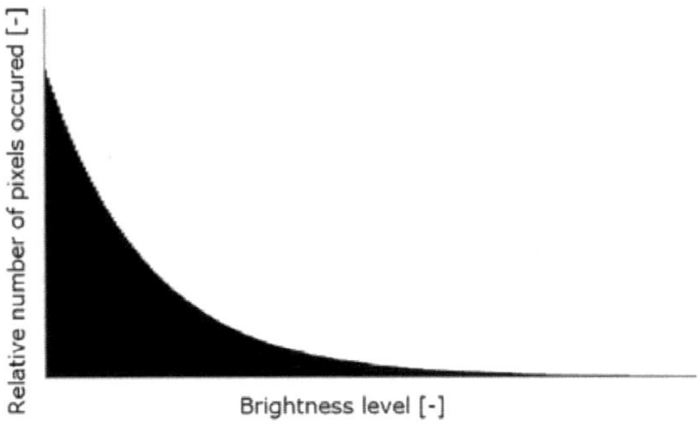

Figure 2. Optimal histogram for switching arc.

2.2. Optimal setting of the diagnostics equipment

When using high speed video for switching arc diagnostics, one has to set properly not only e.g. field of view, depth of field and sharpness, but also the right exposition. The preliminary right exposition could be easily checked by the histogram of the data obtained. For each situation, a different type of histogram is optimal. Regarding switching arc, it was found that optimal histogram should be close to the Fig. 2 – decaying exponential function.

Due to the high intensity of switching arc radiation, various filters have to be used. To estimate their setting (in EV) the following procedure was found to be useful [3]. From several experiments performed, with currents ranging from 3000 to 7600 A, the useful formulas for preliminary setting of the camera exposition was found. As it results from the experiments, the expected switching arc exposition could be estimated from the following formula (exposition expressed in EV):

$$EV_{\text{arc}} = 37.33 + 1.33 \cdot \Delta I + \Delta EV_{\text{reflection}} \quad (1)$$

In eq. (1) ΔI stands for the difference between the maximum current tested and 5000 A. $\Delta EV_{\text{reflection}}$ stands for influence of background material other than matt black – e.g. a gray middle reflective material could be considered to be equal to 4. Eq. (1) is valid for spectrally neutral transparent viewing window. Thus, for example for test current of 10 kA and gray middle reflective material we have $EV_{\text{arc}} = 42.66$.

The exposition of the camera itself can be expressed by the following equation:

$$EV_{\text{cam}} = \log_2 \frac{100 \cdot f^2}{\text{ISO} \cdot t} \quad (2)$$

In eq. (2) f stands for focal ratio of the lens used (including iris), ISO sensitivity of the camera sensor and t time of the electronic shutter used. E.g. for the camera used: ISO equates to 25 000, t is set to 293 ns, and $f = 16$, gives $EV_{\text{cam}} = 21.74$. The required exposition shift of the filters can be then estimated from the simple formula:

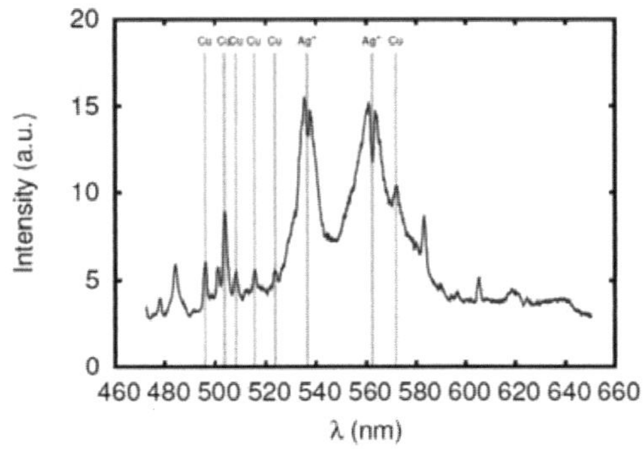

Figure 3. Arc spectrum measured in the experimental model.

$$EV_{\text{filter}} = EV_{\text{arc}} - EV_{\text{cam}} \quad (3)$$

Using (3), we obtain the required filter shift $EV_{\text{filter}} = 20.92$

This can be simply done by neutral density filters only. However, when the particular phenomena like arc roots want to be studied, a combination of bandpass filters and neutral density filters is more convenient. Fig. 3 shows the spectrum of the arc done with the model above in CVVOZE Power Laboratories. One can notice significant line of Ag^+ resulting from the contacts. If we want to enhance arc root tracking, usage of bandpass filter like 540/10 nm is of great advantage – see section 3. Furthermore, bandpass filters are often used together with laser light for background illumination of mainly mechanical events [4].

3. Arc root movement

3.1. Arc root movement automatic evaluation

Due to a lot of experiments performed and a lot of data (individual pictures) within an experiment, it is necessary to evaluate automatically arc roots movement. The program for this purpose was prepared in MATLAB code [5]. The principle of arc root evaluation is briefly explained in Fig. 4 - 6. Fig. 4 shows the original picture extracted from high speed video. By special algorithm, this pictured is converted into black-and-white (binary) regime and for each spot the centre is being found - Fig. 5. Subsequently by another part of algorithm, the individual cathode and anode spots are identified - Fig. 6.

3.2. Arc root movement example

Fig. 7 shows an example of arc root movement along the arc runners from the ignition point towards the quenching chamber. The high speed camera was set to 50000 fps. The distance between arc runners was 65 mm, behind the chamber 3 openings of 9 mm in diameters were used to set up the pressure. Maximum tested current value is 6 kA. It is evident that cathode spot is moving towards the chamber by the action of

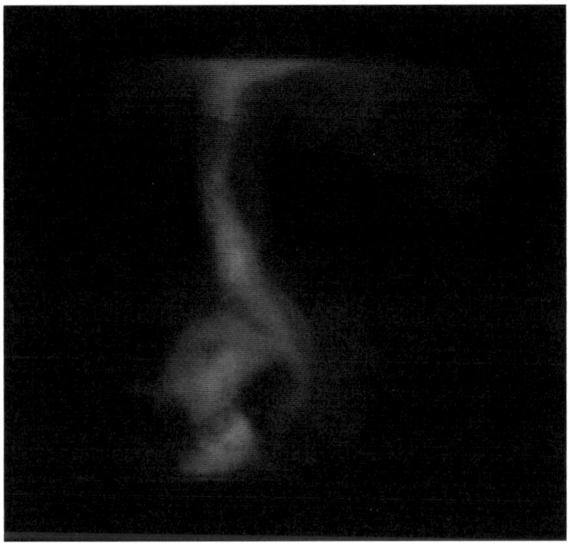

Figure 4. Single picture extracted from high speed video - raw data.

magnetic field with several jumps (marked by cycles), which corresponds to the theory of thermoemission principle of cathode spot creation. The anode spot, on the other hand, moves quite smoothly, which is also expected. There is only one step (marked by a cycle), which is caused by pressure (reflection from the side cover behind the chamber). The position of the anode spot runs ahead of the cathode spot most of the time.

3.3. Influence of parameters on arc root movement

Several influences on arc roots movement were studied. Fig. 8 shows the influence of pressure conditions on average velocity v_{avg} of arc spots. The pressure conditions were changed by changing the array of exhausts on the right side cover (behind the chamber) - change of the holes number and their diameter. The distance between arc runners was 65 mm. The general trend is that with the higher pressure both the spots have lower velocity - it corresponds to the fact, that with higher pressure the aerodynamic resistance against arc movement is higher while the electromagnetic force acting on the arc (it has opposite direction) could be considered constant. The exceptions can be seen for very small pressure (the higher local pressure accelerates the arc) and high pressure (reflections from the side cover causes rapid backward movement mainly of the anode spot which moves more easily). Fig. 9 shows the influence of arc runner distance on arc movement. Both the cathode and anode spots move slower as the distance increases. The main reason is the increase of aerodynamic resistance (higher cross section of the arc from the side point of view). This influence overcomes the small increase of the overall arc force due to electrodynamic forces with increase of the distance.

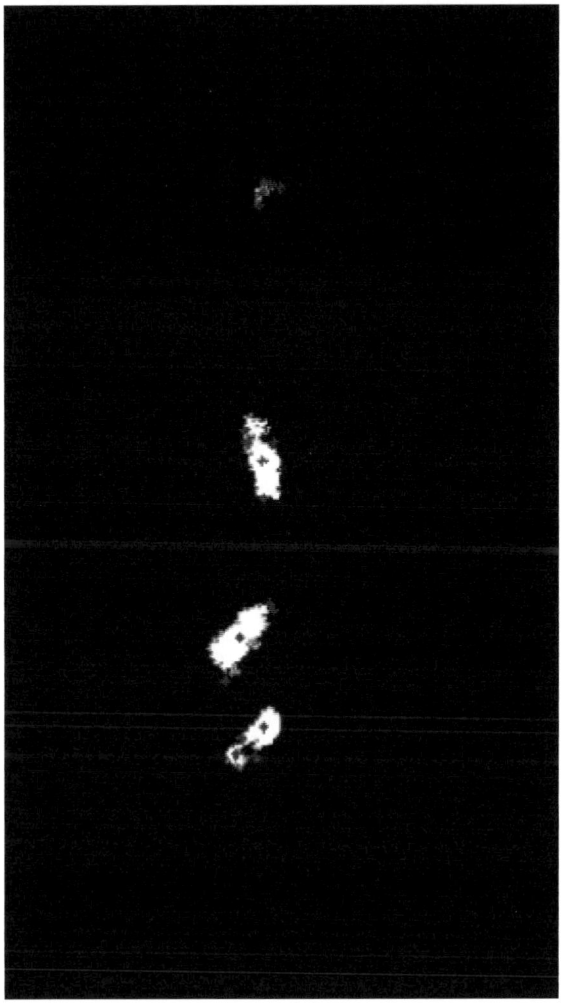

Figure 5. Binary transformed picture.

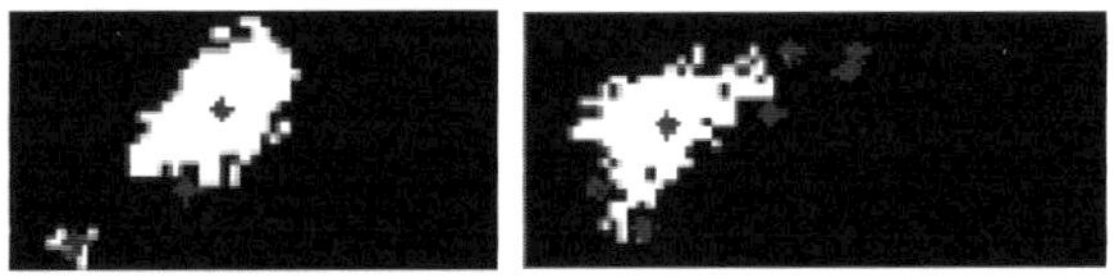

Figure 6. Identification of the center of cathode (left) and anode (right) spot.

4. Conclusions

The paper describes optimal setting of various parameters for high speed video usage to study switching arc under real switching conditions. Special attention was paid to proper selection of combination of neutral density filters and band pass filters to study arc root movement. Based on this setting, the study of arc roots movement and influences affecting the arc root movement were evaluated. For automatic arc root movement evaluation, a special program in Matlab code was prepared. The paper shows the influence of pressure in the chamber and distance between electrodes on arc root velocities. The figures of velocities

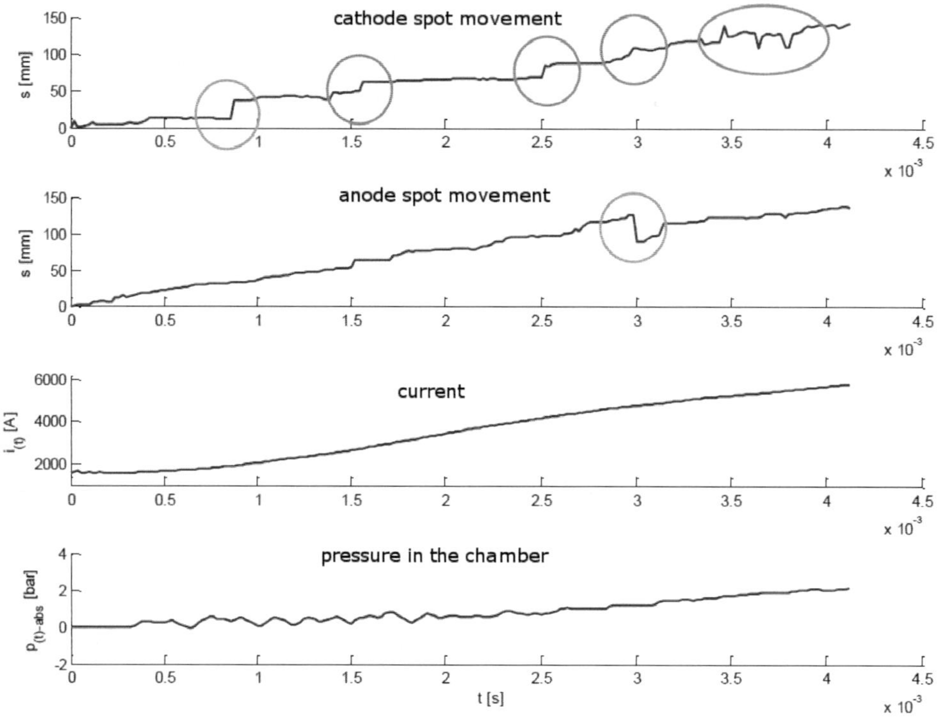

Figure 7. Arc root movement evaluation.

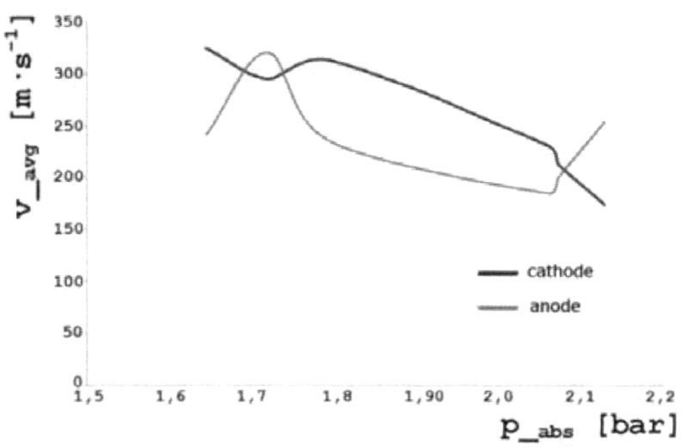

Figure 8. Influence of pressure on arc root movement.

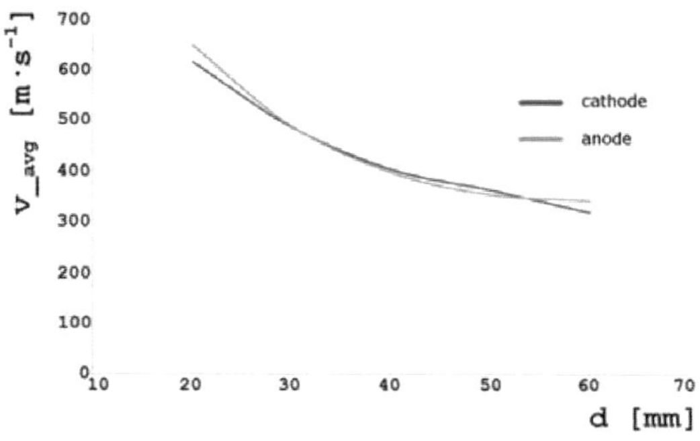

Figure 9. Influence of distance on arc root movement.

and several jumps in characteristics are valid for the experimental model only, but the general trends can be expanded to any other geometry and volume of the quenching chamber.

Acknowledgements

This research work has been carried out in the Centre for Research and Utilization of Renewable Energy (CVVOZE). Authors gratefully acknowledge financial support from the Ministry of Education, Youth and Sports of the Czech Republic under NPU I program (project No. LO1210) and from the Czech Science foundation under project No. GA 15-14829S.

References

[1] M. Lindmayer et al. Low-voltage switching arcs - experiments and modeling. In *Proceedings of the XVth Symposium on Physics of Switching Arc*, volume II, pages 252–267, 2003.

[2] J. Quemeneur et al. Cathode arc root movement: Models comparison. *Plasma Physics and Technology*, 2(2):187–190, 2015.

[3] M. Samohejl. Enhancement of optical diagnostics of the switching arc. Master's thesis, Brno University of Technology, 2016.

[4] J. Qu et al. Application of laser arc imaging technology to observe arc behavior and contact motion. *Plasma Physics and Technology*, 2(2):183–186, 2015.

[5] M. Fendrych. Study of arc root movement in a model of the LV quenching system. Master's thesis, Brno University of Technology, 2016.

doi:10.14311/ppt.2017.3.261
Plasma Physics and Technology **4**(3):261–264, 2017

MODELLING OF RADIATIVE TRANSFER IN AIR ARC PLASMA

M. Bartlova[a,b,*], P. Kloc[a], N. Bogatyreva[b], V. Aubrecht[a], J. Pokorny[b]

[a] *Centre for Research and Utilization of Renewable Energy, FEEC, Brno University of Technology, Technicka 10, 616 00 Brno, Czech Republic*

[b] *Department of Physics, FEEC, Brno University of Technology, Technicka 10, 616 00 Brno, Czech Republic*

[*] bartlova@feec.vutbr.cz

Abstract. The objective of this work is to compare the accuracy of several approximate models of radiative properties for the prediction of radiative transfer in air arc plasma at the temperatures in the range of 300–25 000 K and the pressure of 0.1 MPa. Calculated absorption coefficients are used to generate the parameters of different models. The radiative transfer inside the cylindrically symmetrical air plasma with prescribed temperature profile was studied. The equation of radiative transfer was solved using the P_1 and SP_3 approximations, calculated radiative source term in the energy equation (net emission) was compared with results obtained by spectral integration.

Keywords: radiative transfer, P_1 approximation, SP_3 approximation, mean absorption coefficients.

1. Introduction

Opening of contacts in switching devices results in an electric arc, which has to be extinguished quickly. The switching arc is a very complex phenomenon, and its realistic simulation has to take into account many effects. A crucial role in the energy balance of the switching arc plasma plays the radiation transport of energy. However, the non-linearity of equations describing the radiation field and strong dependency of input parameters on the radiation frequency and properties of the medium make the mathematical plasma models very complicated.

Several approximate methods have been developed to simplify the solution of radiation transfer. The method of spherical harmonics (P_N approximation) provides a tool to obtain an approximate solution of arbitrary high order of accuracy by transforming the equation of radiation transfer into a set of simultaneous partial differential equations [1]. However, for higher order approximations the mathematical complexity increase extremely rapidly. Therefore, P_1 approximation is usually used in association with different spectral approximations. This method is rather popular since it reduces the equation of transfer from a very complicated integral equation to a relatively simple second-order partial differential equation [2, 3]. Furthermore, if overall energy conservation is computed (also a partial differential equation), compatibility of the solution methods is virtually assured. However, the P_1 approximation may be substantially in error in optically thin situations. Another drawback are the boundary conditions, especially when the surface emission dominates over medium emission.

The simplified SP_N method has been developed originally for neutron transport and was extended to radiation transfer [4]. It was shown that SP_3 approximation improves the P_1 approximation especially for high pressures [3]. Unlike P_1 method that requires for each frequency the solution of one elliptic differential equation, the simplified SP_3 approximation leads to the solution of two similar equations. It remains then very attractive in comparison with the full solution of the radiation transfer, as well as in comparison with the standard P_3 approximation.

Various spectral approximations have been studied in this work, including Planck and Rosseland mean absorption coefficients (MAC) for cutting the frequency interval in different number of bands, line limited Planck MAC, and Planck and Rosseland MAC in k-distribution type model [2]. Spectral approximations have been used in P_1 and SP_3 calculations.

2. Absorption coefficients and spectral approximations

Under the assumption of local thermodynamic equilibrium (LTE) and neglecting the scattering, the equation of radiation transfer can be written in the form

$$\mathbf{\Omega} \cdot \nabla I_\nu(r, \mathbf{\Omega}) = \kappa_\nu (B_\nu - I_\nu) \tag{1}$$

where I_ν is the spectral intensity of radiation, $\mathbf{\Omega}$ is a unit direction vector, B_ν is the Planck black-body intensity. The only system dependence quantity in eq. (1) is the absorption coefficient κ_ν.

At high arc temperatures, dissociation of molecules and ionization of atoms occurs in the gas. We assume that plasma is in a state of LTE, its equilibrium composition can be determined by, for example, minimizing Gibbs energy [5]. Neutral atoms and up to triple ions of N, O, Ar elements were considered. Molecules photo-absorption was taken into account for N_2, O_2, NO, NO_2 and N_2O molecular species and their relative ions.

Both continuum and line radiation have been considered in calculation of absorption coefficients. Continuum spectrum is formed by bound-free transitions (photoionization) and free-free transitions (inverse bremsstrahlung). Photoionization cross sections were taken from OPACITY project database [6] or calculated using Coulomb approximation for hydrogen-like species [7]. In the discrete radiation calculations (bound-bound transitions) Doppler, Stark, Van der Waals, natural and resonance broadening of spectral lines were taken into account, semi-empirical formulas [7] were used for calculation of various broadening mechanisms. In the calculation of cross sections for the diatomic molecular band systems we have used Franck-Condon principle [8]. Molecular photo-dissociation cross sections were taken from PHIDRATES database [9]. As can be seen in Fig. 1, the absorption spectrum is very complex and widely varying function of frequency.

One procedure for handling very complicated frequency variable in the equation of radiation transfer is to find an efficient way of averaging the absorption coefficients into frequency bands. Any averaging method depends strongly on the properties of the absorption spectrum, the physical properties of the system (such as typical temperature and size), and the accuracy required. The mean absorption coefficient (MAC) values are generally taken as either the Rosseland MAC $\overline{\kappa}_{\mathrm{R}}$ (based on the assumption that almost all radiation is reabsorbed) or the Planck MAC $\overline{\kappa}_{\mathrm{P}}$ (suitable in optically thin, emission dominated system):

$$\overline{\kappa}_{\mathrm{R}}^{-1} = \frac{\int_{\Delta\nu} \kappa_\nu^{-1} \frac{\mathrm{d}B_\nu}{\mathrm{d}T} \mathrm{d}\nu}{\int_{\Delta\nu} \frac{\mathrm{d}B_\nu}{\mathrm{d}T} \mathrm{d}\nu}, \tag{2}$$

$$\overline{\kappa}_{\mathrm{P}} = \frac{\int_{\Delta\nu} \kappa_\nu B_\nu \mathrm{d}\nu}{\int_{\Delta\nu} B_\nu \mathrm{d}\nu}. \tag{3}$$

There is no universal way of the definition of spectral bands $\Delta\nu$. Generally, the value of the absorption coefficient should not strongly vary over the band. However, increasing number of bands leads to increase of calculation cost. We have used 11 radiation bands with following cutting frequencies (in units of 10^{15} Hz) [3] which are shown in Fig. 1:

$$(0.001, 0.057, 0.296, 0.386, 0.746, 0.986,\\ 1.71, 2.098, 2.64, 2.997, 4.49, 10) \tag{4}$$

In the second approach, the band selection process was based upon numerical optimization with a mean value calculated as a line limited Planck MAC [10]. Using numerical optimization the three-band model was created with cutting frequencies (see Fig. 2)

$$(0.001, 2.938, 3.513, 10) \times 10^{15}\,\mathrm{Hz} \tag{5}$$

Beside of standard Rosseland and Planck MAC (2), (3), the line limited Planck MAC was calculated with the absorption coefficient of spectral lines replaced by

$$\kappa_{\lim} = \frac{1}{L}(1 - e^{-\kappa_{\mathrm{line}} L}) \tag{6}$$

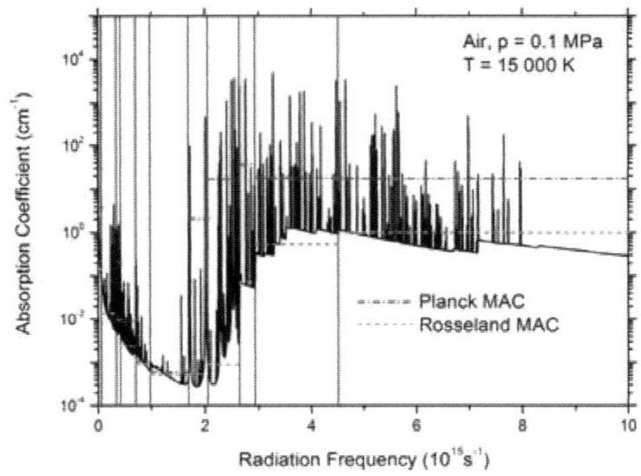

Figure 1. The high resolution spectrum of air plasma compared with standard Planck and Rosseland MAC, 11 frequency bands (eq. (4)).

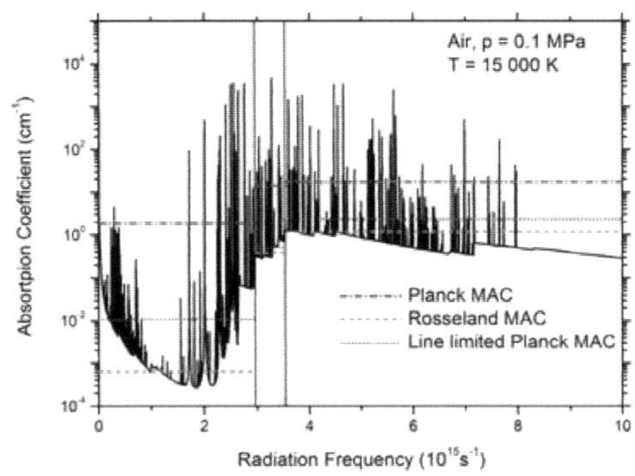

Figure 2. The high resolution spectrum of air plasma compared with standard Planck and Rosseland MAC, and line limited Planck MAC with L=10 mm; 3 frequency bands (eq. (5)).

where L is the characteristic absorption length.

It can be seen in Figs. 1, 2 that the standard Planck MAC strongly overestimate the influence of spectral lines.

Another way how to eliminate the overestimated influence of spectral lines on Planck MAC is the k-distribution type model [2]: in each band frequencies are further divided into four subgroups according to the value of κ_ν (in cm^{-1})

$$\kappa_\nu \in [0, 0.1], [0.1, 1], [1, 10], [10, \infty]. \tag{7}$$

This approach was used for three frequency bands (5), and Rosseland and Planck MAC (2), (3) were calculated for each subgroup.

3. P_1 and SP_3 approximations

In the P_1 approximation, the local intensity is approximated by a series of spherical harmonics, and the series is truncated after the first order terms. The directional equation (1) is transformed into one second-order elliptic partial differential equation for the density of radiation U_ν (zero-order moment of the radiation intensity):

$$\nabla\cdot(-\frac{1}{3\kappa_\nu}\nabla U_\nu)+\kappa_\nu U_\nu=4\pi\kappa_\nu B_\nu \tag{8}$$

where $-\frac{1}{3\kappa_\nu}\nabla U_\nu=\mathbf{F}_\nu$ is the radiation flux. The associated Marshak boundary condition at a black-body diffuse wall take the form

$$\mathbf{n}\cdot\nabla U_\nu=\frac{3}{2}\kappa_\nu(4\pi B_\nu^{\mathrm{wall}}-U_\nu) \tag{9}$$

Using MAC in equations (8), (9) the net emission of radiation is

$$\nabla\cdot\mathbf{F}_{\mathrm{R}}=\sum_j\nabla\cdot\mathbf{F}_j \tag{10}$$

where j denotes the individual frequency bands.

Simplified P_3 equations (SP_3) are an asymptotic correction to standard diffusion theory and asymptotically relate to the slab geometry P_3 equations. In this approximation, from the equation (1) the system of two second-order partial differential equations for two variables $\psi_{1,\nu}$ and $\psi_{2,\nu}$ was derived in [3]

$$\begin{aligned}-\nabla\cdot\frac{\mu_1}{\kappa_\nu}\nabla\psi_{1,\nu}+\kappa_\nu\psi_{1,\nu}&=4\pi\kappa_\nu B_\nu\\-\nabla\cdot\frac{\mu_2}{\kappa_\nu}\nabla\psi_{2,\nu}+\kappa_\nu\psi_{2,\nu}&=4\pi\kappa_\nu B_\nu\end{aligned} \tag{11}$$

where $\mu_i=(3\mp2\sqrt{1.2})/7$, $i=1,2$. These two equations are weakly coupled by the Marshak boundary conditions for a black-body diffuse wall

$$\begin{aligned}\alpha_1\psi_{1,\nu}+\frac{1}{\kappa_\nu}\mathbf{n}\cdot\nabla\psi_{1,\nu}&=-\beta_2\psi_{2,\nu}+\eta_1B_\nu^{\mathrm{wall}}\\\alpha_2\psi_{2,\nu}+\frac{1}{\kappa_\nu}\mathbf{n}\cdot\nabla\psi_{2,\nu}&=-\beta_1\psi_{1,\nu}+\eta_2B_\nu^{\mathrm{wall}}\end{aligned} \tag{12}$$

where

$$\begin{aligned}\alpha_i&=5(34\pm11\sqrt{1.2})/96,\ i=1,2,\\\beta_i&=5(2\mp\sqrt{1.2})/96,\ i=1,2,\\\eta_i&=5\pi(3\pm\sqrt{1.2})/2,\ i=1,2.\end{aligned} \tag{13}$$

Using MAC in equations (11), (12) the net emission of radiation is

$$\nabla\cdot\mathbf{F}_{\mathrm{R}}=\sum_j\nabla\cdot\frac{1}{\kappa_\nu}\nabla(a_1\psi_{1,j}+a_2\psi_{2,j}) \tag{14}$$

with $a_i=(5\mp3\sqrt{1.2})/30$, $i=1,2$.

4. Results

The radiative transfer inside the cylindrical air plasma column at uniform pressure of 0.1 MPa with prescribed temperature profile was calculated using the P_1 and SP_3 approximations with various MAC. In Figures 3–6, the calculated net emission is compared with results of the full spectral integration [10].

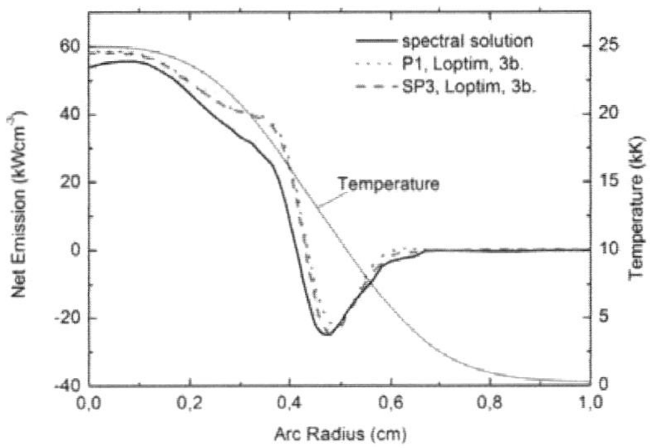

Figure 3. Net emission calculated using full spectral integration [10], and P_1 and SP_3 approximation with line limited Planck MAC, 3 frequency bands (eq. (5)).

Fig. 3 shows the results of P_1 and SP_3 approximations with line limited Planck MAC in three frequency bands (5). The characteristic absorption length is $L_{\mathrm{optim}}=3R_{\mathrm{p}}=1.38\,\mathrm{cm}$. R_{p} is the plasma radius which is defined as a distance from the cylinder axis, where the temperature reaches half point between the maximum and minimum value [10]. Good accuracy can be seen, SP_3 approximation leads to more accurate results in the strongly absorbing region at the plasma edge.

The importance of spectral line limiting is shown in Fig. 4 where net emission calculated with standard Planck and Rosseland MAC in 3 frequency bands (5) is presented. Planck MAC overestimate very strongly the emission of radiation (so much that it is not shown in the figure due to the scale), on the other hand, Rosseland MAC underestimate it. Some improvement can be obtained by combination of Planck and Rosseland MAC - Planck MAC is used in frequency bands with low values of absorption coefficients, while Rosseland MAC in frequency bands with strong absorption. However, the set of three bands does not work well because even the combination of Planck and Rosseland MAC (Rosseland MAC in 1st and 3rd band, Planck MAC in 2nd band) increase the net emission ten times.

A small improvement in the results can be obtained by cutting the frequency interval in more bands, as can be seen in Fig. 5, where 11 bands (4) were used. In the combination of Planck and Rosseland MAC, Planck MAC were used only in 4th and 6th band. Planck MAC alone, or used in more bands lead to strong overestimation of emission.

Relatively good accuracy can be obtained by combination of frequency and κ-based cutting (k-distribution type model (7)), as it is shown in Fig. 6. In combination of Planck and Rosseland MAC, the Rosseland MAC were considered in subgroups with $\kappa>1\,\mathrm{cm}^{-1}$.

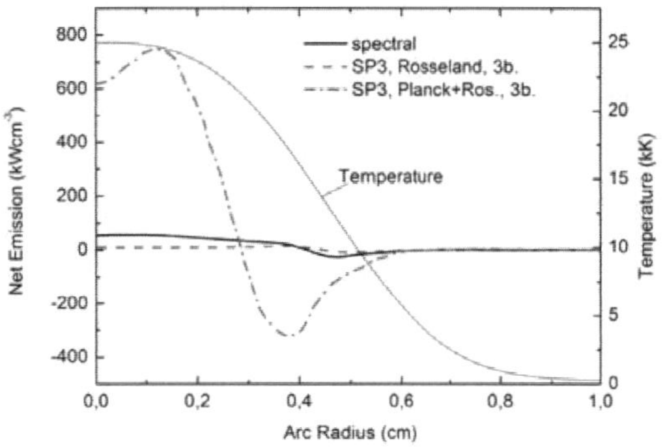

Figure 4. Net emission calculated using full spectral integration [10], and SP_3 approximation with Rosseland MAC and combination of Planck and Rosseland MAC for 3 frequency bands (eq. (5)).

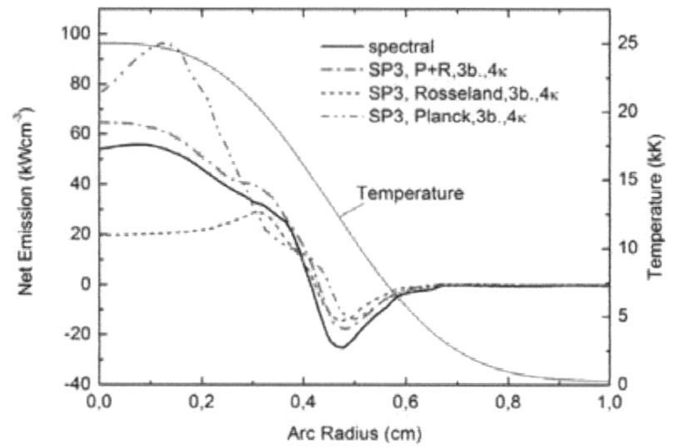

Figure 6. Net emission calculated using full spectral integration [10], and SP_3 approximation with Planck MAC, Rosseland MAC and combination of Planck and Rosseland MAC, k-type model, 3 frequency bands (eq. (5)).

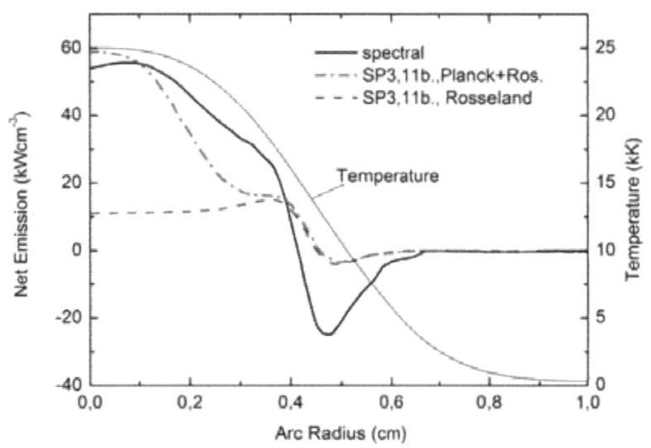

Figure 5. Net emission calculated using full spectral integration [10], and SP_3 approximation with Rosseland MAC and combination of Planck and Rosseland MAC for 11 frequency bands (eq. (4)).

5. Conclusions

The net emission has been calculated in air thermal plasma using P_1 and SP_3 method in association with different spectral approximations. SP_3 method is slightly more accurate than P_1 model. Rosseland MAC underestimate both emission and absorption of radiation, Planck MAC overestimate the influence of spectral lines. Good accuracy can be obtained using line limited Planck MAC and three-band model obtained by numerical optimization. Another suitable approach is the k-distribution type model with combination of Planck and Rosseland MAC. The advantage of this approach is that it does not depend on a certain thickness of the arc.

Acknowledgements

Authors gratefully acknowledge financial support from the Centre for Research and Utilizaton of Renewable Energy under project No. LO1210 and from Czech Science Foundation under project No. 15-14829S.

References

[1] M.F. Modest. *Radiative Heat Transfer.* Oxford: Academic Press, 2013.

[2] H. Nordborg and A.A. Iordanidis. Self-consistent radiation based modelling of electric arcs : I. Efficient radiation approximations. *J. Phys. D: Appl. Phys.*, 41(13):135205, 2008. doi:10.1088/0022-3727/41/13/135205.

[3] B. Peyrou et al. Radiative properties and radiative transfer in high pressure thermal air plasmas. *J. Phys. D: Appl. Phys.*, 45(45):455203, 2012. doi:10.1088/0022-3727/45/45/455203.

[4] E.W. Larsen et al. Simplified P_N approximations to the equation of radiative transfer and applications. *J. Comput. Phys.*, 183(2):652–675, 2002. doi:10.1006/jcph.2002.7210.

[5] O. Coufal. Composition and thermodynamic properties of thermal plasma up to 50 kK. *J. Phys. D: Appl. Phys.*, 40(11):3371–3385, 2007. doi:10.1088/0022-3727/40/11/018.

[6] OSC Technical Information Server. *Opacity Project.* http://opacities.osc.edu.

[7] R. Liebermann and J.J. Lowke. Radiation emission coefficients for sulphur hexafluoride arc plasmas. *J. Quant. Spectrosc. Radiative Trans.*, 16(3):253–264, 1976. doi:10.1016/0022-4073(76)90067-4.

[8] G. Herzberg. *Molecular Spectra and Molecular Structure. I. Spectra of Diatomic Molecules.* New York: D. Van Nostrand Co., 1950.

[9] Southwest Research Institute. *PHoto Ionization/Dissociation RATES.* http://phidrates.space.swri.edu.

[10] P. Kloc, V. Aubrecht, and M. Bartlova. Numerically optimized band boundaries of Planck mean absorption coefficients in air plasma. *J. Phys. D: Appl. Phys.*, 50(30):305201, 2017. doi:10.1088/1361-6463/aa7627.

doi:10.14311/ppt.2017.3.265
Plasma Physics and Technology **4**(3):265–268, 2017

EXPERIMENTAL TEMPERATURE MEASUREMENTS IN MINIATURE CIRCUIT BREAKER

D. Simek*, D. Pecek

Faculty of Electrical Engineering and Communication, Brno University of Technology, Technicka 3058/10, 616 00 Brno, Czech Republic

* xsimek18@stud.feec.vutbr.cz

Abstract. Low voltage switching apparatuses efficiency depends on the speed of movement of the switching arc from the contacts to the quenching chamber. The paper is focused on investigation of this movement of an arc. Measurement of radiation spectra of the electric arc burning inside miniature circuit breaker and moving to quenching chamber are presented. Measured radiation spectra contain suitable atomic spectra lines for calculation purposes. The problems connected with the measurements are discussed. The main part of the paper deals with a calculation of temperature of the investigated plasma of the arc. Atomic lines database of National Institute of Standards and Technology was used as a spectral data source for the calculations.

Keywords: switching arc, miniature circuit breaker, atomic emission spectroscopy, temperature measurement.

1. Introduction

Low voltage circuit breakers are electrical apparatuses widely used for breaking of short-circuit currents and overloads. Those devices are divided into three main groups: miniature circuit breakers (MCB), molded case circuit breakers (MCCB) and air circuit breakers (ACB). Miniature circuit breakers are the smallest breakers of those groups. These devices are the most common devices in the standard switchboards.

One of the crucial parameters of circuit breaker is a breaking capacity, which depends on arcing time of MCB. Long arcing time may cause damages due to a very high temperature of an arc. On the other hand, short arcing time may create overvoltage across the insulation. Arcing time is affected by movement speed of contacts, arc movement from contact area to the quenching chamber and quenching chamber efficiency. Arc movement between the moving contacts of MCB were experimentally observed recently using laser arc imaging technology [1]. Another approach of an analysis of this phenomenon are computer simulations. Some interesting simulations of the contact opening process were presented in [2]. Simulation of arc movement to the quenching chamber including temperature calculation inside MCB under specific conditions was performed [3]. Temperature maximum was found at about 20 000 K in this simulation.

The temperature of an arc is significant parameter of quenching. The main goal of our measurement was to find out the development of the arc temperature during switching off process because of its effects to the speed of quenching. Optical emission spectroscopy was used as a mean of obtaining the arc temperature inside the MCB. MCBs are usually made of plastic, silver plated copper and ferromagnetic materials which are all affected by the arc. Parts whose are in direct contact with an arc are generally made of copper. Therefore, plasma inside miniature circuit breaker often contains copper vapour, which is clearly visible in its radiation spectrum. It allows calculation of the temperature and other important physical properties using atomic lines of copper.

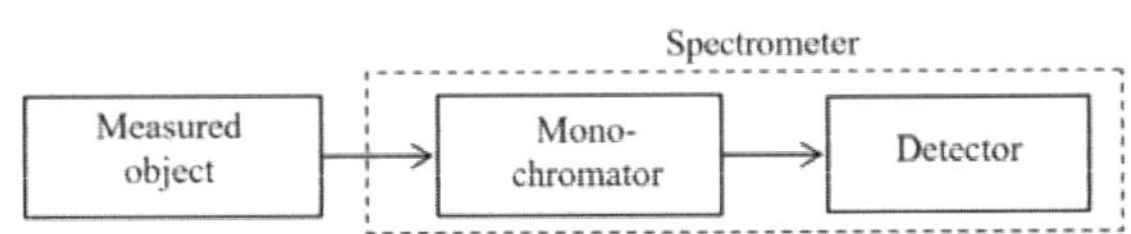

Figure 1. Block diagram of emission spectra measurement.

Block diagram of measuring circuit is shown in Figure 1. Spectrometer was used as measuring instrument. CCD chip is commonly used as a detector.

2. Experimental setup

Measurements were performed in the High current laboratory of the Centre for Research and Utilization of Renewable Energy. 16 MVA three phase asynchronous generator was used as a power supply. Excitation of generator were 13 %. Series of special resistors and coils were used for the current and power factor adjustment. Current flew to the test bench 1 through the 16 MVA high current transformer with conversion rate 4800 V/250 V. Investigated device was miniature circuit breaker with B characteristic, nominal current of 10 A and tripping capacity of 10 kA. Optical fiber was used for necessary separation of measuring devices from the power circuit and transmission of light from a specific position to the input slit of the spectrometer. The Avantes Avaspec ULS-3648TEC was used as spectrometer. Data were stored and processed in laptop using AvaSoft 8.0 software.

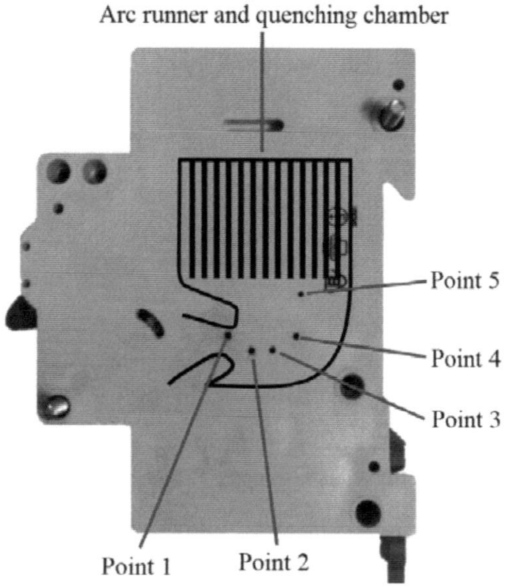

Figure 2. Diagnostic holes positions (investigated points).

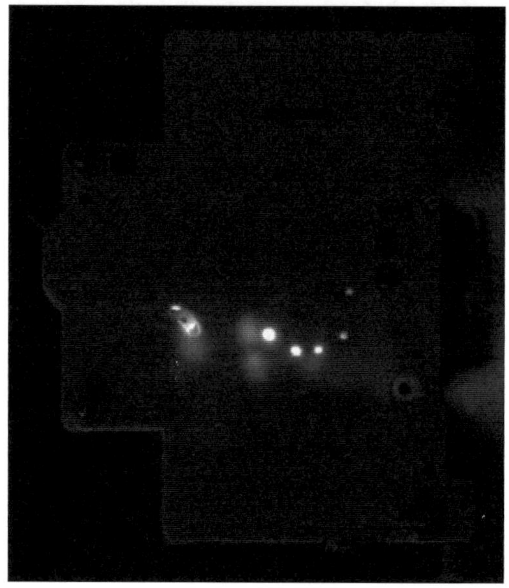

Figure 3. Picture cut out of high-speed video.

The arc was burning inside the circuit breaker, thus small holes were drilled for diagnostic purposes along the arc runner, see Figure 2.

3. Measurement procedure

1. Test circuit were calibrated to these parameters: $U_{ef} = 42\,\text{V}$, $I_{ef} = 2.8\,\text{kA}$, $\cos\varphi = 0.5$, $f = 50\text{Hz}$.
2. Calibration of spectral sensitivity of the spectrometer was carried out.
3. Optical fiber was adjusted to the investigated point 1.
4. Test circuit was switched on and the overload was interrpted by the tested sample.
5. Radiation spectrum was saved into the laptop.
6. Procedure from 3 to 4 was repeated several times.
7. Optical fiber was shifted to next point and spectra were saved.
8. Radiation spectra lines were identified.
9. Relative radiation intensities of suitable lines were calculated.
10. Boltzmann plot was created for each point and the temperature was calculated.
11. Temperatures for each point were plotted into the graph ordered from the nearest to the fearthest distance from the contact pair of the breaker.
12. High speed videos of investigated points were taken in second part of measurements.

4. Results and discussion

The high-speed video were taken by a Photron SA-X2 high-speed camera. One picture of the high-speed video is shown in the Figure 3.

In this case, voltage and current were measured by data acquisition system of the laboratory during the test. Waveforms of voltage and current during the one of tests are presented in the Figure 4.

High-speed video in combination with waveforms allow evaluation of important parameters of the test. Evaluated parameters are presented in the Table 1.

Parameter	**Units**	**Value**
Voltage switching peak	(V)	100
Maximum current	(A)	880
Switching off time	(ms)	2.5
Reaction time	(ms)	close to zero
Lever movement delay	(ms)	0.8
Lever movement duration	(ms)	4.2

Table 1. Important parameters of the test.

Example of radiation spectrum can be seen in the Figure 5. Radiation spectrum of copper is suitable for calculation purposes, because it contains appropriate lines. The spectral lines are intensive and have appropriate distance between each other. Copper was probably evaporated from the arc-runner or contacts.

λ measured	**λ database**	**Element**
510.2 nm	510.6 nm	copper
514.9 nm	515.3 nm	copper
520.5 nm	520.9 nm	silver
521.4 nm	521.8 nm	copper
546.2 nm	546.5 nm	silver
577.9 nm	578.2 nm	copper

Table 2. Assignment of elements.

Measured spectral lines were assigned to elements

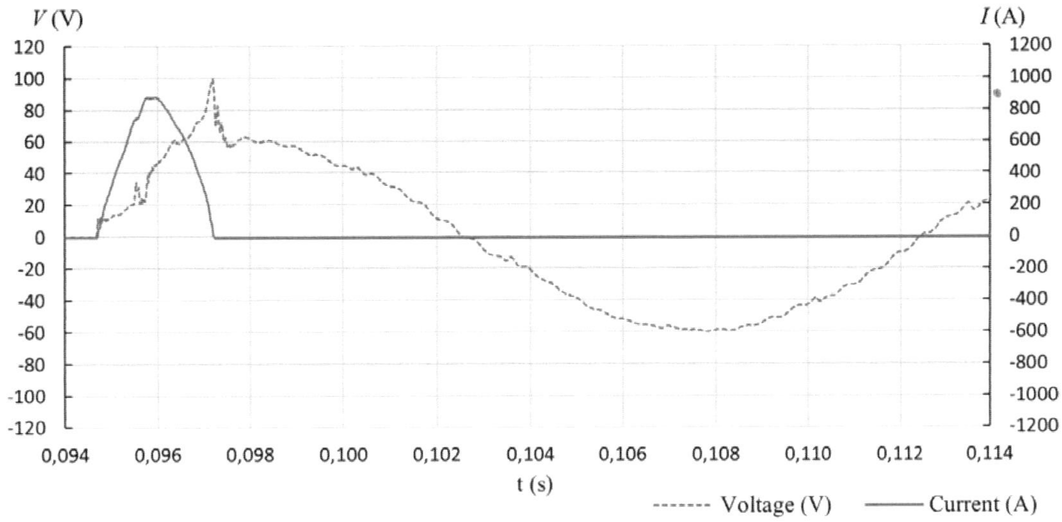

Figure 4. Waveforms of voltage and current during the test.

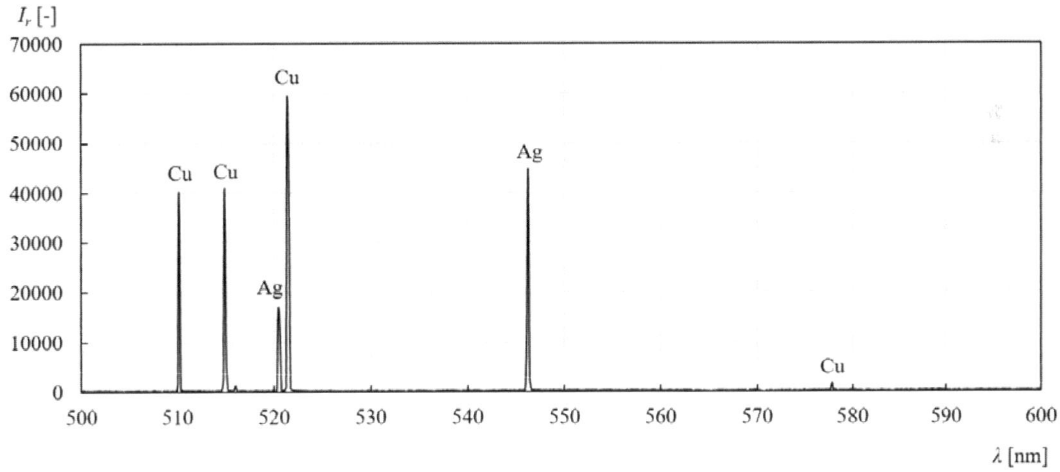

Figure 5. Example of the spectrum measured at the point 5.

using NIST database [4], see table 2 .

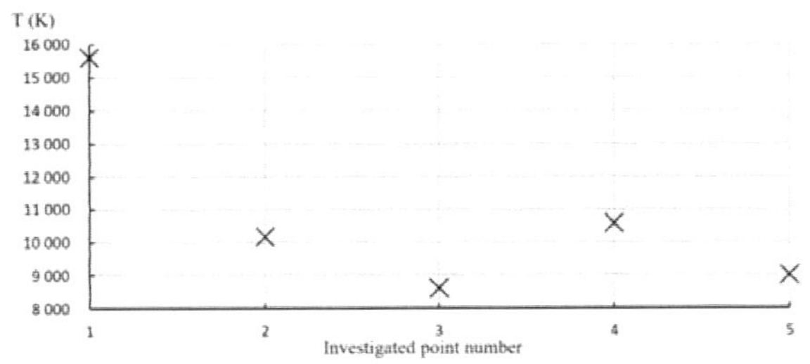

Figure 6. Temperatures measured at investigated points.

Temperatures measured at investigated points are presented in the Figure 6.

The arc temperature was calculated about 15 600 K as maximum temperature near the contact system of the breaker. The temperature sharply decreases with decreasing distance from the contact system to 8 600 K at the point 3. At the point 4, there is an increase of temperature to 10 600 K. At the point 5 temperature decreases to 9 000 K because of splitter plates cooling.

It is important to mention that the temperature calculation may be affected by a few complications. The first one is a spectral lines shape deviation. It depends on spectrometer resolution and measurement conditions. The second problem corresponds with database using. The transition strength and the transition probability, or Einstein coefficient were taken from the NIST database. The NIST database states that the values used for calculations has accuracy of C+ class. That means accuracy more than 82 %.

Another problem is related to the database using again. NIST database cannot cover all experiment conditions of the spectral line measurement. That means that the database contains lines measured under different pressures, humidity, ambient temperatures etc.

5. Conclusions

The aim of the measurement was to find out the temperature of the arc inside miniature circuit breaker during the switching off proccess. Temperatures measured at investigated points are presented in Figure 6.

Atomic emission spectroscopy was used for spectra measurement. High-speed videos helped to find out important times for diagnostic of the breaker behaviour under this conditions. Temperature was calculated from the relative intensity of spectral lines of copper. Calculated values correspond to realistic assumptions, although the calculation contains simplifying assumptions.

Acknowledgements

Authors gratefully acknowledge financial supports from the Ministry of Education, Youth and Sports of the Czech Republic (NPU I programme, project No. LO1210 and faculty project No. FEKT-S-17-4374) and from the Czech Science Foundation under project No. GA 15-14829S.

References

[1] J. Qu, S. Yuan, Q. Wang, and X. Li. Application of laser arc imaging technology to observe arc behaviour and contact motion. *Plasma Physics and Technology*, 2(2):183–186, 2015. URL: `http://ppt.fel.cvut.cz/articles/2015/qu.pdf`.

[2] Y. Wu, M. Rong, Z. Sun, X. Wang, F. Yang, and X. Li. Numerical analysis of arc plasma behaviour during contact opening process in low-voltage switching device. *Journal of Physics D: Applied Physics*, 40(3):795–802, 2007. `doi:10.1088/0022-3727/40/3/016`.

[3] M. Rong, F. Yang, Y. Wu, A. B. Murphy, W. Wang, and J. Guo. Simulation of arc characteristics in miniature circuit breaker. *IEEE Transactions on Plasma Science*, 38(9):2306–2311, 2010. `doi:10.1109/TPS.2010.2050703`.

[4] National Institute of Standards and Technology. Atomic spectra database lines form. URL: `https://physics.nist.gov/PhysRefData/ASD/lines_form.html`.

doi:10.14311/ppt.2017.3.269
Plasma Physics and Technology **4**(3):269–272, 2017

OBJECTIVE FUNCTION FOR NUMERICAL MEAN ABSORPTION BANDS OPTIMIZATION

P. KLOC*, V. AUBRECHT, M. BARTLOVÁ

Centre for Research and Utilization of Renewable Energy, FEEC, Brno University of Technology, Technicka 10, Brno, Czech Republic

* klocpetr@feec.vutbr.cz

Abstract. Mean absorption coefficients (MACs) offer great potential for fast numerical calculation of radiation heat transfer. They are based on replacing complex absorption coefficient spectrum by a handful of frequency bands with a single, temperature dependent value assigned to each band. Accuracy of radiation transfer calculation thus depends on the accurate interpretation of the mean value inside each frequency band as well as on the proper band distribution. Yet finding optimal band distribution is not an easy task often requiring numerical optimization process. This contribution focuses on the parameters of such optimization process, namely selection of an objective function and its effect on the optimal band distribution. It demonstrates, that improper objective functions can produce physically unreasonable artifacts in the calculation of radiation heat transfer. Optimal formulation of the objective function is proposed in this contribution.

Keywords: mean absorption coefficients, numerical optimization, radiation transfer.

1. Introduction

It is a well known fact, that the temperature inside switching arc plasma can reach tens of thousands of Kelvins. At such high temperature levels, radiation transfer plays a very important role in the total energy balance of the arc. An accurate description of the radiation energy transfer is therefore crucial for any numerical simulation of the switching arc.

Only two radiation quantities are necessary to describe radiation energy transfer in the most cases. The first quantity is the divergence of radiation flux. It describes the energy sink or gain inside the plasma volume and must be incorporated into the plasma energy balance equation [1]. This quantity is therefore important for accurate simulation of a thermal plasma volume. The second quantity is tied to the radiation energy transfer at the outer boundary. Escaping radiation can induce plasma composition changes due to outer walls ablation and different material emission into the plasma volume. The amount of radiation reaching the outer walls is best quantified by the radiation flux quantity.

Fast and accurate evaluation of both radiation quantities is thus required for any reasonable numerical simulation of thermal plasma. Unfortunately, the accurate calculations are very computationally demanding due to a very complex nature of the radiation spectrum. Several approximate solutions were developed through the history, including Net Emission Coefficients (NEC) [2] and Mean Absorption Coefficients (MAC) [3]. The MACs show great promise in simplification of radiation transfer calculations, but require careful handling in order to maintain acceptable accuracy [4]. One possible way for achieving reasonable accuracy is using the numerical optimization of the frequency bands distribution [5] or even the mean value inside each band itself [6].

The numerical optimization process relies on the so called objective function, i.e. a function, that is searched by a numerical optimization process for the position of minima. In theory, this objective function can be based on any radiation quantity such as radiation flux or divergence of radiation flux. However, due to the complex nature of the radiation transfer inside plasma it is very hard to predict, whether the outcome of the optimization process is independent of the objective function definition or whether different definitions produce unique results. We try to answer this question by a series of tests presented in this contribution.

2. Model

We wanted to keep the radiation model itself as simple as possible. Therefore, we considered infinitely long cylindrical domain with radius of $R = 1\,\mathrm{cm}$ filled with air plasma at the uniform pressure of 1 bar. A fixed predefined temperature profile is imposed on the calculation domain (see Figure 1) to emulate the plasma column inside the domain. The temperature profile is described by the analytical function

$$T(r) = T_{\max} - (T_{\max} - T_{\min}) \frac{1 - \mathrm{e}^{-n\left(\frac{r}{R}\right)^3}}{1 - \mathrm{e}^{-n}}, \quad (1)$$

which allows a large variety of different shapes. The following parameters were selected in this particular case to approximately represent a free burning arc: $T_{\min} = 300\,\mathrm{K}$, $T_{\max} = 25\,\mathrm{kK}$, $n = 7$.

The divergence of radiation flux as well as the radiation flux itself were evaluated in 50 points along the

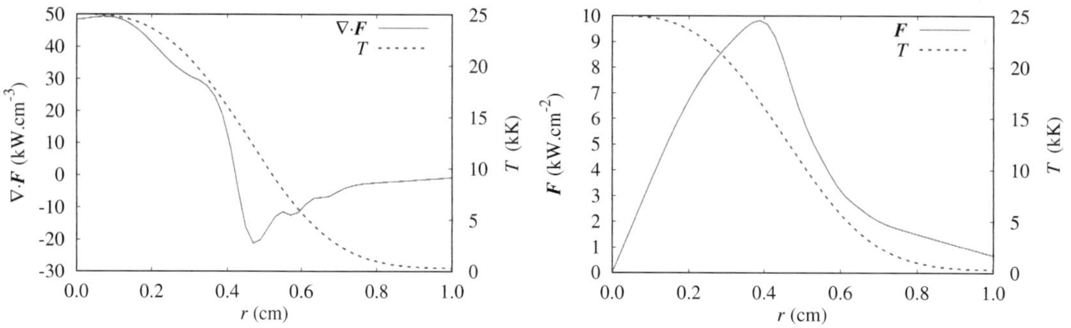

Figure 1. Divergence of radiation flux (left) and radiation flux (right) along the radius of infinitely long cylindrical domain with fixed temperature profile.

cylinder radius (see Figure 1) using a model and material data taken from [7]. Both profiles were calculated using a full spectral resolution of absorption coefficient (an example of absorption coefficient spectrum is in Figure 2) and are referred to as *spectral* solutions through the text or *sp* subscript in equations. The spectral solution serves two purposes. Firstly, it is used to evaluate the accuracy of the approximate solution described in the following paragraph and secondly it is used for definition of the numerical optimization objective function.

In the subsequent step we used a numerical optimization procedure [8] to calculate the optimal band distribution for three-band Planck mean absorption coefficients. The process is similar to the one described in [5]. We used line limiting factor proposed by Nordborg [9] with characteristic plasma length set to 1.5 cm to mitigate the known overestimation of atomic lines by Planck mean absorption coefficient. By employing only three frequency bands we were able to characterize the final band distribution by just two parameters ν_1 and ν_2, which define the boundaries between the bands. The outer boundaries are fixed at 10^{12} Hz and 10^{16} Hz for lower and higher limit respectively.

We defined an universal numerical optimization objective function to test the effect of several different radiation quantities on the mean absorption band distribution. The objective function is written as

$$\Delta f(\nu_1,\nu_2) = \sqrt{\sum_{i=1}^{50} A_i^2 \left(\nabla\cdot\mathbf{F}_i - \nabla\cdot\mathbf{F}_{i,\mathrm{sp}}\right)^2} + \sqrt{\sum_{i=1}^{50} B_i^2 \left(\mathbf{F}_i - \mathbf{F}_{i,\mathrm{sp}}\right)^2} + \sqrt{\sum_{i=1}^{50} C_i^2 \left(G_i - G_{i,\mathrm{sp}}\right)^2}, \qquad (2)$$

where the summation is carried over all the 50 spatial points in which the spectral properties were resolved

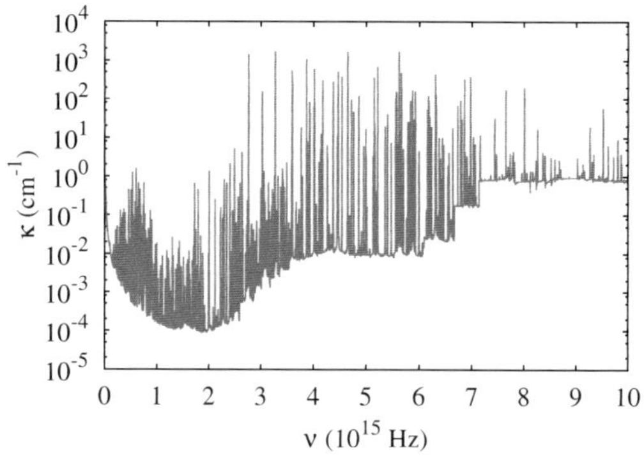

Figure 2. Absorption coefficient of air at 25 kK.

and G_i is the incident radiation defined as

$$G_i = \int_0^\infty \int_{4\pi} I(r_i,\nu)\mathrm{d}\Omega\,\mathrm{d}\nu \qquad (3)$$

with $I(r_i,\nu)$ representing radiation intensity at point r_i and frequency ν. The variables A_i, B_i and C_i are used to modify the objective function according to our needs. In total we calculated eight series of numerical optimization procedure, each series containing minimum of 3 optimization attempts to verify the convergence repeatability. Finally, we evaluated the accuracy of radiation flux and divergence of radiation flux calculated with the optimized three-band mean absorption model by comparing the profiles with the spectral solution.

3. Results

Even though the numerical optimization procedure can operate with any arbitrary value of a objective function, it is often advantageous to limit the objective function to the interval between 0 and 1. To do so, the definition of objective function often rely on the maximum value of the appropriate quantity. In such case, this maximum value is denoted by additional subscript *max* in the text.

Four distinct objective functions were tested in total with each test being described in more details in the following subsections. Generally, we expected all the tested objective functions to perform quite similarly, but the results show quite different picture.

3.1. Divergence of radiation flux

The objective function is represented only by divergence of radiation flux in case of $A_i = 1/\nabla \cdot \mathbf{F}_{\mathrm{sp,max}}$, $B_i = 0$ and $C_i = 0$. With this definition the focus is mainly on the areas where the divergence of radiation flux exhibits high absolute value. The areas on the outskirts of the cylinder as well as the position of the transition between emitting and absorbing regions are considered with lesser significance, thus some degree of deviation can be expected.

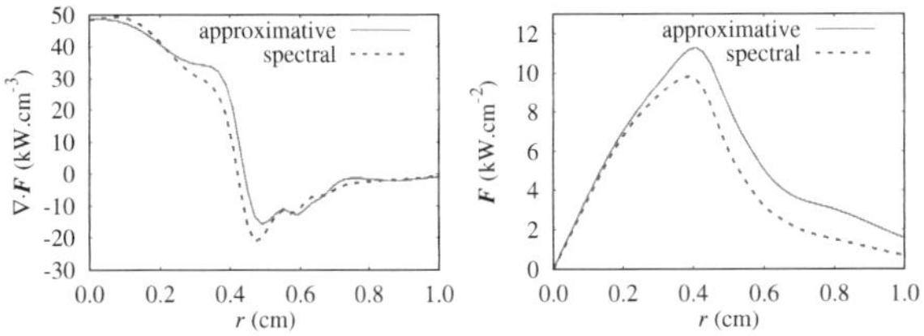

Figure 3. Divergence of radiation flux (left) and radiation flux (right) profiles evaluated by the objective function based upon divergence of radiation flux only.

The calculated optimal band boundaries were found at frequencies $\nu_1 = 2.7591 \cdot 10^{15}$ Hz and $\nu_2 = 3.5528 \cdot 10^{15}$ Hz with corresponding profiles of radiation flux and divergence of radiation flux are shown in Figure 3. One can clearly see, that the divergence of radiation flux is relatively well approximated. Only the position of transition from emitting region to the absorbing one is slightly shifted and the absorption is underestimated by approximately 20 %. However, this inaccuracy is large enough to cause the difference by the factor of 2 in the radiation flux at the domain boundary.

3.2. Radiation flux

One obvious way to improve the radiation flux accuracy is to use the radiation flux itself as the accuracy evaluating quantity. This can be achieved in our test objective function by defining the variables $A_i = 0$, $B_i = 1/F_{\mathrm{sp,max}}$ and $C_i = 0$. This objective function emphasize the area with high values of the flux around $r = 0.4$ cm with lesser focus on the central areas.

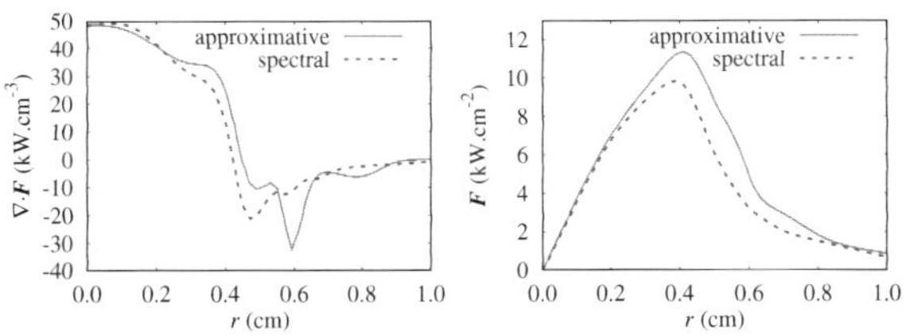

Figure 4. Divergence of radiation flux (left) and radiation flux (right) profiles evaluated by the objective function based upon radiation flux only.

The optimal band distribution differs significantly from the previous test case. The band boundaries are now located at $\nu_1 = 2.3965 \cdot 10^{15}$ Hz and $\nu_2 = 3.0232 \cdot 10^{15}$ Hz. The impact of the changed band boundaries is visible in Figure 4, where the radiation flux profile is quite improved and matches the spectral profile much closer. Especially the value at the domain boundary is resolved quite accurately with the error less than 20 %. Unfortunately this improvement was not achieved by improving the divergence of radiation flux profile. An arbitrary absorption area is created around $r = 0.6$ cm which is responsible for the improvements in the radiation flux profile. Consequently, using these band boundaries would lead to the incorrect evaluation of the energy balance inside plasma.

3.3. Incident radiation

Incident radiation represent another tempting option for objective function. Unlike the previous quantities, incident radiation profile never reaches zero value making its impact more uniform across the calculation domain. In our test objective function we can achieve the pure incident radiation evaluation by setting variables $A_i = 0$, $B_i = 0$ and $C_i = 1/G_{\mathrm{sp,max}}$. Maximum of incident radiation $G_{\mathrm{sp,max}}$ is located at the cylindrical domain axis.

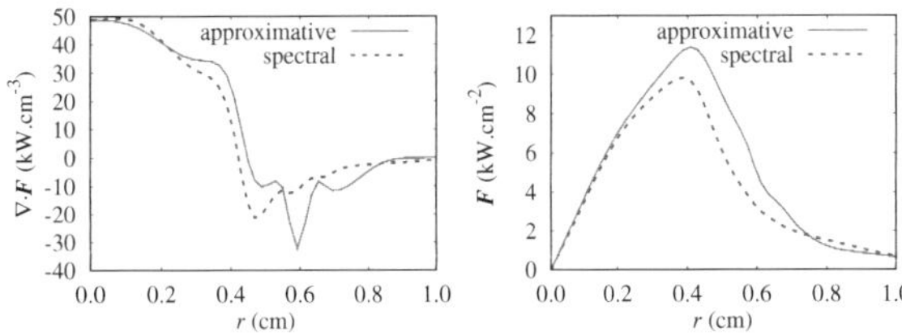

Figure 5. Divergence of radiation flux (left) and radiation flux (right) profiles evaluated by the objective function based upon incident radiation only.

Even though the incident radiation seems like good candidate for radiation objective function, its performance is inferior to the previous cases. The best band distribution band boundaries are located at $\nu_1 = 2.0123 \cdot 10^{15}$ Hz and $\nu_2 = 3.0404 \cdot 10^{15}$ Hz with corresponding divergence of radiation flux and radiation flux profiles captured in Figure 5. The results are quite similar to those obtained with objective function based upon radiation flux. Direct comparison reveals that the absorption part is even more overestimated in the case of incident radiation. This is clearly documented on the radiation flux profile where the approximate mean absorption coefficients solution reaches below the spectral solution in area close to the domain boundary.

3.4. Weighted linear combination

All the previous objective function were based on a single radiation quantity only. However, in many cases the results did not satisfy all the expectations. Each one improved the related quantity, usually at the cost of decreased accuracy in other quantities. The apparent room for improvement is an inclusion of multiple radiation quantities into the objective function. This can be easily achieved with our definition of the objective function by properly modifying the variables A_i, B_i and C_i.

For this particular test, we decided to focus on the most impactful quantities only. Therefore we used the following definition: $A_i = 1/\nabla \cdot \mathbf{F}_{\mathrm{sp,max}}$, $B_i = 1/F_{\mathrm{sp,max}}$ and $C_i = 0$, which ensures, that both radiation flux and divergence of radiation flux are equally weighted in the objective function. It might be advantageous to focus on one of the quantity in the real scenario, but for this test the equal balance is more desired.

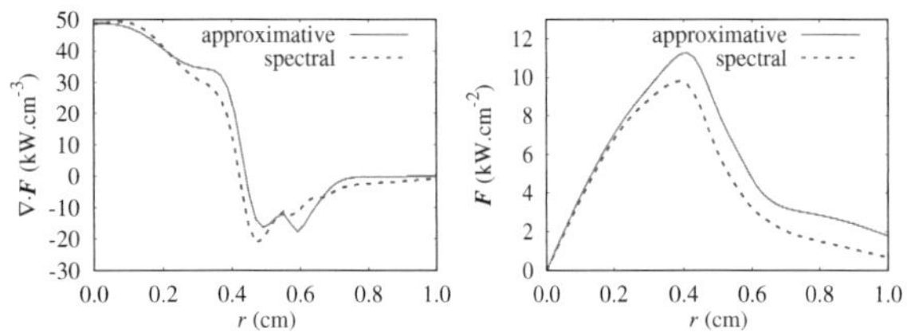

Figure 6. Divergence of radiation flux (left) and radiation flux (right) profiles evaluated by the objective function based upon weighted linear combination of radiation flux and divergence of radiation flux.

The linear combination distribution function results in frequency band distribution similar to the first test case based purely on the divergence of radiation flux. The band boundaries are located at $\nu_1 = 2.9146 \cdot 10^{15}$ Hz and $\nu_2 = 3.5528 \cdot 10^{15}$ Hz with the corresponding approximate profiles shown in Figure 6. The approximate divergence of radiation flux still exhibits some arbitrary absorption areas, but the discrepancy is far smaller than in the case of pure radiation flux objective function. Unfortunately, this does not lead to the significant improvement in the radiation flux at the domain boundary. Rather the radiation flux is improved in the area around $r = 0.7$ cm. The linear objective function therefore seems to be useful in the case when the domain is relatively small and the outer walls are close to the plasma boundary.

4. Conclusions

In this contribution we tested several different objective functions for numerical optimization of mean absorption coefficients frequency band distributions. The obtained results clearly indicate the importance of proper formulation of the objective function. The optimized mean absorption coefficients can establish an artificial absorption area without careful handling of the objective function. On the other hand, the impact of the objective function formulation is minimal in the central parts of the plasma column.

We propose the objective function based upon divergence of radiation flux to be used for numerical optimization, since the radiation source term is important for the plasma energy balance equation. Although, linear combination of radiation flux and divergence of radiation flux can be useful for cases, where the correct evaluation of radiation energy transfer to the outer walls plays critical role or the outer walls are close to the plasma boundaries.

We would like to note, that our conclusion is based on the limited number of tests. Only one temperature profile with a single plasma composition was considered in the tests. More test are required for broader applicability assessment of our conclusions.

Acknowledgements

Authors gratefully acknowledge financial support from the Centre for Research and Utilization of Renewable Energy under project No. LO1210 – "Energy for Sustainable Development (EN-PUR)" and from Czech Science Foundation under project No. 15-14829S.

References

[1] A. A. Iordanidis and C. M. Franck. Self-consistent radiation-based simulation of electric arcs: Ii. application to gas circuit breakers. *Journal of Physics D: Applied Physics*, 41(13), 2008. doi:10.1088/0022-3727/41/13/135206.

[2] J.J. Lowke. Predictions of arc temperature profiles using approximate emission coefficients for radiation losses. *Journal of Quantitative Spectroscopy & Radiative Transfer*, 14(2):111–122, 1974. doi:10.1016/0022-4073(74)90004-1.

[3] H. Z. Randrianandraina, Y. Cressault, and A. Gleizes. Improvements of radiative transfer calculation for sf6 thermal plasmas. *Journal of Physics D: Applied Physics*, 44(19), 2011. doi:10.1088/0022-3727/44/19/194012.

[4] P. Kloc, V. Aubrecht, M. Bartlova, O. Coufal, and Ch. Ruempler. On the selection of integration intervals for the calculation of mean absorption coefficients. *Plasma Chemistry and Plasma Processing*, 35(6):1097–1110, 2015. doi:10.1007/s11090-015-9648-3.

[5] P. Kloc, V. Aubrecht, and M. Bartlova. Numerically optimized band boundaries of Planck mean absorption coefficients in air plasma. *Journal of Physics D: Applied Physics*, 50(30), 2017. doi:10.1088/1361-6463/aa7627.

[6] L. Fagiano and R. Gati. On the order reduction of the radiative heat transfer model for the simulation of plasma arcs in switchgear devices. *Journal of Quantitative Spectroscopy & Radiative Transfer*, 169:58–78, 2016. doi:10.1016/j.jqsrt.2015.10.002.

[7] P. Kloc, V. Aubrecht, M. Bartlova, and O. Coufal. Radiation transfer in air and air-Cu plasmas for two temperature profiles. *Journal of Physics D: Applied Physics*, 48(5), 2015. doi:10.1088/0022-3727/48/5/055208.

[8] L. Changtong and Y. Bo. Low dimensional simplex evolution: a new heuristic for global optimization. *Journal of Global Optimization*, 52(1), 2012. doi:10.1007/s10898-011-9678-1.

[9] H. Nordborg and A.A. Iordanidis. Self-consistent radiation based modelling of electric arcs: I. efficient radiation approximations. *Journal of Physics D: Applied Physics*, 41(13), 2008. doi:10.1088/0022-3727/41/13/135205.

doi:10.14311/ppt.2017.3.273
Plasma Physics and Technology **4**(3):273–276, 2017

A REVIEW OF PROGRESS TOWARDS SIMULATION OF ARC QUENCHING IN LIGHTNING PROTECTION DEVICES BASED ON MULTI CHAMBER SYSTEMS.

A. Chusov[a,*], E.Rodikova[a], M. Murmann[b], H. Nordborg[b], R. Fuchs[b]

[a] *Streamer Electric, St Petersburg, 191024, Russia*
[b] *Institute for Energy Technology, HSR, Rapperswil, 8640, Switzerland*
[*] alexander.chusov@streamer.ru

Abstract. Two distinct modes of follow current suppression were observed in multi-chamber systems (MCS) under lightning overvoltage: Zero Quenching (ZQ) and Impulse Quenching (IQ). Sufficiently lower erosion of electrodes and evaporation of discharge chamber walls makes the IQ more preferable as a mechanism of arc quenching. Since experimental search for best MCS design is both difficult and expensive numerical modeling is considered as a prospective method for geometry optimization. Several steps were made towards development of efficient arc model. This article highlights most important results of arc quenching simulation and current status of arc model development.

Keywords: lightning protection, multi-chamber system, arc discharge, plasma simulations, design optimization.

1. Introduction

Nowadays multi-chamber arresters (MCA) (Fig. 1) are considered as prospective solutions for lightning protection of overhead powerlines [1, 2]. MCA consists of spark gap sequence covered by silicone rubber (Fig. 2). During normal grid operation MCA works as an insulator but when lightning overvoltage is applied it causes cascade of spark gap breakdowns. Each gap when triggered operates as a discharge chamber and conducts lightning current along with current of power frequency.

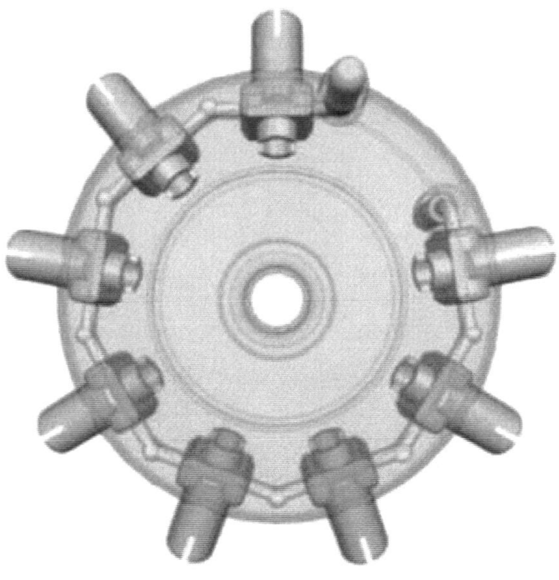

Figure 1. Multi-chamber arrester.

Large current causes strong pressure buildup which in couple with Lorentz force results in outflow of plasma and hot gas leading to arc extinction. Early experiments demonstrated quenching mechanism similar to one employed in circuit breakers - so called Zero Quenching (ZQ): current flows until it passes zero, where cooled arc is quenched (Fig. 3, right). The arcing times are of the order of 10 ms. Large heat loads on electrodes brought by follow current puts an inherent limit on MCA application range. Fortunately another distinct quenching mode - Impulse Quenching (IQ) characterized by sufficiently shorter decay time was discovered in latter experiments. In case of IQ current is suppressed within approximately 0.1 ms (Fig. 3, left), causing significantly less erosion. Further research identified certain MCS geometries capable to perform IQ more efficiently. In order to employ IQ for lightning protection we are looking for optimal design giving best quenching efficiency. Design optimization however was found to be challenging due to difficulty and high expense of tests. To support experimental search numerical simulations of impulse arc quenching in MCA were used to predict better chamber geometry.

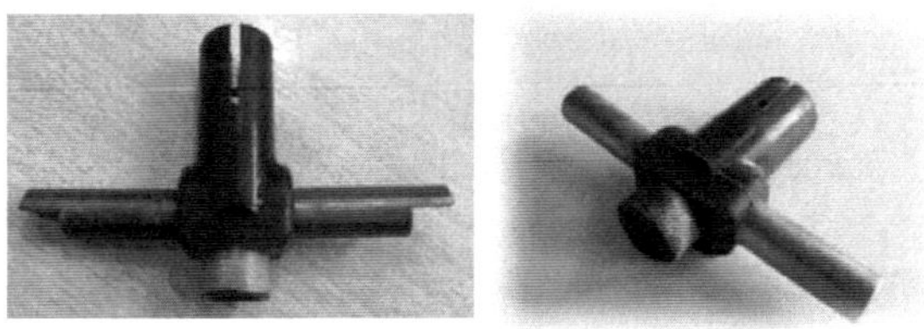

Figure 2. Two examples of MCA discharge chamber design.

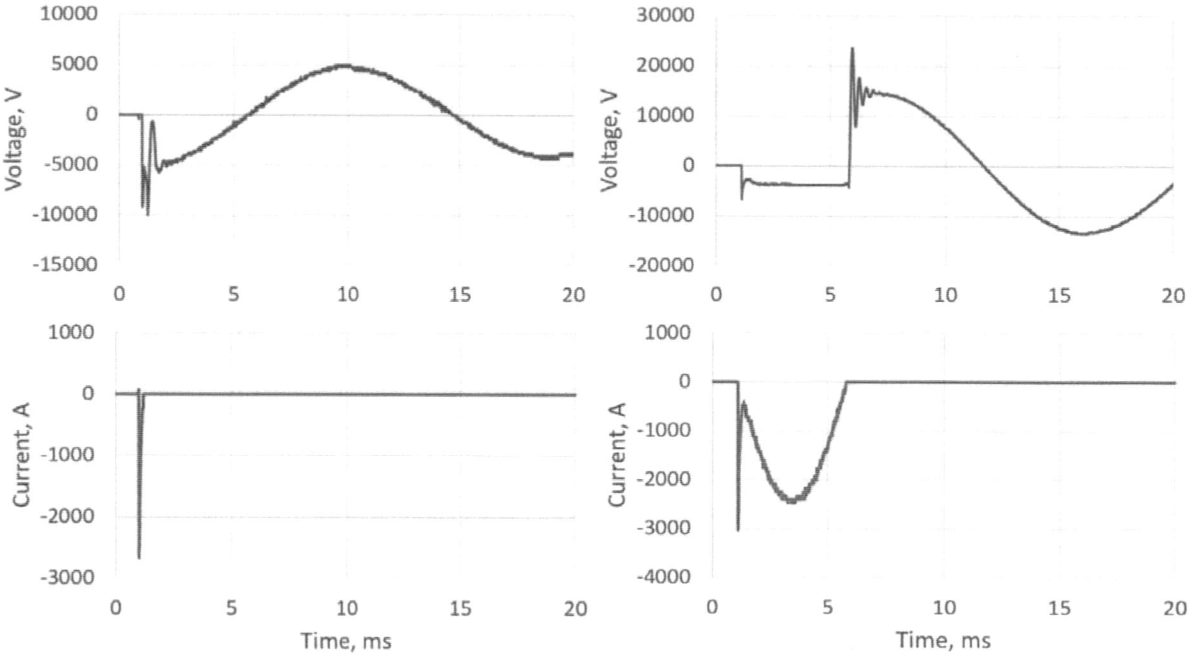

Figure 3. Zero Quenching (ZQ) - right. Impulse Quenching (IQ) - left.

2. Electrical arc simulation

2.1. Basic equations

General approach to arc simulations relies on MHD equations [3]:

$$\frac{\partial \rho}{\partial t} + \nabla \cdot (\rho \mathbf{u}) = 0 \quad (1)$$

$$\frac{\partial (\rho u)}{\partial t} + \nabla(\rho \mathbf{u} \otimes \mathbf{u}) = -\nabla p + \nabla \cdot \tau + [\mathbf{j} \times \mathbf{B}] \quad (2)$$

$$\frac{\partial (\rho h)}{\partial t} + \nabla \cdot (\rho h\ \mathbf{u}) = \frac{\partial p}{\partial t} + \nabla \cdot (\tau \mathbf{u}) + \mathbf{j} \cdot \mathbf{E} + \nabla \cdot (\mathbf{q} + \mathbf{q_{rad}}). \quad (3)$$

Here, ρ represents the mass density, u is the gas velocity, p is the pressure, τ is the viscous part of the stress tensor, and h is the specific enthalpy. The last term of Eq. (3) represents the heat flux, which has been split into a heat conduction q and radiative heat flux q_{rad}. The turbulence is included in the effective viscosity and heat conductivity of the gas. $k - \varepsilon$ turbulence model was used in all simulations ([4]). In order to solve the above set of equations, it is necessary to add state equations,

$$\rho = \rho(p, T) \quad (4)$$

and

$$h = h(p, T). \quad (5)$$

Data for viscosity, thermal and electric conductivity is also needed. The coupling to the electromagnetic field is realized by the Lorentz force $[\mathbf{j} \times \mathbf{B}]$ and the ohmic heating $\mathbf{j} \cdot \mathbf{E}$, where $\mathbf{j}$ is the electric current density,$\mathbf{E}$ is the electric field, and $\mathbf{B}$ is the magnetic induction. The current density is given by

$$\mathbf{j} = \sigma(\mathbf{E} + [\mathbf{u} \times \mathbf{B}]), \quad (6)$$

where σ is the electric conductivity of the plasma. Instead of full Maxwell equations to compute the electric and magnetic fields the magnetostatic approximation was used in this study,

$$\nabla \cdot j = 0 \quad (7)$$

and

$$[\nabla \times \mathbf{B}] = \mu_0 \mathbf{j}. \quad (8)$$

2.2. Material properties

Following the approximation of local thermodynamic equilibrium (LTE), the properties of the plasma only depend on temperature and pressure. It is possible then to precompute all thermodynamic properties and store in the form of lookup tables as described in ([5–7]). For simplicity, electrode erosion and ablation of wall material were neglected, making it possible to consider pure air arc discharge. The pressure dependence was only included for quantities showing strong variation with changing pressure – density $\rho(p, T)$ and the electric conductivity $\sigma(p, T)$ of the plasma.

2.3. Radiation transport

The phenomenon can be modeled by the radiative heat transfer equation for the spectral radiative intensity, which is given by

$$\mathbf{s} \cdot \nabla I_\nu(r, s) = \kappa_\nu [I_\nu^b(T) - I_\nu(r, s)] \quad (9)$$

Here I_ν is the radiative intensity at position r in direction s with frequency ν, κ_ν is the absorption coefficient of the medium, and I_ν^b is the Planck function

$$I_\nu^b(T) = \frac{2h}{c^2} \frac{\nu^3}{e^{\frac{h\nu}{k_B T}} - 1}. \quad (10)$$

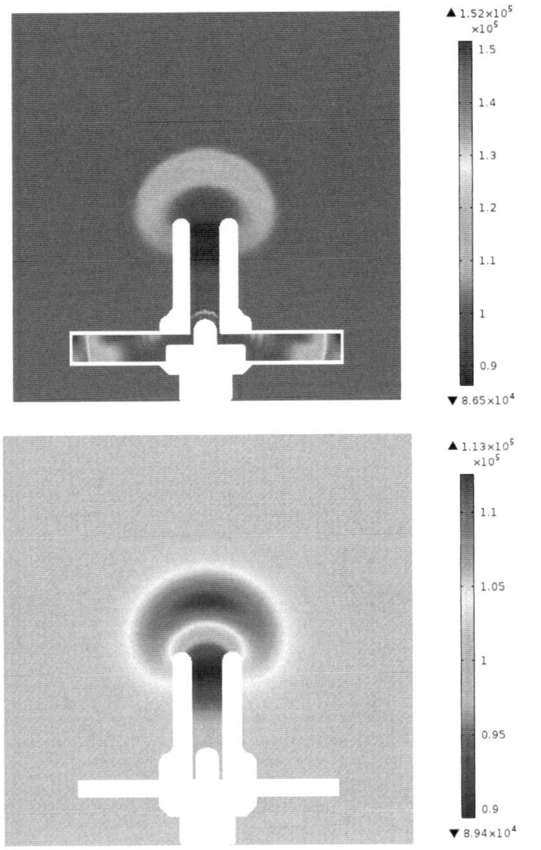

Figure 4. Pressure distribution at time of 100 µs for lightning current of 3 kA. Top - Chamber 1 (hollow electrodes), bottom (solid electrodes) - Chamber 2.

where h is the Planck constant, c the velocity of light, and k_B the Boltzmann constant. This equation is typically solved by the Discrete Ordinate Method (DOM). Since absorption coefficient is strongly frequency dependent, it is necessary to use a multi-band approach with averaged absorption coefficients. The most important effect is that high frequencies are strongly absorbed, whereas low frequencies are not so it was decided to use a simple two-band model. The first frequency band has a wavelength starting from zero up to $\lambda = 120\,\mathrm{nm}$ and an absorption coefficient of $\alpha = 2000\,\mathrm{m}^{-1}$, the second band starts from $\lambda = 120\,\mathrm{nm}$ up to $\lambda = 1\,\mathrm{mm}$ with $\alpha = 50\,\mathrm{m}^{-1}$ [7].

3. Preliminary results.

3.1. Investigation of residual resistance.

As a preliminary step simulations with lightning pulse only applied to chamber (meaning no circuit interaction involved) were conducted for several chamber designs. For further simplification 3D-geometry was replaced by 2D assuming planar flow. Turbulence is neglected as well. Since arc quenching was not simulated directly the perfomance of different chamber geometries was evaluated on the basis of residual resistance assuming the lower the latter the more

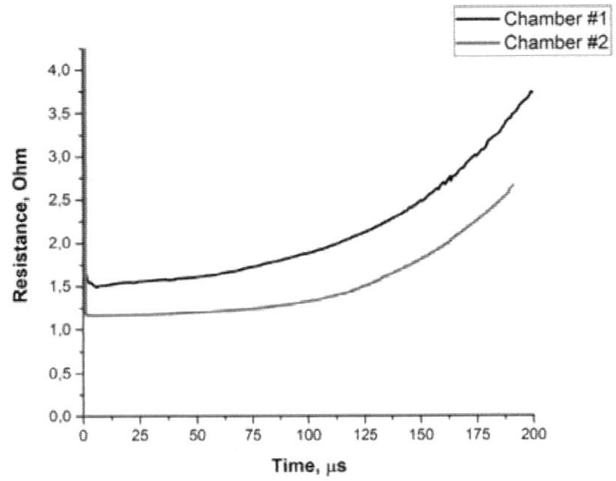

Figure 5. Resistanse curves for two different chambers. Black curve - upper chamber, red curve - lower chamber (see Fig. 4).

chances to prevent follow current to flow. The current pulse was imposed on chamber, its rise time and half-width were set to 8 µs and 50 µs correspondingly while magnitude was set to 3 kA. Results obtained for two previously tested chambers (Fig. 2) are depicted in Fig. 4 and Fig. 5 where the former displays pressure distribution at time about 100 µs while the latter displays resistance curve. The upper chamber in Fig. 4 corresponds to Chamber 1 on Fig. 5 and as it can be seen outperforms Chamber 2 (lower chamber) in terms of resistance rise rate. This result is in accordance with experimental knowledge.

3.2. Arc-circuit interaction

In order to simulate arc quenching (1)–(3) were coupled with circuit equations [7]. The circuit is depicted in Fig. 6, it represents simplified model of power grid. The capacitor C_0 is initially charged by voltage U_c which in fact stands for powerline nominal voltage so appearance of ZQ or IQ is dependent on the value of U_c. Indeed simulations gave different results for charging voltage values of 1 kV and 2 kV (Fig. 7 and Fig. 8 correspondingly). It can be seen from Fig. 7 that in case of 2 kV current vanishes only after approximately 5 ms while in case of 1 kV current hardly ever starts to flow and is interrupted after about 300 µs.

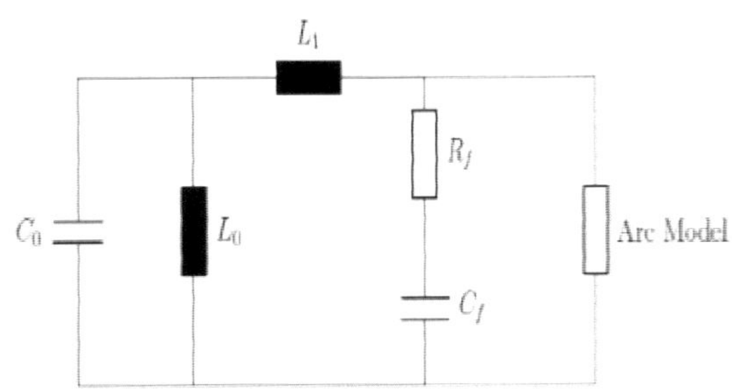

Figure 6. Electrical circuit model used for arc quenching simulations.

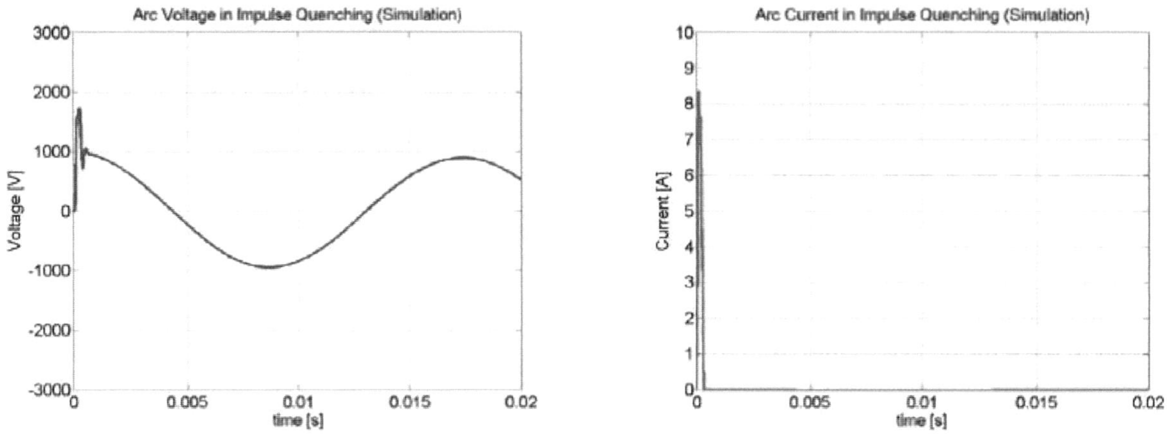

Figure 7. Current and voltage curves for $U_c = 1\,kV$.

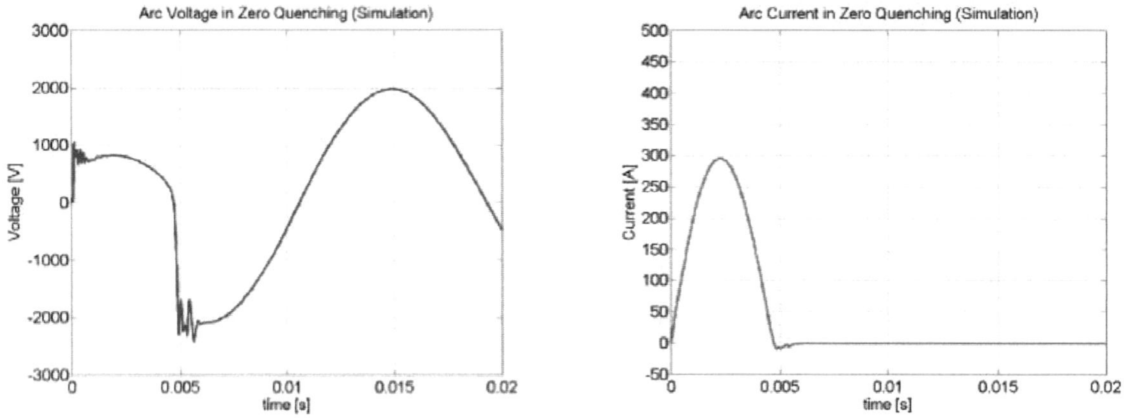

Figure 8. Current and voltage curves for $U_c = 2\,kV$.

4. Conclusions

Simulations of arc quenching in MCS were carried out with the help of specially developed impulse arc model based on MHD equations. Two important results should be highlighted. First, simulation results qualitatively match the experimental data. Second,full arc model with involved arc-circuit interaction demonstrated the transition from ZQ to IQ mode. Those promising achievements allow to consider simulation as a prospective tool for future MCS design optimization.

References

[1] G. Podporkin et al. Overhead lines lightning protection by multi-chamber arresters and insulator-arresters. *IEEE Transactions on Power Delivery*, 26(1):214–221, 2011. doi:10.1109/TPWRD.2010.2076368.

[2] G. Podporkin et al. Development of multi-chamber insulator-arresters for lightning protection of 220 kV overhead transmission lines. *XI SIPDA*, pages 160–165, 2011. doi:10.1109/SIPDA.2011.6088440.

[3] A. Gleizes et al. Thermal plasma modelling. *Journal of Physics D: Applied Physics*, 38(9):R153–R183, 2005. doi:10.1088/0022-3727/38/9/R01.

[4] F. Yang et al. Low-voltage circuit breaker arcs-simulation and measurements. *Journal of Physics D: Applied Physics*, 46(27):273001, 2013. doi:10.1088/0022-3727/46/27/273001.

[5] A. Cressault et al. Properties of air-aluminum thermal plasmas. *Journal of Physics D: Applied Physics*, 45(26):265202, 2012. doi:10.1088/0022-3727/45/26/265202.

[6] P. Kloc et al. Radiation transfer in air and air-Cu plasmas for two temperature profiles. *Journal of Physics D: Applied Physics*, 48(5):055208, 2015. doi:10.1088/0022-3727/48/5/055208.

[7] M. Mürmann et al. Modeling and simulation of the current quenching behavior of a line lightning protection device. *Journal of Physics D: Applied Physics*, 50(10):105203, 2017. doi:10.1088/1361-6463/aa560e.

doi:10.14311/ppt.2017.3.277
Plasma Physics and Technology **4**(3):277–280, 2017

PROGRAM PROCESSING DATABASE DATA FOR CALCULATION OF SPECTRAL LINES WIDTH AND SHIFT IN PLASMA

J. Pokorný

Faculty of Electrical Engineering and Communication, BUT in Brno, Technická 3058/10, 616 00 Brno, Czech Republic

xpokor31@stud.feec.vutbr.cz

Abstract. Electric and magnetic fields cause splitting of energy levels in an atom. Transition of electrons among these levels could be seen as broadening and shift of spectral lines. We recognize various types of effects, the most important is Stark effect. We developed a program for calculations of temperature dependence linear coefficients of Stark broadening and shift of spectral lines. Our results were calcutated for temperatures usual for SF_6 circuit breaker.

Keywords: spectral line broadening, spectral line shift, Stark effect, Van der Waals effect.

1. Introduction

The article is about database data processing to be able to calculate width and shift of spectral lines. There are a lot of information about studying object in the spectral line. Spectroscopic observation of spectral lines brings unique information which is not available by any other physical method. Analysis of spectral line position, width, deformation and intensity brings information about plasma density, temperature, pressure and kind of matter or mixture of matter.

We are able to investigate electric and magnetic properties of an object thanks to Stark and Zeeman effects present in spectral lines. Elaboration and theoretical understanding of demanding physical processes, where information is available thanks to spectral lines, are the only way to gain knowledge about processes where direct measurements are not feasible, ie. processes on distant stars.

2. Spectral lines properties

Hot plasma is described by a degree of ionization. The degree of ionization is determined by ionized to all neutral particles ratio in observed volume. The degree of ionization is an important factor which describes plasma behaviour. The degree of ionization for plasma of once ionized atoms in local thermodynamic equilibrium can be calculated by the equation which was invented by Indian scientist M. N. Saha in the twentieth years of the past century.

Spectral lines in radiation belong to discrete frequencies. The broadening of these lines is caused by natural atomic effects as well as external effects. Next important factor is temperature broadening of spectral lines which is caused by temperature motion of absorbing atoms. There is no line exactly monochromatic, but the line is broadened with maximum at wavelength predicted by quantum mechanics. Broadening and shift of the spectral line could be calculated:

$$\delta = \delta_{\mathrm{S}} + \delta_{\mathrm{R}} + \delta_{\mathrm{V}} + \delta_{\mathrm{N}}, \tag{1}$$

$$d = d_{\mathrm{S}} + d_{\mathrm{V}}, \tag{2}$$

where index S stands for Stark, R for resonance, V for Van der Waals and N for natural broadening or shift repectively.

2.1. Stark broadening and shift

External electric field belongs among external effects which cause spectral lines broadening due to splitting energy layers in atom – Stark broadening. Stark broadening effect could be either in linear or quadratic form. The form of Stark broadening effect depends on external electric field intensity. Stark effect is very significant during plasma arcs when there are high degree of plasma ionization. This phenomenon is used in spectroscopy as the most precise method for measurements of vibrational and rotational states of molecules.

There are several ways how to calculate Stark broadening of spectral line, we can calculate the broadening:

$$\delta_S = F(T,p)\,(P+Q)\,, \tag{3}$$

where

$$P = \sum_{J''} \frac{S\left(J'', J\right)}{2J+1} g_{se}\left(X_{J''J}\right), \tag{4}$$

$$Q = \sum_{J''} \frac{S\left(J'', J'\right)}{2J'+1} g_{se}\left(X_{J''J'}\right) \tag{5}$$

and function F is defined:

$$F(T,p) = 16\left(\frac{\pi}{3}\right)^{2/3} cR_\infty a_0^3 N_{\mathrm{e}} \left(\frac{hcR_\infty}{k_B T}\right)^{1/2}, \tag{6}$$

where Rydberg constant $R_\infty = 109737.3\,\mathrm{cm}^{-1}$, a_0 is the Bohr radius, variable N_{e} is the electron density,

$S\left(J'', J^*\right)$ is the line strength of transition between states with quantum numbers J'', J^*: $J'' \to J^*$, this line strength is available in NIST database [1]. J'', J^* are quantum numbers of total momentum of atom [2].

Stark effect also causes shift of spectral line:

$$d_S = F(T, p)(R - S), \qquad (7)$$

where

$$R = \sum_{J''} \frac{\Delta E_{J''J}}{|\Delta E_{J''J}|} \frac{S\left(J'', J\right)}{2J+1} g_{sh}\left(X_{J''J}\right), \qquad (8)$$

and

$$S = \sum_{J''} \frac{\Delta E_{J''J'}}{|\Delta E_{J''J'}|} \frac{S\left(J'', J'\right)}{2J'+1} g_{sh}\left(X_{J''J'}\right). \qquad (9)$$

Kramers-Gaunt g_{se} and g_{sh} factors are used for necessary corrections when using classical physics approach in areas on the border of quantum mechanics. These factors were published by Gaunt in 1930 and they were made more accurate by Hummer 1988, Hoof 2014 [3].

Stark broadening and shift depend on temperature. There could be calculated linear coefficients a, b resp. A, B for temperature dependence of width resp. shift of spectral line.

2.2. Van der Waals broadening and shift

Van der Waals broadening occurs during neutral particles interactions. Firstly we have to calculate effective main qunatum number:

$$n_k^* = Z\sqrt{\frac{R_\infty}{I_p - E_k}}, \qquad (10)$$

where I_p is ionization potential and E_k is energy of electron state. Then we have to calculate mean matrix elements:

$$\bar{R}_k^2 = \frac{\left(n_k^*\right)^2}{2Z^2} \cdot \left[5\left(n_k^*\right)^2 + 1 - 3l_k\left(l_k+1\right)\right], \qquad (11)$$

where l_k is orbital angular momentum quantum number of electron. Van der Waals broadening is given by:

$$\delta_V = \frac{a_0^2}{2}\left[\frac{9\pi\alpha c}{4}\right]^{2/5} \cdot \left(\bar{R}_{\mathrm{up}}^2 - \bar{R}_{\mathrm{low}}^2\right)^{2/5} \cdot \sum_{p \neq a}\left(\frac{R_\infty}{E_p}\right)^{4/5} \cdot V_{\mathrm{p}}^{3/5} \cdot N_1^{\mathrm{p}}, \qquad (12)$$

where α is the polarisability, $\bar{R}_{\mathrm{up}}^2$ and $\bar{R}_{\mathrm{low}}^2$ are averaged radial matrix elements of upper and lower energy states of the i-th spectral line of species a, V_{p} is the reltive velocity of the perturber and N_1 is the population density of the ground state of the perturbing species [4].

3. Program for computation of spectral lines properties

There are spectroscopic databases available on the internet. I would like to mention some free access databases: american institute NIST database, professor Kurucz database (from Harvard University) and international database The Opacity Project.

We used the NIST database in our program. NIST is commercial database, but some of the spectra of selected substances can be accessed for free for noncommercial purposes. There was once ionized sulphur atoms taken as an example. This sulphur ion is produced during the quenching mechanism in SF_6 circuit breaker.

We used Eclipse environment and C programming language due to speed of calculation and compatibility with Linux and Windows operating systems. Our program loads data from NIST database, it fills internal fields of variables with these data. The program adds increments from equivalent energy levels to broadening and shift of spectral line. There are linear coefficients a, b resp. A, B counted for spectral lines broadening and their shift in temperature range 5000 K – 50 000 K. We use a linear approximation for thermal dependence for width and also for shift:

$$y = kx + q, \qquad (13)$$

where $k = a$ and $q = b$ for thermal dependence of spectral line width, resp. $k = A$ and $q = B$ for thermal dependence of spectral line shift. There are thirteen parameters in the output of the program describing absorption spectral lines, the most important is the width according to Stark and according to Van der Waals.

4. Results

There are various types of broadening and shift of spectral lines. We have taken into consideration external as well as internal factors and sum up their influence. We calculate linear coefficients a, b resp. A, B necessary for calculation of spectral lines broadening resp. shift in the whole temperature range 5000 K – 50 000 K. The chosen temperature range covers the whole range needed for study of spectral effects connected with SF6 circuit breakers functioning. SF_6 molecule begins to react with used construction materials at temperature of 550 K, thermal dissociation of SF_6 gas starts about 2000 K and final temperature before quenching of an arc could exceed 20 000 K according to circuit breaker type [6].

Selected results are presented in table 1 and comparison with literature in table 2. We need to obtain function F, after that Stark broadening and shift of spectral lines can finally be calculated.

5. Conclusion

Observed spectrosocopic microsecond effects during burning and quenching of an arc in SF_6 circuit break-

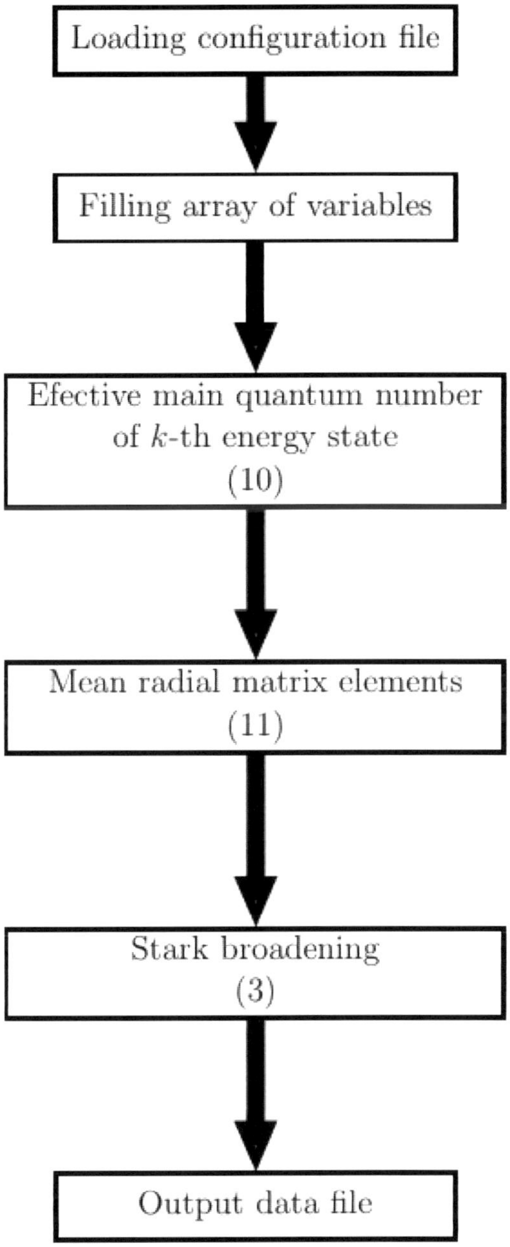

Figure 1. Block diagram of our program.

```
/* lower energy level */
if ( (Euj == El) && (semaphore1 == 0) ) {
  semaphore1 = 1;
  C4ini = C4ini+foscj*pow((lambdaj*1e-7),2.0); /* parameter Stark France */
  Aeml = 6.6702e13*glj*foscj/(guj*lambdaj*lambdaj); /* natural broadening low */
  Dnl = Dnl+Aeml;
  for ( cntK=0; cntK<6; cntK++ ) {
    gJD[cntK] = 0.275664*log(1+0.012854*pow(T[cntK],1.5)/(z*abs(Elj-El)));
    AJD[cntK]=gJD[cntK]*Sj/gl;
    DHni[cntK]=DHni[cntK]+AJD[cntK];
    gsJD[cntK]=0.001391*pow(T[cntK],1.5)/(z*abs(Elj-El));  /* shift */
    BJD[cntK]=gsJD[cntK]*Sj*(Elj-El)/(gl*abs(Elj-El));
    DHd[cntK]=DHd[cntK]+BJD[cntK];
  }
}
```

Figure 2. Example of code from our program. The segment calculates Gaunt factors.

λ nm	width a $s^{-1}K^{-1}$	width b s^{-1}	shift A $s^{-1}K^{-1}$	shift B s^{-1}
267.0604	1.50×10^{-4}	-1.55×10^{-1}	-3.22×10^{-5}	1.94×10^{-1}
386.1224	2.66×10^{-4}	-2.21×10^{-2}	3.93×10^{-5}	-2.37×10^{-1}
421.8370	2.65×10^{-4}	-2.50×10^{-1}	5.12×10^{-5}	-3.08×10^{-1}
470.1519	2.50×10^{-4}	-1.86×10^{-1}	7.12×10^{-5}	-4.29×10^{-1}
545.5371	2.50×10^{-4}	-2.29×10^{-1}	1.68×10^{-5}	-1.01×10^{-1}
560.7707	1.89×10^{-4}	-1.79×10^{-1}	-3.89×10^{-5}	2.35×10^{-1}
564.8587	2.22×10^{-4}	-1.65×10^{-1}	3.16×10^{-5}	-1.90×10^{-1}
630.7227	2.00×10^{-4}	-1.92×10^{-1}	-5.65×10^{-5}	3.41×10^{-1}
631.4431	1.62×10^{-4}	-1.50×10^{-1}	-2.60×10^{-5}	1.57×10^{-1}
727.5214	2.49×10^{-4}	-1.95×10^{-1}	3.29×10^{-6}	-1.98×10^{-2}

Table 1. Linear coefficients a, b resp. A, B for Stark broadening resp. shift of S^+.

Transition array	Multiplet or Term	λ nm	w_{MSE} nm	T 10^3K	N_e $10^{23}m^3$	J	J'	δ_λ nm
	$^4F-{}^4D^0$	560.7707	0.0304	23.5	1.0	9/2	7/2	0.019
$3p^23d-3p^24p$	$^4D-{}^4P^0$	630.7227	0.0963	23.5	1.0	7/2	5/2	0.025
	$^2F-{}^2D^0$	631.4431	0.103	23.5	1.0	5/2	3/2	0.020
	$^4P-{}^4D^0$	545.5371	0.0301	28.5	1.0	5/2	7/2	0.026
$3p^24s-3p^24p$			0.050	32.6	0.7	5/2	7/2	0.019
	$^2P-{}^2D^0$	564.8587	0.0424	27.0	0.67	1/2	3/2	0.016
			0.0374	34.0	1.02	1/2	3/2	0.028

Table 2. Comparison of calculated Stark broadening of spectral lines of S^+ with literature [5].

ers lead to possibilities of theoretical modelling of these effects. Pressure at burning arc has direct influence on spectral lines broadening and also thermal effects in SF6 gas plays an important role. SF_6 gas gradually thermally dissociates about temperature of 2000 K and rapidly cools down the arc. SF_6 circuit breaker has great advantage against air circuit breakers because of lower dissociation temperature. The thermal dissociation of air begins about 7000 K [6].

Emission spectrum changes over time due to gradual thermal heating of circuit breaker components. At the end of burning of the arc there are usually emission lines of eletrodes' material found in the emitted spectrum. We designed a program for calculation of spectral properties of various ions. We calculated data for ionized sulphur atoms in this article, selected results for S^+ ion are in table 1. In table 2, comparison of our results with calculation of other authors [5] is presented. In [5], modified semi–empirical approach is used; the difference with our results occurs due to the fact that in [5] the Stark widths of whole multiplets have been calculated.

Thorough theoretical study of spectra could bring new ways for design of electrodes shapes in order to avoid soon arc cutoff, improve cooling down mechanisms or to find new alloys for electrodes. As a result of these studies should be increased number of working cycles and improved safety of high voltage devices.

Acknowledgements

This work was supported by the Internal Grant Agency of Brno University of Technology, grant No. FEKT-S-17-4626 and by the Czech Science Foundation under project No. 15-14829S.

References

[1] NIST ASD Team. NIST atomis spectra database (version 5.4). [2017-09-01] `https://www.nist.gov/pml/atomic-spectra-database`.

[2] H. R. Griem. Semiempirical formulas for the electron–impact widths and shifts of isolated ion lines in plasmas. *Phys. Rev.*, 165(1):258–266, 1968. `doi:10.1103/PhysRev.165.258`.

[3] P. A. M. van Hoof et al. Accurate determination of the free-free gaunt factor I – non-relativistic gaunt factors. *Monthly Notic. Royal Astr. Soc.*, 444:420–428, 2014. `doi:10.1093/mnras/stu1438`.

[4] R. W. Liebermann et al. Radiation emission coefficients for sulfur hexafluoride arc plasmas. *J. Quant. Spectr. Radiat. Transfer*, 16(3):253–264, 1976. `doi:10.1016/0022-4073(76)90067-4`.

[5] N. Milovanovic et al. Cowan code and Stark broadening of spectral lines of S II, S III and S IV. *AIP Conference Proceedings*, 938(1):258–261, 2007. `doi:10.1063/1.2800140`.

[6] V. N. Maller et al. *Advances in High Voltage Insulation and Arc Interruption in SF_6 and Vacuum.* Pergamon Press, Oxford, U.K., 1981. ISBN 0-08-024726-1.

doi:10.14311/ppt.2017.3.281
Plasma Physics and Technology **4**(3):281–285, 2017

MODELING OF THE INTERACTION BETWEEN THE SWITCHING ARC AND HYDRAULIC DRIVING MECHANISM IN GAS-BLAST CIRCUIT BREAKERS

H. Zhang[a], K. Cao[b], Q. Zhang[b], J.D. Yan[b,*]

[a] *Pinggao Group Co. Ltd., Pingdingshan City, Henan, China, 467001*
[b] *Department of Electrical Engineering, University of Liverpool, Brownlow Hill, Liverpool, UK*
[*] yaneee@liverpool.ac.uk

Abstract. The presence of an arc in a circuit breaker interrupter creates an opposing force to the driving mechanism by changing of the pressure field. This opposing force alters the dynamics of the driving mechanism, the travel characteristics of the moving contact and therefore the switching process. The severity of the influence depends on the structure of the interrupter, the travel profile and also the current waveform, especially the magnitude of the fault current. A 252 kV puffer circuit breaker was used in the present work to study the key factors that contribute to the uncertainty of the predicted contact travel based on coupled simulation.

Keywords: Coupled simulation, driving mechanism, switching arc.

1. Introduction

High voltage circuit breaker is a crucial element in modern power transmission system and its reliability and performance play an important role in the safe operation of the network. It is well known that the performance of a breaker is determined by the design and operational parameters among which the travel characteristics of the moving components (e.g. contact-nozzle assembly) is a key factor that is controlled by the driving mechanism but modified by the arcing process. Despite that much effort has been devoted to arc modelling in high voltage circuit breakers [1–3] little has been reported on the influence of the arc on the dynamics of the driving mechanism. Measured travel curves are normally used in the simulation of high voltage circuit breaker [4]. A detailed analysis of a typical three-level hydraulic driving mechanism is given in [5]. As a continuation of the work done in [5], coupled circuit breaker simulation was attemped in [6]. However, the complex arcing process was approximated by a simple pressure device and assumed pressure variation with time. In the present work, a lumped mechanical model of a hydraulic driving mechanisms has been developed and coupled to a differential arc model in a way as shown in figure 1. The coupling between the two models allows the determination of the travel characteristics of the moving components in a self-consistent manner, considering automatically the effect of pressure field variation in the arcing process. The aim is to answer the following two questions. First, using the lumped model for the driving mechanism, what are the main factors that affect the accuracy of the predicted travel characteristics and how? Secondly, what accuracy can be achieved and what is the applicability of the model parameters?

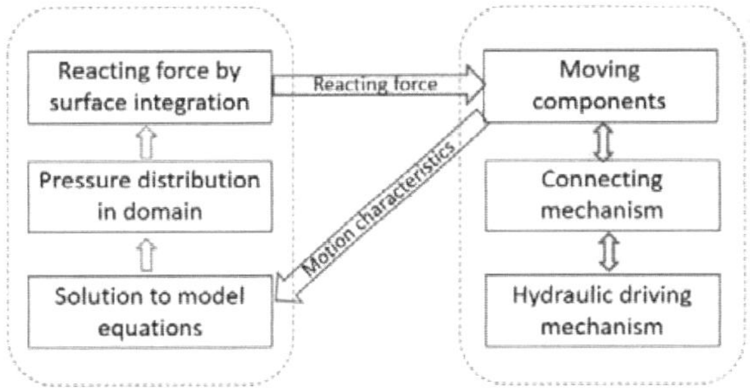

Figure 1. Coupling of the mechanical driving mechanism and the arcing process.

2. Arc model

The gas flow in the interruption chamber is largely unsteady and turbulent with the assumption that the arc is axis-symmetric (2-D). The governing equations (modified Naiver-Stokes equation) of switching arcs can be written in a general form as:

$$\frac{\partial(\rho\phi)}{\partial t} + \nabla \cdot (\rho\phi\vec{V}) - \nabla \cdot (\Gamma_\phi \nabla\psi) = S_\phi \qquad (1)$$

With a comprehensive description of the arc model given in [7, 8], for the sake of simplicity, details regarding the arc model and equation (1) will not be presented in this paper.

The modified N-S equation takes into account all important process and factors during arcing, such as: radiation, ohmic heating, nozzle ablation, electromagnetic effect and turbulence. The arc model is implemented in a commercial computational fluid dynamics (CFD) package, PHOENICS. A typical 252 kV puffer circuit breaker has been chosen as an example, based on which two stes of reference simulation have been conducted with current of 10 and 50 kA.

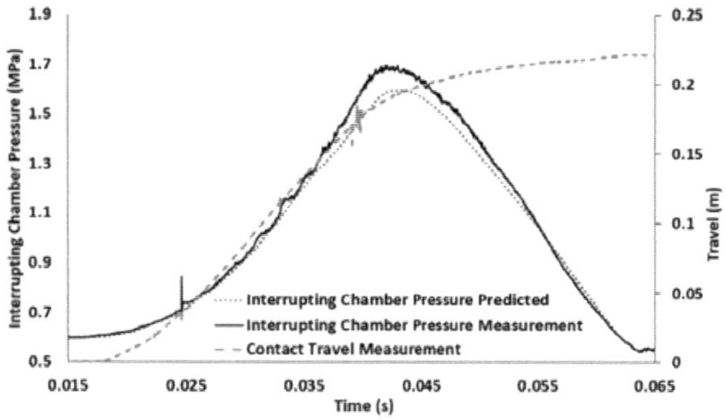

Figure 2. Predicted and measured pressure in the compression chamber of a puffer circuit breaker under 10 kA conditions. Measured contact travel is also given.

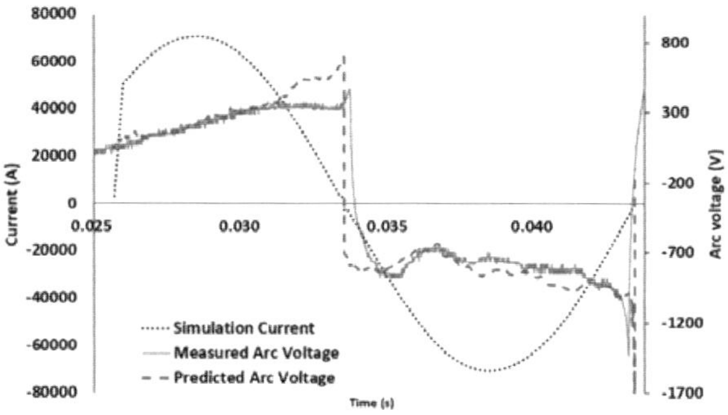

Figure 3. The comparison of simulation and measured arc voltage under 50 kA conditions.

The calculation results are then compared with available measurement. Detailed experimental procedure regarding the measurement of contact travel, arc voltage and interruption chamber pressure is presented in [9]. In the case of 10 kA, a comparison between the measured and simulation arc chamber pressure is provided in figure 2. At 10 kA, the current is relatively low and the arc duration is also short. Thus, the arc has less impact on the pressure distribution in the interruption chamber. As a result, this is an ideal condition to verify pressure predictions caused by compression. On the other hand, at 50 kA, the arc is more stable compared to the low current cases and calculated arc voltage is an important parameter for verifying the arc model. The predicted and measured arc voltage under 50 kA condition are presented in figure 3. The predicted pressure and arc voltage show good agreement with experiment results. The pressure comparison shows that the arc model is capable of predicting the pressure variation in the interruption chamber caused by the moving objects while the arc voltage comparison demonstrates that the arc model is capable of calculating the arc parameters with sufficient accuracy.

3. Hydraulic driving mechanism model

The functional structure of the hydraulic driving mechanism is shown in figure 4. This is a two-level system

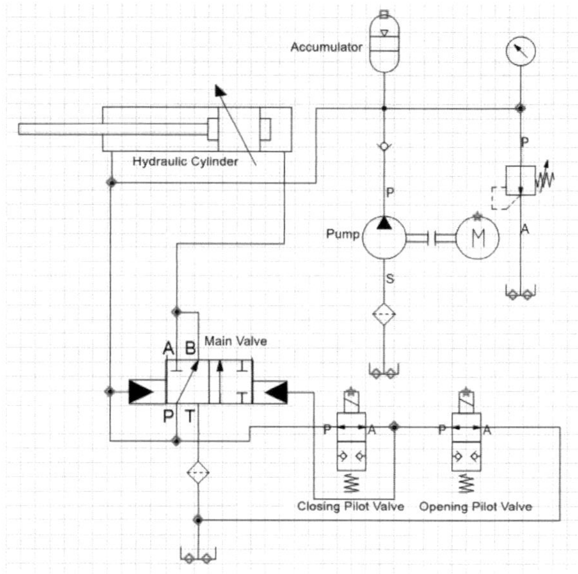

Figure 4. Schematic of the two-level hydraulic driving mechanism upon which the driving mechanism model is based. Only the opening operation is considered. The main components are labled in the diagram.

in the sense that it has two tiers of control valves controlling the operation of the main cylinder i.e. the opening and closing pilot valves and the main valve. The operation of a control valve is a dynamic process, by analyzing the force balance on its control member, this process can be described as:

$$m_i \frac{\mathrm{d}x_i^2}{\mathrm{d}t^2} = F_{\mathrm{s}i} - F_{\mathrm{c}i} - B_{\mathrm{v}i}\frac{\mathrm{d}x_i}{\mathrm{d}t} - F_{\mathrm{r}} \quad i = 1,2,3 \quad (2)$$

where the subscript stands for different levels of hydraulic components (1: pilot valve, 2: main valve, 3: hydraulic cylinder), m represents the mass of the control member (mass of the connecting mechanism is included in the hydraulic cylinder level), B_{v} is the viscous friction coefficient, F_{r} the reacting force (only applies to hydraulic cylinder), F_{s} and F_{c} are the forces on the high pressure (system pressure) and control side of the control member, which can be expressed as:

$$F_{\mathrm{s}i} = A_{\mathrm{s}i}P_{\mathrm{s}i} \quad F_{\mathrm{c}i} = A_{\mathrm{c}i}P_{\mathrm{c}i} \quad i = 2,3$$

where A_{s} and A_{c} are the effective high pressure and control side areas of the control member, P_{s} and P_{c} are the corresponding pressures. Note that the pilot valves are not differential valves, they are operated by electrical actuators. The high-pressure side of any control member can be considered as connected to the accumulator directly since the pressure loss along connecting pipelines is negligible [10]. Therefore, the high-pressure side pressure is equal to the pressure inside the accumulator, which is assumed to remain constant throughout the operation (45 MPa). The pressure of the control side can be calculated using:

Figure 5. Structure of the interruption chamber used in the simulation.

$$\frac{dP_{ci}}{dt} = \frac{\beta}{V_{ci}}\left(A_{ci}\frac{dx_i}{dt} - Q_{i-1}\right) \quad i = 2, 3 \qquad (3)$$

where β is the bulk modulus of the hydraulic oil, V_{ci} the instantaneous volume of the control side chamber and Q_{i-1} is the volumetric flow rate that exits the control side volume. The subscript $_{i-1}$ indicates that the outflow of the current level is always controlled by the previous level component. The flow rate through the control valves is determined by:

$$Q_i = C_{\mathrm{d}i}A_{\mathrm{v}i}\sqrt{\frac{2(P_{\mathrm{c}(i+1)} - P_{\mathrm{b}})}{\rho_{\mathrm{h}}}} \quad i = 1, 2 \qquad (4)$$

where C_{d} is the discharge coefficient of the orifice, P_{b} the back pressure (pressure in the oil reservoir, when compared with the operational pressure of the system, it can be considered as atmospheric pressure), ρ_{h} the density of hydraulic fluid and A_{v} the coressponding orifice area. Equations (2)–(4) constitute the governing equations of the hydraulic driving mechanism. By solving them simultaneously, the travel profile of the moving components (without considering reacting force) can be obtained.

4. Reacting force calculation and coupled simulation procedure

The reacting force applied to the driving mechanism is determined by the net force acted by the working gas on the surface of all moving components. This can be obtained by integrating on the surface of all moving components the elementary forces exerted by the pressure in the direction of movement. Within each simulation time step, an integration is performed and the total net reacting force calculated. This new data is then substituted into equation (2) (for hydraulic cylinder only), and a new displacement for the moving components is subsequently obtained. In this manner, the interaction between the arc and the driving mechanism can be included in the predicted travel during the simulation. Structure of the arc chamber under investigation is shown in figure 5. It is a 2-D axis symmetric representation of the actual arc chamber. Filling pressure inside the chamber is 0.6 MPa, the maximum travel of the moving contact (downstream) is 220 mm and the over-travel is 47 mm.

5. Analysis of travel characteristics

During the operation of the hydraulic driving mechanism, the motion of the mechanical components is

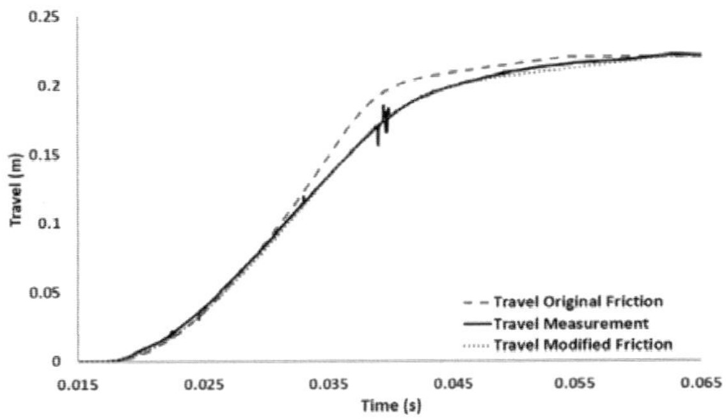

Figure 6. Travel curves for 10 kA case, together with measured travel and travel obtained under same condition with original B_{v3} setup.

closely coupled with the flow of hydraulic fluid, such flow is generally complicated since it involves the acceleration, deacceleration, and compression of the fluid. In addition, there are various friction sources that exist between both fluid-solid and solid-solid interfaces. Therefore, it is inevitable that the lumped parameter model contains a number of uncertainties, among which the most prominent is one the frictional force exerted on the piston-rod assembly inside hydraulic cylinder. The magnitude of this frictional force is determined by material, structure of the cylinder as well as the contact area between piston-rod assembly and hydraulic oil. When, a constant $B_{\mathrm{v}3}$ ($1250\ \frac{\mathrm{N{\cdot}s}}{\mathrm{m}}$) is used in equation (2) to model the frictional force, the travel curve (for 10 kA case) obtained deviates from the measurement as shown in figure 6. Evidently, a constant $B_{\mathrm{v}3}$ is inadequate. Considering the contact area between the rod and hydraulic oil changes during the motion of the piston, it is necessary to divide $B_{\mathrm{v}3}$ into two parts: a constant part that describes the friction between the piston and the rubber sealing rings installed between the piston and cylinder housing and a linearly changing part that accounts for the changing area of solid-fluid interface i.e.:

$$B_{\mathrm{v}3} = a + b_{x_3} \qquad (5)$$

The value of $B_{\mathrm{v}3}$ is calculated based on experimental results. Figure 6 also presents the travel curve obtained using equation(5). It can be seen that the new result is significantly improvemened over the previous one. The maximum error (1.8%) occurred near the end of the travel is within the acceptable limit. As long as the hydraulic driving mechanism under consideration has a similar structure, the lumped driving mechanism model is capable of predicting the travel profile accurately.

Under high current conditions, another important factor that affects the travel is the reacting force. In this case, the arc can raise the local pressure significantly as shown in figure 7 with both measured and predicted compression chamber pressure. As the moving components are only allowed for translation movement in the arc chamber, their area subjected

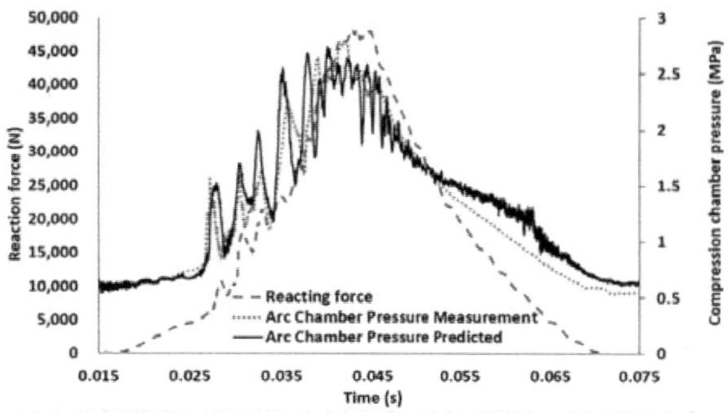

Figure 7. Calculated pressure and reacting force under 50 kA conditions, together with the measured arc chamber pressure. Pressure variations in the figure are recorded at the exit of compression chamber.

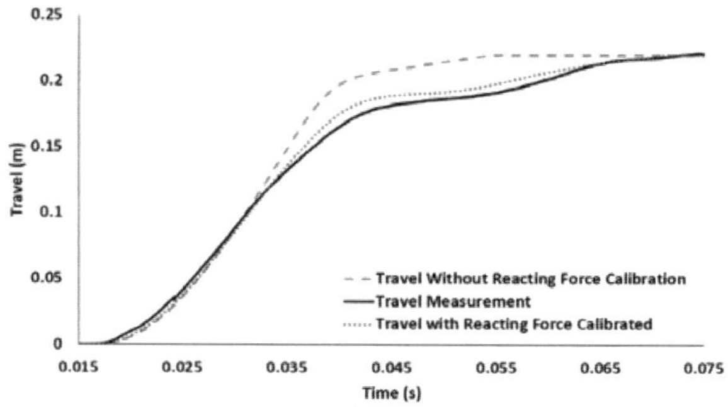

Figure 8. Travel curves for 50 kA case, together with measured travel obtained under same condtion.

to high pressure will remain unchanged throughout the simulation. Thus, the arc will have a much higher impact on the travel compared with the 10 kA case. As showcased in figure 7, although the general profile of the predicted and measured pressure matched up nicely, their instantaneous value still differs. Between 27 ms and 39 ms, the predicted pressure is lower than the measured pressure. As a result, an error naturally exists between the calculated reacting force and its real value. This is further demonstrated by the calculated and measured travel curve as compared in figure 8. It can be observed that a significant difference exists between the two travel profiles. The arc model has underestimated the reacting force as the calculated travel indicates a higher contact speed in the middle portion. To quantify the effect of error in pressure calculation, a dimensionless coefficient is introduced so the total reacting force is:

$$F_\mathrm{r} = B_\mathrm{r} \int P \cdot \mathrm{d}A \tag{6}$$

where F_r is the total reacting force and $\mathrm{d}A$ is the elementary surface area contributing to reacting force that is projected in the direction of movement. B_r is the coefficient used to adjust for the error in pressure calculation, and P is the corresponding local pressure. By comparing with measured travel curve, it is found that the optimum value for B_r is 1.15. The calibrated travel is also shown in figure 8. In this case, the maximum error occurred in the middle portion of the travel profile is 5.8%. It is noteworthy that in the 50 kA case, the maximum reacting force recorded is 47 kN. Considering that the driving mechanism is only capable of outputing 31 kN at most, the reacting force is definitely an important factor when determining the travel profile of the moving components under high current conditions.

6. Conclusion

For no-load and low current cases, the main factor that affects the travel is the frictional force on the cylinder piston. The coefficient for this frictional force should be adjusted using the measured travel as a reference. On the other hand, calculation of pressure distribution in the arc chamber may not always be accurate due to the complex physical processes and geometry. Therefore, the reacting force which essentially quantifies the interaction between the driving mechanism and arcing chamber of a circuit breaker also needs to be calibrated accurately. Despite these uncertainties, the coupled circuit breaker model is capable of describing the operation process under both low and high current conditions. Therefore, it is a valuable tool for circuit breaker design optimization.

References

[1] G. Bizjak, P. Zunko, and D. Povh. Circuit breaker model for digital simulation based on mayr's and cassie's differential arc equations. *IEEE Transactions on Power Delivery*, 10(3):1310–1315, 1995. doi:10.1109/61.400910.

[2] R.E. Blundell and M.T.C. Fang. A simplified turbulent arc model for the current-zero period of a SF_6 gas-blast circuit breaker. *Journal of Physics D: Applied Physics*, 31(5):561–568, 1998. doi:10.1088/0022-3727/31/5/013.

[3] M. Okamoto, M. Ishikawa, K. Suzuki, and H. Ikeda. Computer simulation of phenomena associated with hot gas in puffer-type gas circuit breaker. *IEEE Transactions on Power Delivery*, 6(2):833–839, 1991. doi:10.1109/61.131142.

[4] A. Petchanka, F. Reichert, F. Methling, and S. Franke. CFD arc simulation of a switching-off process in a model chamber. *Plasma Physics and Technology*, 2(1):63–66, 2015. URL: http://ppt.fel.cvut.cz/articles/2015/petchanka.pdf.

[5] B. Xu, R. Ding, J. Zhang, and Q. Su. Modeling and dynamic characteristics analysis on a three-stage fast-response and large-flow directional valve. *Energy conversion and management*, 79:187–199, 2014. doi:10.1016/j.enconman.2013.12.013.

[6] B. Xu, R. Ding, J. Zhang, L. Sha, and M. Cheng. Multiphysics-coupled modeling: Simulation of the hydraulic-operating mechanism for a SF_6 high-voltage circuit breaker. *IEEE/ASME Transactions on Mechatronics*, 21(1):379–393, 2016. doi:10.1109/TMECH.2015.2460351.

[7] K.Y. Park and M.T.C. Fang. Mathematical modeling of SF_6 puffer circuit breakers. I. High current region. *IEEE Transactions on Plasma Science*, 24(2):490–502, 1996. doi:10.1109/27.510015.

[8] K.Y. Park, X.J. Guo, R.E. Blundell, M.T.C. Fang, and Y.J. Shin. Mathematical modeling of SF_6 puffer circuit breakers. II. Current zero region. *IEEE Transactions on Plasma Science*, 25(5):967–973, 1997. doi:10.1109/27.649608.

[9] J.Y. Zhong, Y.J. Guo, and H. Zhang. Pressure and arc voltage measurement in a 252 kV SF_6 puffer circuit breaker. *Plasma Science and Technology*, 18(5):490–493, 2016. doi:10.1088/1009-0630/18/5/08.

[10] H.E. Merritt. *Hydraulic control systems*. John Wiley & Sons, 1967.

doi:10.14311/ppt.2017.2.108
Plasma Physics and Technology 4(2):108–111, 2017

OPTICAL INVESTIGATIONS ON PLASMA TEMPERATURE ESTIMATION IN A MODEL SPARK GAP FOR SURGE CURRENTS

T. Runge[a,*], St. Franke[b], S. Gortschakow[b], M. Kurrat[a]

[a] *Technische Universität Braunschweig, Institute for High Voltage Technology and Electrical Power Systems, 38106 Braunschweig, Germany*

[b] *INP Greifswald, Felix-Hausdorff-Str. 2, 17489 Greifswald, Germany*

[*] t.runge@tu-braunschweig.de

Abstract. In this experimental investigation optical emission spectroscopy is used to characterize the radiation of the plasma in a spark gap during surge. Different approaches are used, compared and discussed in order to estimate plasma temperatures. The measurements were carried out in a narrow gap arrangement based on spark gap technology. This model is tested using 8/20 µs surge currents according to the IEC 62475 with amplitudes of 5 kA and 11 kA.

Keywords: arc, temperature, plasma, spark gap, surge current, surge protective device.

1. Introduction

Class I surge protective devices (SPD) are used in order to protect electrical devices from surge currents caused for example by lightning strikes. Thus following tasks have to be handled by the SPD [1, 2]:

- Plasma ignition caused by occuring surge
- Discharge of the surge to the ground potential
- Ensure of a galvanic isolation

These tasks can be achieved by a SPD based on spark gap technology due to their high energy absorption potential. For the limitation and extinguishing of the follow current in SPDs it is common use to quench the plasma in a narrow gap with gassing chamber walls. The successful extinguishing of the follow current mainly depends on the plasma properties like temperature, pressure and electrical conductivity. Thus one part of understanding the plasma processes and current limitation is the knowledge about temperature behaviour and how to influence it. To investigate the dependence of plasma temperature on different boundary conditions a method has to be found to measure the temperature. In this investigation different methods are presented in order to determine the plasma temperature in a spark gap model at surge currents. Further a comparison of the different results is made to identify the uncertainty of the plasma temperature.

2. Methodology

The measurements are carried out in a narrow gap arrangement based on spark gap technology at surge currents. In order to measure the temperature optical emission spectroscopy (OES) is applied at the spark gap as described in subsection 2.1. Afterwards the different methods to determine the temperature are described in subsections 2.2 to 2.4.

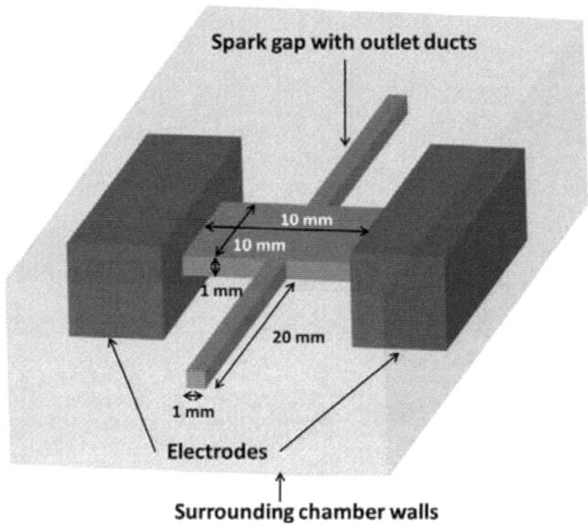

Figure 1. 3-D model of the used spark gap model without SDD [4].

2.1. Experimental Setup

The used spark gap model is already presented in other publications like [3]. A narrow gap is formed as a box with two outlet ducts by two W-Cu electrodes and a chamber wall. The chamber wall is made of Polyoxymethylen (POM) and Polymethylmethacrylat (PMMA) in order to get an optical access into the spark gap (Fig. 1). Surge currents of the form 8/20 µs are applied according to the IEC 62475 with amplitudes of 5 kA and 11 kA. For the ignition of the spark gap a surface discharge device (SDD) is used connected by a varistor. Afterwards the plasma is spreading in the spark gap and a breakdown occurs leading to a current flow through the plasma. The plasma expands with time and is finally forming a nearly homogeneous discharge in the spark gap [4].

For the optical emission spectroscopy two different setups are used. The first one consists of a

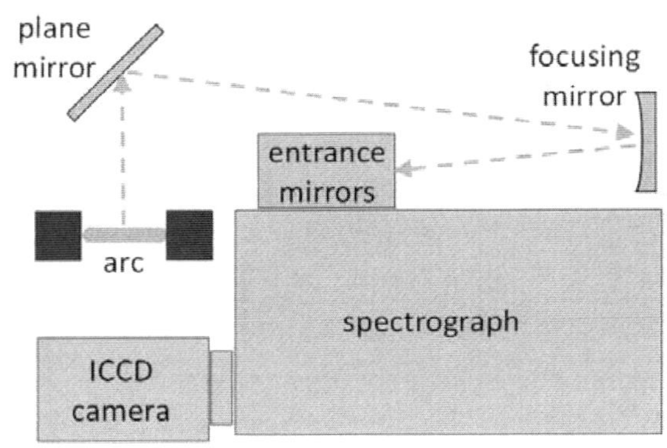

Figure 2. Setup for optical emission spectroscopy of SPD.

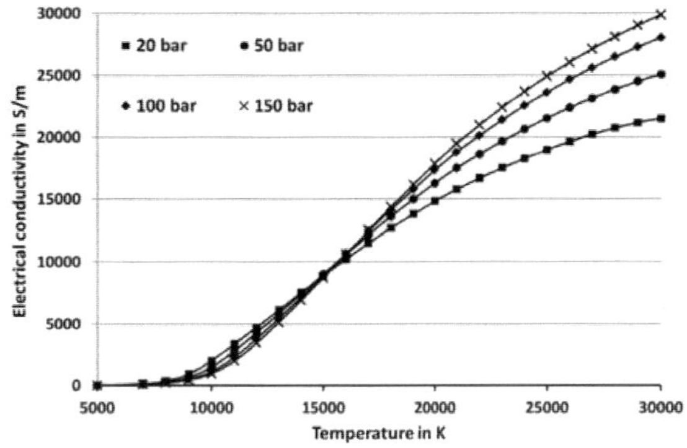

Figure 3. Electrical conductivity of air in dependence of temperature and pressure [6].

compact spectrometer (Avantes AcvaSpec 2048XL) with an entrance slit of 25 µm and a grating with 300 lines per mm on a 75 mm optical bench. Thus a spectrum from 250 nm up to 1000 nm can be recorded with a spectral resolution of about 1.5 nm full width at half maximum. The spectrometer is triggerable and the exposure time is set to 2 µs. To collect the radiation of the plasma a bare fibre with 200 µm diameter is used.

The second spectroscopic setup consists of an imaging spectrograph with 0.5 m focal length (Roper Acton SpectraPro SP2500i) equipped with an intensified CCD camera (Princeton Instruments PI-MAX) to record one image per shot at exposure times of about 1 µs (see Fig. 2). For this purpose the arc is imaged on the entrance slit of the spectrograph (40 µm slit width) by a focusing mirror. Deflecting mirrors are used to guide the radiation. Making use of a grating with 150 lines per mm, a spectral range of about 150 nm can be recorded by the triggered ICCD camera with a spectral resolution of about 0.3 nm full width at half maximum.

The trigger delay was adjusted to 13 µs for the first and 14 µs for the second setup to obtain spectra around current maximum. In both setups an absolute calibration of side-on spectra in units of spectral radiance are performed placing a calibrated tungsten strip lamp (OSRAM Wi 17/G) at the arc position.

2.2. Method I

The presented method I is not based on optical emission spectroscopy. In this case measurements of the electrical conductivity and plasma pressure are used as done in [4] and [5]. Together with data from [6] (Fig. 3) the electrical conductivity and pressure are used to determine the temperature. In this case data for air are used. This assumption can be made due to the negligible influence of POM and PMMA on the electrical conductivity compared to air plasma [7, 8].

2.3. Method II

From spatially and spectrally resolved OES the volume emission coefficient of a spectral line can be observed. In this investigation H_α at 656.279 nm as most marked spectral line in the observed spectrum is used. The so-called single line method makes use of this volume emission coefficient which can also be calculated from the following equation:

$$\varepsilon(T) = \frac{hc_0}{4\pi\lambda_0} \cdot \frac{n_g(T,p)}{Z} \cdot g_u \cdot A_{ul} \cdot \exp\left(-\frac{E_u}{k_b T}\right) \quad (1)$$

with h - Planck constant , c_0 - speed of light in vacuum, λ_0- center wavelength of H_α line, n_g - density of atomic hydrogen, Z - partition function of atomic hydrogen, g_u - statistical weight of the upper level of the transition, A_{ul} - transition probability, E_u - energy of the upper level, k_B - Boltzmann constant, and T - plasma temperature. The hydrogen density n_g can be obtained from local thermodynamic equilibrium (LTE) plasma composition calculations. The atomic data of the transition are taken from Kurucz database [9]. Now the volume emission coefficient is calculated for a selected temperature range at a given total pressure. From a look-up table the plasma temperature is deduced for the experimental emission coefficient.

2.4. Method III

In this method it is assumed that the plasma behaves like a black-body radiator due to a high optical depth caused by a high plasma pressure. With this assumption the temperature can be determined by comparing the measured intensity ratio L_2/L_1 of two wavelengths λ_1 and λ_2 with the intensity ratio of a black-body radiator at the same wavelengths as given in Eqn. (2).

$$\frac{L(\lambda_2,T)}{L(\lambda_1,T)} = \frac{\lambda_1^2 \cdot \left[\exp\left(-\frac{hc_0}{\lambda_1 k_b T}\right) - 1\right]}{\lambda_2^2 \cdot \left[\exp\left(-\frac{hc_0}{\lambda_2 k_b T}\right) - 1\right]} \quad (2)$$

3. Results and Comparison

This section is divided into three subsections: Firstly the calculation of temperatures by the use of measured pressure and electrical conductivity (method I), secondly the determination of temperatures using OES

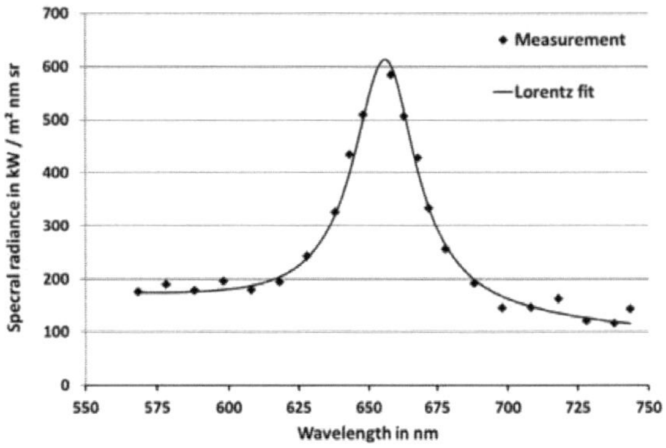

Figure 4. Side-on intensity of H_α line and Lorentz fit including linear offset for current amplitude of 5 kA.

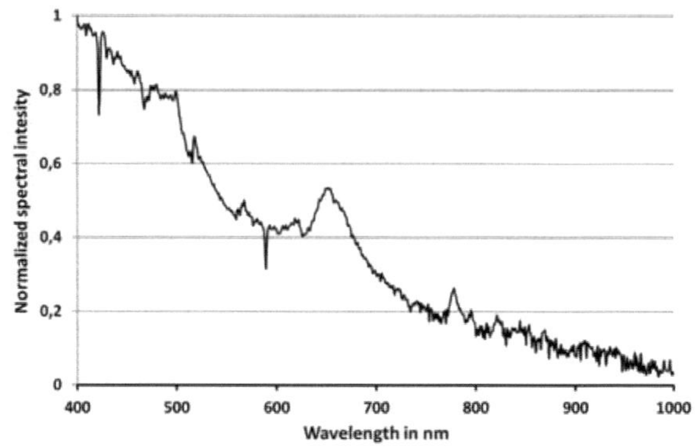

Figure 5. Normalized spectral intensity at a current amplitude of 11 kA.

(method II and III) and thirdly the comparison of the different results.

3.1. Results using method I

In this subsection the estimated temperature is presented using method I. Thus the measured electrical conductivity and pressure are used together with the data of [6] in order to determine the temperature. The measured pressure and electrical conductivity are shown in Tab. 1 together with the calculated temperature for different current amplitudes. They are representing the maximum temperature during surge occurring at current maximum. Next to the calculated temperature an uncertainty of these results are given. This uncertainty is arising from the uncertainties in the measurements of pressure with 10 % and electrical conductivity with 15 % [4, 5].

3.2. Results using method II and III

In order to determine the plasma temperature by OES two different methods are used. For the temperature at a current amplitude of 5 kA method II and for 11 kA method III is applied. Caused by the absorption in the PMMA at wavelengths below 400 nm only the spectral radiance between 400 nm and 1000 nm can be used. Fig. 4 presents the spectral radiance of the plasma at a current amplitude of 5 kA. The detected radiation originates from a region in the center of the chamber of about 1 mm lateral dimension. There appears a strongly broadened H_α spectral line at 656.279 nm with a full width at half maximum of 28 nm, which can be fitted by a Lorentz profile. Assuming a homogeneous plasma along the 1 mm line of sight and integrating over the spectral line profile, taking account of linear background signal, one yields a volume emission coefficient of $2.2 \cdot 10^{10}$ W/m^3 sr.

Before application of Eqn. (1) the hydrogen density n_g has to be determined from plasma composition calculations for pure POM and pure PMMA plasmas at a given total pressure of 90 bar (see Tab. 1). This yields a temperature of 18 600 K in case of a pure POM plasma and a temperature of 18 400 K for a pure PMMA plasma. The experimental uncertainty of the emission coefficient is estimated to be better than ± 40 % leading to an uncertainty in the temperature of less than ± 4 % (± 750 K).

At a current amplitude of 11 kA the H_α spectral line becomes less marked and the spectrum becomes more and more Planck like, as can be seen in Fig. 5. Thus the ratio of measured intensities at different wavelengths L_2/L_1 as given in Eqn. 2 is used to determine the plasma temperature in this case (method III). The ratio of intensities depends on the temperature and the chosen wavelengths. At the use of this method it is important that no marked spectral line is located at the choosen wavelengths. Further two wavelength with a high distance have to be used to get a higher slope in the ratio over temperature and therefore a better resolution. Here the ratio is determined for 525 nm and 900 nm. Due to the signal noise, spectral intervals of ± 5 nm are averaged at the given wavelengths. This yields a measured ratio of 6.5 ± 0.75. A comparison with the calculated ratio using Eqn. (2) is presented in Fig. 6. Thus the temperature is determined to be 23 200 K with a lower boundary of 17 000 K and upper boundary of 36 000 K caused by the uncertainty of the measured ratio. For the comparison in the next subsection the uncertainty is determined by the average of difference between 23 200 K and lower and upper boundary which yields in 9 500 K.

3.3. Comparison of Temperatures

The comparison of the different results for plasma temperature at current amplitude of 5 kA and 11 kA are shown in Tab. 2.

At a current amplitude of 5 kA method I and II are in good accordance among themselves with an acceptable uncertainty. In case of 11 kA current amplitude the results using method I and III differ more. Especially the uncertainty of method III is too high in order to have an informative character.

Current Amplitude	Pressure	Conductivity	Temperature
5 kA	90 bar	15 250 S/m	18 750 ± 1 875 K
11 kA	250 bar	31 815 S/m	29 500 ± 5 900 K

Table 1. Estimated plasma temperature determined by method I.

Current Amplitude	Method I	Method II	Method III
5 kA	18 750 ± 1 875 K	18 500 ± 750 K	—
11 kA	29 500 ± 5 900 K	—	23 200 ± 9 500 K

Table 2. Comparison of plasma temperature using different evaluation methods.

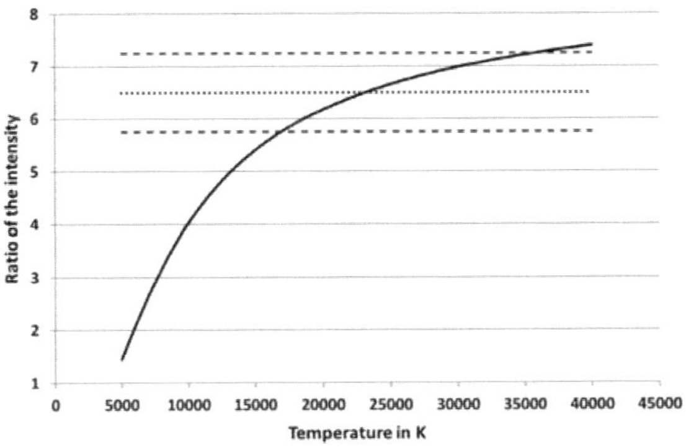

Figure 6. Comparison of the calculated intensity ratio (solid line) with measured intensity ratio (dotted lines) within given uncertainty (dashed lines).

4. Conclusion and Outlook

Different methods are used in order to determine plasma temperatures in a model spark gap at 8/20 μs surge currents according to the IEC 62475 with amplitudes of 5 kA and 11 kA. Method I based on measurements of plasma pressure and electrical conductivity, method II on OES using the volume emission coefficient of H_α spectral line and method III on OES using the ratio of black-body radiance at different wavelengths. At a current amplitude of 5 kA plasma temperature was determined to be about 18 500 K using method I and II. Caused by the uncertainty no clear temperature can be determined at 11 kA. Alternative methods applied to even higher currents are under investigation to obtain temperatures from Planck like spectra. However, especially at a current amplitude of 5 kA it is shown that combined measurements of pressure, electrical conductivity and temperature can be used in order to verify plasma parameters with theoretical data for electrical conductivity depending on pressure and temperature. This may offer an option to proof theoretical plasma transport data by experiments in the presented model spark gap. For this purpose the composition of plasma should be known precisely, which makes further experiments with non-gaseous chamber walls advantageous.

References

[1] B. Schottel and et al. Investigations on the influence of surge current on plasma in a model spark gap. In *Proc. International Conference on Lightning protection*, Shanghai, China, 2014.

[2] S. Ait-Amar. Arc extinguishing method of spd type1. In *Proc. International Conference on Lightning protection*, Uppsala, Sweden, 2008.

[3] T. Runge and et al. Experimental investigation on plasma pressure in an narrow gap for short time currents. In *Proc. International Conference on Electrica Contacts*, Edinburgh, Great Britain, 2016.

[4] T. Runge, T. H. Kopp, and M. Kurrat. Experimental investigations on electrical plasma conductivity in a model spark gap for surge currents. *submitted to Plasma Physics and Technology*, 2017.

[5] T. Runge and et al. Measurement of plasma pressure in a narrow gap for different surge currents. In *Proc. International Conference on Gas Discharges and their Applications*, Nagoya, Japan, 2016.

[6] A. D'Angola, G. Colonna, C. Gorse, and M. Capitelli. Thermodynamic and transport properties in equilibrium air plasmas in a wide pressure and temperature range. *Eur. Phys. J. D*, 46(1):129–150, 2008. doi:10.1140/epjd/e2007-00305-4.

[7] Q. Ma, M. Z. Rong, A. B. Murphy, Y. Wu, and T. J. Xu. Simulation Study of the Influence of Wall Ablation on Arc Behavior in a Low-Voltage Circuit Breaker. *IEEE Trans. Plasma Sci.*, 37(1):261–269, 2009. doi:10.1109/TPS.2008.2007733.

[8] P. André, L. Brunet, W. Bussiere, J. Caillard, J. M. Lombard, and J. P. Picard. Transport coefficients of plasmas consisting of insulator vapours - Application to PE, POM, PMMA PA66 and PC. *Eur. Phys. J.-Appl. Phys.*, 25(3):169–182, 2004. doi:10.1051/epjap:2004007.

[9] R. L. Kurucz and B. Bell. Atomic Line Data. Kurucz CD-ROM No. 23. [online], Smithsonian Astrophyical Observatory, 1995. URL: http://www.cfa.harvard.edu/amp/ampdata/kurucz23/sekur.html.

doi:10.14311/ppt.2017.2.112
Plasma Physics and Technology **4**(2):112–115, 2017

ADVANCED APPROACH FOR RADIATION TRANSPORT DESCRIPTION IN 3D COLLISIONAL-RADIATIVE MODELS

D. Kalanov[a], Yu. B. Golubovskii[a], D. Uhrlandt[b], S. Gortschakow[b,*]

[a] *Saint Petersburg State University, 7/9 Universitetskaya nab., 199034 St. Petersburg, Russia*
[b] *Leibniz Institute for Plasma Science and Technology (INP Greifswald), Felix-Hausdorff-Str. 2, 17489 Greifswald, Germany*
[*] sergey.gortschakow@inp-greifswald.de

Abstract. The description of radiation transport phenomena in the frames of collisional-radiative models requires the solution of Holstein-Biberman equation. An advanced solutuion method for 3D plasma obejcts is proposed. The method is applicable for various line contours in a wide range of absorption coefficients. Developed approach is based on discretization of the arbitrary plasma volume on a Cartesian voxel grid. Transport of photons between the cells is computed using the ray traversal algorithm by Amanatides [1]. Solution of the particle balance equations with computed in advance radiative transfer matrix is demonstrated for various typical arc shapes, like e.g. free-burning arc and cylindric arc. Results are compared with corresponding calculations using previously developed approaches. As the method is suited for finite geometries and allows for a strict solution of the radiation transport equation, applicability ranges of previous approximations can be specified.

Keywords: radiation transport, resonance radiation, plasma modelling, arcs.

1. Introduction

Trapping of line radiation is a fundamental problem of gas discharge physics. Over the years a Holstein's approximation of effective transition probability [2] was used to evaluate the influence of radiation trapping. This approximation describes the photon escape taking absorption into account rather roughly and fails to describe the correct spatial redistribution of resonance atoms. The description of radiation transport phenomena in the frames of collisional-radiative models requires the solution of Holstein-Biberman equation [3]. Various solution methods of this equation were proposed over the years (see, for example, [3, 4]). One of promising solution approaches is the matrix method [3]. A kernel of the integral operator is represented as a matrix, which can be successfully coupled with self-consistent plasma models [5, 6]. Despite the low computational cost and simple integration into plasma-chemical schemes, this method, as well as an another matrix approach [7], has certain limits.

The main drawback is related to the plasma domain geometry. Depending on the shape of a plasma source, a discretization scheme should be specified. In mentioned appoaches the scheme was chosen based on the source symmetry (plane-parallel, infinite and finite cylinder). An analytical conversion of the kernel to the convenient form was performed, where each type of symmetry had a unique kernel representation. Finite domains require extensive mathematical treatment. Furthermore, the case of an asymmetric domains was beyond the scope of the method. In addition, inhomogeneity of absorption spatial profile significantly complicated the derivation of suitable radiation transport matrix.

In present work an advanced matrix method is described. It provides a solution approach for the Holstein-Biberman equation on a uniform cartesian grid, which is universal for various 3D domains. The method allows for the strict integration over the frequency having spatial integrals fully discretized. Results of computations are compared to data obtained with previously developed methods and demonstrated for typical arc plasma domains.

2. Description of the method

2.1. Basic equations

The Holstein-Biberman equation for the case of stationary plasma reads

$$W(\vec{r}) = N(\vec{r}) \cdot A - \int_V N(\vec{r}\,') \cdot A \cdot K(\vec{r}, \vec{r}\,') \mathrm{d}^3 r'. \quad (1)$$

Here $N(\vec{r})$ is the population of the excited level (density of excited atoms), $W(\vec{r})$ represents the excitation source, and A is the spontaneous emisson probability. The kernel $K(\vec{r}, \vec{r}\,')$ describes the absorption in the point $\vec{r}$ of photons emitted in the point $\vec{r}\,'$:

$$K(\vec{r}, \vec{r}\,') = \frac{1}{4\pi} \int_0^\infty \mathrm{d}\nu \frac{\varepsilon_\nu(\vec{r}\,')\kappa_\nu(\vec{r})}{|\vec{r} - \vec{r}\,'|^2} \exp\left(- \int_{\vec{r}\,'}^{\vec{r}} \kappa_\nu(\xi) \mathrm{d}\xi \right). \quad (2)$$

Under the assumption of complete frequency redistribution the line profiles of emission and absorption

ε_ν,κ_ν fulfil the criteria

$$\int_0^\infty \varepsilon_\nu \mathrm{d}\nu = 1, \qquad \kappa_\nu = \kappa_0 \frac{\varepsilon_\nu}{\varepsilon_0}.$$

Here ε_0 and κ_0 are the values in the line center.

For the sake of simplicity further derivation will be performed for the case of homogeneous absorption over the volume. Therefore, the kernel becomes the representation

$$K(\vec{r},\vec{r}\,') = \frac{1}{4\pi}\int_0^\infty \mathrm{d}\nu \frac{\varepsilon_\nu \kappa_\nu}{|\vec{r}-\vec{r}\,'|^2} \exp\left(-\kappa_\nu |\vec{r}-\vec{r}\,'|\right). \tag{3}$$

To make the problem discrete the integral operator in (1) will be replaced by a matrix dividing the entire volume into a number af small finite cells: $V = \sum_{i=1}^{M} \Delta V_i$, where M is a number of cells. Parameters in plasma are assumed to be constant within the subvolume. The equation (1) then becomes

$$W(r_k) = \sum_{i=1}^{M} N(r_i) a_{ik}, \qquad a_{i,k} = A \cdot \left(\delta_{i,k} - \int_{\Delta V_i} K(r_k, r')\mathrm{d}^3 r' \right). \tag{4}$$

In order to calculate the integrals in (4) a certain discretization scheme has to be chosen.

2.2. Calculation of the matrix elements

In the present approach a universal discrete scheme based on an uniform cartesian 3D-grid (or "voxel grid") is used. An extension of the approach for the case of the non-uniform grids is also possible.

To calculate the distances between the points $\vec{r}$ and $\vec{r}\,'$, a fast voxel traversal algorithm by Amanatides and Woo [1] was implemented. This algorithm casts the ray from one voxel to another and counts all voxels that have been crossed. The distance between the points $\vec{r}$ and $|\vec{r}-\vec{r}\,'| = \rho$ can be associated with indices i,k for each cell:

$$\rho_{i,k} = \Delta l \sqrt{(x_i - x_k)^2 + (y_i - y_k)^2 + (z_i - z_k)^2} \qquad i,k = 1..M. \tag{5}$$

Here x,y,z denote integer voxel coordinates, M is a total number of voxels, $\Delta l = \sqrt[3]{\Delta V}$ is a length of the voxel edge relative to the source volume. Considering the homogeneous absorption only the knowledge of i and k is necessary. In order to account for the inhomogeneity the coordinates of all crossed voxels should be used.

Introducing a dimensionless distance $\widetilde{\rho}_{i,k} = \rho_{i,k}/\Delta l$, the general form of matrix coefficient reads:

$$a_{i,k} = A \cdot \left(\delta_{i,k} - \Delta V \int_0^{\widetilde{\rho}_{i,k}} K(\widetilde{\rho})\mathrm{d}\widetilde{\rho} \right). \tag{6}$$

In the case of homogeneous absorption ($\kappa_\nu(r) = \kappa_\nu$) the integral in exponent (2) is reduced to $\kappa_\nu \rho_{i,k}$. The discretization of the kernel requires consideration of two different cases.

$i \neq k$: in this case $\rho \neq 0$ and (6) can be rewritten as

$$a_{i,k} = -A \cdot \frac{\Delta V}{4\pi} \int_0^\infty \frac{\varepsilon_\nu \kappa_\nu}{\rho_{i,k}^2} \exp\left(-\kappa_\nu \rho_{i,k}\right) \mathrm{d}\nu \tag{7}$$

For the case $i = k$: in $\rho = 0$ the integral has a singularity. It can be eliminated by the conversion to spherical coordinates considering $\rho_{i,i} \to \Delta r$:

$$a_{i,i} = A \cdot \left[1 - \frac{1}{4\pi} \int_0^\infty \left(4\pi \int_0^{\rho_{i,i}} \frac{\varepsilon_\nu \kappa_\nu}{\rho'^2} e^{-\kappa_\nu \rho'} \rho'^2 \mathrm{d}\rho' \right) \mathrm{d}\nu \right] = = A \cdot \left[1 - \int_0^\infty \left(\int_0^{\Delta r} \varepsilon_\nu \kappa_\nu e^{-\kappa_\nu r} \mathrm{d}r \right) \mathrm{d}\nu \right]; \quad \Delta r = \sqrt[3]{\frac{3\Delta V}{4\pi}}. \tag{8}$$

Finally, the matrix coefficients in the case of homogeneous absorption ($\kappa_\nu(r) = \kappa_\nu$) are given by

$$a_{i,k} = A \cdot \begin{cases} -\frac{\Delta V}{4\pi} \int_0^\infty \frac{\varepsilon_\nu \kappa_\nu}{\xi_{i,k}^2} e^{-\kappa_\nu \Delta l} \mathrm{d}\nu, & i \neq k; \\ 1 - \int_0^\infty \left(\int_0^{\Delta r} \varepsilon_\nu \kappa_\nu e^{-\kappa_\nu r} \mathrm{d}r \right) \mathrm{d}\nu, & i = k. \end{cases} \tag{9}$$

Considering the case of the Lorentzian line shape and spatially homogeneous absorption the simplified expressions for matrix coefficients can be derived to speed up the computation. The kernel can be rewritten in following form

$$K(\vec{r},\vec{r}\,') = -\frac{1}{4\pi |\vec{r}-\vec{r}\,'|^2} \left.\frac{\mathrm{d}T}{\mathrm{d}\rho}\right|_{\rho=|\vec{r}-\vec{r}\,'|}, \tag{10}$$

where $T(\rho)$ is a Biberman's transmission factor:

$$T(\rho) = \int_0^\infty \varepsilon_\nu e^{-\kappa_\nu \rho} \mathrm{d}\nu = e^{-\frac{\kappa_0 \rho}{2}} I_0\left(\frac{\kappa_0 \rho}{2}\right). \tag{11}$$

Here $I_{\alpha=0,1,2...}$ denotes the modified Bessel function of the first kind. Then, the expression for the kernel reads

$$K(\rho) = \frac{1}{4\pi\rho^2} \cdot \frac{\kappa_0}{2} e^{-\frac{\kappa_0 \rho}{2}} \left[I_0\left(\frac{\kappa_0 \rho}{2}\right) - I_1\left(\frac{\kappa_0 \rho}{2}\right) \right]. \tag{12}$$

Substituting of Eq.(12) into Eq. (6) gives

$$a_{i,k} = A \cdot \begin{cases} -\frac{\kappa_0 \Delta V}{8\pi \rho_{i,k}^2} e^{-\frac{\kappa_0 \rho_{i,k}}{2}} \left[I_0\left(\frac{\kappa_0 \rho_{i,k}}{2}\right) - I_1\left(\frac{\kappa_0 \rho_{i,k}}{2}\right) \right], \\ e^{-\frac{\kappa_0 \Delta r}{2}} I_0\left(\frac{\kappa_0 \Delta r}{2}\right), \quad i = k. \end{cases} \tag{13}$$

The benchmark tests show that the computational time for elements decreases up to a factor of five comparing to numerical integration over the line contour.

2.3. Solution algorithm

A domain with a volume V is discretized on a uniform cartesian grid using voxelizer code [8]. Characteristic linear size of the voxelized mesh is determined by the maximum number of voxels along the x-dimension n_{dim}.

The matrix coefficients are computed in two steps: calculation of the distances traveled by photons $\rho_{i,k}$ and integration of the kernel 2) for each pair of voxels. In order to optimize the calculation of distances, ray tracing algorithm [1] was modified for parallel computing using a graphical processing unit (GPU). Each thread of GPU can handle the operations only with scalars, therefore, the vectors with coordinates of each (r, r')-pair are prepared. The number of voxel pairs being simultaneously processed (the array chunk size) is determined by the size of the graphics card memory. Then, the chunk with computed distances is transferred to the CPU where the coefficients (9) are calculated and stored in a file. After the loop over all chunks the saved data are collected in the matrix which represents the coupling between all cells in plasma.

When the radiation transport matrix is ready, the system of linear equations representing the Holstein-Biberman equation can be solved for a given excitation source in the plasma domain. This equation system can be also easily included into a multicomponent collisional-radiative model, which considers the coupling with particle transport equations for other plasma species.

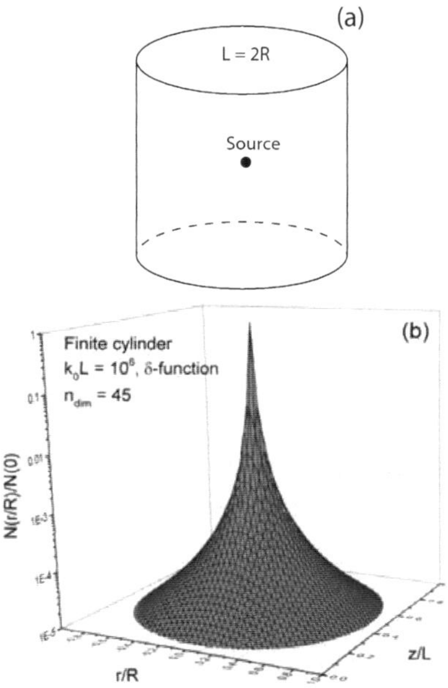

Figure 1. (a) Scheme of the finite cylinder domain with a point excitation sourcein the center, (b) Spatial distribution of the density of resonance atoms in case of point source in the finite cylinder. Blue grid - ray tracing method, orange mesh - asymptotic matrix method.

3. Results

To validate the approach a finite cylinder geometry of size $L = 2R = 1$ with a point excitation source located in the volume center was used. In previous publication [9] a similar problem was considered for the case of the high-opacity asymptotics for a Lorentzian line contour. The kernel of the radiation transport operator was integrated over cylindrical layer cells. The size of corresponding linear equation system was n_{dim}^2. The same problem is solved by ray tracing method without the use of the asymptotics. The resulting linear equation system has the size of the order of n_{dim}^3. Figure 1 demonstrates the schematic representation of the geometry as well as the solution of Eq. (4) using both methods for the case with an optical depth $\kappa_0 L = 10^6$. Linear mesh size for ray tracing method is $n_{\text{dim}} = 45$. As it becomes obvious from Fig. 1 both approaches are in an excellent agreement. Similar results have been obtained for various values of the absorption coefficent as long as the later remained high.

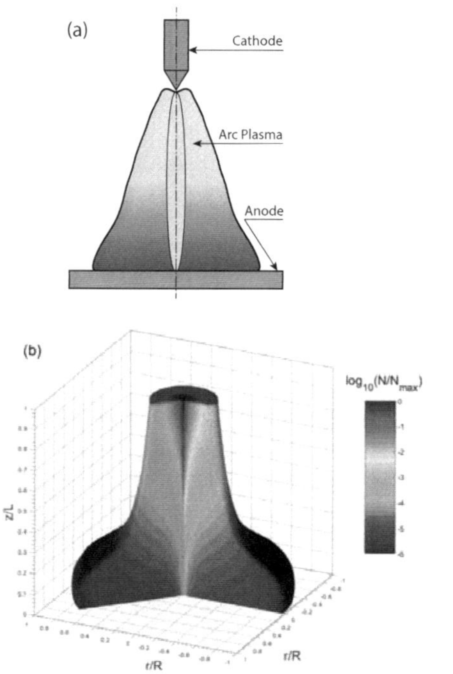

Figure 2. (a) Schematic picture of a free-burning arc. (b) Normalized solution for the case of $n_{dim} = 41$, $k_0 L = 10^6$.

The main advantage of the approach is the possibility to solve the radiation transport equation in an arbitrary 3D domain. As examples of non-trivial geometry a "bell"-like shape of a free-burning arc (welding arc) and two hemispherical electrodes (switching arc) were chosen. The size of the grid in both cases was $n_{\text{dim}} = 41$ and the optical depth was set to $\kappa_0 L = 10^6$. The excitation source is approximated by a function $W(r) = 10^{-ar}$ along discharge axis.

The first example is illustrated by a Figure 2. Figure

2a present a schematic picture of the free-burning arc. Figure 2b demonstrates the solution of the Holstein-Biberman equation in a discretized 3D-domain.

A typical cylindrical circuit breaker geometry is presented in Figure 3a. The solution is demonstrated by Figure 3b. One could see the effects of radiation screening by electrodes. To include these effects one needs to exclude the coupling between unit volumes where photons cannot travel along the straight line. For this purpose, the check is performed whether the ray trace only allowed voxels or not.

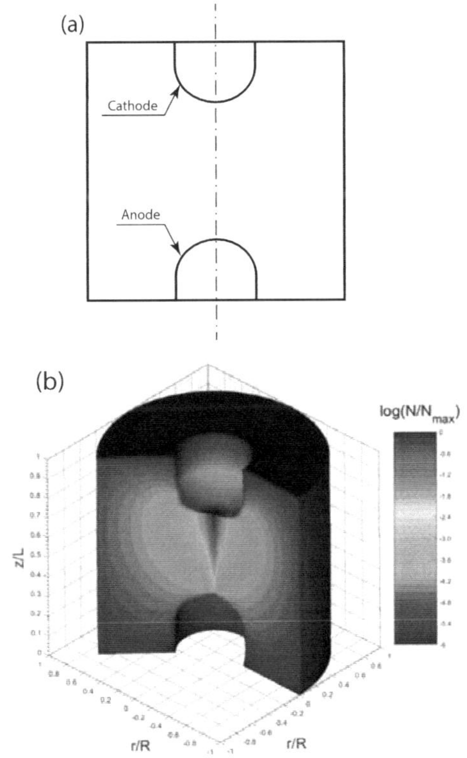

Figure 3. (a) Schematic picture of a switching arc geometry. (b) Normalized solution for the case of $n_{dim} = 41$, $k_0 L = 10^6$.

4. Conclusions

In the present work an advanced matrix method based on the ray tracing algorithm for the solution the Holstein-Biberman equation is proposed. The method is applicable for arbitrary 3D-domains discretized on a uniform Cartezian mesh. The transformation of the radiation transport operator was performed for the Lorentzian line shape and high opacity. The method can be extended for a wide range of absorption coefficients and is in general not dependent on the line shape. In the case of Lorentzian line shape and high opacity limit the matrix coefficients can be simplified to reduce the computation time.

The ability to solve the radiation transport equation in complex domains is illustrated by two examples: free-burning welding arc and cylindrical switching arc. The results demonstrate expected qualitative distribution of the radiation known from the experiments. Developed approach is compared with the previous matrix method for the case of a finite cylinder geometry with a point excitation source located in the center. The results of both approaches are in excellent agreement for the case of large absorption coefficients.

The radiation transport matrices computed with the ray tracing method can be successfully integrated into self-consisent multicomponent plasma models of various complexity.

Acknowledgements

Authors gratefully acknowledge computational resources provided by Computer Center of Saint-Petersburg State University. The work of D.K. was also supported by German-Russian Interdisciplinary Science Center (G-RISC) P-2016b-17.

References

[1] J. Amanatides and A. Woo. A Fast Voxel Traversal Algorithm for Ray Tracing. *Eurographics*, 87(3):3–10, 1987. doi:10.1.1.42.3443.

[2] T. Holstein. Imprisonment of resonance radiation in gases. *Physical Review*, 72(12):1212–1233, 1947. doi:10.1103/PhysRev.72.1212.

[3] Yu.B. Golubovskii, S. Gorchakov, and D. Uhrlandt. Transport mechanisms of metastable and resonance atoms in a gas discharge plasma. *Plasma Sources Science and Technology*, 22(2):32pp, 2013. doi:10.1088/0963-0252/22/2/023001.

[4] A.F. Molisch and B.P. Oehry. *Radiation trapping in atomic vapours*. Oxford University Press, 1998.

[5] Yu.B. Golubovskii and V.A. Maiorov. The influence of resonance radiation transport on the contraction of a glow discharge in argon. *Plasma Sources Science and Technology*, 24(2):25027, 2015. doi:10.1088/0963-0252/24/2/025027.

[6] Yu.B. Golubovskii, D. Kalanov, S. Gortschakow, M. Baeva, and D. Uhrlandt. Excited atoms in the free-burning Ar arc: treatment of the resonance radiation. *Journal of Physics D: Applied Physics*, 49(47):475202, 2016. doi:10.1088/0022-3727/49/47/475202.

[7] J.E. Lawler, G.J. Parker, and W.N.G. Hitchon. Radiation trapping simulations using the propagator function method. *Journal of Quantitative Spectroscopy and Radiative Transfer*, 49(6):627–638, 1993. doi:10.1016/0022-4073(93)90006-4.

[8] D. Morris and K. Salisbury. Automatic preparation, calibration, and simulation of deformable objects. *Computer methods in biomechanics and biomedical engineering*, 11(3):263–279, 2008. doi:10.1080/10255840701769606.

[9] Yu.B. Golubovskii, A. Timofeev, S. Gorchakov, D. Loffhagen, and D. Uhrlandt. Population of resonance and metastable atoms in a cylindrical volume of finite size. *Physical Review E - Statistical, Nonlinear, and Soft Matter Physics*, 79(3):25–27, 2009. doi:10.1103/PhysRevE.79.036409.

doi:10.14311/ppt.2017.2.116
Plasma Physics and Technology **4**(2):116–119, 2017

CALCULATION OF PRESSURE RISE AND ENERGY OF HOT GASES DUE TO HIGH ENERGY ARCING FAULTS IN THE METAL-CLAD SWITCHGEAR

M. Iwata[a,*], T. Miyagi[a], Y. Goda[a], K. Shirai[b]

[a] *Central Research Institute of Electric Power Industry (CRIEPI), Electric Power Engineering Research Laboratory, 2-6-1 Nagasaka, Yokosuka-shi, Kanagawa-ken 240-0196, Japan*

[b] *Central Research Institute of Electric Power Industry (CRIEPI), Nuclear Fuel Cycle Backend Research Center, 1646 Abiko, Abiko-shi, Chiba-ken 270-1194, Japan*

[*] m-iwata@criepi.denken.or.jp

Abstract. This paper presents the 3-D CFD calculation results of the pressure rise due to the High Energy Arcing Faults (HEAFs) in the metal-clad switchgears. The calculations were performed considering the came-off of the roof panel that was observed in the arc tests. The calculated pressure development approximately agreed with the measured one. Furthermore, the energy of hot gases exhausted from the broken roof panel was calculated to investigate the thermal effect of hot gases.

Keywords: switchgear, internal arc, pressure rise, energy of hot gases.

1. Introduction

High Energy Arcing Faults (HEAFs) can be caused by the failure within electrical installations due to either a defect or an exceptional service condition or maloperation [1]. The HEAFs lead to the temperature rise and the pressure rise in the installations and then the exhausting of hot gases from the installations. As a result, the successive fire incident owing to the ignition of flammable materials may occur inside and/or outside the installations.

In Japan, the Tohoku District-off the Pacific Ocean Earthquake attacked the Onagawa Nuclear Power Station of Tohoku Electric Power Co., Inc. on March 11, 2011 and the successive fire incident due to the HEAFs occurred in the electrical cabinet. In order to clarify the pressure rise in the non-arc proof electrical cabinets due to HEAFs and understand the critical condition for the successive fire occurrence in the cabinets, a series of 3-phase internal arc tests with 6.9 kV switchgears were performed [2]. In the tests, the arc current I_{arc} was around 20 kA, the arcing duration tarc was 0.1–2 s, and the electric arc energy W_{arc} was 3–45 MJ.

In this paper, the calculations of the pressure rise due to the HEAF in the switchgear were carried out using a CFD (Computational Fluid Dynamics) method. The calculated pressure rise was compared with the measured one. The propagation of the hot gases in the switchgear and the energy of the hot gases exhausted from the switchgear through the broken roof panel were also investigated.

2. Calculation method and conditions

In the 3-D calculations, a CFD program (CFD-ACE+) was applied to solve the full set of basic hydrodynamic equations (i.e., conservation equations of mass, momentum and energy). Those conservation equations in terms of partial differential equations were solved at each location in the switchgear iteratively, while the energy input, a portion of the arc energy, was set locally. The general form of the conservation equations can be written as:

$$\frac{\partial}{\partial t}(\rho\,\phi) + \nabla\cdot(\rho\,\mathbf{v}\,\phi) = \nabla\cdot(\Gamma\,\nabla\,\phi) + S \qquad (1)$$

where ϕ represents the conserved variables under consideration, ρ the gas density, $\mathbf{v}$ the gas velocity, Γ the transport coefficient of ϕ and S the source term for ϕ [3].

Figure 1 shows the schematic drawing of the switchgear used in the tests. The switchgear has two cable compartments (CCs), a bus compartment (BC) and two circuit-breaker compartments (CBCs). The compartments are separated by the metal plates. For the ventilation in the switchgear, the metal plates of the rear wall as well as the metal plates between the CC and BC have some rectangular slits. Figure 2 shows the sketch of the upper CC considered in the calculations. The volume of the CC was 1.31 m^3. In order to simplify the calculations, the pressure and the temperature of the outside of the upper CC were 0.1 MPa and 293 K, respectively. As the arc was ignited in the upper CC, the energy input W_{input} was applied locally in the CC. W_{input} was given as k_p times W_{arc}. k_p is a fraction of the W_{arc} leading to the pressure rise. k_p changes depending on the arc current and the arcing electrode material, e.g. k_p decreases from 0.6 to 0.53 gradually with increasing arc current from 1 kA to 12.5 kA in the case of copper electrodes [4]. The values of k_p estimated in other publication [5] is in the same range as those mentioned above. When estimating the values of k_p, the energy balance

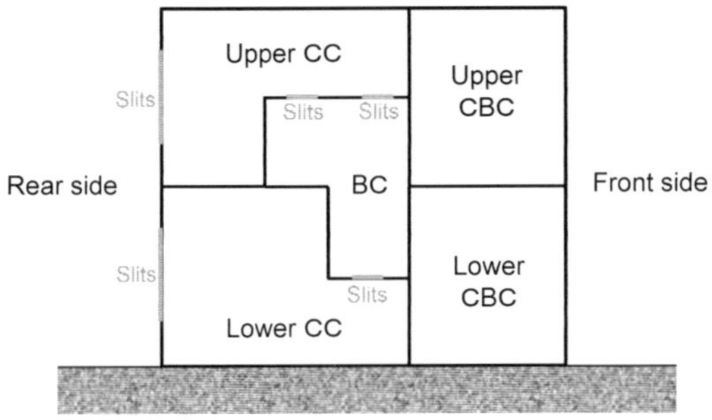

Figure 1. Lateral view of the compartments in the switchgear [8]. [CC: Cable compartment, BC: Bus compartment, CBC: Circuit-breaker compartment]

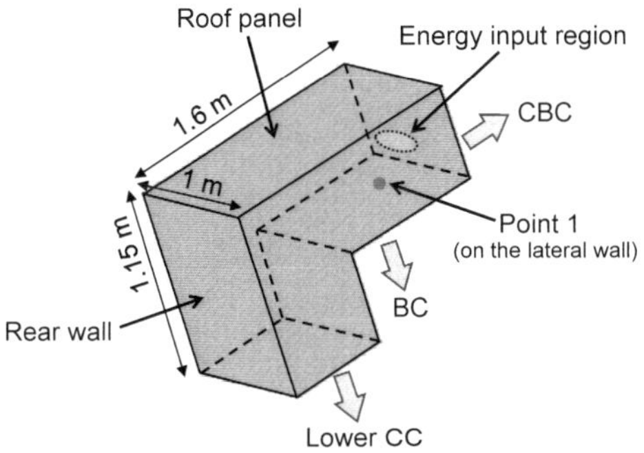

Figure 2. Geometrical arrangement considered in the calculations (upper CC). [8]

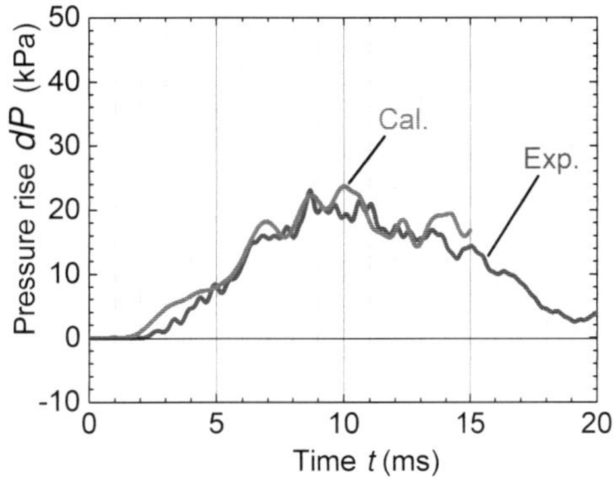

Figure 3. Calculation and experimental pressure developments at point 1 in the upper CC.

in the closed container was investigated [4–7], where the "input" energy in the container was the electric arc energy and the energy of chemical reactions like oxidation of electrode vapor in the case of exothermal reactions. Conversely, the output energy in the container was the energy causing pressure rise in the container and the energy losses due to arc radiation energy and the energy portion for electrode melting and evaporation.

The pressure rise in the CC was measured at point 1 in the test. The gas inside was air with an initial pressure and temperature of 0.1 MPa and 293 K, respectively. I_{arc} was 21 kA, t_{arc} 0.1 s (100 ms), W_{arc} 3.1 MJ, the current frequency 50 Hz. k_p was taken as 0.53, because the arcing electrode material was copper in the tests, and k_p did not change so much when the arc current increased from 4 kA to 12.5 kA [4]. The metal plate with the slits was simply considered as a medium of gas flow resistance and of heat resistance/capacitance in the same way as in [9]. The walls of the CC were assumed to be adiabatic, because the arcing duration was short (0.1 s), and the CC was embedded in ambient air, which was a good heat-insulating medium.

During the test, the tightening bolts of the roof panel of the upper CC broke owing to the pressure rise in the CC and the roof panel came off. In the calculations, it was assumed that a part of the roof panel began to open when the pressure rise reached about 5 kPa, which was estimated using the ultimate tensile strength (420 N/mm^2) and the cross-sectional area (14.2 mm^2) of the bolt, and the area of the roof panel. In our previous paper [8], as a first step, the opened area of the roof panel was constant (0.064 m^2). In this paper, the opened area increased gradually (0.036–0.295 m^2) with time by considering the motion of the roof panel. The opened area was estimated by using the following method. Supposing that the roof panel is rigid and the roof panel comes off with the rear eyebolts as the supporting points, the motion of the roof panel due to the internal pressure rise in the upper CC was calculated by using the motion equation. The gap between the roof panel and the upper CC was calculated, and then the opened area of the roof panel was estimated by using the gap area.

3. Calculation results

3.1. Pressure rise

Figure 3 shows the calculated and measured pressure developments of point 1 in the CC. The measured pressure rise dP increased and reached the maximum value of about 23 kPa at time t of about 8.7 ms, and dP showed a tendency to decrease although the arc continued to burn in the CC. In this paper, the calculation was carried out until $t = 15$ ms, where dP decreased gradually. The calculated pressure development agreed well with the measured one.

Figure 4 shows the 3-D pressure distribution in the CC. With time t, dP near the energy input region increased and the pressure wave propagated to another region, e.g. the rear of the CC. Also, the hot gases in the CC went out mainly through the opened area of

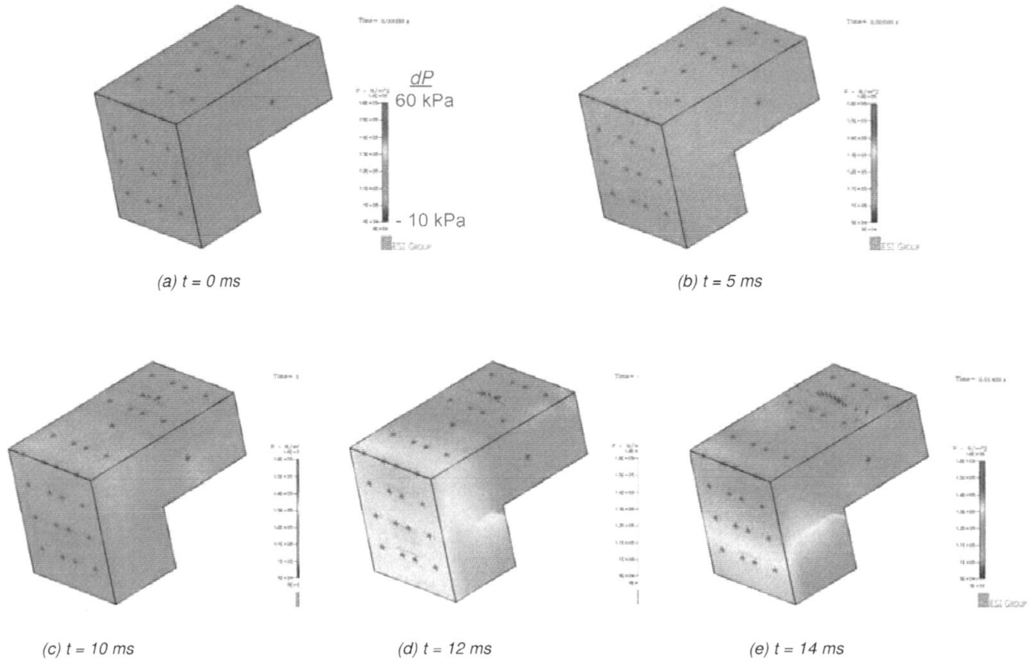

Figure 4. Time dependence of 3-D pressure distribution in the upper CC.

the roof panel after t of about 4.4 ms when dP of the roof panel reached about 5 kPa.

3.2. Energy of hot gases exhausted from roof panel

When investigating the thermal effect on personnel and flammable materials near the electrical equipment in which fault arcs occur, it is necessary to estimate the energy of hot gases exhausted from the equipment. In this section, the energy flow density EFD of hot gases exhausted from the opened area of the roof panel was calculated using the same method as in [10]. The energy flow density EFD of hot gases is given by

$$EFD = H_{\mathrm{ex}}\,\rho_{\mathrm{ex}}\,v_{\mathrm{ex}} \tag{2}$$

where H_{ex}, ρ_{ex}, and v_{ex} are the specific enthalpy, gas density, and velocity of the gas at the exit of the opened area of the roof panel, respectively.

Figure 5 shows the energy flow density EFD of hot gases exhausted from the broken roof panel. EFD increased rapidly at time t of about 4.4 ms when the roof panel began to open owing to the pressure rise in the CC.

Figure 6 shows the measured arc energy W_{arc} and the calculated energy W_{ex} of the exhausted hot gases. The energy W_{ex} increased gradually because both EFD and the opened area increased with time.

Figure 7 shows the ratio k_{ex} of W_{ex} to W_{arc}. The ratio k_{ex} increased approximately in proportion to the opened area A_{op} of the roof panel. This result suggests that the ratio k_{ex} can rise more as the opened area increases. It is necessary to consider the development

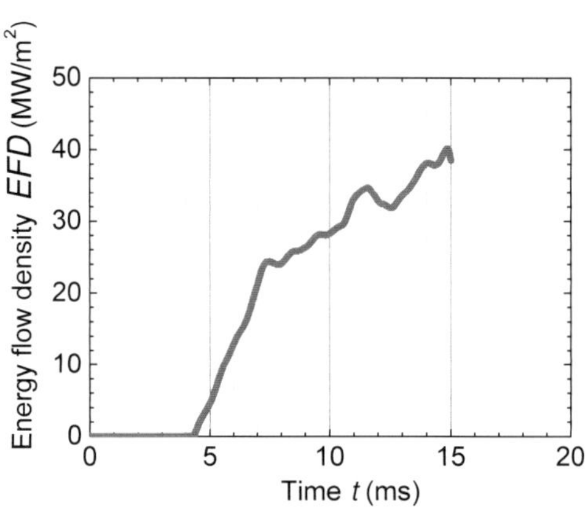

Figure 5. Calculated energy flow density of hot gases exhausted from the opened area of roof panel.

of the energy of hot gases exhausted from electrical equipment when investigating the thermal effect of hot gases on personnel and flammable materials near the equipment.

4. Conclusions

This paper described the 3-D CFD calculation results regarding the pressure rise due to HEAFs in the metal-clad switchgears and the energy of hot gases exhausted from the switchgears thorough the broken roof panel. The energy input took place locally in the upper cable compartment of the switchgear. The gas inside was air with an initial pressure of 0.1 MPa. The arc current I_{arc} was 21 kA, the t_{arc} 0.1 s (100 ms),

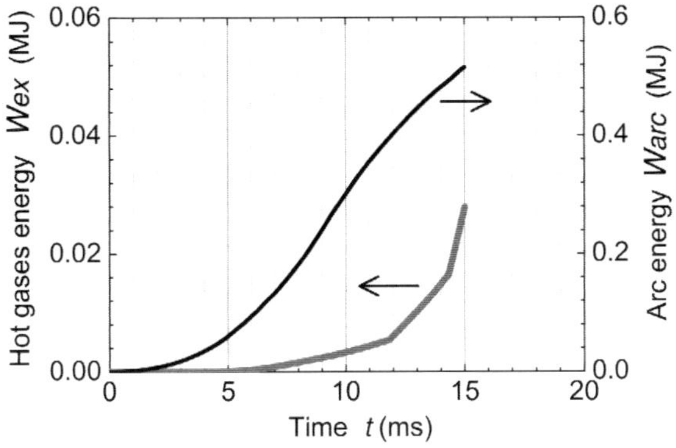

Figure 6. Calculated energy of the exhausted hot gases and measured arc energy.

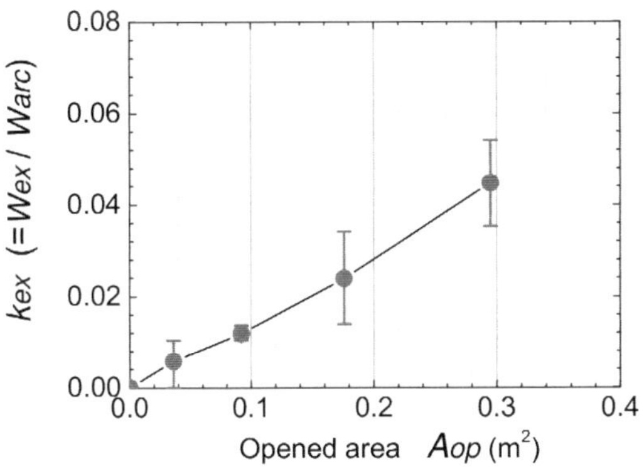

Figure 7. Dependence of the ratio $k_{ex} = (W_{ex}/W_{arc})$ on the opened area of roof panel.

W_{arc} 3.1 MJ. The calculations were performed considering the propagation of the hot gases from the cable compartment to the outside and the came-off of the compartment roof panel that was observed in the test. As a result, the calculated pressure developments of the cable compartment agreed well with the test results. In addition, the calculated energy of hot gases exhausted from the opened area of the roof panel increased gradually because both the energy flow density of hot gases and the opened area increased with time. The above properties of hot gases must be considered when investigating the thermal effect of hot gases on personnel and flammable materials near the switchgears.

References

[1] IEC Std.62271-200. *AC metal-enclosed switchgear and controlgear for rated voltages above 1 kV and up to and including 52 kV*, 2011. `https://archive.org/details/gov.in.is.iec.62271.200.2003`.

[2] K. Shirai, Y. Suzuki, Y. Goda, M. Iwata, and T. Miyagi. High energy arcing faults fire tests using high voltage metal-clad switchgear components (1) preliminary tests for fire evaluation. In *2013 Fall Meeting of the Atomic Energy Society of Japan*, 2013.

[3] S.V. Patankar. *Numerical Heat Transfer and Fluid Flow*. Hemisphere Publishing Corp., 1980.

[4] M. Iwata, K. Anantavanich, and G. Pietsch. Influence of arc current on fraction kp of electric arc energy leading to pressure rise in a closed container. In *Proc. 17th Int. Conf. on Gas Discharges and their Applications*, pages 189–192, 2008.

[5] X. Zhang, G. Pietsch, and E. Gockenbach. Investigation of thermal transfer coefficient by the energy balance of fault arcs in electrical installations. *IEEE Trans. Power Delivery*, 21(1):425–431, 2006. `doi:10.1109/TPWRD.2005.852274`.

[6] T. Miyagi, S. Tanaka, M. Iwata, T. Tadokoro, T. Ohtaka, and T. Amakawa. Influence of arc current on energy balance due to high current arc in a closed chamber. *IEEJ Trans. Power and Energy*, 128(12):1561–1568, 2008. `doi:10.1541/ieejpes.130.232`.

[7] S. Tanaka, T. Miyagi, T. Ohtaka, M. Iwata, T. Amakawa, and Y. Goda. Influence of electrode material on pressure-rise due to arc in a closed chamber. *IEEJ Trans. Power and Energy*, 130(2):232–238, 2010. `doi:10.1541/ieejpes.128.1561`.

[8] M. Iwata, T. Miyagi, Y. Goda, Y. Suzuki, and K. Shirai. Numerical approach to pressure rise due to short-circuit fault arc in the metal-clad switchgears. In *Proc. 20th Int. Conf. on Gas Discharges and their Applications*, pages 171–174, 2014.

[9] M. Iwata, S. Tanaka, T. Miyagi, T. Amakawa, and G.J. Pietsch. Simulation of pressure rise due to fault arc in a closed container with a perforated metal plate. In *Proc. IEEJ Technical Meeting*, pages ED–11–70/ SA–11–43/ SP–11–16, 2011.

[10] M. Iwata, S. Tanaka, T. Miyagi, T. Amakawa, and G.J. Pietsch. Influence of perforated metal plate on pressure rise and energy flow due to internal arcing in a container with a pressure-relief opening. *IEEE Trans. Power Delivery*, 29(3):1292–1300, 2014. `doi:10.1109/TPWRD.2014.2305161`.

doi:10.14311/ppt.2017.2.120
Plasma Physics and Technology 4(2):120–123, 2017

EXPERIMENTAL STAND FOR INVESTIGATIONS OF INSULATOR DEGRADATION AND ELECTRODE EROSION IN HIGH-CURRENT BREAKER

A.V. Budin[a], M.E. Pinchuk[a,b,*], V.V. Leontev[a], A.G. Leks[a], N.K. Kurakina[a,b], A.A. Kiselev[a], J.V. Simakova[b,c], V.Ya. Frolov[b]

[a] *Institute for Electrophysics and Electric Power of Russian Academy of Sciences, Dvortsovaya nab., 18, St.-Petersburg, 191186, Russia*
[b] *Peter the Great Saint-Petersburg Polytechnic University, Polytechnicheskaya 29, St.-Petersburg 195251, Russia*
[c] *NE Open Joint Stock Company, Partizanskaya , 21, St.-Petersburg, 195248, Russia*
[*] pinchme@mail.ru

Abstract. An experimental stand for studies of electric arc, electrode erosion and insulator degradation processes in high-current circuit breakers and some preliminary experimental data is described. The setup includes a discharge chamber, a capacitive energy storage with capacitance of 0.11 F, voltage up to 10 kV, and all necessary diagnostic techniques. The stand is designed for modeling current pulse with amplitude of 3–150 kA and duration of the first half period of 1.0–3.0 ms during the process of disconnecting the ring and the pin contacts. The arc is cooled by transverse gas blowing at pressure in the chamber of 0.5–3 MPa. Acquired experimental data can be used for verification of the modelling results of the heat transfer processes in the discharge chamber. At the stand, advanced composite materials based on carbon and iron-copper pseudoalloy are studied.

Keywords: high-current high-pressure gas discharge, electrode erosion, insulator degradation, high-voltage test techniques, high-current circuit breaker, high-voltage engineering.

1. Introduction

Design and development of high-voltage high-current gas circuit breakers and new technologies for manufacturing their contact systems is rather uneasy and expensive problem. Now instead of traditional technologies there come new ones, including the creation of new composite materials, such as laser weld deposition [1, 2], and plasma technologies [3, 4].

In the present work, the description of the experimental stand with a modelling discharge chamber intended for research of arc processes, electrode erosion and degradations of insulation materials is presented. The stand has been created in addition to the available one [5] for research and test of lightning protection devices in air power lines [6] and for continuation the research of erosive characteristics of new perspective composite materials [7, 8].

2. Discharge chamber

The experimental setup is intended for studies on blown arcs with current amplitude up to 300 kA and pulse pressure up to 1 MPa. In the design of the discharge chamber flexible change of the configuration is incorporated. So, for example, an puffer-type electrode unit or an transverse gas blowing through an arc are stipulated.

Figure 1 shows the variant of the discharge chamber with a transverse gas blowing.

The housing of the chamber (1) is a thick-walled vessel made of the stainless steel, inside it there is a mobile acetal resin piston (2). The anode holder (3) is mounted on the piston. The studied material sample is fixed on anode end (4). The opposite end of the anode holder is connected with the ground conductor line by means of flexible copper wires laid inside of polycarbonate tube (5). The orificed cathode (6) is fasten on the current input (7), isolated from the chamber case and connected with the high voltage conductor line.

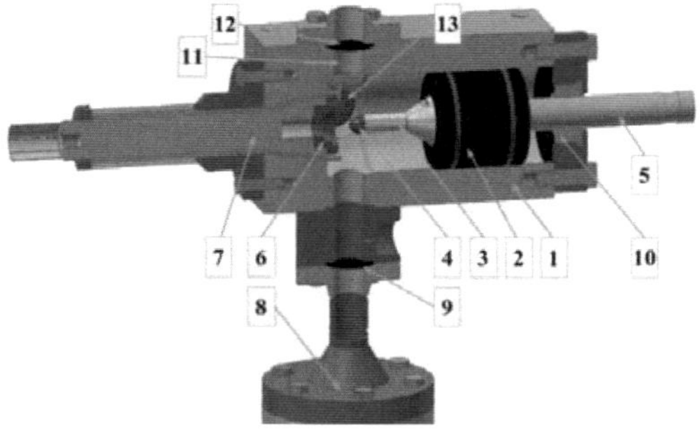

Figure 1. Discharge chamber (all designations in the text of article).

Movement of the piston occurs by means of compressed gas which is pumped into a cylindrical volume (8) and enters the chamber volume after the break of the diaphragm (9). The cavity behind the piston also plays the role of a shock-absorber since air is squeezed out from it through a narrow slit between the tube (5) and the fluoroplastic cartridge (10).

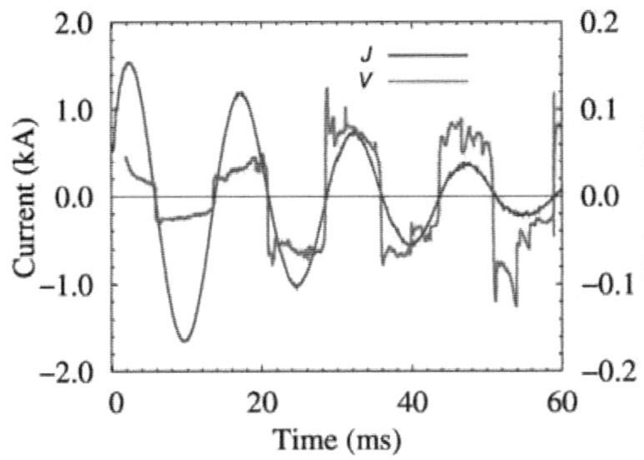

Figure 2. Current (J) and voltage across arc gap (V) at frequency of 75 Hz.

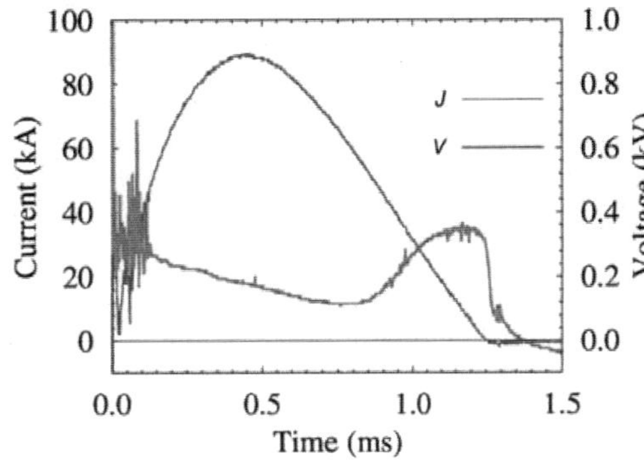

Figure 3. Current (J) and voltage across arc gap (V) for high current amplitudes.

Coaxially with an inlet aperture for gas puffing, the exhaust outlet (11) also closed by a diaphragm (12) is placed, which is allowed to limit the maximum pressure in the chamber volume and to model an arc blowout at transverse gas blowing through an arc.

Perpendicularly to the chamber axes, two coaxial diagnostic windows (13) are located, through which optical diagnostics is performed. The construction of the windows allows for mounting of various materials for test of thermal and radiative arc action. Specially constructed channel in the chamber housing is used for pulse pressure measurements in the discharge volume.

The energy source is the modular capacitive system with maximal storage energy of 6 MJ, maximal charging voltage of 10 kV and capacity of 0.11 F. The storage system is divided into six identical independent modules which can be switched on a load both simultaneously, and independently with delays. The commutation of each module is carried out by a triggered air switch (trigatron). The current is measured by Rogovskiy coil. For the measurement of the voltage drop on an arc a high-ohmic resistive divider with a transformer decoupling is used, which high-voltage end is connected to the current input (7 in figure 1), and the low-voltage end – to a junction of flexible copper wires going from the anode with the ground conductor line.

The study of arc quenching is feasible in several current modes.

At current amplitude up to 10 kA, it is possible to pass up to 20 half-cycles of a current with frequency of 75 Hz through electrode system (figure 2). Higher current amplitudes (up to 300 kA) correspond to a current mode in one or two half-cycles of the discharge (figure 3) with a half-cycle duration of 1–3 ms. The duration depends mainly on number of used modules and connected inductive reactor. During the first current half-cycle of the current (1–3 ms), the contacts move away on the distance of 3–4 cm.

The chamber construction is described in more details in [9].

3. Available diagnostics

At the stand, various high-speed diagnostic methods are presented, which are now used for detailed studies of arc processes. The scheme of the diagnostic equipment arrangement is shown in figure 4. Additional details about the diagnostic methods and equipment can be found in [10, 11].

For a pulse pressure measurements in the discharge volume the membrane piezoelectric gauge T-500 is used. It allows to register a pressure magnitude during all discharge process with a time resolution of $\sim 5\,\mu s$.

To study temperature conditions of electrodes, two identical high-speed monochromatic pyrometers with calibrated semiconductor sensors and a time resolution of $\sim 1\,\mu s$ are used. In figure 5, the brightness temperature of the cathode vs time is presented for current amplitude of 40 kA and electrodes made of copper-chromium 50/50 alloy.

For a high-speed dual angle photographing of the arc burning and the piston movement, a 9-frame ICCD camera K011 (BIFO Moscow, frame exposition up to 100 ns) and a 16-frame rotatory mirror CCD camera CORDIN-530 (USA, frame exposition up to 0.2 µs) are used. These cameras allow to carry out high-speed imaging of the processes by a duration up to several milliseconds with a high temporal and spatial resolution. In figure 6, the frames of a high-speed imaging of an arc with current amplitude of 90 kA between the disconnecting copper electrodes are shown.

The direct shadow method is realized with a probing of the discharge volume by a continuous laser expanded beam. The shadow images is registered by a camera CORDIN-530. A schlieren technique configuration is also possible.

Spectral measurements are carried out with the imaging spectrograph LOT-Oriel MS257 and CCD-camera with a mask DU-420-UV-FK (UK, Andor Tech.). On the CCD-matrix the metal mask [12] is put, that allows to register several consecutive spectra with a time resolution up to several microseconds.

In figure 7, the spectra of the arc radiation in three consecutive periods of the current with amplitude of

Figure 4. The scheme of the diagnostic arrangement on the experimental stand: 1 – discharge chamber; 2 – high-speed pirometer; 3 – high-speed ICCD camera Bifo K-11; 4 – high-speed rotatory mirror CCD camera Cordin-530; 5 – imaging spectrograph LOT-Oriel MS257i with masked CCD Andor DU-420-UV-FK; 6 – collimating optics and 7 – laser for shadow and schlieren techniques.

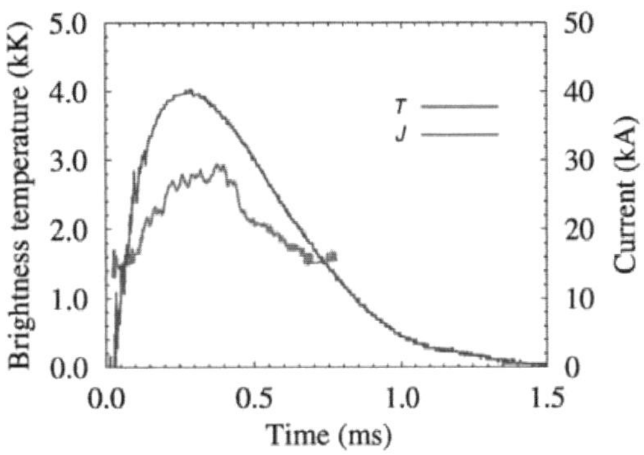

Figure 5. Brightness temperature of the cathode (T) and current (T).

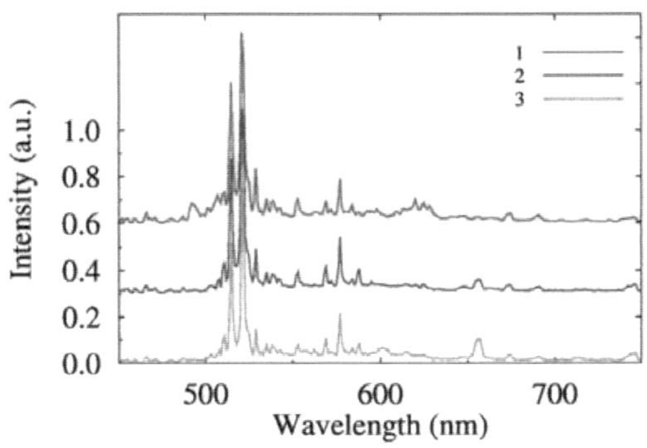

Figure 7. Arc spectrum for current amplitude of 2 kA and frequency of 75 Hz: 1 – first half-cycle of the current, 2 – second, 3 – third.

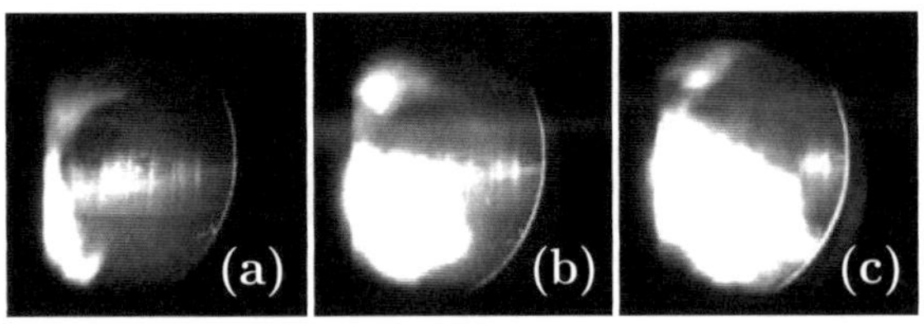

Figure 6. High-speed frames of arc with current amplitude of 90 kA between divergent copper electrodes in time moments of 500 *(a),* 550 *(b) and* 600 μs *(c) after discharge ignition.*

2 kA between electrodes with copper-graphite covering deposited by means of plasma sputtering [3] and fluoroplastic sample installed in the field of discharge action are shown. A change in the spectrum corresponds to a change of the structure and parameters of the arc plasma during the discharge development.

4. Conclusions

The experimental installation for a physical modelling and study of processes in high-voltage high-current gas breaker is created. At the stand, the complex of various methods of high-speed diagnostics for detailed investigations of discharge processes is realized.

Acknowledgements

This work was supported in part by the Russian Foundation for Basic Research, Project Nos. 16-08-00767a.

References

[1] I. V. Shishkovsky, I. A. Yadroitsev, and I. Yu. Smurov. Manufacturing three-dimensional nickel titanium articles using layer-by-layer laser-melting technology. *Technical Physics Letters*, 39(12):1081–1084, 2013. doi:10.1134/S1063785013120250.

[2] V.V. Vasil'tsov, M.G. Galushkin, I.N. Ilichev, A.I. Misurov, and V.YA. Panchenko. Layer-wise

metal-powder laser surfacing: Analytical theory and experiment. *Engineering Journal: Science and Innovations*, 2012. doi:10.18698/2308-6033-2012-6-227.

[3] V.Ya. Frolov, G.K. Petrov, B.A. Ushin, D.V. Ivanov, and S.G. Zverev. Air-plasma technologies of spraying of coatings. In *Plasma Physics and Plasma Technology VIII International Conference. Minsk, Belarus, September 17-21, 2012*, pages 608–611, 2012. URL: http://ifanbel.bas-net.by/pppt-7/Proceedings.rar.

[4] V. Frolov, D. Ivanov, and M. Shibaev. Plasma installation for TiO_2 fine powder production: Mathematical simulation. *Plasma Physics and Technology*, 3:57–61, 2016.

[5] A. V. Budin, M. E. Pinchuk, V. E. Pilschikov, A. G. Leks, and V. V. Leont'ev. An experimental stand for investigating protective devices for high-voltage overhead lines. *Instruments and Experimental Techniques*, 59(5):673–677, 2016. doi:10.1134/S0020441216040163.

[6] M. E. Pinchuk, A. V. Budin, I. I. Kumkova, and A. N. Chusov. Studying energy evolution in the discharge chamber of a multichamber lightning protection system. *Technical Physics Letters*, 42(4):395–398, 2016. doi:10.1134/S1063785016040222.

[7] A. V. Budin, M. E. Pinchuk, V. E. Kuznetsov, and F. G. Rutberg. The influence of the production technology of iron-copper composite alloy on its erosion properties in a high-current high-pressure arc. *Technical Physics Letters*, 40(12):1061–1064, 2014. doi:10.1134/S1063785014120050.

[8] V.B. Kovshechnikov, N.I. Litvinov, G.V. Nakonechnyi, R.V. Ovchinnikov, and A.V. Surov. Study of the erosion resistance of copper-iron and copper-nickel electrodes. *Russian journal of non-ferrous metals*, 46:36–40, 2005.

[9] A.V. Budin, M.E. Pinchuk, V.E. Kuznetsov, V.V. Leont'ev, and N.K. Kurakina. Experimental stand for investigations of arc and erosion processes in high-voltage powerful breaker. *Instruments and Experimental Techniques*, 60(6), 2017. in press.

[10] Victor A. Kolikov, Mikhail E. Pinchuk, Anatoly G. Leks, and Philipp G. Rutberg. High-speed diagnostic pulsewise-periodic of electric discharge in water. *Proc. SPIE*, 6279:62795C–62795C–7, 2007. doi:10.1117/12.725381.

[11] M.E. Pinchuk, A.A. Bogomaz, A.V. Budin, and P.G. Rutberg. Electrode plasma jets in powerful pulsed discharge in high-pressure gas. *Plasma Science, IEEE Transactions on*, 42(10):2434–2435, Oct 2014. doi:10.1109/TPS.2014.2316064.

[12] Andor Technology. http://www.andor.com.

doi:10.14311/ppt.2017.2.124
Plasma Physics and Technology **4**(2):124–128, 2017

ANALYSIS OF ARC PROCESSES IN MULTI-CHAMBER ARRESTER FOR LIGHTNING PROTECTION AT HIGH-VOLTAGE OVERHEAD POWER LINES

Iu.V. Murashov[a], V.Ya. Frolov[a,*], D. Uhrlandt[b], S. Gortschakow[b], D.V. Ivanov[a], A.D. Sivaev[c]

[a] *Peter the Great St.Petersburg Polytechnic University, Polytechnicheskaya, 29, 195251, Saint-Petersburg, Russia*
[b] *Leibniz Institute for Plasma Science and Technology, Felix-Hausdorff-Str. 2, 17489 Greifswald, Germany*
[c] *Streamer Electric Company, Nevsky Prospect, 147, 191024, Saint-Petersburg, Russia*
[*] eie@spbstu.ru

Abstract. Nowadays multi-chamber arresters are widely distributed as devices of lightning protection of overhead power lines. A mathematical modelling of processes in the discharge chamber of multi-chamber arrester is necessary to carry out in order to improve its breaking capacity. A three-dimensional mathematical transient model of thermal, gas-dynamic and electromagnetic processes taking place in the discharge chamber of multi-chamber arrester is presented in the article. Basic assumptions, model equations, a computational domain and the boundary conditions are described. Plasma turbulence is taken into account. The results of the calculation i.e. distributions of plasma temperature and overpressure in the discharge chamber at different time points are shown. The analysis of the results was carried out. It is shown that the presence of cavities in the electrodes design promotes electric arc extinction in the discharge chamber of multi-chamber arrester.

Keywords: lightning protection, electric arc extinction, arrester .

1. 1. Introduction

Protection of overhead power lines and equipment of electrical substations from a direct lightning strike requires special lightning protection devices. A new promising method for such protection is the use of multi-chamber surge arresters [1, 2] with electrodes placed into dielectric chamber (see Figure 1).

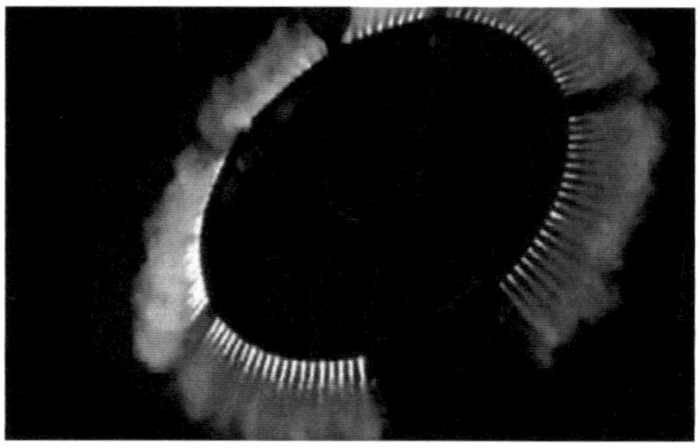

Figure 1. Arc discharge in multi-chamber arrester

Recently, a series of theoretical and experimental studies of such arresters was performed. The goal of those investigations was a choice of electrode and discharge chamber materials that provide the best arc extinguishing conditions and a long service life of the device [3]. Also studies of the electrical strength and electrical conductivity of plasma near the chamber outlet during the pulse current through the discharge chamber were made [4]. Based on the results of those studies a non-stationary three-dimensional mathematical model of plasma processes in a discharge chamber of a multi-chamber arrester was created [5–8].

This article is a continuation and development of theoretical investigations that has been carried out in the papers [5–8].

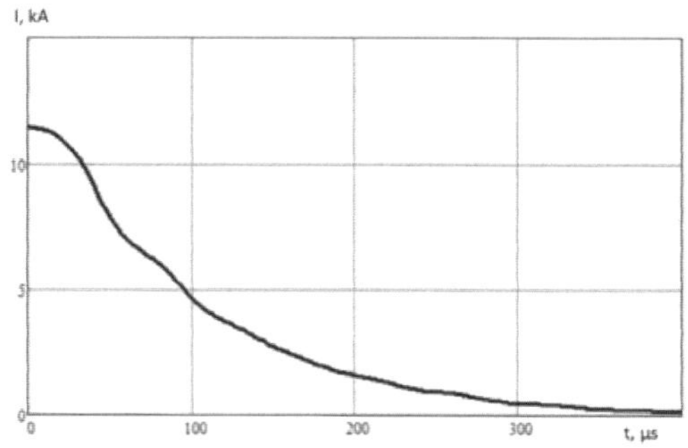

Figure 2. Discharge current dependence on time

2. 2. Mathematical model and initial data

A three-dimensional non-stationary mathematical model has been developed for modeling the processes in the discharge chamber of the multi-chamber arrester. The model includes the energy equation, the motion equation, the continuity equation, the electromagnetic equation on the basis of Maxwell's system. The details of used mathematical model was presented in the paper [8]. The main differences between this model and the model presented in [8] are plasma turbulence and the use of polygonal meshes.

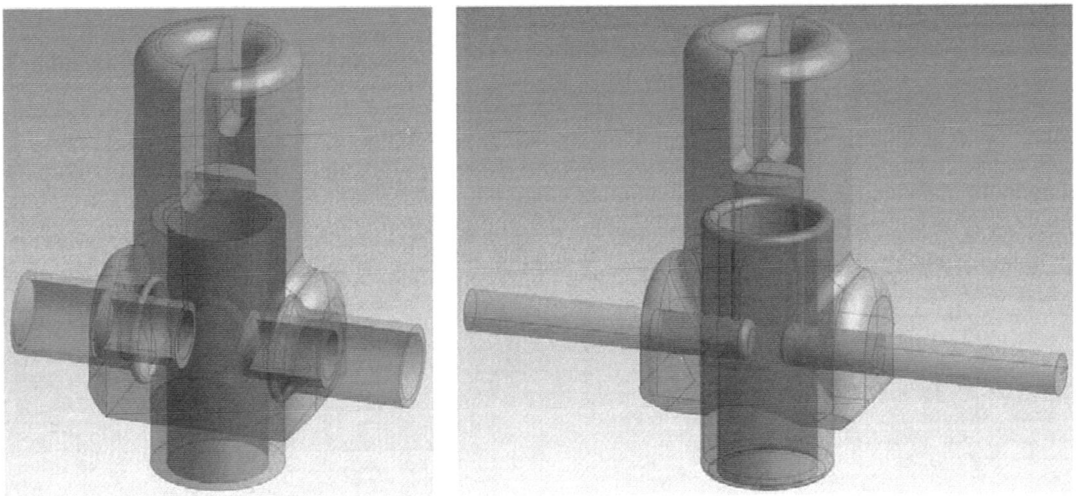

Figure 3. Investigated designs of the discharge chamber

The turbulence of plasma flow was taken into account by the SST-model which is a combination of $k - \epsilon$ and $k - \omega$ turbulence models: $k - \epsilon$ model are used for calculation in the free flow zone while $k - \omega$ model are used in the zones near walls.

The use of polygonal mesh accelerates the convergence of the solution in comparison with the conventional triagonal mesh.

An electrical potential difference is defined on electrodes surfaces as boundary condition for the electromagnetic problem so that a discharge current through the chamber corresponds to the experimentally obtained dependence at each moment of time (see Figure 2).

The properties of gas (plasma) was calculated based on erosion of the electrodes and ablation of material of walls of the discharge chamber [5].

The solution was performed using the software *ANSYS Fluent* that is used by many researchers to investigate different plasma processes [9–11].

Two cases of the discharge chamber design shown in Figure 3 was calculated. The main difference between cases is the design of the electrodes. A design with tube electrodes forming cavities connected to the main volume of the discharge chamber is shown in the left part of Figure 3 . A design with rod electrodes (without cavities) is shown in the right part of Figure 3 .

3. 3. Results

Results of simulation i.e. distributions of plasma temperature at different points of time are shown in Figure 4 and Figure 5 for two cases of the discharge chamber design.

The obtained results show that:

1) In the case of a discharge chamber with tube electrodes the plasma jet reaches the exit of the chamber gap at time point of 10 µs approximately, and the chamber outlet – at time point of 30 µs;

2) In the case of a chamber with rod electrodes the plasma jet reaches the exit of the chamber gap at time point of 5 µs approximately, and the chamber outlet – at time point of 15 µs. It means that motion of plasma to the outlet is approximately twice as high as in the case with tube electrodes;

3) The maximum temperature at the outlet of the discharge chamber is 15–20 thousand K, in the first 100–120 µs the plasma jet increases in size, then its dimensions decrease;

4) The high-pressure zone reaches the outlet of the discharge chamber earlier than the high-temperature zone. It means that a plasma heated to a high temperature moves towards the exit and "pushes" cold gas before itself;

5) The presence of a longitudinal cut in the cylindrical part of the discharge chamber leads to the ejection of a part of plasma in this direction and to a decrease in the velocity and temperature of the plasma jet;

6) Analysis of the simulation results shows that the presence of cavities joined to the main volume of the discharge chamber by small openings leads to the fact that at the initial stage of discharge, when pressure inside the discharge chamber exceeds pressure inside the cavities, plasma begins to spread into these cavities. The pressure inside the cavities is increased as long as it begins to exceed the pressure in the discharge chamber, which is lowered due to propagation of the discharge towards the outlet of the discharge chamber. It occurs in about 140–150 µs after the start of discharge. Thereafter, the gas accumulated in the cavities with lower temperature than the discharge core, begins to move in the opposite direction, i.e. from cavities to the discharge chamber. After 160 µs it leads to cooling of the central part of the discharge, which undoubtedly contributes to the extinction of the electric arc. This conclusion is confirmed by the results of experimental observations.

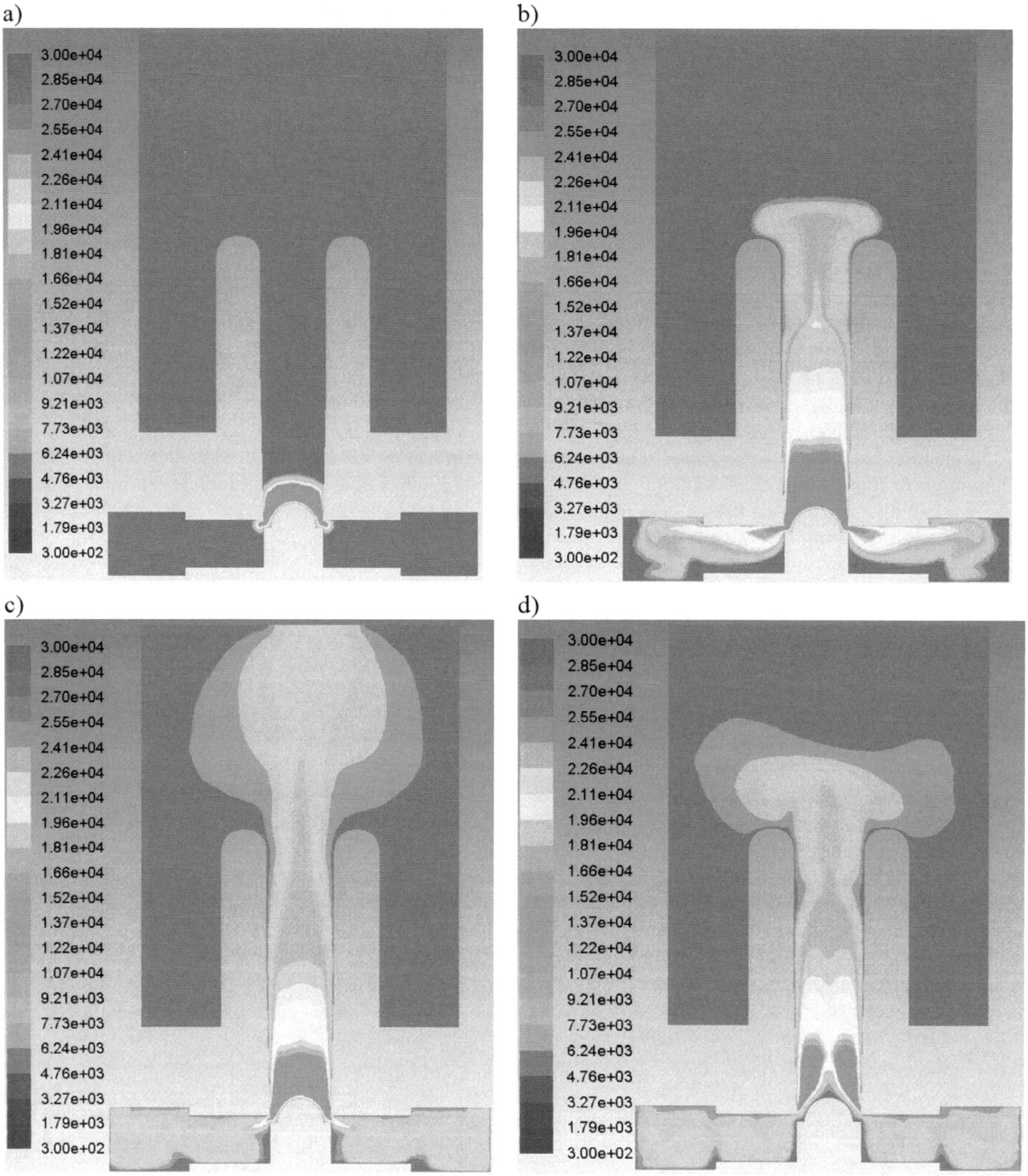

Figure 4. Distributions of temperature (T, K) in the discharge chamber with tube electrodes at different points of time: a – 2 μs; b – 30 μs; c – 140 μs; d – 210 μs

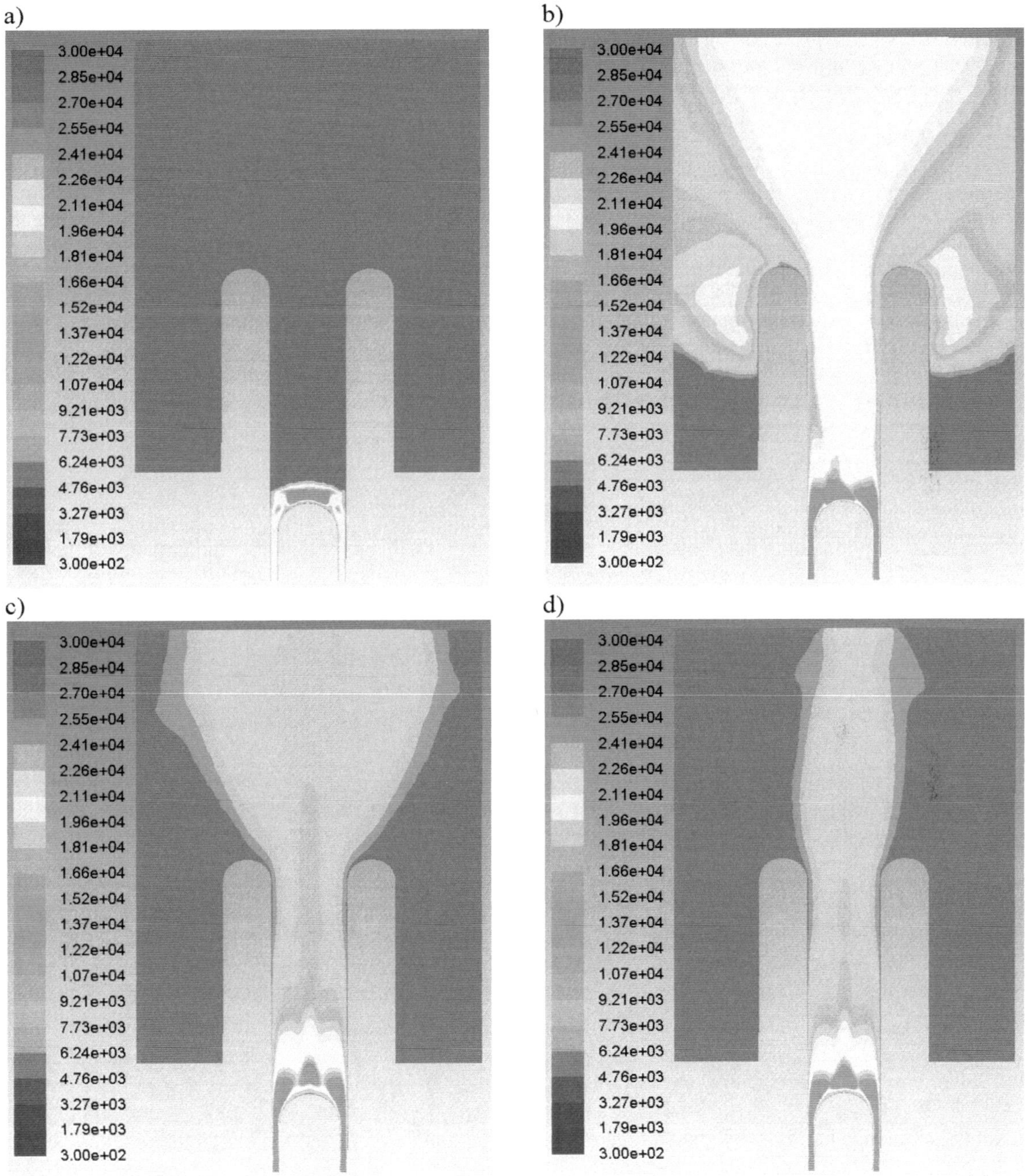

Figure 5. Distributions of temperature (T, K) in the discharge chamber with rod electrodes at different points of time: a – 2 μs; b – 30 μs; c – 140 μs; d – 210 μs

4. 4. Conclusions

The developed mathematical model allows to adequately simulate the processes occurring in the discharge chamber of the multi-chamber arrester and in the surrounding space.

Further studies will be devoted to taking into account in the model the following effects: deformation of the chamber due to high pressure, transport of plasma radiation, deionization processes near the zero of the pulse current.

Acknowledgements

This research has been supported by The Russian Science Foundation.

References

[1] G.V. Podporkin. Development of long flashover and multi-chamber arresters and insulator-arresters for lightning protection of overhead distribution and transmission lines. *Plasma Physics and Technology*, 2(3):241–250, 2015.

[2] Streamer, Electric Company Inc. *Official site*. `http://www.streamer-electric.com/`.

[3] R. Kozakov, A. Khakpour, S. Gorchakov, D. Uhrlandt, D. Ivanov, I. Murashov, G. Podporkin, and V. Frolov. Investigation of a multi-chamber system for lightning protection at overhead power lines. *Plasma Physics and Technology*, 2(2):150–154, 2015.

[4] V.Ye. Pilshchikov, V.Ya. Frolov, and R.A. Belsky. Investigation of the electrical strength and electrical conductivity of plasma near the chamber outlet during the pulse current through the discharge chamber. *Proceedings of V Russian Conference on Lightning Protection*, pages 243–246, 2016.

[5] V.Y. Frolov, D.V. Ivanov, I.V. Murashov, and A.D. Sivaev. Calculation of the composition of plasma of an arc pulsed discharge in a multi-chamber arrester. *Technical Physics Letters*, 41(4):310–313, 2015. `doi:10.1134/S1063785015040069`.

[6] V.Y. Frolov, D.V. Ivanov, I.V. Murashov, and A.N. Chusov. Numerical simulation of plasma processes in a discharge chamber of a multi-chamber surge arrester for lightning protection. *Proceedings of V Russian Conference on Lightning Protection*, pages 334–337, 2016.

[7] A.N. Chusov, G.V. Podporkin, M.E. Pinchuk, D.V. Ivanov, Iu.V. Murashov, and V.Ya. Frolov. Modeling of discharge in multi-chamber systems. *Proceedings of V Russian Conference on Lightning Protection*, pages 351–357, 2016.

[8] V. Frolov, D. Ivanov, I. Murashov, and A. Sivaev. Mathematical simulation of processes in discharge chamber of multi-chamber system for lightning protection at overhead power lines. *Proceedings of the 2016 IEEE North West Russia section young researchers in electrical and electronic engineering conference, ELCONRUSNW 2016*, pages 562–565, 2016. `doi:10.1109/EIConRusNW.2016.7448245`.

[9] V.Ya. Frolov, A.V. Lopota, and D.V. Ivanov. Radio frequency inductively coupled plasma torch for nanomaterials production. *Nauchno-tehnicheskie vedomosti SPbGPU*, 117:146–151, 2011.

[10] V.Y. Frolov, D.V. Ivanov, and M.A. Shibaev. Modeling the plasmachemical synthesis of nanopowdered materials using a combined plasmatron. *Technical Physics Letters*, 40(8):676–679, 2014. `doi:10.1134/S1063785014080185`.

[11] V. Frolov, D. Ivanov, and M. Shibaev. Plasma installation for TiO_2 fine powder production: mathematical simulation. *Plasma Physics and Technology*, 3(2):57–61, 2016.

doi:10.14311/ppt.2017.2.129
Plasma Physics and Technology 4(2):129–132, 2017

QUANTIFYING UNCERTAINTIES IN MEAN ABSORPTION COEFFICIENTS FOR A WALL-STABILIZED ELECTRIC ARC

R. Fuchs*, H. Nordborg

Institute for Energy Technology, HSR University of Applied Sciences Rapperswil, Oberseestrasse 10, 8640 Rapperswil, Switzerland

* roman.fuchs@hsr.ch

Abstract. The quality of arc simulations depends significantly on radiation modeling. Uncertainties due to physical parameters and modeling errors should be systematically quantified. We solve the energy balance equation for a wall-stabilized arc using the P1 model, i.e. without a prescribed temperature profile. We derive the linearized equation and assess the arc voltage sensitivity. This method allows us to optimize the definition of mean absorption coefficients consistently and at low computational costs.

Keywords: sensitivity, mean absorption coefficient, P1 model, optimization, arc simulation.

1. Introduction

Electric arcs and their numerical modeling in industrial devices lead to a diverse set of questions raising many separate research topics. One of them is radiation, since electrical arcs dissipate enormous amounts of thermal energy leading to elevated temperature levels. Therefore, radiation is almost always an important energy transfer mode in electrical arcs.

The basic relations of radiation are given by the radiative transfer equation (RTE)

$$\hat{s} \cdot \nabla I_\nu = \kappa_\nu (B_\nu - I_\nu), \tag{1}$$

which is stated here for the simplest case of a non-scattering medium in local thermal equilibrium (LTE).[1] It describes the change of radiative intensity I_ν along direction $\hat{s}$ that is due to emission, described by the Planck function B_ν and given in equation (5) below, and absorption of the medium. Solving the RTE is computationally demanding since the radiative intensity is a function of location, direction, and frequency. The absorption coefficient κ_ν and its impact on the computational complexity of the RTE are discussed subsequently.

To meet the limited computer resources available to application engineers, the complexity of the RTE must be reduced to simpler models that account appropriately for the radiative thermal energy transport. The method of net emission coefficients (NEC) [2] is often used and has been developed to acceptable accuracy [3]. Its main advantage lies in the marginal computational costs incurred at run time of a plasma simulation, since the radiative thermal source term is tabulated in advance. However, evaluating the radiative heat flux on a nearby surface requires more sophisticated methods. One option is the P1 model, which is a first-order approximation of the RTE in spherical harmonics and frequently used in industrial arc simulations. This model leads to additional differential equations that are of the same structure as the conservation equations for mass, momentum, and energy.

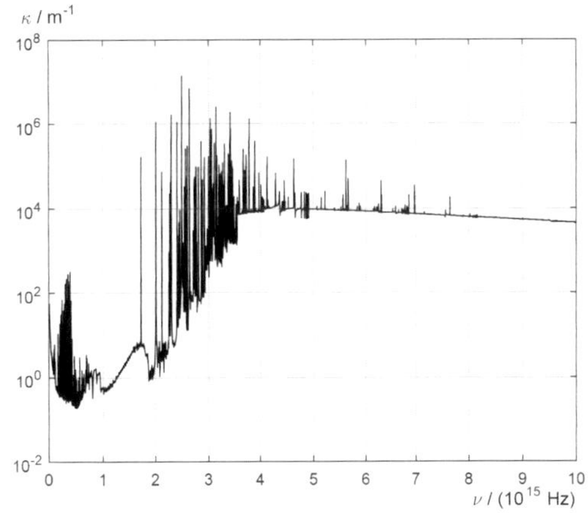

Figure 1. Absorption spectrum of air at $T = 10\,kK$. [4]

We also have to consider the material data. The most important quantity in radiation modeling is the spectral absorption coefficient κ_ν, which depends on the radiation frequency ν, gas pressure, temperature, and gas composition. The complex structure of the absorption spectrum (see figure 1) is due to the atomic structure of the gas (see, e.g., [1]). It does not allow for a naive usage of this data, since an appropriate resolution of the frequency domain would lead to excessive computational costs. Despite the large variations in the value of the absorption coefficient, the frequency domain is split into a small number of intervals or bands and mean absorption coefficients (MAC) are computed for each of them. Since each frequency band acts as a gray body, this approach is termed as the multi-banded gray P1 model.

The main problem of the averaging step is how the absorption lines shall be treated that span over multiple orders of magnitudes. Classically, the Planck and Rosseland average are considered which are weighted averages of κ_ν and κ_ν^{-1}, respectively. The former is dominated by the peak values of the absorption lines while the latter almost ignores them. It is clear that the MAC depend on the definition of the frequency band boundaries that are often chosen heuristically.

The raw data itself comes with uncertainties that affect the MAC values and propagate to the final results. Their sensitivities can be quantified by running simulations repeatedly with slightly edited input data, but comes with major computational costs. For small changes in the MAC, however, we will show that the same information can be obtained much quicker by deriving the linearized equations and solving for the first-order effects on the temperature profile and the quantities depending on it. Such modeling errors are often studied using a fixed temperature profile. From a physical point of view it makes more sense to solve the energy conservation equation, since any change in the radiation model results in an updated temperature field and, consequently, other quantities such as arc voltage.

In this paper, we quantify modeling errors in the mean absorption coefficient that are due to the averaging procedure or inherent to the raw data itself. We iteratively solve the energy balance equation for a wall-stabilized arc to ensure energy conservation. Ultimately, we present the linearized equations as a validated and computationally efficient methodology to determine first-order effects on the arc voltage.

2. Model

2.1. Energy conservation equation

We consider a wall-stabilized arc, i.e. a cylindrical plasma column of radius R, large aspect ratio, and a fixed wall temperature. Moreover, we assume that the plasma is in local thermal equilibrium (LTE) so that the pressure p and electric field E are constant across the plasma column. Further, convective heat transfer is considered weak enough to be negligible. This results in a temperature profile $T(r)$ that only depends on the arc radius r. In this case, the energy conservation equation consists of heat conduction, Ohmic heating, and a radiative heat sink U,

$$\mathrm{div}(-\lambda\,\mathrm{grad}(T)) = \sigma E^2 + U; \tag{2}$$

here, $\lambda(T)$ and $\sigma(T)$ denote the thermal and electrical conductivities. The electric field is given as the ratio of electric current I to the plasma conductivity $S[T]$, which is a functional of the temperature profile:

$$E = \frac{I}{S[T]}, \quad S[T] = \int_0^R \sigma(T(r))r\,\mathrm{d}r. \tag{3}$$

In this model, we consider the current as a constant.

The radiative heat sink U is given by the divergence of the total radiative heat flux. We use the multi-banded gray P1 model to account for radiative heat transfer. Hence, we split the spectral domain into a finite number of intervals $D_i = [\nu_{i-1}, \nu_i]$. The mean absorption coefficients $\kappa_i = \langle\kappa\rangle_i$ are obtained using the Planck and Rosseland average

$$\langle\kappa\rangle_i^{Pl} = \frac{\int_{D_i} B_\nu \kappa_\nu\,\mathrm{d}\nu}{\int_{D_i} B_\nu\,\mathrm{d}\nu}, \quad \langle\kappa\rangle_i^{Ro} = \frac{\int_{D_i} B'_\nu\,\mathrm{d}\nu}{\int_{D_i} B'_\nu \kappa_\nu^{-1}\,\mathrm{d}\nu}, \tag{4}$$

respectively. Therein, $B_\nu(T)$ denotes the Planck function

$$B_\nu(T) = \frac{2h}{c^2}\frac{\nu^3}{\exp(\frac{h\nu}{k_B T}) - 1}, \tag{5}$$

B'_ν is its derivative with respect to temperature, c denotes the speed of light in vacuum, and h, k_B are the Planck and Boltzmann constants, respectively. The P1 model approximation results in an expression relating the radiative heat flux $\vec{F}_\nu$ and the irradiation function G_ν by:

$$\vec{F}_\nu = \frac{-1}{3\kappa_\nu}\mathrm{grad}(G_\nu). \tag{6}$$

This also holds for band-averaged quantities, so that G_i is obtained as the solution of the linear problem

$$\mathcal{L}_i(G_i) := \mathrm{div}\left(\frac{-1}{3\kappa_i}\mathrm{grad}(G_i)\right) + \kappa_i(G_i) = \kappa_i 4\pi B_i(T) \tag{7}$$

with $B_i(T)$ denoting the band-integrated Planck function. The radiative heat sink is then given by

$$U = (-1)\sum_i \kappa_i(4\pi B_i(T) - G_i), \tag{8}$$

which follows from the RTE (1) by integration over the solid angle.

2.2. Linearized equation

Equations (2) and (8) show that any change in the absorption coefficient results in changes to the temperature profile. For small variations, we can derive the linearized energy conservation equation with respect to increments in the MAC ($\delta\kappa_i$) and temperature (δT). This first-order expansion provides a relation which we write as

$$C_{th}(\delta T) - M_I(\delta T) = \delta U, \tag{9}$$

with $\delta U = \delta U(\delta\kappa_i, \delta T)$ summarizing the linear effects in the radiation model. The details are derived in the remainder of this section.

The linearized heat conduction is given by

$$\begin{aligned} C_{th}(\delta T) &= \mathrm{div}\left(-\lambda(T)\,\mathrm{grad}(\delta T)\right) \\ &\quad - \mathrm{div}\left(\lambda'(T)\,\mathrm{grad}(T)\delta T\right), \end{aligned} \tag{10}$$

with $\lambda'(T) = \frac{\mathrm{d}}{\mathrm{d}T}\lambda(T)$. The linearized Ohmic heating at constant current results in

$$M_I(\delta T) = \left(\sigma'(T) - 2\frac{\sigma(T)}{S[T]}\frac{\delta S[T]}{\delta T(r)}\right)\frac{I^2}{S[T]^2}\delta T \quad (11)$$

where $\frac{\delta S[T]}{\delta T(r)}$ represents the functional derivative of the conductivity, $\sigma'(T) = \frac{\mathrm{d}}{\mathrm{d}T}\sigma(T)$, and the radiative heat sink increment is expanded to

$$\delta U = (-1)\sum_i (4\pi B_i(T) - G_i)\delta\kappa_i + \kappa_i(4\pi B_i'(T)\delta T - \delta G_i). \quad (12)$$

The linear increment in the irradiation function δG_i is obtained as the solution of

$$\delta G_i = \mathcal{L}_i^{-1}(M_{\kappa,i}\delta\kappa_i + M_{th,i}\delta T) \quad (13)$$

with

$$M_{\kappa,i}(\delta\kappa_i) = \mathrm{div}\left(\frac{-1}{3\kappa_i^2}\,\mathrm{grad}(G_i)\delta\kappa_i\right) + (4\pi B_i(T) - G_i)\delta\kappa_i, \quad (14)$$

$$M_{th,i}(\delta T) = \kappa_i 4\pi B_i'(T)\delta T. \quad (15)$$

In summary, the terms can be rearranged so that the increment in the temperature profile is given as a linear function of those in the MACs,

$$\delta T = Z^{-1}\sum_i Y_i\delta\kappa_i, \quad (16)$$

with Z and Y being linear operators given by

$$Z := C_{th} - M_I + \sum_i \kappa_i 4\pi B_i'(T) - \kappa_i\mathcal{L}_i^{-1}M_{th,i}, \quad (17)$$

$$Y_i := \kappa_i\mathcal{L}_i^{-1}M_{th,i} - (4\pi B_i(T) - G_i). \quad (18)$$

Finally, we find an expression for the linear increment in the electric field with respect to a temperature increment:

$$\delta E = (-1)\frac{I}{S^2[T]}\frac{\delta S[T]}{\delta T(r)}\delta T. \quad (19)$$

2.3. Simulation conditions

We use a domain radius of $R = 5\,\mathrm{mm}$ discretized with 200 cells and cell-centered quantities. The wall temperature is fixed at 300 K. The energy conservation equation (2) is iteratively solved using a relaxation update with a constant current of $I = 50\,\mathrm{A}$ until the maximal temperature update is smaller than 1 K.

The radiation frequency bands are defined as N uniformly spaced intervals $D_i = [\nu_{i-1}, \nu_i]$, $i = 1, \ldots, N$, with $\nu_0 = 10^{10}\,\mathrm{Hz}$ and $\nu_N = 6 \times 10^{15}\,\mathrm{Hz}$. An additional frequency band is defined $D_{N+1} = [\nu_N, 10^{16}\,\mathrm{Hz}]$. Material data and spectral absorption coefficients are taken for air at $p = 10\,\mathrm{bar}$.[4]

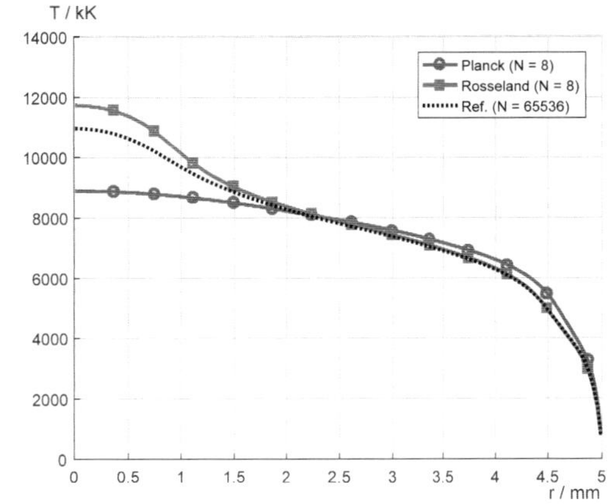

Figure 2. Temperature profiles at 50 A *total current.*

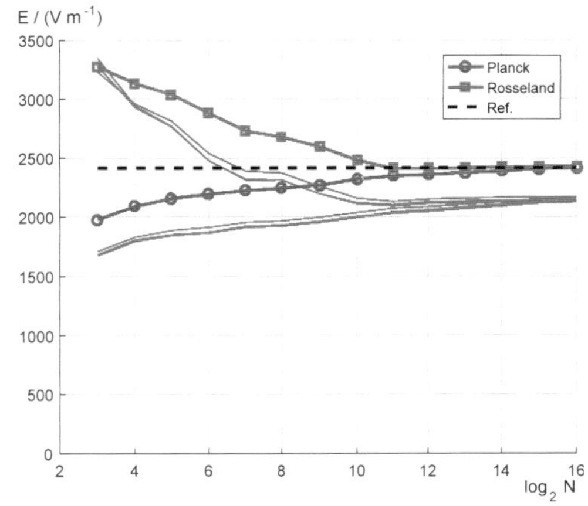

Figure 3. Electric field at 50 A *total current as a function of spectral resolution. The reference value is taken for the finest spectral resolution. Gray lines:* 100, 150, *and* 200 A.

3. Results and Discussion

Figure 2 shows temperature profiles of the wall-stabilized arc obtained with Planck and Rosseland average using 8 bands. We also plot the reference solution with $N = 65536$ bands, which shows an arc center temperature of 10970 K and a noticeable core region with a higher temperature gradient than in the outer parts. Near the wall, the temperature drops quickly to the prescribed wall temperature. In comparison to this, the Rosseland average leads to a more pronounced core region with a higher arc center temperature but qualitatively similar temperature profile. In contrast, the Planck average shows a diffusive temperature profile with an arc center temperature much lower than the reference solution, and a sightly increased temperature in the outer parts. These differences disappear gradually with finer spectral resolutions and the temperature profiles converge to the reference solution.

This data is explained by the properties of the averaging methods. Since the Planck average is dominated by the peak values of the absorption lines, it resembles an optically thick plasma, and the radiative heat transfer has the same effect as an increased thermal conductivity. On the other hand, the Rosseland average yields an optically thin material and radiative energy is transported to the walls.

Figure 3 shows the electric field in variation of the spectral resolution and for several arc currents. We see that, at low spectral resolutions, the Planck average yields a higher electric field than the Rosseland average. This is due to the lower temperature levels and lower electrical conductivity. We also note that the curves converge to the reference solution, with the Planck average being rather constant for spectral resolutions $N > 2^{10}$. This figure shows that accurate results are only obtained with a finely resolved absorption spectrum.

We now turn to the question, which parts of the spectrum are most significant. Figure 4 and 5 show the relative sensitivity of the electric field obtained by the linearized equation for the two averaging methods. We used an uncertainty of 1% in the MAC for each band individually. The same data is also obtained with direct computations for $N \leq 256$, i.e. editing the MAC values and running the simulation until convergence, but at much higher computational costs. The results coincide almost exactly and validate the linearized method. Only minor differences are noted for the solution using $N = 8$ bands and Rosseland average: however, they are at higher radiation frequencies where the sensitivities are orders of magnitudes lower and therefore negligible. This method has also been applied to finely resolved spectral data with drastically reduced computational effort.

The sensitivity of the electric field is limited by 1‰ for the 8-banded solution. The sensitivity curves scale with the interval length; in fact, considering the relative sensitivity of the electric field per interval length, i.e. $\Delta E/(E_{ref}\Delta\nu)$, results in a characteristic curve. We also see that a higher number of spectral bands allows to resolve the sensitivities of the absorption lines. The sensitivity curve of a finer resolved solution is bounded almost everywhere by the coarser ones. The Planck average yields notable lower sensitivities at frequencies above 2.5×10^{15} Hz and band resolutions $N \leq 256$.

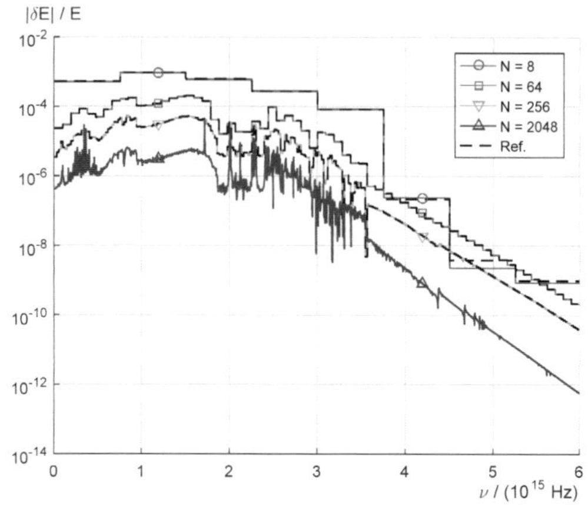

Figure 4. Relative sensitivity of the electric field at 50 A *using Rosseland average and uncertainty of 1%.*

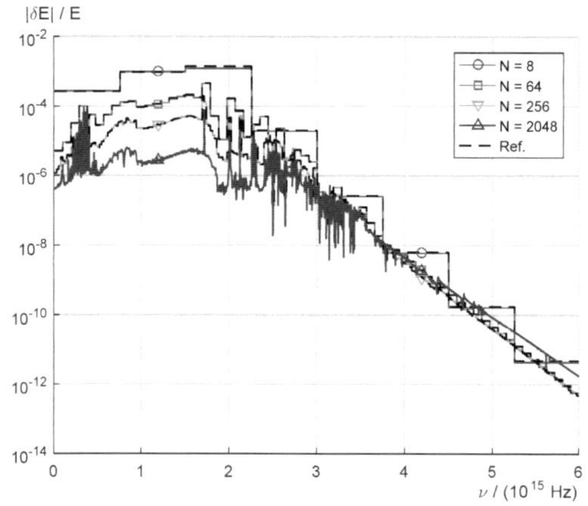

Figure 5. Relative sensitivity of the electric field at 50 A *using Planck average and uncertainty of 1%.*

4. Conclusions

We presented a simple model to study effects of band-averaged MACs on the temperature profile and the electric field of a wall-stabilized arc. The linearized method provides a systematic methodology to assess the sensitivity of the frequency bands at low computational costs, and helps to define MAC inside each band. Hence, frequency bands of low sensitivity can be identified and may be merged or grouped into coarser ones. For future work, the method is easily extended to non-uniform intervals. Preliminary tests with a renormalization length [5] are promising and are currently being investigated.

Acknowledgements

We gratefully thank P Kloc for providing the material data and absorption coefficients.

References

[1] M. F. Modest. *Radiative heat transfer.* Elsevier Science, San Diego, third edition, 2013.

[2] J. J. Lowke. Predictions of arc temperature profiles using approximate emission coefficients for radiation losses. *J Quant Spect and Rad Transfer*, 14(2):111–122, 1974. doi:10.1016/0022-4073(74)90004-1.

[3] H. Z. Randrianandraina, Y. Cressault, and A. Gleizes. Improvements of radiative transfer calculation for SF 6 thermal plasmas. *J Phys D: Appl Phys*, 44(19):194012, 2011. doi:10.1088/0022-3727/44/19/194012.

[4] P. Kloc, V. Aubrecht, M. Bartlova, and O. Coufal. Radiation transfer in air and air-Cu plasmas for two temperature profiles. *J Phys D: Appl Phys*, 48(5):055208, 2015. doi:10.1088/0022-3727/48/5/055208.

[5] H. Nordborg and A. A. Iordanidis. Self-consistent radiation based modelling of electric arcs: I. Efficient radiation approximations. *J Phys D: Appl Phys*, 41(13):135205, 2008. doi:10.1088/0022-3727/41/13/135205.

doi:10.14311/ppt.2017.2.133
Plasma Physics and Technology **4**(2):133–136, 2017

SYNTHESIS OF NANOSILVER IN PLASMA-LIQUID MICRODISCHARGE SYSTEM

V. Chernyak[a,*], D. Hamazin[a], O. Kolomiiets[a], O. Prysiazhna[a], A. Goriachko[a], A. Trohimchuk[b], O. Legenchuk[b], V. Lendiel[a], I. Fedirchyk[a]

[a] *Faculty of Radio Physics, Electronics and Computer Systems, Taras Shevchenko National University of Kyiv, 64/13 Volodymyrska Street, 01601, Kyiv, Ukraine*

[b] *F. D. Ovcharenko Institute of Biocolloidal Chemistry NAS Ukraine, 42 Acad. Vernadskoho Blvd., 03142, Kyiv, Ukraine*

[*] chernyak_v@ukr.net

Abstract. This paper presents the results of investigation of the plasma-liquid system with the secondary discharge supported by atmospheric pressure microdischarge in the vortex Ar flow. The plasma treated fluids were aqueous solutions of $AgNO_3$. The microplasma discharge was powered by a DC supply. The plasma channel behaviour was characterized by photo/video recording, also plasma was studied using emission spectroscopy technique. The working liquid and firm products created after the treatment were studied also.

Keywords: microdischarge, secondary discharge, Ag nanoparticles, plasma-liquid system.

1. Introduction

For a long time, an active study of the interaction between plasmas and liquids (solutions, in particular) were performed. Classic systems have two electrodes, one of which is immersed in a liquid [1]. There is another approach in which liquid plays a function of an electrode in a secondary discharge. Thus, the principal generator of discharge can serve different types of discharges. Today atmospheric pressure microdischarges are one of the promising discharge types [2]. Microplasma is characterized by high current density and relatively low gas temperature that makes it handy for the electrochemical applications of aqueous solutions.

Creation of plasma involving liquids becomes increasingly important in industrial applications [3]. The issues of water purification using plasma, the usage of plasma treated liquids in agriculture, the synthesis of nanosized particles, the reforming of hydrocarbons and others still remain open. Microplasma creates a unique environment for nanomaterial synthesis, which restricts their agglomeration and allows the formation of crystalline materials by selective heating. The key advantage of synthesis processes based on micro-plasma is the fact that all chemical processes happen at atmospheric or higher pressures and so the collision frequency will significantly increase, which may be favorable for the formation of particles [4]. It should be held in mind that the nanoparticles in the solution can become charged and make self-organized structures through interaction as a result of the Coulomb and Van-der-Waals forces.

The paper presents the results of studies of the plasma-liquid system with the secondary discharge supported by the microdischarge in the vortex Ar flow. The fluids were aqueous solutions of $AgNO_3$.

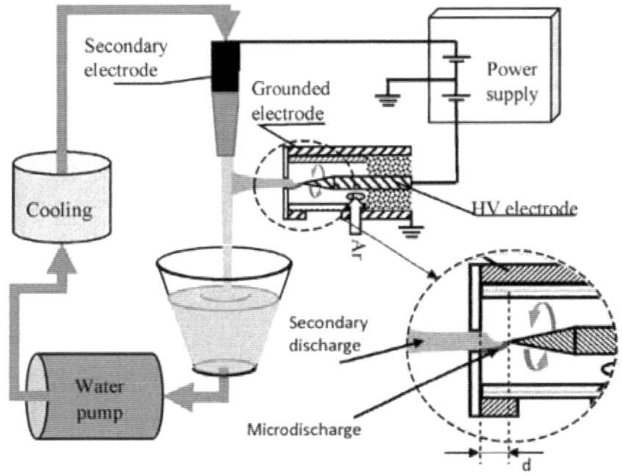

Figure 1. Schematic representation of studied plasma-liquid system with secondary discharge supported by microdischarge.

2. Experimental setup

A schematic representation of the studied plasma-liquid system with secondary discharge supported by atmospheric pressure DC microdischarge is shown in Fig. 1.

The source of microdischarge is axisymmetric plasma generator of microplasma jet. The peculiarities of the system and its detailed description one can find in [5]. Discharge power supply provides the output voltage of 7 kV (cathode). External grounded copper electrode had water cooling (anode). The electrodes of microdischarge were made from copper. The distance between the cathode and orifice of the anode ("d" in fig. 1) was 1 mm. Generated plasma was carried out by vortex flow of working gas (Ar) from discharge gap through a hole in the anode (d = 1 mm) in open air space. The gas flow was G = 1 L min^{-1}.

The secondary discharge was implemented by supplying the electric potential to a hollow graphite cylinder which has working fluid flowing through it. The cylinder is marked as a secondary electrode in Fig. 1. The discharge that burns between the working fluid and the microdischarge is the secondary discharge.

The study of the system parameters was carried out at 10 mA current of microdischarge and 5 mA secondary discharge current. The working liquid was $AgNO_3$ with NH_4OH (hydrate of ammonia) in excess. The concentration of Ag^+ ions in the solution was 50 mM l^{-1}. 100 ml of this working liquid was treated for 15 min. The working liquid was diluted by distillate (1:2) for the absorption spectrum investigation.

Plasma system parameters were studied using optical emission spectroscopy method. Also, parameters of liquids and products, which were formed in the solution after plasma treatment were investigated by absorption spectroscopy and non-contact atomic force microscopy in air.

3. Results and discussion

It is known that the atmospheric pressure microdischarge is similar to glow discharge at low pressure. As was noted in [6], in the case where Ar was used as working gas for microdischarge at atmospheric pressure microplasma jet length was less than 3 mm. However, plasma region can be increased by secondary discharge generation supported by microdischarge. Photo and video recordings confirmed that the secondary discharge in argon looks like a set of individual channels fast moving in space [5]. With the decrease of camera exposure time, it was determined that the system has one channel, which quickly changes its position. Furthermore, the colour of the discharge jet varies along its length from white and blue at the source electrode area to purple at the surface of the liquid.

The emission spectra inside and outside the microdischarge system and absorption spectra of processed solutions were recorded using CCD-based spectrometer Solar TII (S-150-2-3648 USB) (operating in the wavelength range of 200–1080 nm). Emission spectra of microplasma inside microdischarge system contain atomic Ar lines and molecular OH bands, spectra outside the system also contain NO, N_2. Spectra are shown in Fig. 2.

Experiments show that microdischarge plasma is highly non-isothermal both inside (T_v*(OH)=1000±500 K and T_r*(OH)=500±250 K) and outside (T_v*(N_2)=3000±500 K and T_r*(N_2)=1500±250 K) MD system. The fact that plasma component composition dependends on gas atmosphere and does not contain electrode material, in addition to the non-isothermality of plasma shows that secondary discharge is of glow type.

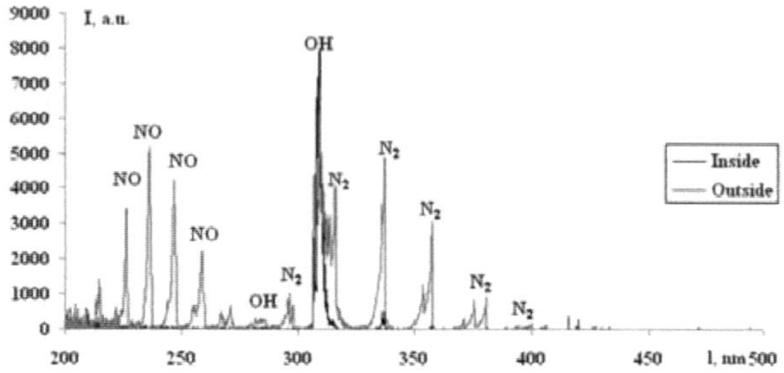

Figure 2. The emission spectra of plasma-liquid interface system with secondary discharge based on MD plasma.

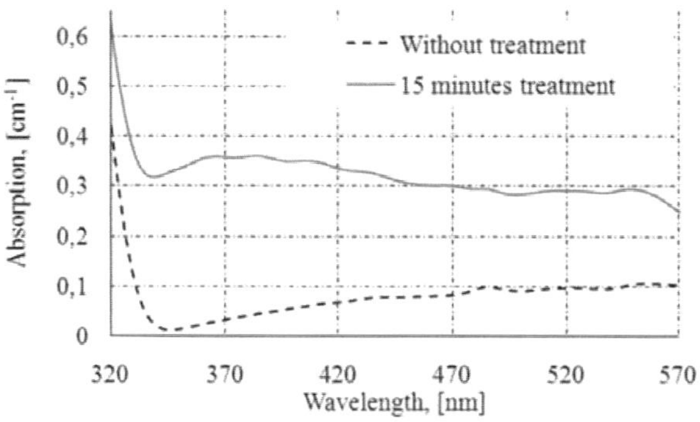

Figure 3. The absorption spectrum of the processed solution ($AgNO_3$+ NH_4OH) after 15 min processing.

The study of the absorption spectra is the simplest way to determine the presence of nanoparticles in the colloid solutions. This type of solutions features the absorption maximum, which corresponds to the plasmon resonance. The location and shape of the maximum depend on the material of the particles, their size and shape. According to [3, 7], the colloid solutions of Ag nanoparticles have maxima at 380 nm and 420 nm.

In this work, the solution of $AgNO_3$ and NH_4 was treated for 15 min. After the plasma treatment of the solution with the secondary discharge, the colour of the investigated liquid significantly changed and became darker. The absorption spectra of processed solutions (Fig. 3). The solid line corresponds to the absorption spectrum of the solution after the treatment. The dashed line shows the absorption spectrum of the untreated solution. The rise of the spectrum that begins at 340 nm corresponds to NO_3^- ion and H_2O_2. The spectum of treated solution shows the presence of weak maximum at 380 nm, which can be caused by the presence of Ag nanoparticles in the solution. In addition, the continuous absorption spectrum is present in the whole studied range of wavelength. This absorption can be caused by the strong scattering of the radiation in the solution due to the presence of the massive particles. These particles can appear as a result of coagulation.

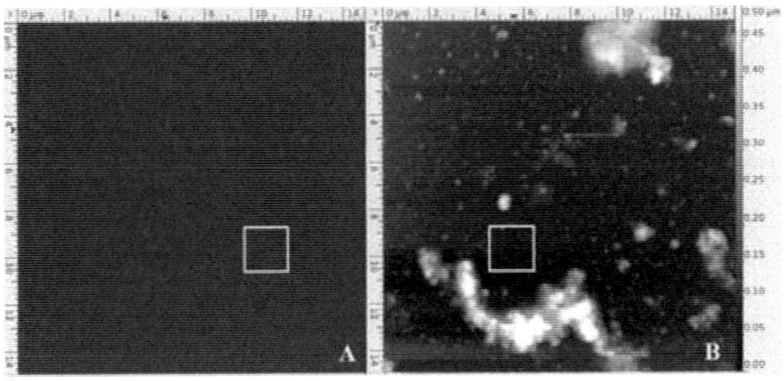

Figure 4. Microimages of clean substrate (A) and nanoparticles stranded substrate (B) field of view size 15×15 μm.

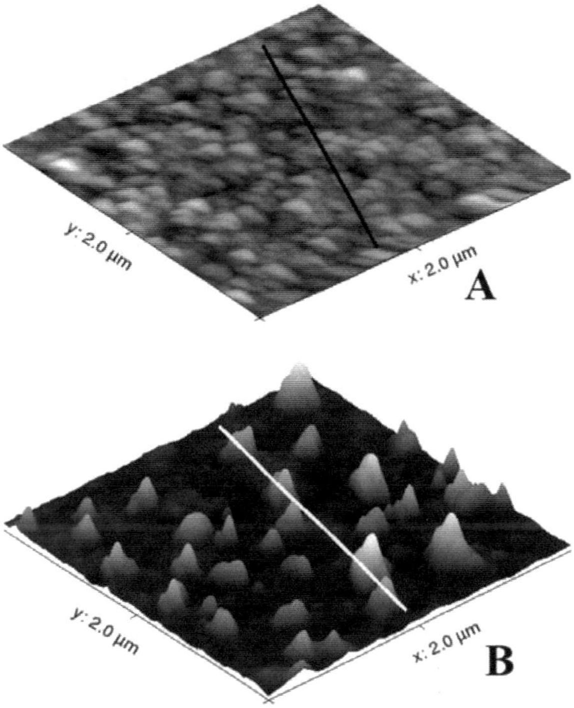

Figure 5. Microimages of clean substrate (A) and nanoparticles stranded substrate (B) field of view size 2×2 μm.

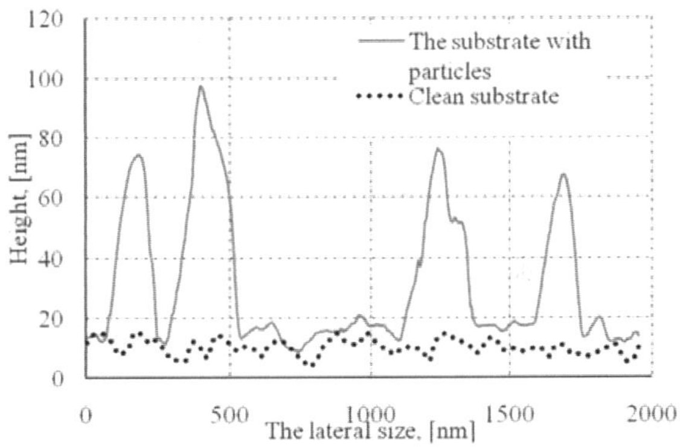

Figure 6. Cross-sections along the lines in Fig. 5. Black dots – clean substrate, Gray lines – nanoparticles stranded substrate.

The darkening of the solution can be observed after a single plasma treatment. This can be explained by the formation of sub-macroscopic sized metal even after a single treatment of working liquid. For more data, we analyzed substance which was formed in the liquid after the plasma treatment. The atomic force microscopy was conducted on the drop of treated liquid that was placed on a silicon wafer covered with a layer of cobalt silicide. Before the study, the liquid samples were dried out. To analyze the samples, we used a Smena NT-MDT microscope. Images of the surface obtained by this method are shown in Fig. 4.

Fig. 4 A shows the image of a clean wafer. Fig. 4 B shows the area covered with the solution. The figure shows that in the second case the size of the particles present on the wafer is 100 nm and more. This massive structures potentially appeared as a result of coagulation of a large number of particles. White squares in Fig. 4 are the areas that have been studied with more spatial resolution. The results are shown in Fig. 5. Along the lines in Fig. 5, cross-sections have been built. These sections are shown in Fig. 6.

The dashed chart in Fig. 6 corresponds to the surface profile of the clean wafer. The solid curve shows the profile of the surface with the presence of nanoparticles. The surface roughness is approximately 10 nm. The characteristic height of the nanoparticles is on the order of 100 nm and their lateral size is approximately 200 nm. However, the lateral size can be overestimated due to the specifics of atomic force microscopy.

Judging from the images the particles of a microscopic size and nano-sized particles with a characteristic size from tens to hundreds of nanometers are present on the substrate.

4. Conclusion

Experiments have shown that treatment of ammoniac silver nitrate solution with non-isothermal plasma of plasma-liquid system with secondary discharge supported by atmospheric pressure direct current microdischarge leads to silver nanoparticles formation. The size of the nanoparticles ranges from 10 to 100 nm. Besides the nanoparticles, the microparticles are observed on the substrate. Presumably, they are the result of the coagulation of individual nanoparticles.

References

[1] C. Richmonds and R. M. Sankaran. Plasma-liquid electrochemistry: Rapid synthesis of colloidal metal nanoparticles by microplasma reduction of aqueous cations. *Applied Physics Letters*, 93(13):13–15, 2008. doi:10.1063/1.2988283.

[2] A. Fridman and G. Friedman. *Plasma Medicine*. First printing. John Wiley & Sons, Ltd., United Kingdom, 2013.

[3] S. Ghosh, B. Bishop, I. Morrison, R. Akolkar, D. Scherson, and R. M. Sankaran. Generation of a direct-current, atmospheric-pressure microplasma at the surface of a liquid water microjet for continuous plasma-liquid processing. *Journal of Vacuum Science &*

Technology A: Vacuum, Surfaces, and Films, 33:021312, 2015. doi:10.1116/1.4907407.

[4] D. Mariotti and R. M. Sankaran. Microplasmas for nanomaterials synthesis. *Journal of Physics D: Applied Physics*, 43(32):323001, 2010. doi:10.1088/0022-3727/43/32/323001.

[5] D. Hamazin, V. Chernyak, O. Solomenko, O. Prysiazhna, E. Martysh, A. Trohimchuk, O. Legenchuk, and V. Lendiel. Atmospheric pressure secondary microdischarge system with vortex gas flow. *Problems Of Atomic Science And Technology*, 106(6):195–198, 2016.

[6] O. Solomenko, O. Prysiazhna, V. Chernyak, V. Lendiel, D. Hamazin, E. Martysh, D. Kalustova, and I. Prysiazhnevych. Investigation of a microdischarge system with the vortex gas flow. *Ukrainian Journal of Physics*, 61(11):960–967, 2016. doi:10.15407/ujpe61.11.0960.

[7] W. H. Chiang, C. Richmonds, and R. M. Sankaran. Continuous-flow, atmospheric-pressure microplasmas: a versatile source for metal nanoparticle synthesis in the gas or liquid phase. *Plasma Sources Science and Technology*, 19(3):1–8, 2010. doi:10.1088/0963-0252/19/3/034011.

doi:10.14311/ppt.2017.2.137
Plasma Physics and Technology **4**(2):137–140, 2017

POST-ARC CURRENT MEASUREMENT IN MECHANICAL CIRCUIT BREAKERS FOR HVDC APPLICATIONS

T. SCHULTZ

High Voltage Laboratory, ETH Zurich, Physikstrasse 3, 8092, Zurich, Switzerland
tschultz@ethz.ch

Abstract.
Post-arc currents contain valuable information on the breaker's performance. In this paper, the dimensioning procedure for a measurement system for such currents, based on a diode clamped shunt resistor is presented. Based on synthetic low and high power tests, the performance of a mock-up is evaluated. Finally, the post-arc current of a model gas circuit breaker, designed for use in DC applications, is measured.

Keywords: post-arc current, mechanical circuit breaker, small current measurement system.

1. Introduction

Mechanical gas circuit breakers are widespread in HVAC networks and may also be used as the mechanical component of HVDC switchgear. Due to the crucial role in the safe and reliable operation of electrical grids, a detailed investigation of their interruption behavior is of importance. The critical moment for interruption is the time around current zero. Measures like the arc resistance at or shortly before current zero as well as the post-arc current can be used to estimate the interruption capability of the tested breaker [1]. Hence, a precise measurement of the arc current around zero provides important information.

The main challenge for the measurement of post-arc currents is their low amplitude and short duration (air / SF_6: ≤ 20 / 1 A, ≤ 25 / 1 µs) compared to the fault current. In the past, a number of measurement systems have been introduced for this task [2–5]. These are based on different principles, such as Rogowski coils, resistive shunts in combination with pyrotechnic or semiconductor switches or the Faraday effect.

With recent improvements in semiconductor technology and digital data acquisition systems (DAQ), it has become easier and considerably more economic to assemble a post-arc current measurement system. This is especially interesting for the investigation of mechanical circuit breakers in HVDC breaker topologies. The smaller fault current level and drastically reduced arcing times allow for new, simpler constructions of such measurement systems. However, realizing a high bandwidth measurement system with sufficient accuracy in the desired low-current range still requires a sophisticated selection of components and design process.

2. Experimental setup

In this section, the realization of a small current measurement system (SCMS) as well as the used low and high power test circuits are described.

2.1. Measurement system

To precisely measure currents in the milliampere or single digit ampere range before and after a zero crossing, measurement systems are needed that limit the output signal during high current phases. One approach are systems based on a parallel connection of shunt resistors and anti-parallel diodes (cf. Figure 1) [4].

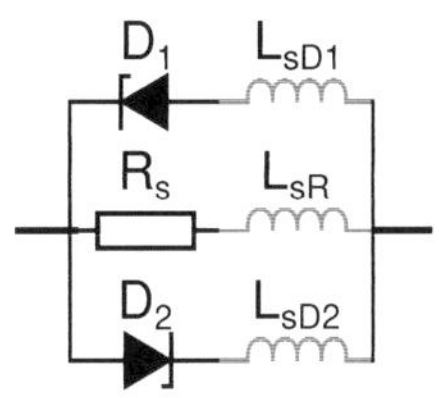

Figure 1. Measurement system consisting of a shunt (R_s) and two anti-parallel diodes (D_1, D_2) as well as corresponding paraisitic inductance (grey).

The anti-parallel connected diodes do not conduct significant amounts of current below a certain threshold voltage (forward voltage V_F). Consequently, the voltage across the measurement system can be considered proportional to the total current in a defined range $\pm I_0$ with:

$$I_0 = V_F / R_s \qquad \text{and} \qquad I_D(V_F) \leq x \cdot I_0 \,, \qquad (1)$$

where R_s is the selected shunt resistance, I_D is the forward biased diode current and x is the desired accuracy, e.g. 1%.

Once this voltage is exceeded, i.e. the current rises to a higher level, the forward biased diode will take over a larger and larger share of the current. This results in a drastically decreased rise of voltage, effectively limiting the output voltage of the measurement system. Due to this effect, the input range of the DAQ can be adjusted accordingly and the range around current zero can be recorded with increased accuracy.

Shunt resistance and diode threshold voltage can be selected to match the desired measurement range. The static voltage-current characteristic defines both the width of the measurement range ($\pm V_F$) as well as the clipping capability (forward slope resistance r_f, peak forward current). Additionally, the whole system needs to be suitable for high frequencies for a fast commutation of the current from and into the shunt (stray inductance, reverse recovery, package inductance). Often, a trade-off has to be made, as high power diodes are not suitable for high frequencies and vice versa. Due to high requirements regarding bandwidth and low output voltage of the system, it is of crucial importance to consider parasitic components and non-ideal behavior of diodes.

As the total current carrying capability of the system can be increased by a parallel assembly of multiple units, Schottky and fast recovery diodes are preferable to high power diodes. However, to avoid thermal runaway of parallel connected diodes, a considerable safety margin is important. If the application requires conducting very high currents, additional high power diodes in parallel may be necessary [4].

2.2. Test circuits

The presented measurement system can be realized in various ways, especially due to the large variety of available diodes. Hence, using separate testing methods for selecting suitable components and testing the complete system under realistic conditions is beneficial.

2.2.1. Low power testing

For a first evaluation of different types of diodes and system assemblies, a pre-charged, low voltage LC-circuit is used. It is tuned to a resonance frequency in the double digit kilohertz range and delivers currents that exceed the selected linear range of the assembly by at least a factor of five. Using this test setup, the maximum rate of rise of current can easily be adjusted to identify the dynamic limit performance of different types of diodes and assemblies. Due to the damping of the circuit, multiple current zero crossings with decreasing rates of rise of current (ROROC) are observed in each measurement. This allows to determine the performance of the system for different conditions. Once a suitable assembly is identified, it can easily be scaled up to the desired current level using parallel connections.

The main drawback of this test setup is that it is not possible to create currents with an amplitude similar to the expected currents of a circuit breaker test. Consequently, the diodes cannot be tested with the low-ohmic shunts that are used in real applications of the measurement system.

2.2.2. High power testing

Once a suitable diode configuration is identified, a verification of a scaled up version with a low-ohmic shunt and parallel connections of multiple diodes has to be performed. This can be done in a circuit breaker test bench, if the parameters of voltage and current can be adjusted sufficiently.

For this paper, a resonant circuit with a pre-charged capacitor is used (Fig. 2). After charging the capacitor C to the desired level, the test can be initiated by closing $AUX\ SW$. The current shape can be influenced by choosing proper values for L, C and the pre-charge voltage. The transient recovery voltage across the DUT is controlled by adjusting R_{TRV} and C_{TRV}. For further details and a description of the used model circuit breaker, refer to [6].

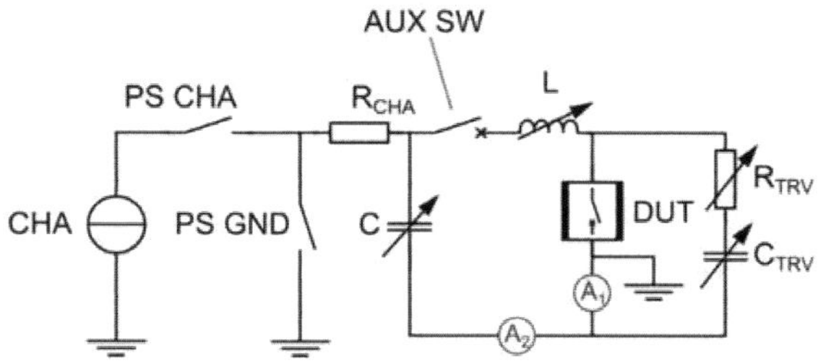

Figure 2. Schematic of circuit breaker test bench. A_1: SCMS, A_2: Reference (Pearson 4418).

3. Determining performance limits

3.1. Low power performance

Fig. 3 shows the signals of the reference current probe (Pearson 110) and the voltage across a SCMS test assembly, obtained during a low power test. The dotted line in Fig. 3 (b) represents a theoretical setup with perfect clipping behavior for currents above I_0.

For the SCMS test system, both the limited output signal during high current as well as the linear relation between current and output voltage during low current phases can clearly be seen. Due to the damped periodic test current, changes in the behavior for different ROROCs around zero can be observed. In this case, the tested combination of diodes and resistors shows a linear behavior for all tested ROROCs with a stable linear range of approximately $\pm 300\,mV$. However, the clamping behavior for higher currents is not very pronounced.

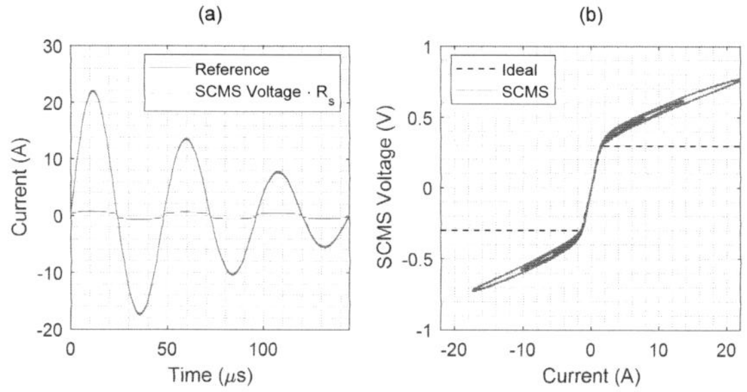

Figure 3. Example of reference current probe and SCMS signals as a function of time (a) and as a function of each other (b).

Using the described setup, tests have been conducted with a number of different assemblies including a variety of diode types (Schottky, FRED, different packages). In the following, the impact of undesired, non-ideal behavior of circuit elements is investigated. The measurements to illustrate them are obtained with modified setups to pronounce their impact on the behavior of the circuit.

3.1.1. Impact of parasitic inductance

Fig. 4 illustrates the measured voltage drop across the SCMS for two assemblies as a function of the reference current measurement. While the diodes used are identical for both setups, leads have been lengthened to artificially increase the parasitic inductance in the diode paths (a) and in the shunt path (b).

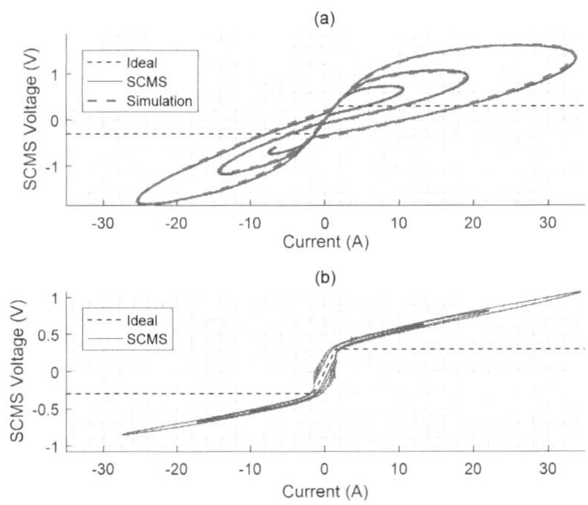

Figure 4. SCMS signal, using lengthened leads to increase inductance in (a) diode and (b) shunt paths.

The impact on the SCMS behavior can clearly be observed when comparing with Fig. 3 (b). In general, parasitic inductance causes an additional voltage drop in the respective branch, depending on both sign and absolute value of the ROROC. This leads to a decreased commutation speed out of and into the respective branch.

In case of the diode branch, for currents approaching zero this reduces the linear range of the measurement system as the current commutation into the shunt is delayed. In contrast, after zero crossing (i.e. the conditions of post-arc currents), the linear range is increased due to a slower commutation into the diode branch. The behavior of the circuit can be reproduced with good accuracy in a simplified simulation, only consisting of the RLC test circuit, simple diode models (V_F, r_F) and stray inductance (cf. Fig. 4 (a)). In this example, the parasitic inductance is considerably below 1 µH. When measuring a post-arc current, care has to be taken that the measurement range for the current approaching zero is large enough to extract ROROC or arc resistance shortly before zero. Hence, only the peak value of rate of rise of fault current before zero ($\mathrm{d}i/\mathrm{d}t_\mathrm{M}$) is of concern:

$$L_{\mathrm{D}\,1,2} < \frac{V_\mathrm{F}}{\mathrm{d}i/\mathrm{d}t_\mathrm{M}} \tag{2}$$

For the shunt resistor branch, the situation is quite different. As can be seen in Fig. 4 (b), the additional voltage drop impacts the linear measurement range. The combination of relatively low output voltage and high rates of change in current make this branch sensitive to the influence of stray inductance. For an error below 1% of the measurement range, the parasitic inductance limit can be in the nanohenry range. This requires high quality resistors and a suitable geometry of the assembly.

3.1.2. Impact of diode properties

In addition to stray inductance, semiconductor properties play an important role for the performance of the measurement system. On the one side, the static current-voltage characteristic of the diodes defines the linear measurement range as well as the clipping capability for high currents (cf. section 2). A series connection can increase the measurement range, however undesired dynamic effects increase correspondingly. On the other side, the bandwidth of the SCMS is influenced by the dynamic behavior of the selected diodes.

The commutation of current from the diode branches into the shunt is impacted by reverse recovery. As explained in section 3.1, this problem can be avoided with the use of either Schottky or fast recovery diodes with recovery times in the low two digit nanosecond range.

Forward recovery on the contrary impacts the commutation from the shunt branch into the diode branches, as the forward biased diodes take some time to become completely flooded with charge carriers. The effect can be seen in Fig. 5, where distinct peaks in the voltage across the SCMS are visible.

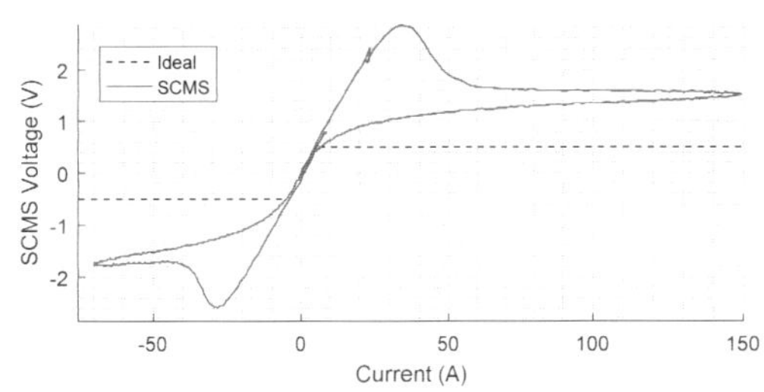

Figure 5. SCMS signal, using a diode with pronounced forward recovery characteristic as a function of reference current.

After recovery, the voltage drops back to the clamping level. While this effect does not impact the current measurement in the linear range, the over voltage due to forward recovery can reach significant levels. To remain in the measurement range of the DAQ and avoid damages to the recording equipment, diode data sheets should be consulted to estimate worst case forward recovery voltages.

3.2. High power performance

After evaluating different components with low power testing, a demonstrator SCMS was assembled to verify that findings can be transferred to high power levels. The system consists of 4 pairs of anti-parallel diodes, each with a 500 A surge current limit, and a 20 mΩ coaxial shunt resistor.

Results of a first high frequency, high current test are illustrated in Fig. 6. As can be seen from subfigure (a), even for high ROROCs, parasitic inductance in the diode branches (connection, package) is not a problem (cf. Fig. 4 (a)). For rates of change of current up to about 12 kA/ms, forward recovery is also negligible. In the zoomed area (subfigure (b)), only a slight overshoot can be seen for the higher ROROC.

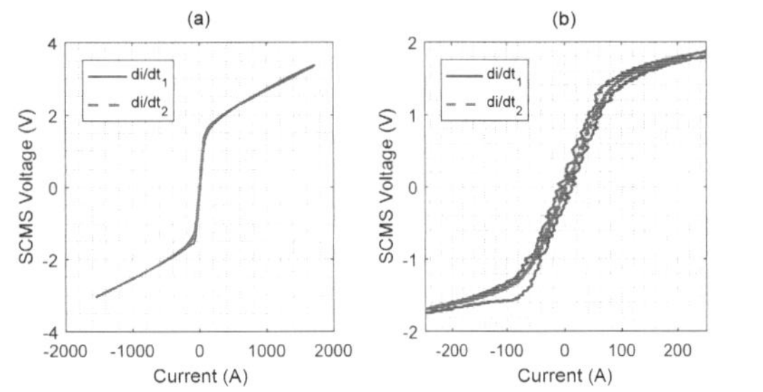

Figure 6. High power testing, peak ROROC: di/dt_1*:* 11.8 kA/ms, di/dt_2*:* 1.9 kA/ms, *(b) detail of (a).*

The limiting factor for performance in this case is the stray inductance in the shunt path. For 11.8 kA/ms, a hysteresis of the curve can be observed (cf. Fig. 4). Using this, the stray inductance can be estimated to about 28 nH. With a linear range of about 1.2 V, the mock-up can be used up to 2.1 kA/ms with less than 5% error, illustrating the scalability of the system.

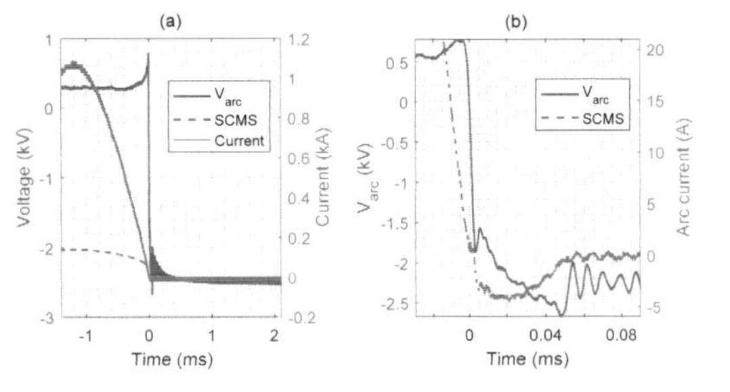

Figure 7. Post arc current measurement with SCMS

Finally, the demonstrator was used in an interruption test of a model gas circuit breaker. The breaker is used to investigate the behavior of mechanical switches as part of HVDC circuit breaker topologies [6]. To model the current as seen by the mechanical circuit breaker during interruption of DC faults, a high frequency oscillatory circuit is used. The breaker is designed for a medium voltage application and uses air at 6 bar as cooling medium.

Fig. 7 shows arc voltage and current (SCMS and reference current probe) of the complete measurement (a), as well as a close-up, illustrating the current zero crossing (b). A post-arc current with an Amplitude of about 4 A and duration of roughly 50 µs can clearly be seen. The signal is not filtered and still contains the capacitive current from the oscillation of circuit breaker stray capacitance and circuit inductance. However, as can be seen on the right-hand side of Fig. 7 (b), this is relatively small.

4. Conclusion

In this paper, dimensioning criteria for a shunt and diode based small current measurement system, suitable to record post-arc currents in high voltage circuit breakers, are discussed. Influencing factors are illustrated with measurements and impacts on measurement of post-arc currents are highlighted. Finally, a demonstrator is used to show the suitability of the dimensioning process.

The presented system has been scaled for the investigation of ultra-fast mechanical breakers, intended for DC systems. Hence, due to longer arcing times and higher fault current amplitudes, further up-scaling would be necessary for a use in HVAC systems. Sufficient bandwidth and sensitivity for measuring post-arc currents in SF_6 can be archived, however, verification tests are recommended.

To conclude, if basic rules are followed and parts are carefully chosen, a diode clamped shunt provides a high-precision, low-cost post-arc measurement system.

Acknowledgements

The work presented in this paper was supported by the Swiss Federal Commission for Innovation and Technology within the SCCER-FURIES.

References

[1] R.P.P. Smeets et al. Performance evaluation of high-voltage circuit breakers by means of current zero analysis. In *Transmis. and Distrib. Conf. and Exhibition APAC*, pages 424–429, 2002. doi:10.1109/TDC.2002.1178398.

[2] V. Vokurka et al. New device for measuring post-arc currents in circuit breakers. *Review of Sci. Instruments*, 58(6):1087–1095, 1987. doi:10.1063/1.1139611.

[3] C. Guilloux et al. Measurement of the post arc current of HV circuit breakers: application to short circuit tests with ITRV. *IEEE Trans. Power Del.*, 8(3):1148–1154, 1993. doi:10.1109/61.252639.

[4] M. Barrault et al. Post-arc current measurement down to the ten milliamperes range. *IEEE Trans. Power Del.*, 8(4):1782–1788, 1993. doi:10.1109/61.248285.

[5] R.P.P. Smeets et al. A new high-resolution high-frequency current-zero measuring system. In *Proc. of ERA Conf. on High Voltage Measurement and Calibration*, pages 1.2.1–1.2.12, 1998.

[6] T. Schultz et al. Interruption Capability Investigations of a Model Gas Circuit-Breaker for HVDC Switching Applications. In *Int. Conf. on Gas Discharges and their Applications*, pages 173–176, 2016.

doi:10.14311/ppt.2017.2.141
Plasma Physics and Technology **4**(2):141–144, 2017

NONEQUILIBRIUM PLASMA CHEMISTRY FOR CONCEPT OF SUSTAINABLE DEVELOPMENT

V. Chernyak*, O. Nedybaliuk, O. Tsymbaliuk, I. Fedirchyk, K. Chunikhina, E. Martysh, V. Iukhymenko, Iu. Veremii, I. Prysiazhnevych, O. Prysiazhna, V. Prysiazhnyi

Faculty of Radio Physics, Electronics and Computer Systems, Taras Shevchenko National University of Kyiv, 64/13 Volodymyrska Street, 01601, Kyiv, Ukraine

* chernyak_v@ukr.net

Abstract. This work is devoted to the exploration of the compatibility of the hybrid plasma-catalytic conversion of liquid hydrocarbons into syngas with the concept of sustainable development. The results of experimental investigations indicate the high efficiency of plasma-catalytic conversion of ethanol to syngas and the small amount of waste (a few percent of feedstock weight). The results of the simulation of the kinetics using ZDPlasKin code for thermochemical and hybrid plasma-catalytic conversion.

Keywords: sustainable development, non-equilibrium plasma chemistry, rotating gliding discharge, conversion.

1. Introduction

The scale on which modern industry uses materials and energy not only led to the substantial depletion of fossil resources but also caused the accumulation of the large amount of wastes, which are hazardous to the environment. In 1980, International Union for Conservation of Nature, United Nations Environment Program and World Wildlife Fund addressed this problem by developing the World Conservation Strategy for Sustainable Development. In September of 2015, the UN proposed 17 goals of sustainable development, which were adopted by 70 countries [1].

The minimization of the use of fossil fuels and their replacement with the renewable energy sources is one of the most important sustainable development goals for ecology, society, and economics. The switch to the renewable biomass can lead to an ecologically beneficial reduction in the emission of carbon, chemicals, and liquid fuels. However, it is widely recognized that the use of first-generation biomass raw materials, such as corn or edible oilseeds, is not a sustainable option in the long run because it competes directly or indirectly with the production of food. Therefore, the European Union is aimed at the use of second-generation biomass raw materials, such as lignocellulose, waste oils and fats [2].

A major barrier to increasing the scale of biofuels production lies in the disadvantages of two main ways of processing lignocellulosic biomass: thermochemical and biotechnological. The biotechnological method of biofuel production is based on using the microorganisms for the processing of biomass into the desired product. This technology is used successfully to obtain ethanol from plant sugars and starches, but it faces significant challenges in the transition to the more complex raw materials, such as lignin and cellulose, and while attempting to produce more sophisticated biofuels than ethanol [3]. At this stage of technological development, the performance of the biotechnological processing of lignocellulosic feedstock is low and is not able to provide the required rate of biofuel production growth.

The modern thermochemical technology of lignocellulosic biomass processing requires much higher temperatures than biotechnological methods (over 700 °C for pyrolysis and 1000 °C for gasification). A significant disadvantage of the thermochemical technologies is low energy efficiency. It is also worth noting that the use of high temperatures not only complicates the design of processing devices and requires the increase of safety requirements, but also decreases the selectivity of the raw materials conversion and leads to the emergence of byproducts. Therefore, the traditional thermochemical (temperature > 700 °C) and traditional plasma-chemical (thermochemical conversion using plasma torches; temperature > 1000 °C) processing of biomass lead to the appearance of the unwanted byproducts and wastes. In the case of catalytic conversion, a high content of impurities in the initial raw biomass negatively affects the yield and longevity of catalysts.

As of today, the discussion is open on the development of principles behind the construction of new technologies for the conversion of substances, which are known as "green chemistry". This is a departure from the traditional evaluation of the effectiveness based on the chemical yield towards the cost-effectiveness evaluation based on the lack of hazardous waste and toxic and/or hazardous substances. "Green chemistry" has to transform the raw materials (preferably renewable), exclude hazardous waste, and use no toxic agents in the production of chemical products and in their application [4]. There should be the development of processes involving H_2, O_2, CO, CO_2, NH as a direct source of atoms H, O, C and N in the production of fine chemicals. The quantitative measure of the environmental acceptability of chemical technology is E factor, which is defined as the ratio of waste weight to the weight of a target product. Waste includes everything that is not a target product [4].

This work is devoted to further experimental research of hybrid plasma-catalytic system with the gas discharge generator of the wide-aperture flow of non-isothermal plasma. Some features of the kinetics in plasma of a wide-aperture rotating gliding discharge are investigated using numerical modeling. The comparison between the model of chemical kinetics during thermochemical reforming and the results of experimental research is provided.

2. Methods and experimental setup

Partial oxidation reforming of hydrocarbons is the most energy-efficient and economical reaction pathway for use in the conversion methods that utilize electric energy. This is especially important for the plasma-chemical reforming methods that use electric energy. The reason for this is that in modern power plants the production of one unit of electrical energy requires three units of chemical energy. Fig. 1 shows the schemes of two hydrocarbon conversion approaches present in plasma-chemistry: plasma reforming and plasma-catalytic reforming.

In plasma approach, the hydrocarbon and oxidant are introduced directly into the discharge area after which the activated mixture is injected into the reaction chamber. Such approach leads to the appearance of the exothermal reactions in the discharge area, which increase the rotational temperature of heavy plasma components and cause the decrease of plasma non-isothermality and drop in the yield of desired products [5]. In plasma-catalytic approach, only a part of the oxidant is introduced into the discharge area and the activated oxidant is injected in a form of radicals into the reaction chamber, which has the separate injection of hydrocarbon. This approach results in the generation of the non-isothermal plasma, which provides the increased selectivity of desired product. Discharge can be used to activate a part of the oxidant while the rest of the oxidant is introduced into the reaction chamber together with the hydrocarbon. The plasma-catalytic approach uses the active species (OH, H, O), which are produced inside plasma from the non-toxic and/or safe reagents, to initiate the chain reactions of renewable biomass conversion at a low temperature (≈250–350 °C).

A discharge chamber, which is attached to the reaction chamber [6], is designed for the plasma-catalytic approach to the hydrocarbon conversion. Plasma is generated using wide-aperture rotating gliding discharge [7, 8]. The discharge is powered using DC power source. Fig. 2 shows the discharge chamber, gas flow directions, and wide-aperture rotating gliding discharge.

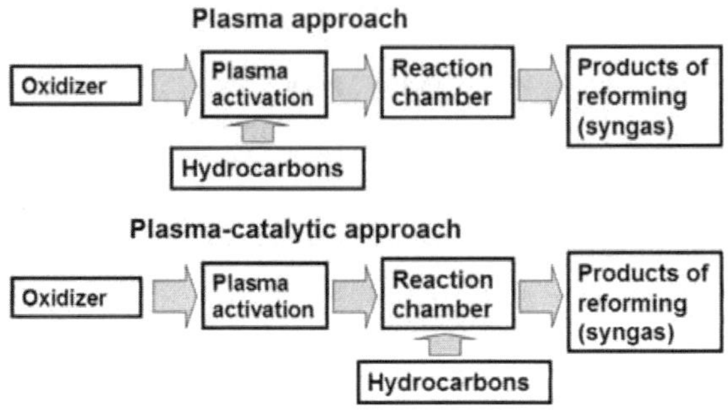

Figure 1. Schematic representation of plasma and plasma-catalytic approaches to hydrocarbon conversion.

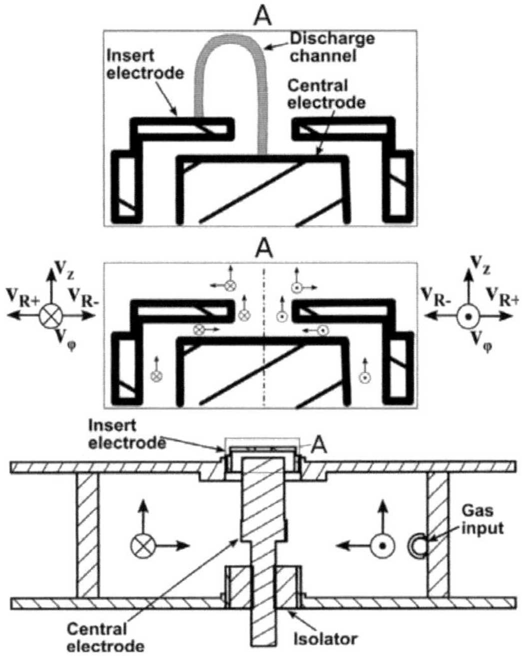

Figure 2. Scheme of discharge chamber, directions of gas flows and wide-aperture rotating gliding discharge during plasma-catalytic reforming of hydrocarbons.

96 % ethanol was used as a model hydrocarbon during the plasma-catalytic reforming of hydrocarbons, atmospheric air, which was supplied using a compressor, was used as an oxidant. The system for plasma-catalytic reforming of hydrocarbons into the synthesis gas was designed for the output power of 25 kW and tested using 53.3 g min^{-1} (25 kW) ethanol flow. The experimental laboratory studies of hybrid plasma-catalytic reforming of ethanol were conducted at lower output power. Optical emission spectroscopy was used for the study of plasma. Gas chromatography was used for the study of gas-phase products.

3. Experimental results

The efficiency of the reforming system in terms of hydrogen production can be evaluated from its hydrogen

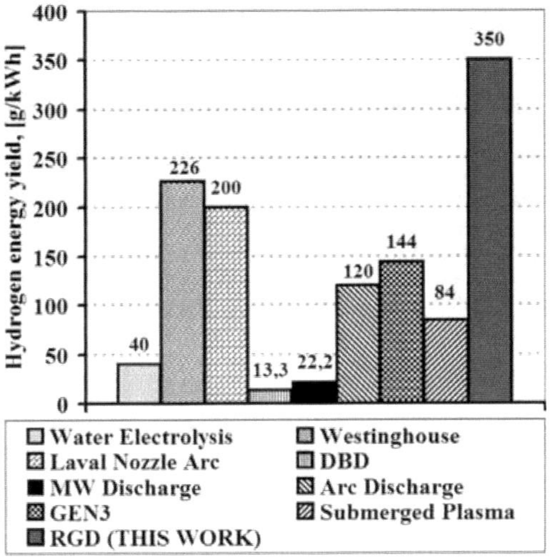

Figure 3. The hydrogen energy yield of reforming of ethanol for different methods of hydrogen production using electrical energy.

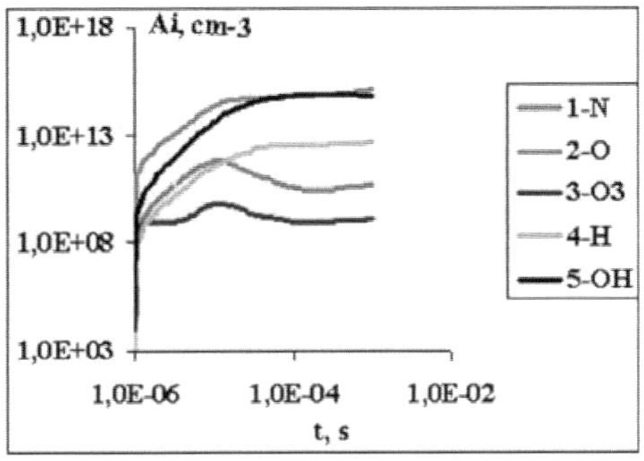

Figure 4. Typical time dependencies of the concentrations of N, O, O_3, H, OH at 2500 K, 400 V cm^{-1} for 0.9Air/0.1H_2O gas.

energy yield. It corresponds to the flow of produced H_2 divided by the electric power spent on plasma generation and is commonly expressed in g kWh^{-1}. Hydrogen energy yield provided in this work corresponds to 4.3 g min^{-1} (2 kW) ethanol input and 24 W input into plasma generation. Based on H_2 concentration in reforming products provided by gas chromatography, the flow of produced H_2 was $\approx$ 8.4 g h^{-1}. Resulting hydrogen energy yield was $\approx$ 350 g kWh^{-1}.

Fig. 3 shows the comparison between the hydrogen energy yields and conversion efficiency of the hybrid plasma-catalytic reforming of ethanol (this work) and other methods of hydrogen production that utilize electrical energy: Water Electrolysis [9], Westinghouse [10], Laval Nozzle Arc [11], DBD [11], MW Discharge [11], Arc Discharge [12], GEN3 [13], Submerged Plasma [14].

The hydrogen energy yield of hybrid plasma-catalytic ethanol reforming is the highest in comparison with the other methods [9–14]. The ratio of the reactor power output to its volume is approximately 100 kW l^{-1}. The hybrid plasma-catalytic approach can be used to solve the existing problems of the traditional renewable biomass reforming methods.

4. Results of numerical modeling

The kinetics of plasma-chemical processes in the plasma of wide-aperture rotating gliding discharge and chemical kinetics of ethanol conversion into synthesis gas were calculated via ZDPlasKin program code [15]. ZDPlasKin allows the simulation of plasma kinetics from the preliminary set of input parameters of the studied system (temperature, pressure, the initial concentration of reagents etc.). During the calculation fulfilled for the plasma of moist air, the database used for the calculation comprised of 900 chemical and electron-molecular reactions involving 83 components. Time sampling step was 10^{-12} s. The time interval for the calculation of changes in the concentration of plasma components at the time evolution was 10^{-3} s. It is significantly higher than the time of the exposure of gas to the current channel of gas discharge considering the experimental data: the rate of gas flow was $\approx 10^3$ cm s^{-1}, wide-aperture rotating discharge is a transverse discharge with the diameter of current channel $\approx 10^{-1}$ cm, respectively the time of gas exposure to the current channel is $\approx 10^{-4}$ s.

This list of reactions is based on a mechanism proposed in [16]. Because the experimental data confirms that the rotational temperature, determined using optical emission spectroscopy, is high enough ($\approx$ 2500 K), all the reaction of the oxidation of atmospheric nitrogen were included to the list of reactions. The mechanism was complemented by the reactions, which involved such species as N_2O, NO_2, NO_3, N_2O_5, N_2O^+, NO_2^+, N_2O^-, NO_2^- NO_3^-, the rates of chemical and electron reactions were taken from [17]. According to the experiment, the electric field in the plasma zone varied in the range of 100–500 V cm^{-1}, which necessitated the addition of dissociation reactions by electron impact of such components as H_2O, NO_2, and HO. At the same time, in this type of conversion, the air with the addition of H_2O is the plasma gas, which is why the list of reactions involving H_2O and its derivatives was added to this mechanism. The list of reactions involving H_2O was composed using [18].

Typical time dependencies of the concentrations of N, O, O_3, H, OH, which are calculated for the plasma of gas discharge – [Ai](t), are shown in Fig. 4. Provided dependences [Ai](t) show that during the experiment the setting time of the fixed values [Ai] of O, H, and OH radicals corresponds to $\approx 10^{-4}$ s, and the degree of the dissociation of molecular oxygen O_2 and water H_2O is $\approx 10^{-2}$.

During the calculation of chemical kinetics in the reactor zone of the experimental system [7], the database contained 130 chemical reactions involving 30 components [19]. Time sampling step was 10^{-6} s. The time interval for the calculation of changes in the concentration of plasma components was 10^{-3} s. The time

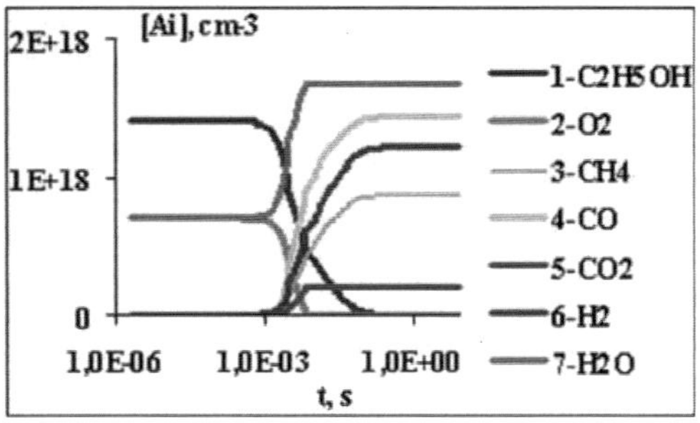

Figure 5. Time dependence of concentrations for the thermochemical conversion of ethanol at 1300 K for $1C_2H_5OH/0.5O_2/0.5H_2O/2N_2$ initial mixture.

interval for the calculation of changes in the concentration of the gas components at the outlet of the reactor zone (Fig. 5) was equal to 10 seconds.

Time dependence of concentrations for the thermochemical partial oxidative conversion of ethanol indicates the existence of two conversion stages: with oxygen and without oxygen. Fig. 5 shows that the ratio between H_2 and CO changes in time.

Product concentrations were calculated for 523–823 K temperature range. Syngas ratio depended on the temperature and changed from 1.6 to 0.7.

5. Conclusions

1. Plasma-catalytic approach is an emerging alternative to the traditional conversion technologies. Plasma-catalytic approach uses active species generated in the plasma from non-toxic and/or non-hazardous reagents to initiate the chain reactions of renewable biomass conversion at low temperature ($\approx$ 500–600 K).

2. Plasma-catalytic approach confirms the concept of sustainable development with the production of small amounts of waste.

3. Studies of chemical kinetics revealed differences in the mechanism of the conversions that result in significant changes in syngas ratio ($[H_2]/[CO]$).

4. Partial oxidative conversion of ethanol has two conversion stages: with oxygen and without oxygen.

5. Temperature dependences of the conversion products concentrations point at the change of syngas ratio caused by the rise of CO concentration with the increase of reforming temperature. In addition, CH_3CHO production correlates with sharp oxidant (O_2) concentration decrease in the system, indicating that reforming is conducted through the chain reactions.

References

[1] SDGs: Sustainable Development Knowledge Platform. https://sustainabledevelopment.un.org/sdgs.

[2] R. A. Sheldon. Green chemistry, catalysis and valorization of waste biomass. *J Mol Catal A Chem*, 422:3–12, 2016. doi:10.1016/j.molcata.2016.01.013.

[3] S. I. Veettil, L. Kumar, and A. A. Koukoulas. Can Microbially Derived Advanced Biofuels Ever Compete with Conventional Bioethanol? A Critical Review. *BioResources*, 11(4):10711–10755, 2016. doi:10.15376/biores.11.4.Veettil.

[4] R. A. Sheldon et al. *Green Chemistry and Catalysis*. Wiley-VCH Verlag GmbH & Co. KGaA, 2007.

[5] V. Ya. Chernyak et al. The Problem of Chemically Active Plasma Nonequilibrium. Review. *Plasma Phys Technol*, 2(3):226–232, 2015.

[6] O. A. Nedybaliuk et al. Plasma-catalytic reforming of liquid hydrocarbons. *Probl At Sci Technol*, 95(1):235–238, 2015.

[7] V. Ya. Chernyak et al. Plasma-liquid system with rotational gliding arc with liquid electrode. In *20th Symp Phys Switch Arc 2013, FSO 2013, 2-6 September, 2013*, pages 112–115, Nove Mesto na Morave, Czech Republic, 2013.

[8] V. Ya. Chernyak et al. The Rotating Gliding Discharge in Quartz Tube and Open Half-Spce at Atmospheric Pressure. *Plasma Phys Technol*, 2(2):116–119, 2015.

[9] Current (2009) State-of-the-Art Hydrogen Production Cost Estimate Using Water Electrolysis (NREL/BK-6A1-46676). Technical report, Golden, Colorado, USA, 2009. http://www.osti.gov/servlets/purl/1218930/.

[10] C. Ducharme. *Technical and economic analysis of Plasma-assisted Waste-to-Energy processes*. PhD thesis, Columbia University, 2010.

[11] J. Mizeraczyk and M. Jasiński. Plasma processing methods for hydrogen production. *Eur Phys J Appl Phys*, 75(2):24702, 2016. doi:10.1051/epjap/2016150561.

[12] G. Petitpas et al. Ethanol and E85 Reforming Assisted by a Non-thermal Arc Discharge. *Energy & Fuels*, 24(4):2607–2613, 2010. doi:10.1021/ef100022r.

[13] K. Hadidi, L. Bromberg, D. R. Cohn, et al. Hydrogen-Rich Gas Production from Plasmatron Reforming of Biofuels. In *World Renew Energy Congr VIII Expo*, Denver, CO, USA, 2004.

[14] L. Bardos, H. Baránková, and A. Bardos. Production of Hydrogen-Rich Synthesis Gas by Pulsed Atmospheric Plasma Submerged in Mixture of Water with Ethanol. *Plasma Chem Plasma Process*, 37(1):115–123, 2017. doi:10.1007/s11090-016-9766-6.

[15] S. Pancheshnyi, B. Eismann, G. J. M. Hagelaar, and L. C. Pitchford. Computer code ZDPlasKin, 2008. http://www.zdplaskin.laplace.univ-tise.fr.

[16] M. Capitelli et al. *Plasma Kinetics in Atmospheric Gases*. Springer Series on Atomic, Optical, and Plasma Physics. Springer Berlin Heidelberg, 2000.

[17] A. Flitti and S. Pancheshnyi. Gas heating in fast pulsed discharges in N_2-O_2 mixtures. *Eur Phys J Appl Phys*, 45(2):21001, 2009. doi:10.1051/epjap/2009011.

[18] L. Morgan. Plasma Chemistry Modeling - KINEMA Research and Software LLC. http://kinema.com/plasma-chemistry-modeling/.

[19] A. N. Tsymbalyuk et al. Influence of the gas mixture temperature on the efficiency of synthesis gas production from ethanol in a nonequilibrium plasma. *Tech Phys*, 58(8):1138–1143, 2013. doi:10.1134/S1063784213080252.

doi:10.14311/ppt.2017.2.145
Plasma Physics and Technology 4(2):145–148, 2017

INFLUENCE OF ABLATION ON DIFFERENTIAL ARC RESISTANCE

L. S. J. Bort*, V. Freiermuth, C. M. Franck

High Voltage Laboratory, Eidgenössische Technische Hochschule, Zürich, Switzerland

* lbort@ethz.ch

Abstract. The influence of ablation on the du/di behavior of an arc in a model gas circuit breaker was examined. Specifically the transition from a state without ablation in the nozzle towards states with ablation was of interest, since prior work indicated that for high currents the voltage becomes constant or du/di gets even positive if ablation is present. Measurements with different blow pressures and rectangular DC currents of varying amplitude were compared, using PMMA-nozzles and dry air as blowing gas. Ablation was measured by weighing the nozzle, scanning the cross section, and using a coordinate measuring machine. The results agreed well, and confirmed that higher pressure shifts the du/di curve towards more favorable values.

Keywords: axially blown arc, ablation, passive oscillation, MRTS, HVDC CB.

1. Introduction

Unlike in AC, the interruption of DC currents requires the creation of an artificial current zero crossing. Today in DC transfer switches, such as metal return transfer switches (MRTS), the passive oscillation circuit is used. An LC-circuit is connected in parallel to a gas circuit breaker. If the arc has a negative differential resistance du/di, the damping of the oscillation in the LC-arc-loop is negative, and eventually will lead to current zero in the breaker [1]. If the arc voltage is not falling with increasing current, passive oscillation fails and additional circuitry is needed which actively brings the current to zero[2]. Previous work at HVL [3] and others [4] indicated that a high amount of ablation leads to du/di being positive, but did not analyze it in detail.

2. Setup

The setup used was previously described in [3], [5], [6]. It consists of a model gas circuit breaker, which uses gas bottles for a freely adjustable blow pressure and two contact pins of which one is movable to draw the arc. Dry air was used as a blowing gas. The inlet pressure is variable, the exhaust pressure is always fixed at 1 bar. Since no gas handling is required to access the parts, it is possible to remove and reinstall the nozzle in a matter of minutes, which made it possible to determine the ablation after each test in reasonable time.

For the measurements presented in this paper, three identical nozzles were produced. Each was used for a series of experiments of one pressure. In the following, these series are shown as 4.5 bar abs., 8.4 bar abs. and 12.7 bar abs., which refers to the inlet pressure of the nozzle. PMMA was chosen because it is transparent, making it possible to observe the arc. To allow separation along the axis of cylindrical symmetry, each nozzle consisted of two identical parts. This

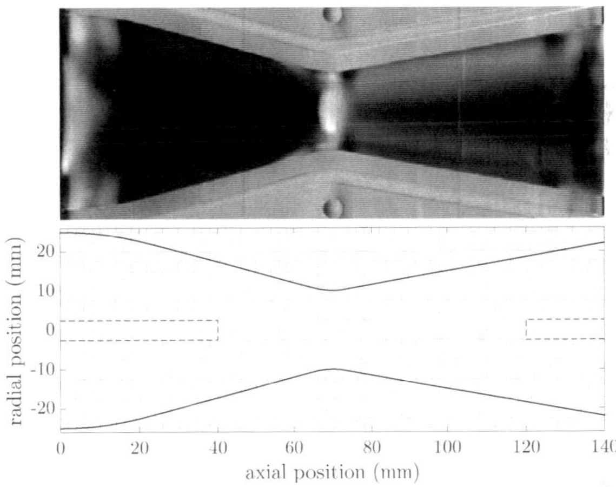

Figure 1. Top: Image of one nozzle half as captured by flatbed scanner. The contour can be seen, as well as the grooves for the sealing cord and two of the six holes for dovel pins. Bottom: Extracted nozzle contour (solid), and position of contacts (dashed). Gasflow was from left to right, cathode was left, anode right.

was achieved by milling the contour into two rectangular blocks of PMMA, which were bolted together. To ensure tightness, a rubber seal in a groove was used, and the alignment was guaranteed by dovel pins. Separating the nozzle allowed scanning the contour line in between experiments to track the changes in geometry, as described in sec. 3.

The nozzle geometry is shown in fig. 1, and identical to the one presented in [5]. The inlet directly attaches to the inlet pressure volume and has a diameter of 50 mm. Both the converging and the diverging part are conical, with half angles of 15° and 10° respectively. The throat was 70 mm after the inlet and was 20 mm in diameter initially. The contacts are copper-tungsten rods with 5 mm diameter, which were at the indicated axial positions of 40 mm and 120 mm.

3. Method

The ablation of nozzle wall material was determined using three different methods. First, the two halves where weighed with a scale, with which the mass could be determined with ±10 mg accuracy, giving ±20 mg certainty for the complete nozzle.

Secondly, the two halves were scanned with an off the shelf flatbed scanner. Assuming cylindrical symmetry, the change of the contour and the density of PMMA (1.188 mg mm^{-3}) can be used to calculate the ablated mass. Additionally, this method shows the axial location of the ablation. The scanner used had a resolution of 4800 dpi, which gives a theoretical resolution of 5 µm per pixel. The gray scale images were aligned, cropped and converted to a binary image first, and the contour of the wall was extracted afterwards using MATLAB. Figure 1 shows one example image before processing and the result.

As a third method, the inner contour of the nozzle was determined by a coordinate measurement machine (CMM) *Mitutoyo KN815*, which is usually used to compare workpieces to the specifications. This machine determined the radius every 5 mm along the axial direction, by measuring 8 points along the circumference. Since the machine was not available at the laboratory, it was only feasible to measure each nozzle before use, and two of the three after the complete series of experiments. The accuracy of these measurements was given by the roundness of the circles, which was about 20–40 µm before and 30–100 µm after all the experiments.

To create the test currents, the FPDCS current source as described in [6] was used. As shown in figure 2, each experiment consisted of two phases: During the pre-current phase, only a small current is flowing while the downstream contact moves from the closed position to the open position. This way, an arc is established. Towards the end of this phase, the current is slightly increased and the valves of the blowing system are opened. As soon as the gas flow has reached a steady state and the contact has stopped its movement at the final position, the main current phase is started. During this time, the current is increased and held at its set value for 10 ms. All results that are shown in section 4 are calculated only from the values of the main current phase.

After each measurement, the nozzle halves were weighed and the contour was scanned with the flatbed scanner, to determine the ablation. Two measurements per current setting were done to ensure repeatability, while keeping the total number of experiments - and therefore ablation - for each nozzle low, to ensure the geometry is as similar as possible for all tests. For the same reason, the measurements were done in ascending order, starting at 300 A up to 1400 A. The measurements with 100 A and 200 A however were done afterwards, when it became clear that they are desired to complete the data set. Since the ablation values of 300 A were already at the edge of the measurement range, the 100 A and 200 A tests were done with an adapted current shape. The inductance of the current source was increased, to reduce ripple and prevent the current to go to zero. Additionally the duration of the main current phase was increased from 10 ms to 40 ms, to increase the amount of ablation towards measurable values. During the complete measurement series, the throat diameter increased by a maximum of 10%, which did not change the arc voltage measurably. This was confirmed by repeating the first test at the very end.

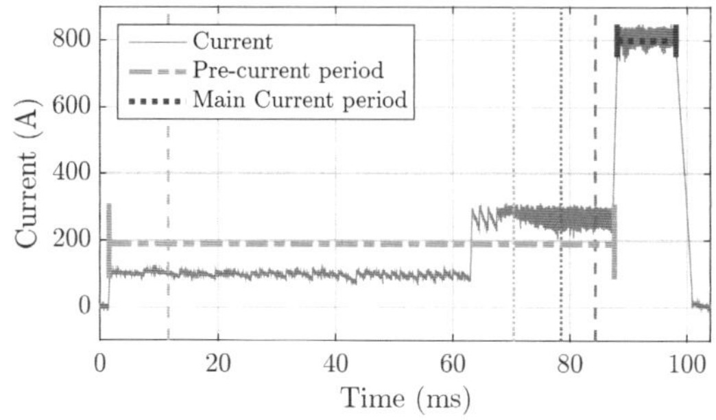

Figure 2. Example current waveform. Start and end of contact movement are indicated with vertical dashed lines, start and stabilization of blowing gas flow with vertical dotted lines. The 85 ms of pre-current were identical for all experiments, followed by 10 ms of main current, 800 A in this case.

4. Results

According to the considerations described in [5] and [7], the voltage and current data of the high current phase was evaluated using the median value instead of the mean, to be more robust against fluctuations. The error bars in figures 3 and 4 are indicating the 15.9 and 84.1 percentiles. Those percentiles give meaningful results even if the voltage fluctuations are not gaussian, and equal the standard deviation if they are. Figure 4 shows the du/di behavior is more favorable for higher blow pressures. For 4.5 bar, the voltage curve is falling up to about 400 A, and is slightly rising until it becomes flat above. For 8.4 bar the rising part is less pronounced and for 12.7 bar the rising part is not observed anymore. The curve gradually transitions into a flat regime around 800 A to 1000 A. Additionally, the overall arc voltage and arc power is higher for higher blow pressures, as shown in fig. 3.

Table 1 shows the mass that was ablated during the entire measurement series, for all three pressures. The CMM measurements agree very well with the weight measurement whereas the flatbed scanner results overestimate the mass loss by about 10%. Figure 7 shows the nozzle geometry change of the 12.4 bar series, as determined by the 29 measured circles of the CMM, and the scan. It can be seen that upstream of the throat (< 70 mm), the scan overestimates the ablation, whereas downstream it shows less in comparison with

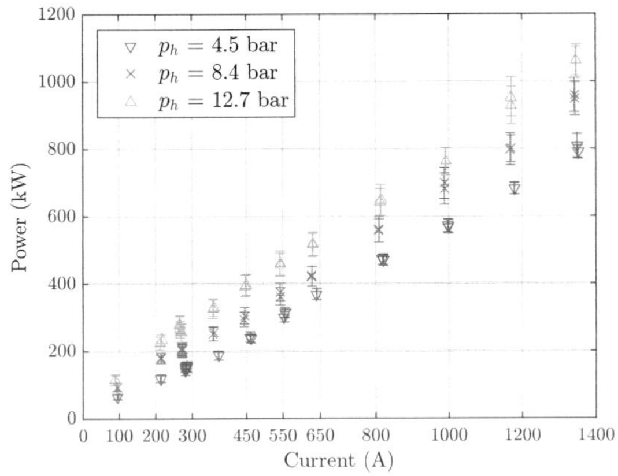

Figure 3. Arc power during the main current phase as function of current. Errorbars indicate the 15.9 and 84.1 percentiles.

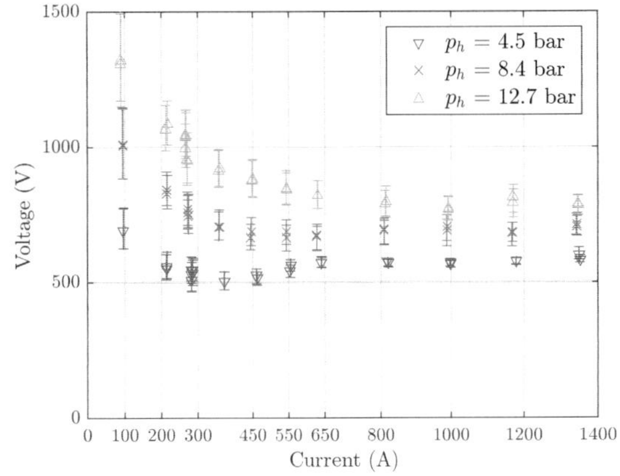

Figure 4. Arc voltage as function of current. The markers show the median, the errorbars indicate the 15.9 and 84.1 percentiles.

	Scale	Scan	CMM
4.5 bar	5.43 g	5.56 g	n/a
8.4 bar	4.05 g	4.41 g	4.14 g
12.7 bar	3.38 g	3.72 g	3.35 g

Table 1. Sum of the ablated mass of all the experiments for each nozzle, determined by weighing, the scans and the CMM. Scale inaccuracy is at least ±20 mg.

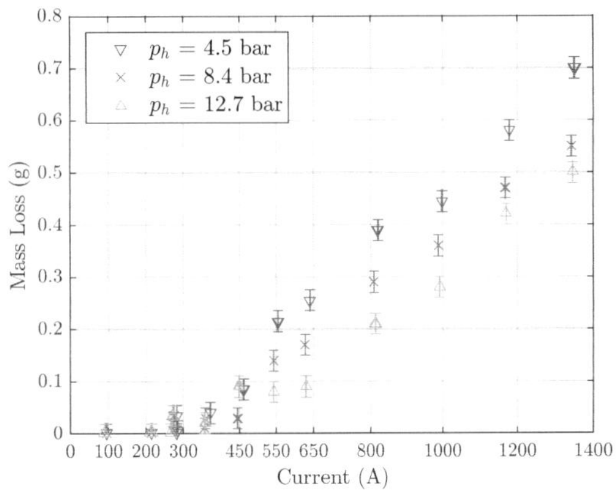

Figure 5. Ablated mass as a function of current. Initially all experiments with currents from 300 A to 1400 A used 10 ms of main current. Additional experiments with 300 A and below were performed, using longer periods of main current, but ablation was still too low to be measured accurately.

the CMM. This means the overestimation in the upstream section has to be higher than 10%. The data for the individual currents was only recorded with the scale and the scan, and indicates that the scanning method has an uncertainty around ±100 mg. Since the scale had an accuracy of ±20 mg, the values from weight measurements were used for figures 5 and 6. Figure 5 clearly shows that the mass loss due to ablation is increasing with current after a certain threshold around 350 A, and higher blow pressure reduced the amount of ablation for all currents.

A more general picture is obtained if the ablated mass is normalized to the total energy the arc dissipated during the high current phase, as shown in figure 6. This especially allows a comparison of tests with low current but long duration with tests of high current and short time. Experiments with 300 A for 40 ms are included, which had a similar specific ablation. Figure 6 shows even more clearly, that for the lower blow pressures the ablation increases sharply above a threshold current. Qualitatively this is similar to the results of Seeger [8]. Comparing the results quantitatively is not possible though, due to the different materials and gases.

5. Discussion

The scanning technique produced valuable results, even though the theoretical accuracy was not achieved. It was not possible to detect the edge of the contour on the scans with single-pixel resolution, since the edge is visible as a brightness gradient of roughly 20 px, or 100 µm. To improve the edge detection accuracy, painting the surface with different marker pens was considered, but all inks that were tried did not increase the detection accuracy by much. Other materials or changes to the background or lighting might be able to improve this technique. Additional to the edge detection uncertainty, the contour determined by the scan significantly overestimated the change of the radius around the throat compared with the CMM, as shown in figure 7. This is attributed to the fact that the gap where the two nozzle halves are joined is not perfectly gas-tight. The tests with 800 A and above showed erosion on the mating surface, up to to the sealing cord groove, which was likely caused by hot gas entering the gap. This erosion most probably eroded the contour edge as well, which means the assumption of cylindrical symmetry does not hold perfectly.

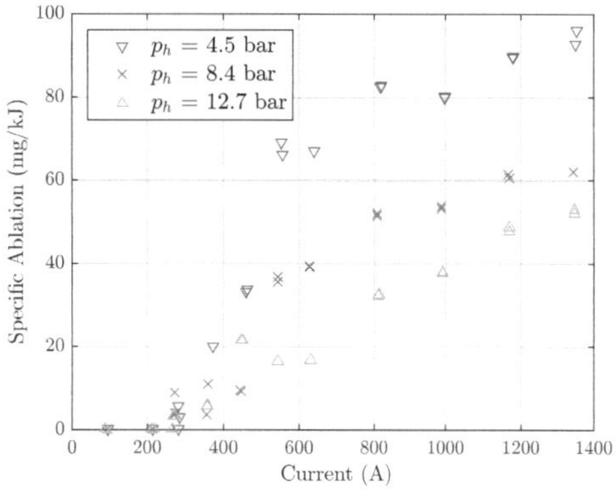

Figure 6. Ablated mass normalized to the energy dissipated by the arc during the main current phase, as function of current. Below 350 A, the ablation rate is about as low as the measuring inaccuracy, above that it rises with current.

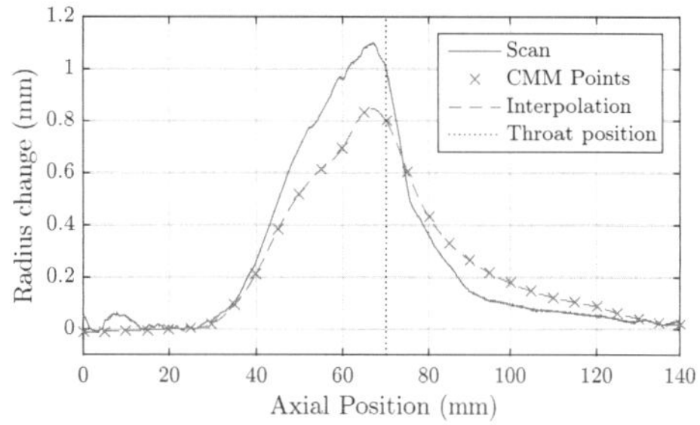

Figure 7. Total change of geometry (see fig 1) due to ablation. Determined by the flatbed scans and the coordinate measurement machine, after the entire measurement series for the 12.4 bar nozzle.

The results of the scans and the CMM (see fig. 7) both show the ablation as a function of axial position. Ablation starts at the position of the upstream contact (40 mm), increases towards the throat position (70 mm) and sharply decreases afterwards. In the diverging part of the nozzle, from 80 mm up to the downstream contact at 120 mm the amount of ablation is much smaller than in the upstream section. One possible reason for this could be higher arc power in the upstream section. Alternatively, the percentage of the dissipated energy arc that contributes to ablation compared with other cooling mechanisms is higher upstream. Higher blow pressures leads to higher arc power and lower ablation rate at the same time. This means, with increasing blow pressure the overall share of the ablation as cooling mechanism is reduced. However the present data does not allow for a quantification or explanation of this effect, since too many factors are still unknown. One of them is the combustion of PMMA with the oxygen of blowing air. According to [9], combustion of 1 g of PMMA yields a similar energy than what is required to vaporize it. This means, it is not possible to determine the energy balance of the ablation properly as long as air is used.

6. Conclusions

The methods described in section 3 worked with an accuracy as expected. Higher weighing accuracy is hard to achieve since it involves including effects like changes in PMMA density due to humidity, or changes in buoyancy due to changes in air pressure. The scanning method could probably be improved by enhancing the scan quality and the edge detection. The results shown in this paper agree well with previous findings [3], which indicate that increasing blow pressure is shifting the $u(i)$ curve towards more favourable conditions for passive oscillation breakers. Additionally it can be seen qualitatively that the threshold current above which ablation starts (300 A to 500 A according to figure 5) coincides with the point where the $u(i)$ changes to unfavorable (flat/rising) conditions (see fig. 4). If this is combined with the fact that the share of ablative cooling is increasing in the same region, the onset of significant ablation is directly correlated with the $u(i)$ curve getting flat.

Acknowledgements

The authors would like to thank ABB Switzerland for financial support, as well as E. Panousis of ABB CRC Switzerland for the frequent discussions.

References

[1] B. Bachmann et al. Development of a 500kV Airblast HVDC Circuit Breaker. *IEEE Trans. on Power Apparatus and Systems*, PAS-104(9):2460–2466, 1985. doi:10.1109/TPAS.1985.318991.

[2] T. Schultz et al. Circuit Breakers for Fault Current Interruption in HVDC Grids. In *VDE-Fachtagung Hochspannungstechnik*, 2016. doi:10.3929/ethz-a-010795297.

[3] M.M. Walter. *Switching Arcs in Passive Resonance HVDC Circuit Breakers*. PhD thesis, ETH Zürich, 2013. doi:10.3929/ethz-a-010112102.

[4] L. Muller. Modelling of an ablation controlled arc. *J Phys D: Appl Phys*, 26(8):1253–1259, 1993. doi:10.1088/0022-3727/26/8/015.

[5] L.S.J. Bort et al. Effects of nozzle and contact geometry on arc voltage in gas circuit-breakers. In *2016 IEEE Intern. Conf. on High Voltage Eng. and Appl.*, 2016. doi:10.1109/ICHVE.2016.7800700.

[6] A. Ritter et al. Five years of pulsed current testing for HVDC switchgear. In *2016 IEEE Intern. Conf. on High Voltage Eng. and Appl.*, 2016. doi:10.1109/ICHVE.2016.7800658.

[7] H. Schmid et al. Measuring a Small Number of Samples, and the 3v Fallacy: Shedding Light on Confidence and Error Intervals. *IEEE Solid-State Circuits Magazine*, 6(2):52–58, 2014. doi:10.1109/MSSC.2014.2313714.

[8] M. Seeger et al. Experimental study on PTFE ablation in high voltage circuit-breakers. *J Phys D: Appl Phys*, 39(23):5016–5024, 2006. doi:10.1088/0022-3727/39/23/018.

[9] B.T. Rhodes et al. Burning rate and flame heat flux for PMMA in a cone calorimeter. *Fire Safety Journal*, 26(3):221–240, 1996. doi:10.1016/S0379-7112(96)00025-2.

doi:10.14311/ppt.2017.2.149
Plasma Physics and Technology **4**(2):149–152, 2017

POLYMERIC ABLATION INDUCED BY FREE BURNING ARCS IN AIR

M. Bator*, R. Bianchetti, P. Suetterlin

ABB Corporate Research Center Switzerland, Segelhofstrasse 1K, 5405 Baden-Daettwil, Switzerland

* matthias.bator@ch.abb.com

Abstract. We investigate the influence of switching arcs on different polymers and their interaction. We describe a set of experiments on a simplified model geometry typical for low voltage switchgear. In a broad range of experimental conditions and parameters such as arc current, polymeric material or contact material, the voltage, the mass loss and the corresponding pressure build-up are examined. From this raw data, we deduce the arc influence on the ablation process as well as the feedback on some arc plasma properties.

Keywords: free burning arcs, low voltage switchgear, plasma properties, polymeric ablation, thermoplastics.

1. Introduction

Low voltage switchgear and the corresponding development processes matured over decades and today are highly optimized. Thus, most development often concentrates on optimization of very specific details such as arc chamber geometries, contact materials and very often the exact composition of plastic parts. Interestingly, when testing new polymers, the approach is often to test a wide range of materials and settle with the best as the understanding of the underlying interaction between the arc and ablated polymers is often lacking.

The goal of the investigation presented in this paper is to improve exactly this understanding. Two simplified laboratory setups were used, one with two parallel polymeric plates with an arc ignited between two cylindrical electrodes, and the other with a polymeric cylinder in order to constrict the arc axially. We investigated the mass loss, the arc voltage and the pressure build up in dependence on the current as well as the polymeric and contact material. Specifically PA6, PA66, PTFE and PE were investigated in pure forms as well as containing additives. The electrodes consisted of either copper or steel.

2. Experimental Setup

The setup consisted of a capacitor bank with 32 mF charged up to 950 V. It was discharged through an inductance and the DUT (device under test) with a pseudo 50 Hz pulse. A second variation used an additional buck converter allowing for an actively controlled current discharge of the capacitor through our DUT (see Figure 1). In these cases, the capacitor bank was used in a 16 mF configuration and was typically charged up to 1600 V. Two different geometries (see were used with both setups.

In the setup including the buck converter, current control was achieved by switching an ABB HiPak IGBT via a feedback PI control (see Figure 1). As

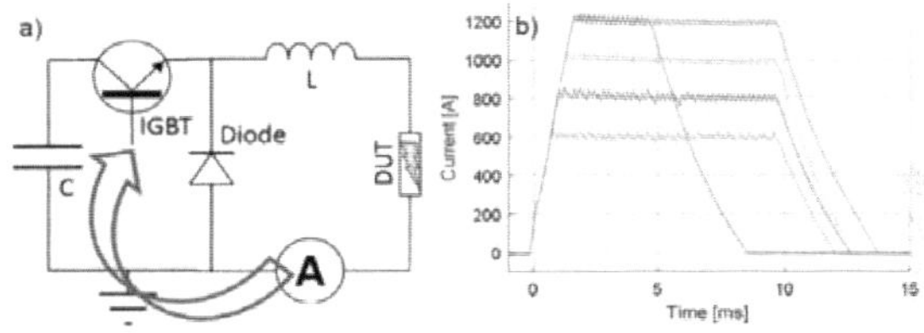

Figure 1. a) Schematics of the active current control setup. A pre charged capacitance (C) is discharged via an IGBT and an inductor (L) through the DUT. The current (A) is used to control the semiconductor. b) Currents (0.6, 0.8, 1.0 and 1.2 kA) set using the setup in a). In one instance, the control loop was stopped after 5 ms; the others were stopped after 10 ms.

long as the current was below the set value, the IGBT was on and the current rose with a time determined by the used inductance (1.5 mH in our case) and the voltage difference between arc and the capacitor bank. After reaching the set current, the ABB HiPak IGBT is switched off and the current decreases as energy is dissipated in the arc. As a result, a small ripple (determined by amongst others the inductance and the control frequency) is visible on the current trace, which typically was in the range of ±20 A on a 1 kA set current. Either after a set time or when the capacitor voltage drops below the voltage of the DUT no modulation of the current occurs and it drops to zero.

2.1. Parallel Plate Geometry

As mentioned, two different geometries were used as DUT, both utilizing a thin wire for ignition. In the first, two polymeric plates were arranged in parallel with one electrode mounted on the right between the two plates (see Figure 2) allowing for optical access from the side perpendicular to the arc direction. The second electrode is mounted to the left of the plates just outside the picture shown in Figure 2. The plate surface measures 20 x 40 mm with an adjustable distance in this case set to 12 mm. This setup resem-

Figure 2. a) Side view of the parallel plate geometry. An electrode is mounted on the right between the two polymeric plates to be investigated. The second electrode, on axis with the first is not shown on the left side. The plates were separated by 12 mm. b) CAD model of the geometry.

bles the typical configuration present in a low voltage breaker with an arc squeezed in between two polymeric walls.

2.2. Nozzle Type Geometry

In the second geometry, a tube replaced the polymeric plates. Here, the electrode is mounted gastight inside the tube on the right (see Figure 3) such that gas flow is only allowed towards the left. The second electrode is again mounted outside of the polymeric region (visible on the left side of Figure 3 a)). The channel within the tube has a length of 45 mm with a variable inner diameter of 5, 7 or 10 mm. The closed geometry allows for an additional pressure measurement inside the tube. Differently to the parallel plate setup, optical access is only given at the outlet of the tube and no longer along the whole arc.

2.3. Optical Measurements

A high-speed camera setup is arranged perpendicular to the arc direction in order to visualize the plasma under different conditions. Typical exposure times were in the range of a few µs. This allowed to either visualize a detailed time evolution of the plasma or give a good overview of its distribution. Using bandpass filters allowed to better distinguish emission from specific sources such as the H_α- or specific metallic lines.

3. Results

The two setups were used to investigate different aspects of the polymer-arc interaction. The parallel plate setup was used to optically investigate the spatial distribution of the arc and how it is influenced by degassing of the polymers. In the cylindrical setup, we concentrated on the pressure and arc voltage build-up during the arcing as well as on the integrated mass loss of the polymeric tubes.

3.1. Plasma Distribution

The main interest of the investigation was the interaction of the plasma with the polymeric walls. In a first step, we investigated how the arc arranges itself within the parallel polymeric plates. This is done by doing time-resolved optical imaging of the arc using the high-speed camera. In a first series, neat PA6 (neat in this context meaning commercially available pure compounds) plates were arranged 12 mm apart. The applied current was a pseudo 50 Hz pulse with 3 kA peak value.

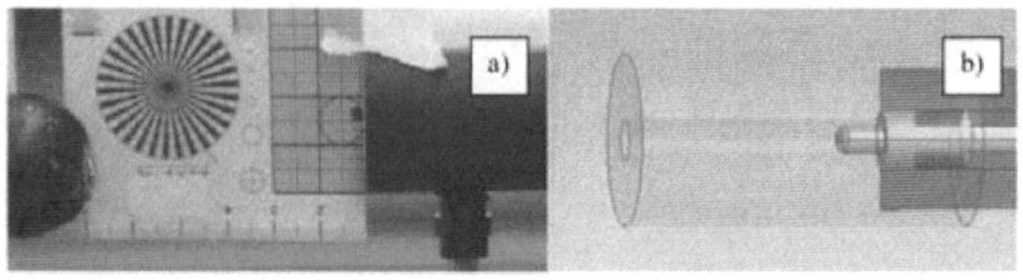

Figure 3. a) Side view of the nozzle geometry. An electrode is mounted on the right inside a hollow polymeric cylinder. The right side is designed to be air tight, while the left is open towards the ground electrode (also shown). b) CAD model of the geometry. The electrode tip has a diameter of 5 mm.

The ignition wire explodes ca. 0.1 ms after current initiation filling the volume in front of the contact with hot metallic plasma and residues of the wire. In this early phase, up to ca. 0.5 ms (see Figure 4, left column), the hot plasma is distributed asymmetrically and touches the polymeric walls (within our resolution of ca. 0.1 mm). The behavior changes quickly as the current increases; the arc becomes brighter and more constricted. After ca. 1.5 ms, the arc generally stabilizes into a symmetric, constricted shape with dark regions between the bright center and the polymeric walls. In all measurements, a sharp edge between the central arc plasma and the dark region around it is observed. A light glow is visible at the polymeric walls (see Figure 4 left column 5.0 ms). Optical emission spectroscopy results (not discussed in this paper in detail) show that the spectral distribution in this region is identical to the center of the arc comprising mostly ionized metal lines. This leads to the conclusion that the light is most likely reflected and/or scattered from

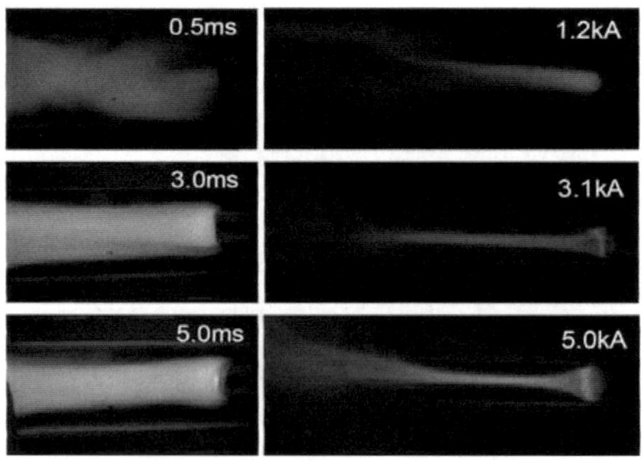

Figure 4. Left panel: Time sequence of a single discharge with neat PA6 walls and a 50 Hz discharge with a 3 kA peak current. Right panel: High-speed frames of an arc discharge with copper electrodes using a bandwith filter at 514.5 nm (3 nm FWHM) exposed for 4 µs at peak current. The walls used are neat PTFE.

the central arc and not emission from evaporating polymers. The exact shape of the arc is not determinable in this setup due to the camera axis being centered between the plates and perpendicular to the arc direction. We assume that the initially cylindrical arc is squeezed into a flattened distribution between the plates.

With changing current and polymeric material, the general behavior does not change. In all measurements, the arc becomes more constricted in the center region, and most likely distributes wider along the optical axis of the camera with increasing currents as shown in Figure 4 (right column). These pictures were taken with an additional bandpass filter around 514.5 nm with a width of 3 nm to enhance the observed radiation coming from atomic emission lines of copper. Furthermore, the wall material in this column is PTFE and not PA6 as in the images on the left. Interestingly, no emission is observed between the polymer and the bright arc regardless of the current. Even additional spectroscopy and absorption measurements (not presented here) do not allow for a classification of the "dark zone" between the polymeric walls and the central arc. A disadvantage of the parallel plate setup and in particular the open sides along the optical axis is the build-up of free-floating fumes in front of the plasma as can be observed towards the top left of the 5.0 kA high speed picture in Figure 4. It covers the plasma, as well as the dark region and therefore falsifies the optical impression of the distance of the arc to the polymeric walls. This effect is considerably stronger using PA6 walls.

The two take-away observations are that the polymer gases are optically not accessible due to most likely being too cold to generate radiation. And secondly that a strong spatial constriction of the arc is observed which considerably increases with higher currents and therefore larger degassing of the polymer.

3.2. Mass Loss and Pressure Build-up

In the second set of experiments, using the nozzle setup, the arc constriction was as apparent as before. At the nozzle opening, the bright plasma region in the center was surround with a dark region towards the walls of the polymeric tube. Still, using this setup the focus of the investigations shifted from the optical investigation to the pressure build-up and voltage during the arcing as well as the mass loss of the polymer tube.

The pressure build-up is an important indicator for the interruption process as the ablated polymers interact in complex ways with the arc. In general, it is considered advantageous to increase the amount of ablated polymer mostly due to its effect of cooling the arc and thus leading to higher arc voltages. Furthermore, the flow can be used either to control the movement of the arc or to diffuse the arc in specific nozzle configurations. Depending on the exact polymeric material used, an additional benefit is the

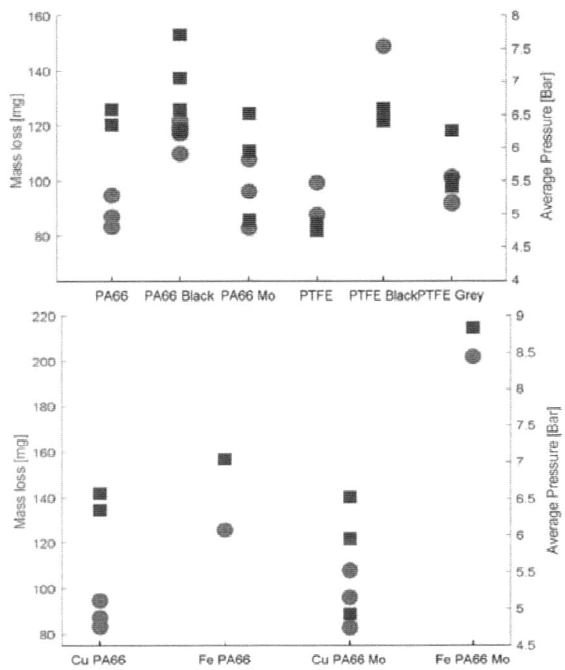

Figure 5. Measured total mass loss (red circles, left axis) and average pressure build-up (blue squares, right axis) for different nozzle materials using a controlled 1.2 kA discharge during 10 ms and different material combinations.

reduced conductivity compared to a metallic plasma and a possibly higher dielectric stability after current zero. Still, there are also disadvantages related to a large amount of hot polymeric gases. Too high pressures might lead to an explosive destruction of the device, in particular at higher currents when the outgassing is scaled up by the current. Furthermore, some polymeric materials, such as e.g. PTFE can lead to higher conductivity in regions with large concentrations [1] leading to either re-strikes outside of the arc chamber during the interruption or to re-ignitions after current zero.

Therefore, it is often necessary to take the decision for a polymeric wall material on a per-device base leading to additional work and uncertainties in the development process. A better understanding of the ablation process on a very general base would help to reduce the necessary testing and ideally enable a better predictability of polymer choices.

In the following we will discuss the findings using combinations of copper, steel, and silver electrodes with neat PA66, "black" PA66 (PA66 with <1% of carbon black), neat PTFE, "grey" PTFE (PTFE with molybdenum disulfide), neat PE and "black" PE (all commercially sourced). The findings are shown in Figure 5. The top panel shows the mass loss and average pressure for different polymers using copper electrodes, while the bottom panel compares PA66 and "black" PA66 using copper and steel electrodes. In both cases, we did not use a 50 Hz discharge, but a

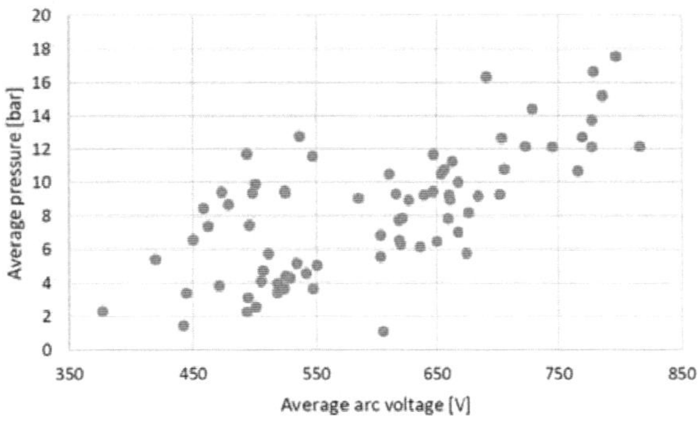

Figure 6. Measured average pressures versus averaged arc voltage in all recorded discharges using the nozzle type setup and the 50Hz synthetic circuit.

controlled 10 ms long current (as shown in Figure 1). The values are averaged between 3 and 8 ms where the variation is the lowest and the influence of the ignition process is no longer present.

We observe a clear trend towards higher ablation ratios for the "black" and "grey" variants of PA66 and PTFE. It has to be mentioned that the spread in identical measurements could be explained by slightly leaky arrangement in particular after several shots when the inner diameter widened slightly. Carbon black is typically used to enhance stability versus UV radiation [2]. Due to the high temperatures in the arcing plasma, usually more than 50% of the radiation reaching the polymeric walls has a wavelength below 300 nm [3]. A stronger absorption in the top-most layers therefore leads to an increase of the ablation processes. Similarly, the higher ablation when using steel electrodes can be explained. The higher net emission compared to copper [4] would result in a larger amount of radiation reaching the walls and thus lead to larger ablation. Furthermore, more ablated metal in addition can change the interaction with the polymeric walls. Interestingly, the pressure in PA66 stays mostly the same for neat PA66, black PA66 and PA66 Mo, while the mass loss is measurably higher for the black version. A possible explanation is that this comparison only includes ablation into the gaseous phase, while the generation of particles and droplets, which can strongly vary between different polymers, typically has less influence on the plasma properties but strongly influences the mass loss. Hence, one could assume that in the case of the black PA66 the stronger absorption leads to a slight increase in gaseous ablation but a stronger increase in generation of particles and droplets. Comparing PTFE and black PTFE on the other hand shows a measurable increase in both, the mass loss and the average pressure. Therefore, in this case one can conclude that the higher absorption is a more direct translation into a larger generation of polymeric gas.

In order to observe the general correlation between the pressure and arc voltages, all measurements using a 50 Hz discharge are shown in Figure 6. The values are taken at the peak currents, but include all material combinations as well as different current settings. Even considering this wide variety and the findings discussed before, a general linear trend is observed. Placing an additional floating electrode at the exit of the nozzle we found that ca. 60-70% of the voltage drop appears in the nozzle. In order to explain the increase of the arc voltage with the pressure, we have to consider the findings from the parallel plate setup regarding how the current influences the arc shape and the extent of the dark layer dominated by relatively cold polymeric gases. The increase of the current leads to higher ablation rates and therefore higher pressures. At the same time, the hotter arc is being constricted in a smaller space. The electrical conductivity is very low at temperatures below 5 kK, grows linearly until ca. 20 kK where it finally flattens out [5]. Therefore, the slightly higher temperature in the core of the arc is not able to compensate the exponentially lower conductivity in the dark regions of "cold" polymer gases leading to an increase of the arc voltage in conditions where a high pressure is expected.

4. Conclusions

We investigated the interaction between polymeric ablation and an electrical arc. We could clearly show the influence the ablated polymer has on the shape of the arc and its flow patterns. Furthermore, we could show how dominant the dark region of relatively cold polymeric gases is not only on the shape but also on the electrical properties of the arc, in particular the arc voltage. Together with the findings that the ablation of polymers can be significantly increased using compounds that incorporate e.g. carbon black the results can be used to further optimize prediction models in switchgear development. The differences in ablation when using steel electrodes compared to copper electrodes further show that the influence of the splitter plates in a real breaker on the ablation properties should not be neglected.

References

[1] Y. Okano et al. Quenching performance of air arc discharge with polymer insulators. *20th Int. Conf. on Gas Disch. and Their Appl.*, 745(A39):199–202, 2014.

[2] R.J. Young and P.A. Lovell. *Introduction to Polymers*. Third Edition. CRC Press, 2011.

[3] ABB. Internal communication/publication.

[4] Y. Cressault and A. Gleizes. Thermal plasma properties for Ar-Al, Ar-Fe and Ar-Cu mixtures used in welding plasma processes: I. net emission coefficients at atmospheric pressure. *J. Phys. D: Appl. Phys*, 46:415206, 2013. doi:10.1088/0022-3727/46/41/415206.

[5] S. Franke et al. Temperature determination in copper-dominated free-burning arcs. *J. Phys. D: Appl. Phys.*, 47:015202, 2014. doi:10.1088/0022-3727/47/1/015202.

doi:10.14311/ppt.2017.2.153
Plasma Physics and Technology 4(2):153–156, 2017

INVESTIGATION OF AN ABLATION-DOMINATED ARC IN A MODEL CHAMBER BY OPTICAL EMISSION SPECTROSCOPY

R. Methling[a,*], A. Khakpour[a], S. Wetzeler[b], D. Uhrlandt[a]

[a] *Leibniz-Institute for Plasma Science and Technology (INP), Felix-Hausdorff-Str. 2, 17489 Greifswald, Germany*
[b] *Institute for High Voltage Technology, RWTH Aachen University, Schinkelstrasse 2, 52056 Aachen, Germany*
[*] methling@inp-greifswald.de

Abstract. A switching arc in a model chamber is investigated by means of optical emission spectroscopy. Ignition wire is applied to initiate an arc of several kiloampere between tungsten–copper electrodes. Radiation emitted by the arc plasma is absorbed by a surrounding PTFE nozzle, leading to an ablation–dominated discharge. Video spectroscopy is carried out using an imaging spectrometer combined with a high–speed video camera. Carbon ion and fluorine atom line emission from the heating channel as well as copper, oxygen and nitrogen from ignition wire and ambient air are analyzed with focus on the low–current phases at the beginning of discharge and near current zero. Additionally, electrical parameters and total pressure are recorded while the general behavior of the discharge is observed by another video camera. Considering rotational symmetry of the arc the corresponding radial emission coefficients are determined. Finally, radial temperature profiles are calculated.

Keywords: ablation, switching arc, spectroscopy.

1. Introduction

High voltage circuit breakers are essential elements in modern power grids which ensure safe power flow [1]. In self–blast circuit breakers the pressure build–up in a heating volume, necessary for arc quenching, is attained by the ablation of material from the nozzle wall due to intense arc radiation. One focus of circuit breaker development is the substitution of the strong greenhouse gas sulfur hexafluoride by carbon dioxide. The current zero (CZ) crossing and the period immediately after current interruption are of high importance for the interruption performance of the circuit breaker. In these time intervals several physical effects occur, such as flow reversal in the heating channel, translation from an ablation–controlled to an axially blown arc, the extinction of the arc and a continued evaporation of nozzle material after current zero due to the preceding thermal stress. The deeper understanding of these effects, their transient behavior and their effects on the dielectric recovery of the gap between the contacts after current zero are key issues for further development of high voltage switchgear, either based SF_6 or alternative gases.

The complicated geometry of real circuit breakers usually does not offer direct optical access to the arc plasma. Thus, model chambers have to be applied to investigate the plasma parameters under more or less similar conditions, comprising solutions with movable pin and tulip electrodes [2] or with fixed electrodes ignited by explosion of thin wires [3] and with non–rotationally symmetric nozzles [4]. In this work two pin electrodes of fixed distance are placed in a long PTFE nozzle. A switching arc in a model chamber is investigated by means of optical emission spectroscopy (OES). In a first step, only the nozzle without the chamber is applied. Here, it is of special interest how far the arc is dominated by the PTFE ablation at low current phases, i.e. during the current increase when the discharge is initiated and when approaching CZ. Beside temperature distribution the influence of ignition wire and the surrounding gas (ambient air) are investigated.

Figure 1. Experimental setup with PTFE nozzle. The observation window made of 2 mm thick quartz glass is positioned in the center of the PTFE nozzle. W–Cu electrodes at both sides are protected from exhaust plasma by ceramic caps.

2. Experimental

A photograph of the chamber setup is shown in Figure 1. The PTFE (polytetrafluoroethylene) nozzle has a length of 126 mm and an outer diameter of 50 mm. The inner diameter is 12 mm over a length of 30 mm along the main part of the arc, then increasing within the next 10 mm to a final diameter of 15 mm. Current flow from a high–current generator is started by a triggered spark gap. An about 0.2 mm thick copper ignition wire is applied to initiate an arc discharge between tungsten–copper electrodes of 40 mm distance and 10 mm diameter. Outside the chamber the ends of the electrodes are protected from exhaust plasma by ceramic caps. The pressure inside the nozzle is measured by means of a piezo–resistive pressure sensor of type 6052C combined with a charge amplifier 5073 (Kistler). The connection port for the pressure sensor can be seen on top of Figure 1. Sinusoidal current waveform of about 100 Hz frequency is applied. The maximum pressure of about 8 bar is measured close to peak current of 11 kA at 4.6 ms.

The intense radiation emitted from the arc plasma is absorbed by the surface of a surrounding PTFE nozzle where it causes photo–ablation of the wall material [5]. Part of the ablated material is exhausted as vapor by axial flow, the other part enters the arc and is heated to plasma temperature by absorption of radiation coming from the arc interior, leading to a discharge that is dominated by the ablation [6]. Both arc plasma and vapor create an overpressure which causes an axial expansion flow towards the ends of the tube. To enable an optical investigation of the arc cross section, a vertical slit is inserted in the middle of the nozzle. Transparent windows made of 2 mm thick quartz plates in the chamber wall are applied to seal the nozzle. The windows transmission is measured before and after each shot. However, it turned out that deposition of ablated material is not critical as long as the windows are cleaned regularly and exchanged every few shots.

Using a combination of 0.5 m–imaging spectrometer and a high–speed video camera (Y4 series of Integrated Design Tools), time-dependent video spectroscopy is carried out. The red dot in Figure 1 marks the central observation position of the OES. The spatial resolution is 28 µm/pixel. The spectral resolution is 0.8 nm using a grating applied for overview spectra with 150 lines/mm and a slit width of 50 µm. The temporal resolution is given by the maximal number of frames per time. It depends mainly on the number of vertical lines of the video camera. For an observation of the complete slit of the nozzle as needed for determination of radial profiles, 600 vertical lines have to be used yielding a frame rate of 7500 fps (frames per second). The exposure time is 2 µs. Reducing the number of vertical lines of the video camera and skipping the spatial resolution, however, allows higher temporal resolution of 100 kfps which is equal to a period between the spectra of only 10 µs. The exposure time is chosen as high as possible (8 µs) in order to obtain highest sensitivity during low–current phases although overexposure occurs during high–current phases. A second high–speed video camera (Y6 series of Integrated Design Tools) is applied for the observation of the general behavior of the discharge.

Ionic as well as atomic line emission from the heating channel are analyzed. Considering rotational symmetry of the arc the corresponding radial emission coefficients are determined. Finally, radial temperature profiles are calculated with temporal resolution.

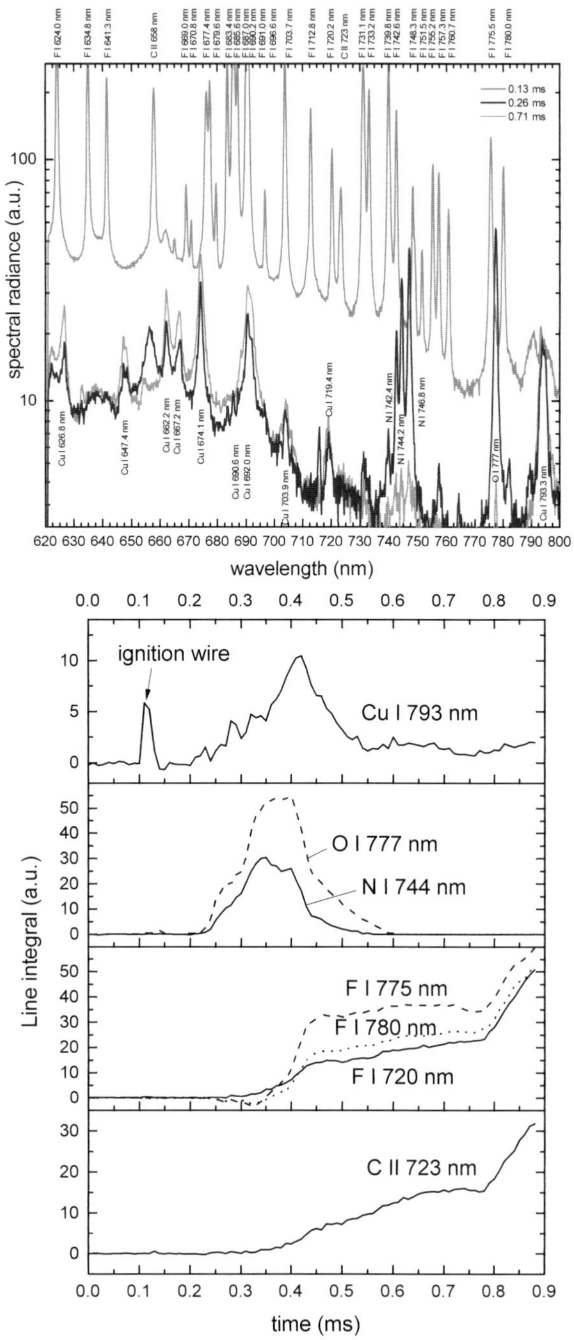

Figure 2. Spectrum variation during initialization. Top: Three spectra measured after 130, 260, and 710 µs are dominated by line emission of Cu I during wire explosion, N I and O I during heating up the arc column, and F I, C I, and C II from nozzle surface during ablation–dominated arc phase, respectively. Bottom: Temporal distribution of spectrally integrated line intensities of selected lines.

3. Results and discussion

Series of two-dimensional spectra containing spatial as well as spectral information are obtained by the video spectroscopy. In the following only the central position is regarded to analyze the temporal evolution

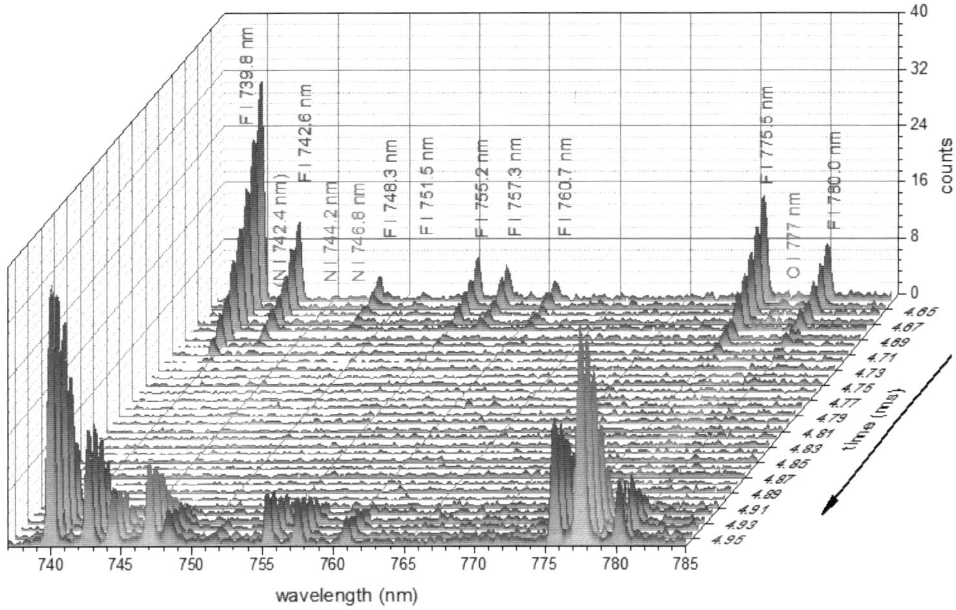

Figure 3. Temporal evolution of spectra around current zero showing influence of ambient air.

of the spectrum during the discharge with 100 kfps. An example of three spectra acquired during the initial phase of one shot is shown in the upper part of Figure 2. The major lines are labeled with PTFE material above and the rest inside the spectrum. For better distinction, logarithmic scaling of the spectral radiance is applied. In the lower part, the temporal distributions of spectrally integrated line intensities of selected lines are plotted.

The arc discharge is initialized by the explosion of a thin copper wire due to Joule heating. Thus, mainly corresponding lines of atomic Cu I radiation can be found in the spectra. The first spectrum with detectable Cu I radiation can be observed after 110 µs (with respect to the point of time when the current flow is started) at current of 1.1 kA. It is followed by a local maximum of Cu I line emission around 120 µs with slightly broadened lines and by weaker spectra at 130 and 140 µs. It should be noted that the intensity of both oxygen and nitrogen atomic lines around 777 and 745 nm, respectively, is very weak and keeps low for the next about 100 µs, cf. lower part of Figure 2. The arc discharge is dominated by the metal vapor of the exploding wire and the influence of ambient air surrounding the discharge plasma within the PTFE nozzle is low.

In the following time of current rise, the Cu line emission rises up again. Now the metal vapor can not originate from the ignition wire any more but is caused by erosion from the tungsten–copper electrodes. In parallel, the atomic line emission from the triplet O I at 777 nm and the three N I lines at 742.4 nm, 744.2 nm, and 746.8 nm increases considerably. That phase starts about 230 µs (100 µs after explosion) and lasts about 150 µs (current increases from 2 to 3 kA). The radiation energy from the arc discharge heats up the nozzle surface and wall material is evaporated. Hot plasma is blown out of the nozzle at both ends as can be observed in video images of the second camera. Thus, the plasma composition is changed, decreasing the amount of nitrogen and oxygen as well as copper. Additionally, a cooling of the arc occurs by increased amount of cold wall material. As a consequence the radiation from air components decreases. The emission from the oxygen line starts before and lasts longer than the ones of nitrogen lines in accordance with their lower energies of the upper levels (O I 10.7 eV, N I 12.0 eV).

Starting around 400 µs (about 3 kA), significant carbon and fluorine line emission can be observed in the spectra. After nearly stable intensity of some lines (F I 775 nm) and relative low increase of others between 400 and about 800 µs, the emission increases very fast. That delay of several hundred µs might result from the time needed to inject the wall material into the arc and to heat it up to temperatures necessary for intense line emission. The temporal evolution of the three F I lines shown here is quite comparable since they have rather similar upper line levels around 14.7 eV. The slower increase of the carbon ionic line C II at 723 nm might be due to the higher level of 18.0 eV.

During the ignition phase it takes more than half but less than one millisecond until the arc plasma is completely dominated by the material ablated from the nozzle wall and the influence of the surrounding gas can be neglected - at least in case of air under ambient conditions of one bar. In the following the influence of the air should be investigated around current zero (around 4.8 ms). Therefore, several successive spectra are plotted in Figure 3. Note that only a part of the spectral range is shown for better distinction of lines. The major lines of atomic fluorine

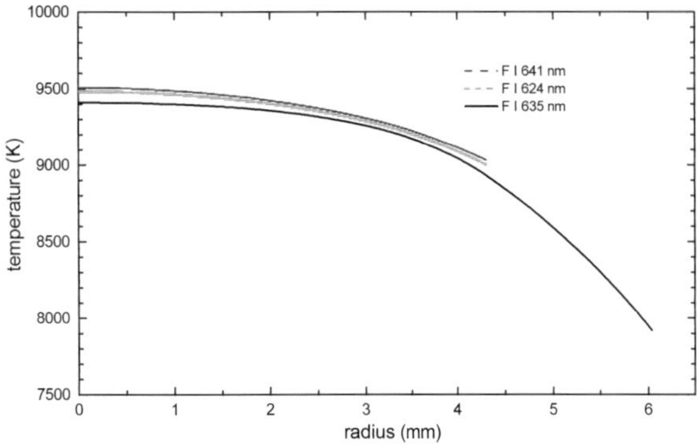

Figure 4. Radial temperature profile 0.4 ms before CZ obtained from three atomic fluorine lines.

as well as oxygen and nitrogen are labeled. Before CZ, only the fluorine lines are observed, cf. spectra between 4.65 ms and 4.71 ms. During a "dark" phase ranging from about 100 µs before until 100 µs after current zero, no emission can be detected due to the low energy input. Here, other spectral techniques are needed for investigation of the residual plasma, e.g. intensified cameras for emission spectroscopy or absorption measurements. However, in the following time starting about 100 µs after CZ (4.9 ms), the sensitivity of the applied video spectroscopy setup is capable to detect enough radiation for further analysis. Now the oxygen triplet O I at 777 nm as well as the three N I nitrogen lines around 744 nm are observed beside the fluorine lines, showing again an influence of ambient air. It was found that also in the second half wave the emission from air components can be detected for the following 500-600 µs.

For determination of radially resolved plasma temperatures, two-dimensional spectra with 600 vertical lines have to be applied. The temporal resolution is reduced to 7500 fps, i.e. increasing of the time between successive spectra to 133 µs. Thus, there remains just one "dark" spectrum at CZ (4.8 ms) for which no line but only noise is detected. The two spectra immediately before CZ, i.e. at 4.66 and 4.53 ms, show distinct emission from all fluorine atomic lines under investigation. Nevertheless, reliable temperature determination is still not possible due to limited line intensity. For the spectrum measured at 4.40 ms (0.4 ms before CZ), two-dimensional spectral fits are carried out. Assuming local thermodynamic equilibrium and rotational symmetry of the arc, Abel inversion of the fitted spectra is done delivering radially resolved emission coefficients. Finally, radial temperature profiles are determined as shown in Figure 4. Three F I lines are applied with quite similar results, including a rather flat temperature profile with a maximum 9500 K and a decrease by 500 K within more than 4 mm. Between radial position of 5 and 6 mm the temperature decreases rapidly to values around 8000 K.

4. Conclusions and outlook

A circuit breaker model is applied to investigate ablation–dominated arc with peak currents of about 11 kA. The radiation of the electrical arc is analyzed by optical emission spectroscopy for a spatially resolved determination of plasma temperature. The focus of our investigations is on the stepwise analysis of the physical properties and processes during the interruption process from the high–current–phase up to some milliseconds after current zero. One question is to which extent and up to which point of time before the arc extinction, the radiation emission of the arc leads to reliable results. Another question is how long the discharge plasma is dominated by ablated material and how far it is influenced by the surrounding ambient when the current is close to the CZ point of time.

In future, additionally to the emission spectroscopy that can also be carried out using more sensitive intensified CCD cameras. Furthermore, laser absorption shall be applied in order to determine the density of colder gas after CZ, such as resulting from the nozzle evaporation. Beside the optical investigations capacitive sensors will be used for measuring the spatially resolved distribution of the arc resistance over the complete time range. The experiments will be accompanied by simulations.

Acknowledgements

R. Methling was supported by grant UH 106/13-1 of the Deutsche Forschungsgemeinschaft.

References

[1] M. Seeger. Future perspectives on high voltage circuit breaker research. *Plasma Physics and Technology*, 2(3):271–279, 2015.

[2] R. Methling, St. Franke, D. Uhrlandt, S. Gorchakov, F. Reichert, and A. Petchanka. Spectroscopic study of arc temperature profiles of a switching-off process in a model chamber. *Plasma Physics and Technology*, 2(2):163–166, 2015.

[3] R. Kozakov, M. Kettlitz, K.-D. Weltmann, A. Steffens, and C. M. Franck. Temperature profiles of an ablation controlled arc in PTFE: I. spectroscopic measurements. *Journal of Physics D: Applied Physics*, 40(8):2499–2506, 2007. doi:10.1088/0022-3727/40/8/013.

[4] D. Eichhoff, P.G. Nikolic, and A. Schnettler. Basic study on the dynamic behaviour of asymmetrically blown arcs in non–rotationally symmetric nozzle arrangements. In *Proceedings of XXth Symposium on Physics of Switching Arc*, pages 126–129. Brno University of Technology, 2013.

[5] C. B. Ruchti and L. Niemeyer. Ablation controlled arcs. *IEEE Trans. Plasma Sci.*, 14:423–434, 1986. doi:10.1109/TPS.1986.4316570.

[6] M. Seeger, L. Niemeyer, T. Christen, M. Schwinne, and R. Dommerque. An integral arc model for ablation controlled arcs based on CFD simulations. *Journal of Physics D: Applied Physics*, 39(10):2180–2191, 2006. doi:10.1088/0022-3727/39/10/029.

doi:10.14311/ppt.2017.2.157
Plasma Physics and Technology 4(2):157–160, 2017

ANALYSIS OF LAST DEVELOPMENT RESULTS FOR HIGH VOLTAGE CIRCUIT-BREAKERS USING NEW G^3 GAS

P. Robin-Jouan[a,*], K. Bousoltane[a], Y. Kieffel[a], J.Y. Trepanier[b], R. Camarero[b], S. Arabi[b], G. Pernaudat[b]

[a] *General Electric Grid Solutions, France*
[b] *Ecole Polytechnique de Montréal, Québec*
[*] philippe.robin-jouan@ge.com

Abstract. Among many alternative gases proposed to replace SF_6 as insulating gas, g^3 (green gas for grid), fluoronitril based component officially introduced at CIGRE 2014, can be now used for insulation projects and new environmental friendly circuit breakers. This gas mixture g^3, presents a reduction of the global warming potential by 98% compared to SF_6 gas and shows quite good dielectric withstand capability. Many new investigations about thermal and chemical behavior have been done and are precised in this paper. Last switching test campaigns will be presented for 145 kV applications. Simulation tools should be updated to be applicable to these projects. Developments and comparisons with last test results will be analyzed.

Keywords: SF_6 alternative gases, CO_2, Circuit Breaker, g^3.

1. Introduction

SF_6 is today extensively used in high voltage gas-insulated switchgear (GIS) and live tank circuit breakers. The positive properties of SF_6 are well known for dielectric withstand and quenching but even if manufacturers were able to reduce the SF_6 mass and the leakage rate of the different switchgears in the past years, a replacement of this gas has become into focus because of its enormous greenhouse potential. Using g^3, we target not only for a significant reduction of the global warming potential (98%), but at the same time we maintain the full ambient temperature range as reached with SF_6, same dimensions and footprint are preserved for a given voltage class as with SF_6; operating pressure values are limited to values already known in GIS and LT activities. The paper focuses on investigations on 145 kV circuit breakers GIS and LT using simulation tools and test results.

2. Reminder about g^3 characteristics

2.1. Flueronitriles as alternative solution

GE Grid Solutions, in partnership with the 3M™ Company, investigated many new compounds and mixtures including fluoroketones, fluoronitriles, hydrofluoroolefins and fluorocarbons. Fluoroketones have shown utility as dielectric fluids. However, compounds such as C5 or C6 fluoroketone have relatively high boiling points resulting in insufficient vapor pressure at low temperatures. This makes them unsuitable for many high-voltage applications. Hydrofluoroolefins (HFOs) have been seen as potential substitutes for SF_6. These molecules have unsaturated carbons and are partially fluorinated and hydrogenated. As such, their dielectric strengths stand at about 0.8 times SF_6. However, they have been found to be flammable in certain situations, causing carbon dust and deposits on the insulator in the event of gas decomposition due to arcing or sparking that prevents this gas from being used in high voltage equipment. Perfluorocarbons offer dielectric strengths that are of interest, but their GWPs are typically in the range 5000 to 12000, eliminating them from consideration. An indepth study was made of the promising chemical family of fluoronitriles. Upon conclusion of this study, a specific compound, namely heptafluoro-iso-butyronitrile $(CF_3)_2$-CF-CN, was selected and commercialized as 3M^{TM} Novec^{TM} 4710 Dielectric Fluid [1]. Gas mixtures of this fluoronitrile with CO_2 were found to be an optimal solution for disconnector and circuit breaker applications. This specific gas mixture called g^3 (green gas for grid) was discovered in the frame of this partnership and is based on 4 %mol or 6 %mol Novec^{TM} 4710 fluid concentration depending upon the minimum operating temperature and rated pressure required for the equipment [2, 3]. g^3 has a low toxicity level and is not classified as CMR (carcinogenic, mutagenic, reproductive toxicant) and requires no specific or additional labels compared to SF_6. The qualification of the new gas mixture was given according to International Standards and the Technical Guide to validate alternative gas for SF_6 in electrical equipment from T&D Europe [4].

2.2. g^3 characteristics

The GWPs of the gas mixtures have been calculated in accordance with the EU 517/2014 F-gas regulation, Annex IV [2]. The GWP for g^3 at 4 %mol and 6 %mol is 327 and 462, respectively. When compared to SF_6 at 23,500, these represent a 98% or greater reduction. The characteristics and performance of g^3 were examined through an extensive battery of laboratory tests.

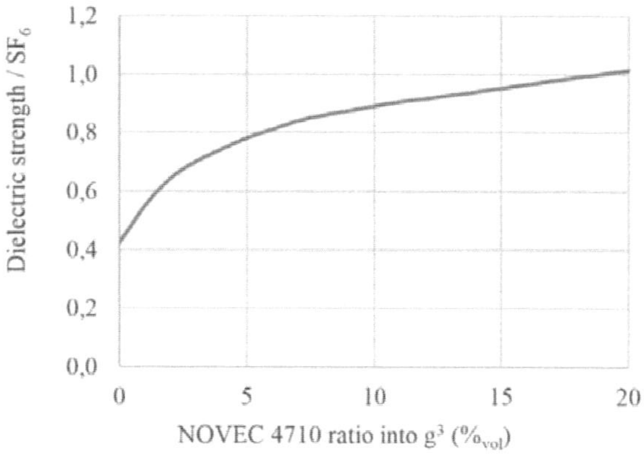

Figure 1. Dielectric strength master curve for Novec^{TM} 4710 into g^3.

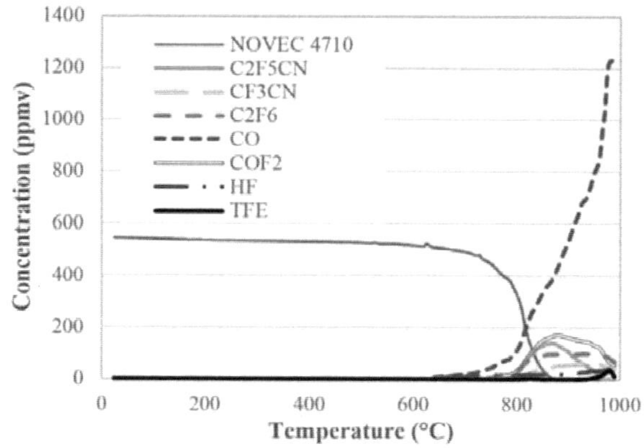

Figure 2. Decomposition products from thermal degradation of Novec^{TM} 4710.

For instance, the power frequency dielectric strength of g^3 was measured on a 145 kV GIS using Novec^{TM} 4710 fluid concentrations ranging from 0 to 20% by volume (Figure 1). At 4 %mol, the dielectric strength of g^3 is roughly 25% less than SF_6 under the same condition and pressure. This means that the overall pressure of g^3 needs to be adapted to compensate for the reduced dielectric strength of the gas mixture [3]. In addition, the thermal stability of the Novec^{TM} 4710 fluid component of the g^3 mixture was evaluated by exposing dilute gas mixtures to elevated temperatures. A mixture of approximately 500 ppmv of Novec^{TM} 4710 fluid in CO_2 was passed through a tube furnace at 1 liter/minute with temperatures increasing to 1000 °C. The fluoronitrile did not display measurable thermal decomposition until temperatures approached 800 °C (Figure 2). SF_6 decomposes at approximately 950 °C under similar conditions. The effluent of the furnace was monitored using Fourier Transform Infrared Spectroscopy and gas chromatography-mass spectrometry. Carbon monoxide was the principal degradation product detected. Above 800 °C, fluorinated species (COF_2, C_2F_5CN, CF_3CN and C_2F_6) were detected. However, all of these compounds decreased in concentration as the temperature neared 1000 °C. The following compounds were monitored but not observed: CF_4, C_3F_8, C_4F_{10}, HFP, HFC-227ea [5, 6].

3. 1D Simulation tools

The development of chambers with g^3 implies to have numerical tools to optimize the design of the chamber with reduced costs. At the early stages of the projects, we mostly use macroscopic tools to predict pressures, mass flows and arc voltage. Our macroscopic tool, called AMASIS based on an AMESIM technology (Siemens PLM Software), performs a calculation in less than 5 minutes and gives relevant information like pressure or mass flow values. The advantage is to give possibility to compute a lot of chamber designs and to take decisions based on it. It is also compatible with our mechanical models to have a full coupling between the mechanical behavior and the breaking. AMASIS has an energetic approach of the arc. Details are given in [7]. The arc volume variation as main result is based on the thermodynamical balance:

$$\frac{\mathrm{d}V_{\mathrm{arc}}}{\mathrm{d}t} = \frac{1}{\rho h}\Big[k \cdot P_{\mathrm{Joule}} - P_{\mathrm{Radiation}} - P_{\mathrm{Convection}} + \left(\frac{\mathrm{d}P_{\mathrm{arc}}}{\mathrm{d}t} + \frac{\mathrm{d}\rho h}{\mathrm{d}T_{\mathrm{arc}}}\frac{\mathrm{d}T_{\mathrm{arc}}}{\mathrm{d}t}\right) \cdot V_{\mathrm{arc}}\Big] \tag{1}$$

with:

- V_{arc}: Arc volume
- P_{arc}: Arc Pressure
- T_{arc}: Arc Temperature
- ρ: Density of the arc
- h: Enthalpy of the arc
- k: Coefficient
- P_{Joule}: Joule power
- $P_{\mathrm{Radiation}}$: Radiation power
- $P_{\mathrm{Convection}}$: Convection power

The calculated arc volume is defined between the contacts, radiates energy that cause ablation; it also interacts with the fluid and exchanges energy. Our model considers multispecies calculations, including CO_2/C_2F_4 mixtures. One typical sketch example is given Figure 3.

The calculation results are good, especially for the thermal volume overpressure and arc voltage measurements(Figure 4). AMASIS has several limitations inherent to any macroscopic tool. It does not give a precise idea of the gas temperature around the arcing contacts but permit to reduce significantly the number of 2D or 3D numerical simulations which are still required for precise investigations [8].

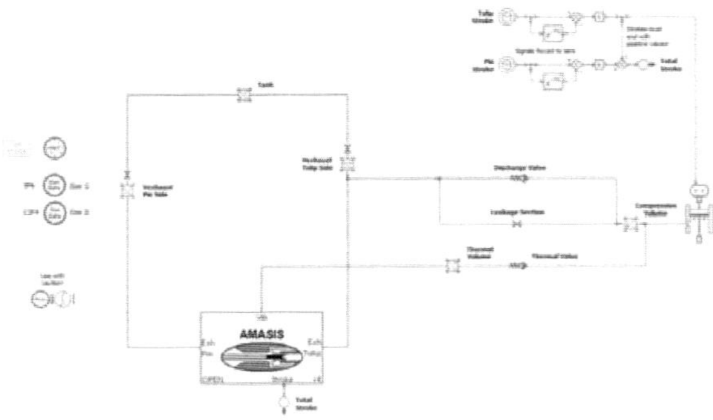

Figure 3. Simple AMASIS sketch of a chamber.

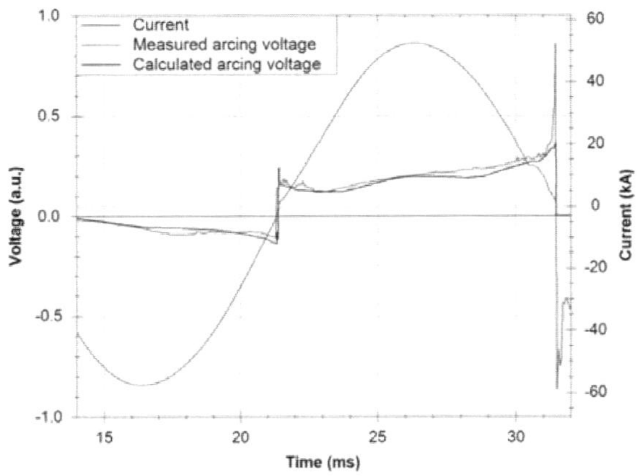

Figure 4. L90 type test result for 18.2 ms arcing time.

4. Hybrid circuit-breaker simulations

Due to the high thermal behaviour of g^3, 3D information is required to improve and optimize exhaust volumes of our breakers. Consequently, 3D simulation should be processed on a complete switchgear. This kind of simulation is not possible due to the long calculation time required even on high performance computer. In order to overcome this problem, we developed the hybrid simulation. This approach is based on a division of the interrupting unit in several areas. Each part could be simulated on the most adapted tool. The AMASIS results, issued from a full interrupting unit simulation, can be applied inside the thermal channel, as an input data in another tool (typically Ansys Fluent) to simulate the 3D flows in the heating or exhaust volumes. In the chamber, complex arc physic modeling has to be taken into account. However, in the exhaust, the principal phenomenon is the hot gas propagation. Connexion between arc modeling area and exhaust volumes is realized as a one way coupling which is acceptable in such application. In arc chambers, the parts are usually 2D axisymmetric shaped (nozzle, cap, pin…) and the exhausts are 3D shaped. So, the results of 1D or 2D simulations with arc considering equivalent sections give good data for simple 3D flow simulations in exhaust volumes. With this kind of hybrid simulation, we can implement and evaluate specific designs in order to improve the gas

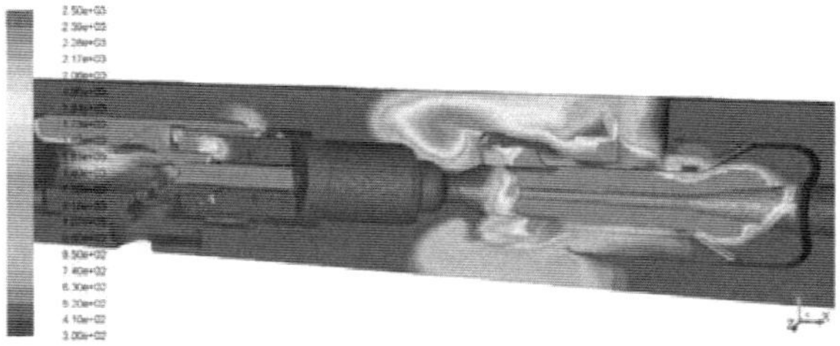

Figure 5. 3D flow simulation in exhaust volumes.

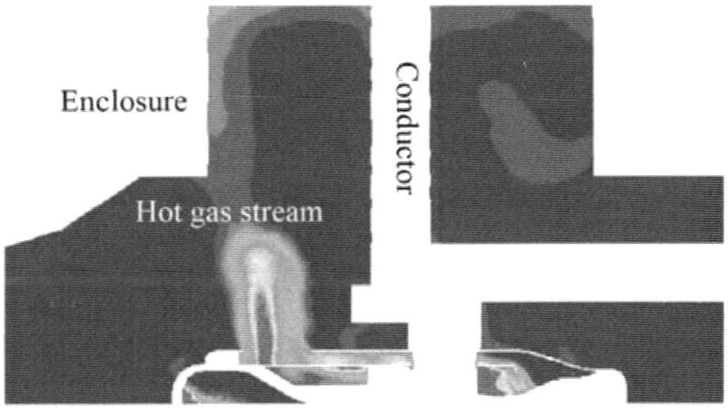

Figure 6. Evaluation of the hot gas stream to the enclosure.

mixing.

The Figure 5 shows a typical 145 kV circuit-breaker geometry with an example of input data, pressure and temperature, as a function of time on the left input side. These data are used to build the final adapted input data for real gas required by the coupling and insuring a compatibility of the data and the physics at the interface. Figure 6 represents the temperature field at one exit of the exhaust volume. This kind of result allows us to evaluate the risk to have a flashover to the enclosure or over an insulator. Thus, with sequential modifications of the chamber, all the tests required by the standards (between 10% to 100% of the default current) have been successful [9].

5. Temperature rise tests

Temperature rise tests validate the thermal behavior of the circuit breaker for the nominal current. We fixed on the interrupting unit a lot of thermocouple sensors to record the temperature all along the contacts (bolted or sliding). The test is valid when the temperature does not increase 65 °C. We performed these tests on the same test set up with g^3, pure CO_2 and SF_6 at several currents, all the sequences tested are listed in the Table 1. The results of sequences 1,2 3 and 6 were presented in [9].

Test results are very similar for the 3 gases: the maximum temperature increase has not been reached; a temperature increase of 2 °C has been measured for the CO_2 test compared to the g^3 test. An increase of 2 °C has been observed for the g^3 test compared to the SF_6 test. Under some conditions, Utilities want to be able to have higher values of current in their

Sequence	Gas	Current $[A]$
1	CO_2	3150
2	g^3	3150
3	SF_6	3150
4	SF_6 temporary over current	from 2867 A to 5985 A
5	g^3 temporary over current	from 2867 A to 5985 A
6	CO_2	3150

Table 1. Temperature rise test sequences

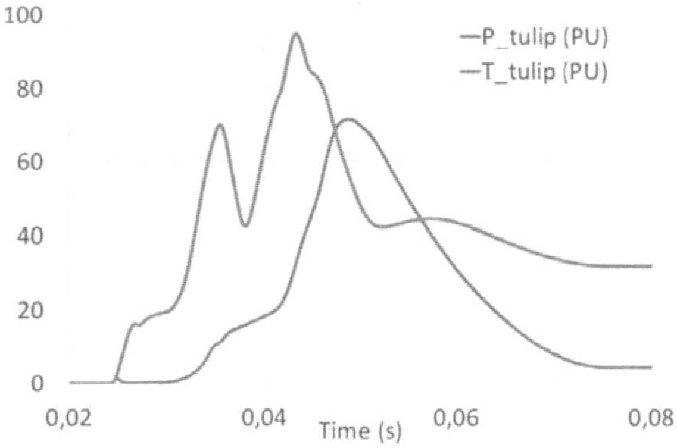

Figure 7. Temperature and pressure as an input data.

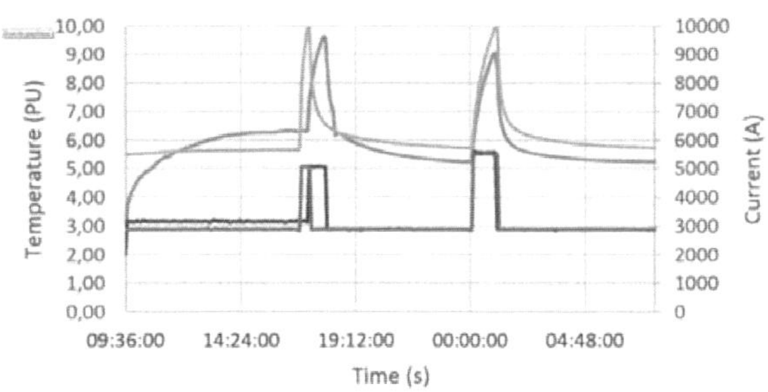

Figure 8. Temperature rise results during the over current tests.

substation for a limited duration. This special requirement has also been tested with g^3 and we compare the performance to SF_6 (sequence 4 and 5). The Figure 8 shows that results obtained on the main contacts with SF_6 and g^3 are very close.

6. Conclusion

The new alternative g^3 gas have been reminded with some new complements about gas stability and toxicity. Different simulations tools are being developed and typically an hybrid method has been presented: this approach has the advantage to run fast simulations and to optimize independently each area of the interrupting unit. Test results proved that g^3 can now be considered as a solution to replace SF_6 as insulation but also considering breaking, temperature rise and in general all performances usually required by utilities. Currently several products exist in AIS and GIS arrangement that passes all the type test. The next step for g^3 is to go for higher short circuit currents such as 50 kA and 63 kA.

References

[1] Y. Kieffel et al. SF_6 alternative development for high voltage switchgears. In *Cigre*, 2014. Cigre Paper D1-305, Paris.

[2] The European Parliament and of The Council. *Regulation (EU) N° 517/2014 on fluorinated greenhouse gases*, April 2014.

[3] D. Gautschi et al. Application of a fluoronitrile gas in gis and gil as an environmental friendly alternative to SF_6. In *Cigre*, 2016. Cigre Paper B3-106, Paris.

[4] http://www.tdeurope.eu.

[5] J.G. Owens. Greenhouse gas emission reductions through use of a sustainable alternative to SF_6. In *IEEE Electrical Insulation Conference (EIC)*, Montreal, Canada, 2016. doi:10.1109/EIC.2016.7548658.

[6] K. Pohlink et al. Characteristics of fluoronitrile/CO_2 mixture-an alternative to SF_6. In *Cigre*, 2016. Cigre Paper D1-204, Paris.

[7] S. Silvant, G. Gaudart, P. Huguenot, Y. Kieffel, M. Perret, P. Robin-Jouan, D. Vigouroux, R. Fray, Y. Cressault, P. Teulet, and J.Y. Trépanier. Detailed analysis of live tanks, dead tanks and gis circuit breakers using a new environmental friendly gas. In *Cigre*, 2016. Cigre Session 46.

[8] D. Godin, J.-Y. Trépanier, M. Reggio, X.D. Zhang, and R. Camarero. Modelling and simulation of nozzle ablation in high-voltage circuit-breakers. *Journal of Physics D: Appl. Phys.*, 33(20):2583–2590, 2000. doi:10.1088/0022-3727/33/20/310.

[9] P. Robin-Jouan, K. Bousoltane, and Y. Kieffel. Dernières avancées sur les développements de l'appareillage ht utilisant le nouveau gaz d'isolation g^3. In *Colloque sur les Arcs Electriques*, Nancy, 2017.

doi:10.14311/ppt.2017.2.161
Plasma Physics and Technology 4(2):161–164, 2017

ANALYSIS OF ELECTROMAGNETIC PROCESSES INSIDE THE ARC INTERRUPTING SYSTEM OF A HIGH-CURRENT CIRCUIT BREAKER

I. Murashov[a,*], V. Frolov[a], D. Ivanov[a], A. Kvashnin[b]

[a] *Department "Electrical Power Engineering and Equipment", Peter the Great Saint-Petersburg Polytechnic University, 29 Polytechnicheskaya, 195251 Saint-Petersburg, Russia*

[b] *NE Open Joint Stock Company, 21 Partizanskaya, 195248 Saint-Petersburg, Russia*

[*] iuriimurashov@gmail.com

Abstract. Description of electromagnetic processes inside an arc interrupting system of high-current circuit breaker is presented in article. Non-stationary mathematical model of the circuit-breaker operation (short-circuit currents mode) is developed. The main regularities of the displacement of the electric arc inside an arc chute are established. The influence of the movable contact position on the conditions for the displacement of the electric arc is presented. Also, the influence of the arc chute material on the motion and interrupting of an electric arc (alternating and direct currents) is established.

Keywords: circuit breaker, mathematical model, electrical switch, short circuit.

1. Introduction

The main device for protecting low-voltage distribution power grids from overload and short-circuit currents is an circuit breaker. Circuit breakers are received the widest application in alternating current circuits. There are several theories of electric arc extinction and current interruptions, which are based on the theories of Slepian and Cassie [1, 2].

The operation of the circuit-breakers is determined and by other factors besides the the electric arc extinction, for example, by the thermal state of the contacts during a long and short-term discontinuity under current load.

The subsequent rejection of the movable contact (accompanied by the formation and arc extinction) depends on the organization of the magnetic damping of the arc.

The consequences of a current transfer high speed (≈ 400–500 A/µs) from main contacts to arc-extinguishing contacts can lead to failure of the device due to strong erosion of electrodes and even their welding, violating its functional capacity for current switching. The existing drive devices speed and the pulling force of the arc-extinguishing contacts allows the commutation process to be completed in the same half-cycle when the current passes through zero at the beginning of the arc contacts dilution in the growing section of the short-circuit current.

Maximum current value can reach half the amplitude of the expected short-circuit current during the current rises in the transient mode, as follows from the above. This makes it possible to estimate and take into account in the analysis of the main-contact system heating with the current transfer speed 400–500 A/µs. It shows necessary to ensure the arc-extinguishing contacts pressing increase (not leading to their welding). It means the rate of pressure increase should outrun the rate of current rise, which causes the contact pad heating up to the melting temperature.

Ensuring the current transition from the main contacts to arc-extinguishing contacts without electric arc is the main task of the rotary mechanism for the dilution of movable contacts. There is a weakening of the main contacts (silver-based lining) pressing and increasing the arc-extinguishing contacts (copper contacts) pressing during the current flows from the main to the arc-extinguishing contacts.

2. Theoretical investigations

The parameters of the arc-extinguishing contacts transient resistance (during their compression depending on the closing speed, the material parameters and the contact surface) can be determined according to the energy conservation equation [3]:

$$m\left(\frac{\mathrm{d}h}{\mathrm{d}t}\right)^2 + kh^{\frac{5}{2}} = \frac{mv^2}{2} + \frac{1}{2}Ph \qquad (1)$$

where $k = \frac{8}{15}\sqrt{r_c\Theta}$ - coefficient taking into account the shape and elastic properties of the contacts; r_c - equivalent radius of the contact surface curvature; Θ - coefficient taking into account the contact material, Poisson's ratio and elastic modulus of moving and stationary contacts [3]; P - force applied to contacts; h - rapprochement of contact bodies; m - mass of movable contact; v - speed of the contacts approach at the moment of their contact. The maximum approach of contacts corresponds to the moment when $\frac{dh}{dt} = 0$. In this case, equation (1) takes the form:

$$kh^{\frac{5}{2}} = \frac{mv^2}{2} + \frac{1}{2}Ph \qquad (2)$$

It is possible to determine the maximum approach of contacts h when they are compressed by solving this equation. At the same time, the compression time of the electrodes: $t = \frac{h}{v}$.
The current transfer from the main contacts to arc-extinguishing contacts without electric will be realized in the absence of melting of a large mass of electrodes in the transient mode of the current, as well as a complete transition of the current to the arc-extinguishing contacts until the moment of throwing the movable contact by the rotary mechanism.Otherwise, if the current from the main contacts does not have time to go to the arc-extinguishing contacts, the arc will ignite on the main contacts. This will lead to an increase in erosion of the main contacts.
The change in the contact resistance of the arc-extinguishing contacts during compression by [3] will be:

$$R = \frac{\rho_0\left(1+\frac{2}{3}\alpha T_{\mathrm{mel}}\right)}{2\sqrt{\frac{h}{2A_s}f_c}} \tag{3}$$

where ρ_0 - specific electrical resistance of electrode material; α - temperature coefficient ($0.0042\,\mathrm{K}^{-1}$); T_{mel} - melting temperature of contact material; $A_s = \frac{1}{2R_s}$ - coefficient taking into account the curvature of the contact surface (sphere-plane); f_c - coefficient taking into account the ratio of the contact area at the end and the beginning of compression ($\approx$2.25).

2.1. Calculation of the electrodynamic forces

The following forces act on the arc at the extinction:

- ☐ The electrodynamic force $F_\perp$ due to the location of the arc (conductor with current) between two contacts
- ☐ The electrodynamic force F_μ due to the location of the arc near the ferromagnetic plates
- ☐ The electrodynamic force $F_\sim$ due to the variable component of the short-circuit current

Calculations of the electrodynamic forces are performed for the distance a between contacts. The angle 0° corresponds to the fully open state of the contacts. The angle −6° corresponds to near a state close.

The results of the electrodynamic force $F_\sim$ estimated calculation acting on the electric arc in the circuit-breaker are presented in Table 1 for four variants of contact opening time (2, 3, 4, 5 ms). The electrodynamic forces $F_\perp$ and F_μ don't depend on time. The results of the electrodynamic forces estimated calculation are presented in Table 2. The electrodynamic forces $F_\perp$ and F_μ can be compared with the electrodynamic force $F_\sim$ (see Table 1).

2.2. Mathematical model (heat transfer task)

A steady-state temperature rise is adopted as an excess of temperature during the device operation in a continuous mode; the maximum temperature rise is adopted during operating in a short-time mode at the end of the operating periods. According to Russian standards, the test should be continued until a steady temperature is reached, but not more than 8 hours.In the case of applying numerical methods, it is possible to calculate the steady-state operating mode when the rated current flows through the contact system. The geometry of the existing switching contacts of the main circuit was used in the development of a mathematical model (see figure 1).

t[ms]	α	a[mm]	$F_\sim$[N]
	−6°	21	446
2	−3°	28	576
	0°	36	693
	−6°	21	948
3	−3°	28	1167
	0°	36	1314
	−6°	21	1551
4	−3°	28	1782
	0°	36	1815
	−6°	21	2175
5	−3°	28	2274
	0°	36	2007

Table 1. The electrodynamic force $F_\sim$.

α	a[mm]	$F_\perp$[N]	F_μ[N]
−6°	21	2084	256
−3°	28	2499	342
0°	36	2828	433

Table 2. The electrodynamic forces.

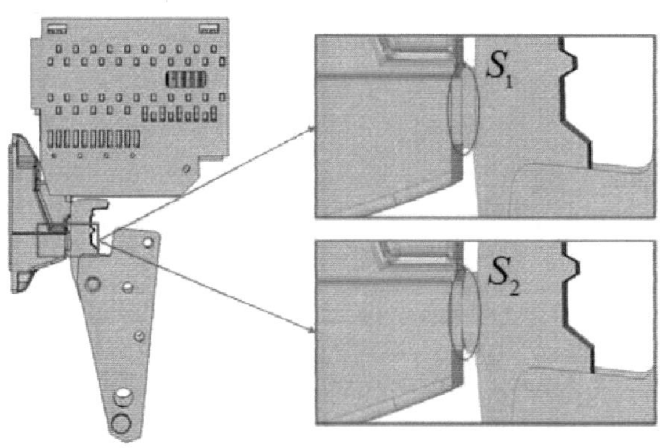

Figure 1. Geometry of contacts.

The area of the contact surface $S_1 > S_2$ (see figure 1), which allows us to analyze the effect of the contact surface area on the contact system heating (it is possible to establish the force of pressing the

contacts in solving a mechanical task).
Basic equations of mathematical model:

$$\rho C_p \frac{\partial T}{\partial t} + \nabla(-\lambda \nabla T) = Q \qquad (4)$$

$$\begin{cases} \overrightarrow{J} = Q_j \\ \overrightarrow{J} = (\sigma + \epsilon_0 \epsilon \frac{\partial}{\partial t})\overrightarrow{E} + J_{\text{ex}} \\ \overrightarrow{E} = -\nabla V \end{cases} \qquad (5)$$

Heat removal from the contact surface (convective component):

$$q_0 = h(T_{\text{ext}} - T) \qquad (6)$$

where h - heat transfer coefficient [4], the numerical value of the coefficient depends on the properties and parameters of the convective flow. The simulation results in the case of a contact surface $\approx 340\,\text{mm}^2$ are shown in figure 2).

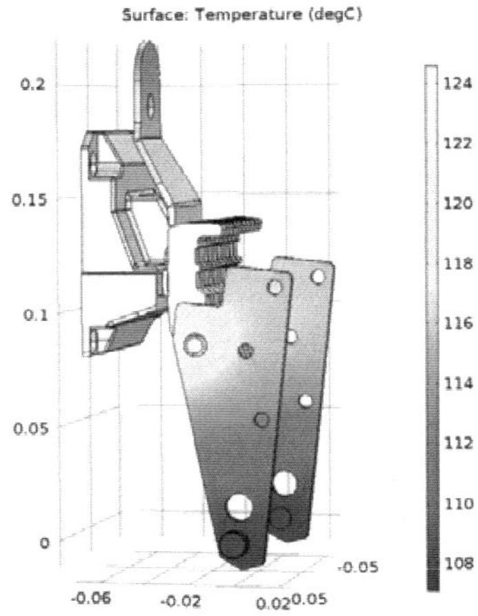

Figure 2. Temperature distribution.

The maximum value of the temperature is 124.6 °C, according to the results of experimental studies, the maximum value is 127 °C.

2.3. Mathematical model (electrostatic task)

From the experimental results it was dedicated that the arc interrupting system is not fully engaged during arc extinguished. It is necessary to provide a high electric field strength in the contact area of the interrupting system to ensure the formation of attachment on the contact of the arc interrupting system and intensive retraction of the arc into the interrupting grid [5],which value will exceed the value of electric field strength near the movable contact (see figure 3). A preliminary analysis of the electric field strength is made on the basis of simulation results of a two-dimensional electrostatic mathematical model of a circuit breaker. The model is based on the existing geometry of the circuit breaker.
The mathematical model is based on the Gauss law for an electric field:

$$\begin{cases} \overrightarrow{E} = -\nabla V \\ \nabla(\epsilon_0 \epsilon_r \overrightarrow{E}) = \rho_V \end{cases} \qquad (7)$$

The results of numerical simulation are shown in figure 4.

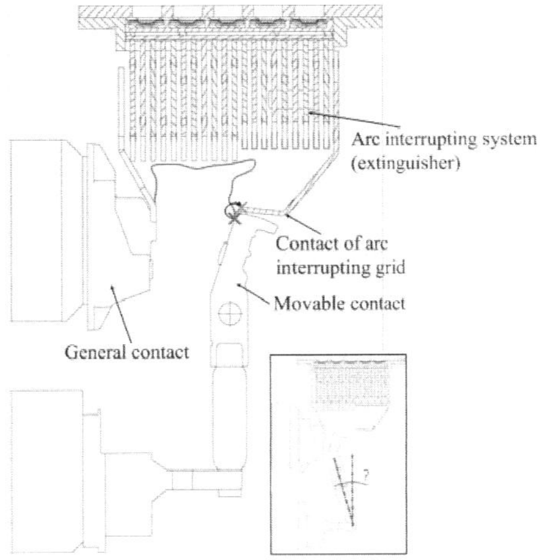

Figure 3. Geometry of contact system.

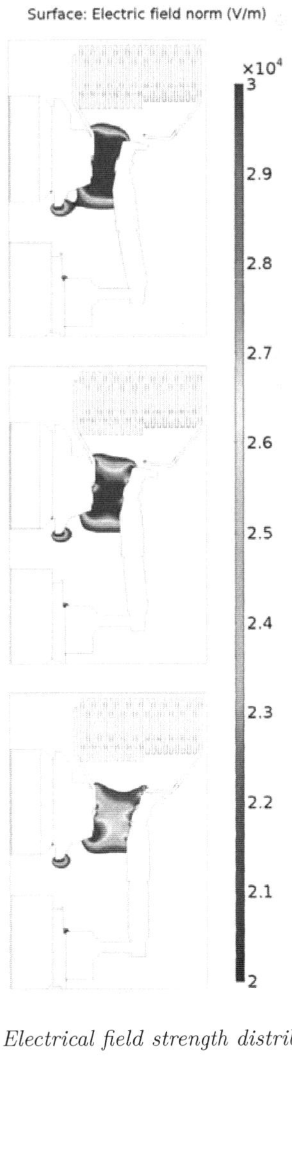

Figure 4. Electrical field strength distribution.

3. Conclusions

The change in the scalar potential inside the ferromagnetic grid is greater in the case the contact of the arc interrupting grid has a zero electric potential. This means that conditions are created for stretching the arc over the entire length of the ferromagnetic grid.
The electric field strength has high values in the contact area of the arc grid with the greatest contact dilatation, which causes strong emission of electrons from the cathode surface and guarantees the attachment at the contact of the arc grid, which leads to an intensive retraction of the arc into the arc interrupter.
The main forces acting on the arc are $F_{\perp}$ (prevails in the interval of contact opening time 2–3 ms) and $F_{\sim}$ (prevails in the interval of contact opening time 4–5 ms).
The maximum temperature does not exceed 125 °C with a long current of 4 kA, which corresponds to the national standards for contact materials used (it follows from the simulation results).
Optimization of the arc interrupting system design based on the simulation results of the electrostatic task.

References

[1] U. Bakshi and M. Bakshi. *Protection and Switchgear.* Technical Publications, 2009.

[2] Stewart S. *Distribution Switchgear.* Institution of Engineering and Technology (IET), 2004.

[3] G. Kukekov and V. Frolov. *Transients in contact-thyristor devices.* Energoatomizdat, 1988.

[4] Comsol Multiphysics. *AC/DC Module. Software for Computational Electromagnetics Modeling.* `https://comsol.com/acdc-module`.

[5] A. Krasko and E. Ponomarenko. *Electric discharges in gases. External insulation of overhead lines and switchgears. Internal insulation.* Minsk: BNTU, 2011.

doi:10.14311/ppt.2017.2.165
Plasma Physics and Technology **4**(2):165–168, 2017

INVESTIGATION OF THE RECOVERY BEHAVIOUR OF A SMALL SWITCHING GAP AFTER CURRENT INTERRUPTION

D. BÖSCHE*, M. ALIJA, M. HILBERT, M. KURRAT

Institute for High Voltage Technology and Electrical Power Systems - elenia, TU Braunschweig, Schleinitzstrasse 23, 38106 Braunschweig, Germany

* dirk.bösche@tu-bs.de

Abstract. The recovery behaviour of switching arcs in a small gap was investigated experimentally. Therefore a simple mechanical model switch without an arcing chamber was designed. For the investigation the current carring contacts are separated. After various arcing times, the current is interrupted by an electronic switch. At a fixed contact distance, the switching gap was tested with variable voltages to investigate the dielectric strength. Based on these results the recovery behaviour was studied using the Weibull distribution function.

Keywords: plasma temperature, switching gap, arc, current zero, reignition, Weibull distribution.

1. Introduction

Switches (e.g. grounding or disconnecting) and surge protective devices (SPD) are common as protection units, which both use gaps and arcs as switching elements[1–3]. The switching task is carried by the arc, that has to be extinguished quickley[1, 2, 4–6]. To prevent reignition of the arc, a high dielectric strength of the switching gap after current interruption is required[1–3]. Hence, the dielectric recovery behaviour of the residual gas column in the switching gap is of importance. Based on the breakdown mechanims in ionized gases, a first treatment on the recovery behaviour of low voltage switches is done in [1, 2, 7–10]. During the recovery process positive residual volume charges are left in front of the cathode of the switching gap. In the residual gas column a region with low ionization probability is established. The dielectric breakdown behaviour of this region can be characterized by the Townsend Mechanism[7–9]. The ionization of the residual gas column depends on the temperature and the electric field[1, 2, 7–10]. Within a temprature range of 300 K to 3000 K of the residual gas column, the density decreases and the effective ionization coefficient rises mainly dependent on dissociation processes[11, 12]. Regarding to the recovery behaviour this can be associated with a higher reignition probability. Especially the recovery behaviour of switching gaps with metal contacts can be modeled as a dielectric breakdown in high ionized gases[13]. Therefore the properties and behaviour of the residual gas column in the switching gap are important for an understanding of the recovery behaviour. Towards first investigations on the recovery behaviour of low voltage switches the dielectric strength of a switching gap is studied in this paper.

2. Measurement Setup

In this section, the designed measurement setup for the basic investigation of the recovery behaviour of the switching gap will be presented.

2.1. Electrical Test Circuit

In order to investigate the dielectric strength of a switching gap after various arcing times T_{Libo}, a test circuit, that is shwon in Figure 1 was developed. The 18 mF capacitor $\mathrm{C_C}$ provides the desired test current, which is adjusted by the resistor R roughly. It can be fine-tuned by varying the charging voltage $\mathrm{U_C}$ and the charging resistor $\mathrm{R_L} = 300\,\mathrm{k\Omega}$. To reduce the influence of the arc on the load current, the voltage of $\mathrm{C_C}$ is only adjusted within the range of 1 kV to 1.6 kV. The test voltage $\mathrm{U_V}$ is stabilized by a 40 µF capacitor $\mathrm{C_V}$ which can prevent any significant voltage drop when the voltage is applied onto the switching gap. Since the capacitor $\mathrm{C_V}$ and parasitic inductances may cause some oscillations the 140 Ω damping resistor $\mathrm{R_D}$ is added to the electrical test circuit. $\mathrm{R_D}$ also limits the current in case of a re-ignition of the arc. The archived slew rate of the test voltage is 2.5 kV/µs.

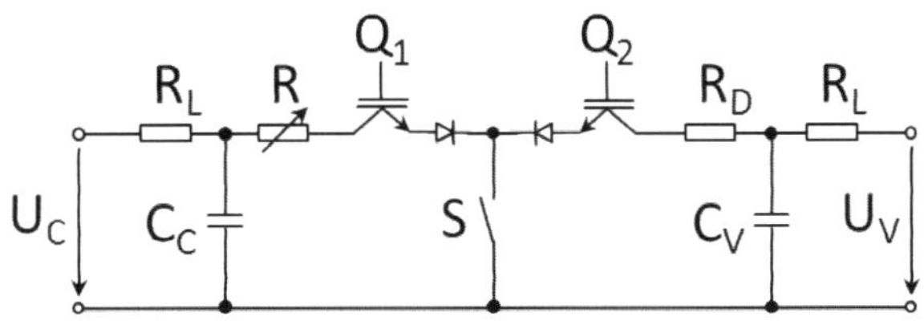

Figure 1. Electrical test circuit to investigate the dielectric strength of the switching gap

2.2. Mechanical Model Switch

The model mechanical switch consists of two contacts which are shown in Figure 2. The contacts are separated using a linear motion. The contact arrangement is placed in an open setup and is only surrounded by two sidewalls. The switching medium is air at atmospheric pressure. The contact plates with (4x3) mm^2 are sligthly spherical. The contact material is Ag/SnO_2. To minimize the influence of contact wear onto the measurements the contact plates were replaced after several switching processes. Quenching units are not implemented to ensure fast pressure equalization.

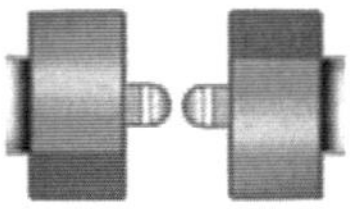

Figure 2. Contact arrangement

The switch is designed to be normally closed and the contact pressure is provided by a coil spring. A pneumatic drive is used to actuate the moving contact. Therefore the switching speed can be adjusted by varying the operating pressure.

3. Measurements and Results

The measurement procedure is described briefly and the results are shown. Based on the evaluated test parameters the analysis of the recovery behaviour is addressed using the Weibull distribution function.

3.1. Measurement Procedure

The measurement procedure is shown in Figure 3. The test circuit setup is to generate a load current of 100 A. The switch is in the closed position when the IGBT Q1 is turned on. The current flows for approximately 5 ms until the contacts are separated. The occurring arc voltage is detected by a separate triggering unit which starts the digital sequence control system. The digital sequence control system ensures exactly the same time sequence for each series of measurement. Depending on the series of measurement the arc burns for $T_{\text{Libo}} = [0.1, 0.5, 1, 1.5]$ ms until the current is interrupted by the IGBT Q1. Exactly 2 ms after contact separation, the IGBT Q2 is turned on which applies the test voltage onto the switching gap.

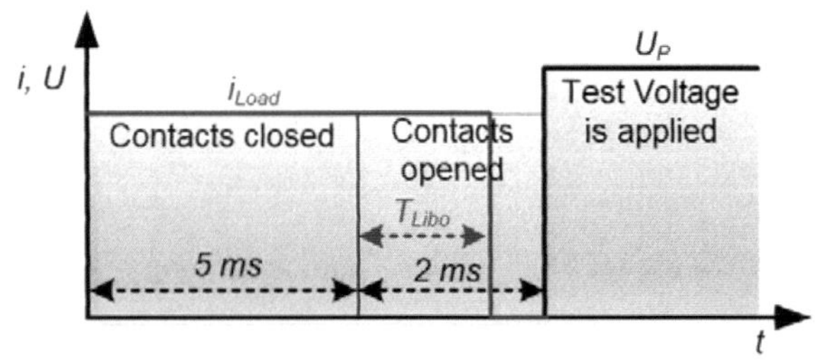

Figure 3. Schematic of the measurement procedure

This time and the operating pressure of the pneumatic system is kept constant to ensure exactly the same contact distance during all series of measurement. The contact distance could not be measured but is estimated to be in the 200 µm range. When the test voltage is applied the contacts are opened. The test voltage U_P was varied in the range of 400 V to 1800 V and 20 tests have been done at each chosen voltage. Depending on the breakdown probability of the last measured step, the voltage was reduced or raised in order to generate the reference points for the Weibull distribution.

3.2. Weibull-Estimation

For the investigation of the recovery-behaviour in the small switching gap a statistical approach based on the Weibull distribution is considered as appropriate[2, 14, 15]. The Weibull distribution function is described by the equation

$$F(U_P) = 1 - e^{-\left(\frac{U_P - U_{P0\%}}{U_{P63.2\%} - U_{P0\%}}\right)^{\beta}} \tag{1}$$

U_P is the test voltage, $U_{P0\%}$ the withstand voltage and $U_{P63.2\%}$ the breakdown voltage. $F(U_P)$ is calculated using the Matlab function *wblfit* and based on the measured data. Due to the fact, that a monotonously increasing behaviour of the input data is requiered the following steps in the calculation procedure are necassery. Assuming that the Weibull distribution function $F(U_P)$ is almost linear within the range of 0.3 to 0.9, only these values (pair of $F(U_P)$ and U_P) in the frequency distribution of the measured data are taken into account [14, 15]. With the selected data pairs, the 63.2 % breakdown-voltage-probability ($U_{P63.2\%}$) is caluclated using the fitting tool *fit* in Matlab in the first step. Based on the approach of linear regression the shape parameter β of the Weibull distribution function is caluclated using the same fitting tool in Matlab in the second step. In the last step the withstand voltage $U_{P0\%}$ of the switching gap is calculated using the Weibull chart.

3.3. Results

Calculation results of the Weibull parameter for a load current of $I_p = 100$ A are given in table 1. Both, U_P and β show a dependency on the arcing time. With a decreasing arcing time the shape paramter β shows a decreasing behaviour. The 63.2 % breakdown-voltage show for an increase of T_{Libo} from 0.1 ms to 0.5 ms and from 1 ms to 1.5 ms no significant change. Whereas for a time step of T_{Libo} from 0.5 ms to 1 ms the 63.2 % breakdown-voltage is reduced. An increase of the arcing time leads to a reduction of the withstand voltage $U_{P0\%}$. Considering the Weibull parameters, calculation results of the Weibull distribution functions are shown in Figure 4. For each function the associated reference points are highlighted as points in the same colour. It can be observed, that an increase of T_{Libo} leads to a shift of the weibull distribution function along the x-axis and hence to a reduction of the dielectric strength of the switching gap. For $T_{\text{Libo}} =$ 1.5 ms the selected reference points show a maximum scattering, whereas for $T_{\text{Libo}} = 1\,\text{ms}$ the scattering is the lowest.

	T_{Libo} in ms			
WB.-Param.	0.1	0.5	1	1.5
$U_{P63\%}$	1222	1239	944	950
$U_{P0\%}$	739	572	431	227
β	9.14	5.96	5.87	3.21

Table 1. Weibull parameters for different arcing times

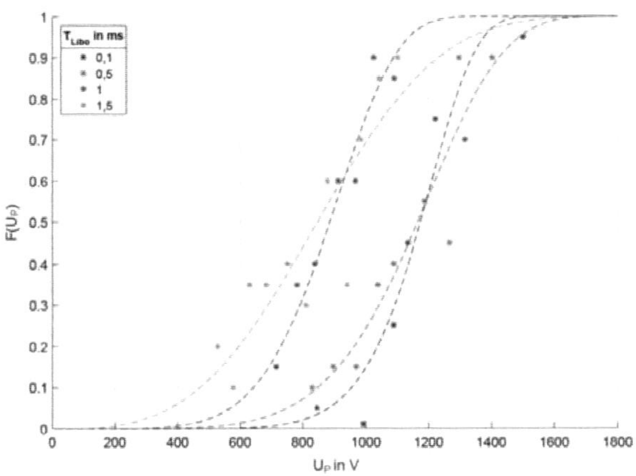

Figure 4. Weibull Distribution functions and reference points

3.4. Discussion

Based on the results the influence of the arcing time on the breakdown behaviour of the switching gap can be observed. Comparing the two time steps of T_{Libo} from 0.1 ms to 0.5 ms and from 1 ms to 1.5 ms, an increase of the arcing time does not reduce the breakdown strength of the switching gap. However comparing the results of $T_{\text{Libo}} = 0.5\,\text{ms}$ and 1 ms, it is obvious that an increased arcing time leads to a reduction of $U_{P0\%}$ and $U_{P63.2\%}$.

Due to the longer arc duration erosion and thereby irregularities on the contact surface are increased. This causes higher local electric fields which decrease the electric strength of the switching gap [3]. But after current interruption a pollution layer is formed onto the contacts which may have the opposite effect. A longer arc duration may cause a thicker pollution layer which even increases the electric strength. The contact surface and the forming process of the pollution layer after current interruption may be of interest for further investigations to estimate the leverage of these two effects on the dielectric strength.

Another influencing factor is the remaining conductivity of the residual gas column. The differential equations of Mayr and Cassie [1, 2, 7] can be used for an estimation. If the dielectric recovery time is several tens larger than the arc time constant of the mechanical model switch the remaining conductivity is negligible. This may be the case with arcing times of 0.1 ms and 0.5 ms and leads to the assumption that those breakdowns where dielectric only. Consequently the dielectric strength of the switching gap with $T_{\text{Libo}} = 1.5\,\text{ms}$ should be significant lower compared to $T_{\text{Libo}} = 1\,\text{ms}$. This assumption is supported by the differential equations which predict that the conductivity decreases exponentially after current interruption. But the test results show a different behaviour. The dielectric strength of the switching gap with $T_{\text{Libo}} = 1\,\text{ms}$ and 1.5 ms are almost identical. Either the arc time constant of the model mechanical switch is exceptional higher or the differential equations cannot be used in this case. This motivates further investigations on the conductivity of the residual gas column.

4. Conclusions

A measurement setup for the investigation of the recovery behaviour of switching gaps is presented. The recovery behaviour is studied basing on the dielectric strength of the switching gap for four different arcing times. Using the Weibull distribution function a statstical analysis of the dielectric strength and the influence of the arcing time is presented and discussed. Furthermore the erosion of the contact surface and the remaining conductivity of the residual gas column are discussed to explain the influence of the arcing time on the dielectric strength. Based on the results further investigations on the recovery behaviour of switching gaps are motivated.

Acknowledgements

The authors would like to thank Robin Schrader for his commitment and support of this work during his master thesis.

References

[1] A. Erk. *Grundlagen der Schaltgerätetechnik*. Springer Verlag, 1974.

[2] M. Lindmayer. *Schaltgerätetechnik*. Springer Verlag, 1999.

[3] P. Slade. *Electrical contacts: Principles and Applications*. Taylor and Francis Group LLC, 2014.

[4] M. Lindmayer, E. Marzahn, A. Mutzke, T. Rüther, and M. Springstubbe. The process of arc splitting between metal plates in low voltage arc chutes. *IEEE Trans. on Comp. and Pack. Techn.*, 2006.

[5] A. Mutzke, T. Rüther, M. Lindmayer, and M. Kurrat. Arc behavior in low-voltage arc chambers. *Eur. Phys. J. Appl. Phys*, 2010. doi:10.1109/TCAPT.2006.875902.

[6] C. Fievet, M. Barrault, P. Chevrier, and P. Petit. Experimental and numerical studies of arc restrikes in low-voltage circuit breakers. *IEEE Trans. on Plasma Science*, 1997. doi:10.1109/27.649604.

[7] W. Rieder. *Plasma und Lichtbogen*. Vieweg Verlag, 1967.

[8] J. Slepian. Extinction of an a.c. arc. *Transactions of the American Institute of Electrical Engineers*, 47(4), 1928. doi:10.1109/T-AIEE.1928.5055155.

[9] M. Schmelzle. *Grenzen der Selbstlöschung kurzer Lichtbogenstrecken bei Wechselstrombelastung.* PhD thesis TU Braunschweig, 1968.

[10] K. Troeger. *Ueber die Wiederverfestigung von Lichtbogenstrecken nach dem Stromnullwerden wandernder Wechselstromlichtbögen in engen Spalten.* PhD thesis TU Braunschweig, 1960.

[11] Y. Tanaka. Prediction of dielectric properties of N_2/O_2 mixtures in the tempreature range of 300–3500 K. *Journal of Physics*, 37, 2004. doi:10.1088/0022-3727/37/6/008.

[12] Y. Tanaka. Influence of copper vapor contamination on dielectric properties of hot air at 300–3500 K in atmospheric pressure. *IEEE Trans. on Diel. and Elec. Ins.*, 12(3):504–512, 2005. doi:10.1109/TDEI.2005.1453455.

[13] G. Timoshenko. Die lichtbogenwiederzündung als durchschlag in stark ionisierten gasen. *Zeitschrift fuer Physik*, 84(11):783–793, 1945. doi:10.1007/BF01330502.

[14] W. Hauschild and W. Mosch. *Statistik fuer Elektrotechniker.* Berlin Verlag Technik, 1984.

[15] A. Küchler. *Hochspannungstechnik: Grundlagen - Technologien - Anwendungen.* Springer-Verlag, 2009.

doi:10.14311/ppt.2017.2.169
Plasma Physics and Technology **4**(2):169–172, 2017

APPLICATION OF RF (13.56 MHZ) ARC DISCHARGE FOR PLASMA CHEMICAL CONVERSION OF VOLATILE FLUORIDES

R.A. Kornev[a,*], P.G. Sennikov[a], V.V. Nazarov[b]

[a] *G.G. Devyatykh Institute of Chemistry of High-Purity Substances of RAS, Nizhny Novgorod, Russia*
[b] *Institute of Applied Physics RAS, Nizhny Novgorod, Russia*
[*] kornev@ihps.nnov.ru

Abstract. The report discusses the use of a new type of plasma source of hydrogen plasma based on a RF (13.56 MHz) arc discharge of atmospheric pressure between two electrodes. This discharge was used for hydrogen reduction of the tetrafluorides of silicon, boron and molybdenum. As a result of the studies, the main regularities of the hydrogen reduction process were established and the main synthesized products were determined. Samples of carbides of silicon, boron and molybdenum were prepared.

Keywords: RF arc discharge, boron fluoride, silicon fluoride, molybdenum fluoride.

1. Introduction

Volatile fluorides of boron ($^{nat}BF_3$), silicon ($^{nat}SiF_4$) and molybdenum ($^{nat}MoF_6$) are the most suitable for the physico-chemical characteristics of the starting materials for preparing their isotopic varieties ($^{10/11}BF_3$, $^{28/29/30}SiF_4$, $^{92/94/95/96/97/98/100}MoF_6$) by the centrifugal method [1]. In the future, already isotopically enriched fluorides can be used to prepare isotopic varieties of B, Si and Mo. The listed isotopes and materials on their basis find application in fundamental scientific research, and also have considerable practical significance. Among the seven stable isotopes of molybdenum, the most important isotopes are ^{98}Mo and ^{100}Mo. Both are used to generate radioactive ^{99}Mo, which turns into ^{99m}Tc, used as a radiopharmaceutical, and the ^{100}Mo isotope can be used in nuclear physics experiments to study double neutrinoless beta decay [2]. It was proposed in [3] to use as starting material molybdenum carbide of a natural composition having a developed surface, which would allow the formed ^{99}Mo atoms to be accumulated with a higher efficiency in comparison with the metal. ^{10}B is mainly used in the control rods of thermal neutron reactors. In order to reduce the size of the control rods while increasing their absorbing capacity with respect to neutrons, the most expedient is the use of highly concentrated ^{10}B in the form of metal ceramic of boron carbide with aluminum oxide [4]. The highly enriched ^{29}Si silicon isotopes in ^{28}Si could be used to develop a quantum computer [5]. An important property of monoisotopic silicon, distinguishing it from a natural analogue, is a higher thermal conductivity which can be used to create various electronic devices [6]. For example, it is known that SiC compound is used to create semiconductor devices of power electronics, and in [7] it is noted that the use of ^{28}SiC significantly improves their performance characteristics. As it has been already mentioned, it is expedient to use their fluorides for preparation of the above-mentioned isotopes and materials. The high cost of enriched fluorides requires minimization of return and non-return losses of the substance at the processing stage. Therefore, the development of highly efficient methods for the isolation of isotopically modified silicon, boron and molybdenum from their fluorides is an urgent task. Plasma chemical methods, in comparison with conventional chemical methods, are most preferable for isolation of simple substances and synthesis of compounds from their thermodynamically stable volatile fluorides. The processes of hydrogen reduction of the fluorides of silicon and molybdenum to elementary Si and Mo in the plasma of RF-IC, RFC discharge and RFAD were studied in [8, 9]. The purpose of this work is to study the processes of plasma chemical synthesis of SiC, B_4C and Mo_3C_2 carbides from volatile fluorides of silicon, boron, molybdenum and methane in RFAD discharge at a pressure close to atmospheric and to study the structure of the prepared samples.

2. Description of the discharge

In our case, the discharge is formed between tungsten electrodes in a hydrogen atmosphere in the pressure range 200–760 Torr and represents an electric arc of alternating current with a frequency of $\omega = 13.56$ MHz (Fig. 1a). In this case, there is a strong heating of the end parts of the electrodes that come into contact with the plasma. The electric current in the gas under given conditions is determined by the characteristics of the external circuit. To determine the total current, two electric shunts of alternating current ($r1 = 2500\,\mu\Omega, r2 = 3750\,\mu\Omega$) were connected in series to the grounded electrical high-frequency circuit (Fig. 1c) . The characteristic value of the total current determined by this technique was $I \sim 5$ A. An estimate of the current was also carried out by estimating the current density of thermoionic emission, knowing

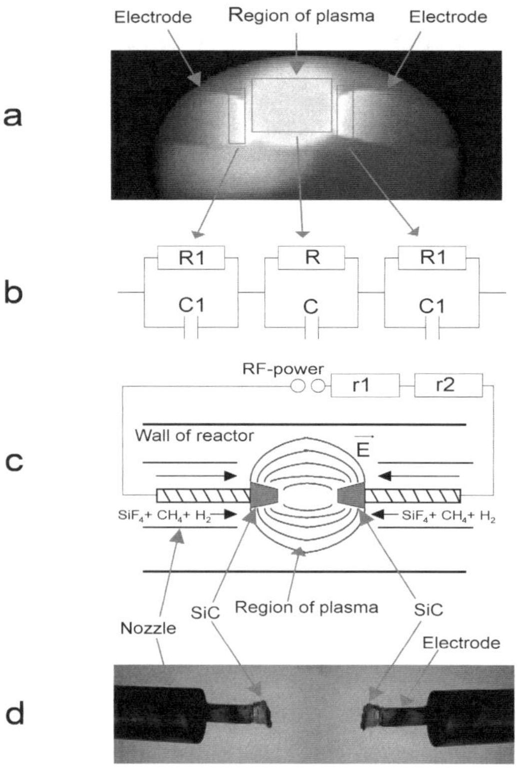

Figure 1. a - is a type of discharge; b - is the equivalent electrical circuit; c - is the scheme of plasma-chemical reactor; d - is the type of precipitated B_4C.

the temperature of the electrodes. The temperature was determined by modeling the thermal processes in the reactor [10] and for the selected experimental conditions was equal to 3500 K. For the electrode material used (tungsten) and the known area of the heated surface, the current density is determined by the formula:

$$j = A \cdot T^2 \cdot exp(-e\varphi/kT) \quad (1)$$

where A for tungsten lies in the interval 40–100, $e\varphi = 4.54\,\text{eV}$ [11]. Knowing the area of the heated electrode surface, the estimated current value is $I \sim 5\,\text{A}$. It can be seen that the experimentally obtained value of the total current coincide with the current value estimated by 1. A double electric layer (DEL) model was used to describe the RF discharge under consideration. Formation of DEL occurs in the area of contact of the metal surface of the electrodes and plasma [11]. According to the adopted model, the equivalent circuit of our discharge is the circuit shown in Fig. 1b. R is the active resistance of the plasma, $C = \epsilon S/(4\pi L)$ is the capacity of the space between the DEL filled with plasma, where S is the characteristic cross-sectional area of the plasma arc, L is the characteristic dimension of the plasma, $\epsilon = -{\omega_{\text{pe}}}^2/\nu_{\text{en}}^2$ is the dielectric constant of the plasma), $R1$ is the active resistance resulting in the heating of the electrodes, connected with the flow of thermoelectrons and electrons that have passed from the plasma through the DEL to the electrodes, $C1 = S1/(4\pi d)$ is the capacity of the DEL, where $S1$ is the area of the electrodes, $d = 3r_D$ (the radius of the Debye) is the characteristic longitudinal size of DEL. The value of R is determined by the formula $R = L/\sigma S$, where L is the characteristic dimension of the plasma, S is the characteristic cross-sectional area of the plasma arc, and $\sigma = {\omega_{\text{pe}}}^2/4\pi\nu_{\text{en}}$ is the conductivity of the plasma [12]. The resistance and impedances of this circuit are calculated according to this formula and are equal to $R = 6\,\Omega$, $Z_C = 10\,\Omega$, $Z_{C1} = 400\,\Omega$, where $Z_C = 1/i\omega C$ is the impedance capacitance C. Since R is comparable with Z_C, there are two components of the electric current in the plasma: the conduction current and the bias current. The value of $R1$ due to the current of thermoionic emission can not be calculated, but judging by the strong heating of the electrodes, $R1 < Z_{C1}$ and the current in the region of DEL is the conduction current, which in our case is 5 A. The role of the bias current is reduced to the transfer of energy through the discharge gap without absorption. Thus, the major part of the energy is released in the near-electrode region, where the deposition of the target product takes place and energy consumption is less as compared with the arc discharge.

3. Experimental

This type of discharge was used to study the processes of hydrogen reduction of fluorides of silicon, boron and molybdenum and synthesis of their carbides at a total reactor pressure of 240 and 760 Torr. Hydrogen was used as the plasma-forming gas. The molar ratio of $H_2/SiF_4/CH_4$ and $H_2/BF_3/CH_4$ was maintained at 9/1.5/1 and the molar ratio of $H_2/MoF_6/CH_4$ equal to 45/1.5/1. The plasma-chemical reactor was a silica glass tube, inside which the rod electrodes were placed equipped with gas flow generators and connected to RF generator (Fig. 1c). During the synthesis of SiC and B_4C, deposition was carried out at the ends of the electrodes in contact with the plasma (Fig. 1d), and in the synthesis of Mo_3C_2, its finely dispersed fraction precipitated on a porous nickel filter. The process of conversion of the fluorides of silicon, boron and molybdenum into their carbides and other possible by-products of the reaction was studied experimentally. The total conversion of fluorides and the presence of gas phase reaction products were determined by IR spectroscopy. The samples were analyzed by X-ray phase analysis, electron microscopy, laser mass spectroscopy, and laser diffraction.

4. Results and Discussion

During the research, in the synthesis of silicon carbide, in gas-phase products at the outlet of the plasma chemical reactor, $SiHF_3$, CF_4, C_2H_4, C_2H_2 and CO_2 were detected. In the synthesis of boron carbide, C_2H_4,

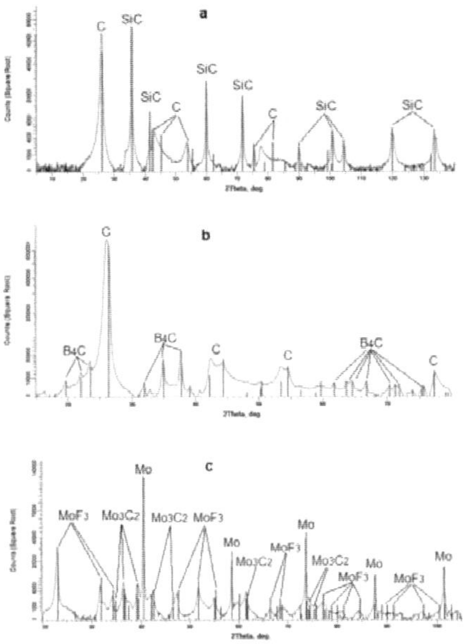

Figure 2. X-ray phase analysis of the samples of silicon carbide (a),boron carbide (b) and molybdenum carbide (c).

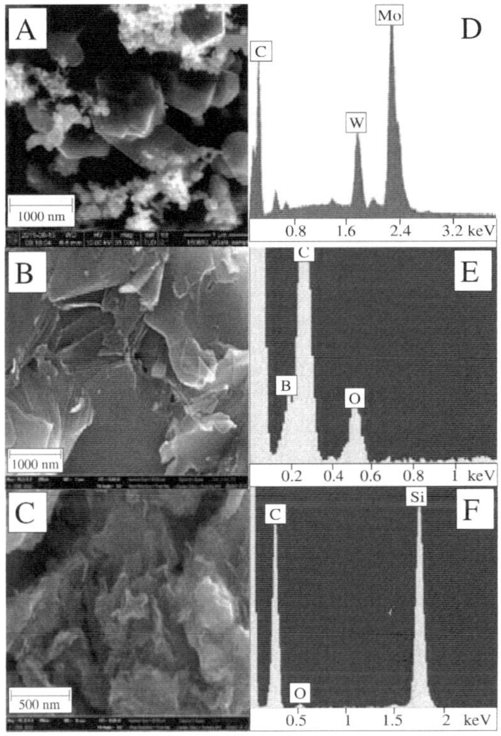

Figure 3. A - Morphology of samples Mo_3C_2, B - B_4C, C - SiC, and their element composition (D), (E) and (F), respectively.

C_2H_2 and CO_2 were also found in the gas phase. In addition, an oily liquid condensed on the walls of the plasma chemical reactor, which represented the complex $[CH_3BF]^-H^+$. It should be noted that both during the synthesis of silicon carbide and the synthesis of boron carbide, the hydrogen fluoride (HF) was not observed in gas phase products. The intensive growth of silicon carbide and boron carbide was observed on electrodes in the form of cylindrical polycrystal, and finely dispersed soot precipitates on the walls of the plasma chemical reactor. According to X-ray phase analysis (Fig. 2a,b), samples of silicon carbides and boron contain up to 40 % of carbon. In the process of synthesis of molybdenum carbide in the exhaust gas mixture, due to small concentration of methane in the initial mixture entering the plasma chemical reactor, the ethylene and acetylene were not observed. The formation of solid-phase products, which include molybdenum carbide, molybdenum and MoF_3, is observed in the volume of the plasma chemical reactor in the form of intense dust formation (Fig. 2c). On the basis of the experimental data obtained, it is possible to assume the basic chemical reactions responsible for the formation of carbides of silicon, boron and molybdenum:

$$SiF_4 + 2CH_4 \rightarrow SiC + CF_4 + 4H_2 \quad (2)$$

$$4BF_3 + 4CH_4 \rightarrow B_4C + 3CF_4 + 8H_2 \quad (3)$$

$$6MoF_6 + 13CH_4 \rightarrow 2Mo_3C_2 + 9CF_4 + 26H_2 \quad (4)$$

According to morphological studies, Mo_3C_2 is formed as a nano-powder with a particle size of 50–70 nm (Fig. 3a). Samples of boron carbide and silicon have a layered structure (Fig. 3b,c). It is possible to say with a high degree of probability that the growth of B_4C and SiC occurs layer by layer. A prerequisite for growth is the initial deposition of a graphite layer on the electrode surface. Graphite precipitates on the surface of the electrode heated up to 3000–3500 K as a result of pyrolysis of methane, forming dendrides. The electric field strength and its gradient on the generated dendrides are high, which allows the SiF_4 or BF_3 molecule to be polarized, and the Si or B layer is deposited. Then, graphite precipitation is required again. The samples grown under these conditions should contain a significant amount of free carbon, which is confirmed by X-ray analysis data.

5. Conclusions

As a result of these studies, the main gas-phase and solid-phase products in the synthesis of the carbides of silicon, boron and molybdenum were determined. The main chemical reactions responsible for the formation of target products – carbides – are proposed. The assumptions are proposed for the mechanism of growth of polycrystalline carbides of silicon and boron on the surface of the electrodes. Samples of carbides of silicon, boron and molybdenum were prepared and their phase, elemental and impurity composition were studied.

Acknowledgements

This work was financially supported by Russian Scientific Foundation Grant 17-13-01027.

References

[1] E. M. Aisen et al. Study of isotope separation of some chemical elements in a gas centrifuge. *Nuclear Instruments and Methods in Physics Research*, 374(1):127–131, 1996. doi:10.1016/0168-9002(96)00065-4.

[2] NUCLEAR ENERGY AGENCY. *The Supply of Medical Isotopes. Review of Potential Molybdenum-99/Technetium-99m Production Technologies.*, 2010. http://www.oecd-nea.org/med-radio/reports/Med-Radio-99Mo-Prod-Tech.pdf.

[3] V. D. Risovany et al. Molybdenum carbide nano-powder for production of Mo-99 radionuclides. *World Journal of Nuclear Science and Technology*, 2(2):58–63, 2012. doi:10.4236/wjnst.2012.22010.

[4] F. Thevenot et al. Boron carbide – a comprehensive review. *Journal of the European Ceramic Society*, 6(4):205–225, 1990. doi:10.1016/0955-2219(90)90048-K.

[5] T. D. Ladd et al. An all silicon quantum computer. *Physical Review Letters*, 89(1):017901(1–4), 2002.

[6] I. C. Kizilyalli et al. Power transistors fabricated using isotopically purified silicon ^{28}Si. *IEEE Electron. Device Lett.*, 26(6):404–406, 2005. doi:10.1109/LED.2005.848111.

[7] T. Noda et al. Silicon isotope enrichment for low activation. *Fusion Engineering and Design*, 41:173–179, 1998. doi:10.1016/S0920-3796(98)00186-0.

[8] P. G. Sennikov et al. A study of silicon tetrafluoride reduction with hydrogen in radiofrequency discharge. *Russian Journal of High Energy Chemistry*, 48(1):49–53, 2014. doi:10.1134/S0018143914010123.

[9] R. Kornev et al. Hydrogen reduction of $^{98}MoF_6$ in RF discharge. *J.Radioanal. Nucl.Chemistr*, 309:833–840, 2016. doi:10.1007/s10967-015-4687-z.

[10] R. Kornev et al. Gas-dynamic and thermal processes in the synthesis of trichlorosilane by hydrogen reduction of silicon tetrachloride in a high-frequency discharge. *Russian Journal of Applied Chemistry*, 87(9):1246–1250, 2014. doi:10.1134/S1070427214090092.

[11] Yu. P. Raizer. *Fizika gazovogo razryada.* Second printing. Nauka, Moscow, Russia, 1987.

[12] D. Yanin et al. Diagnostics of the atmospheric-pressure plasma parameters using the method of near-field microwave sounding. *Technical Physics. The Russian Journal of Applied Physics*, 57(4):468–477, 2012. doi:10.1134/S1063784212040251.

doi:10.14311/ppt.2017.2.173
Plasma Physics and Technology 4(2):173–176, 2017

ROTATING CONSTRICTED SWITCHING ARCS BURNING IN GAS AND IN VACUUM

D. Baron[a], S. Ettingshausen[a], M. Koletzko[c], A. Lawall[b], T. Rettenmaier[c], N. Wenzel[c,*]

[a] *Siemens AG, Energy Management, Medium Voltage & Systems, Frankfurt /Main, Germany*
[b] *Siemens AG, Energy Management, Medium Voltage & Systems, Berlin, Germany*
[c] *Siemens AG, Corporate Technology, Erlangen, Germany*
[*] norbert.wenzel@siemens.com

Abstract. The method of controlling high-current switching arcs by transverse magnetic fields (TMF) forcing the constricted arc to rotate in a contact system is being applied successfully to improve the breaking capability of vacuum interrupters and gas circuit breakers. We describe the behavior of magnetically driven switching arcs in vacuum and in gas environment. We report on experiments using high-velocity videography, magnetic probes, and spectroscopy; they deliver the velocity, the temperature and the voltage of an arc. We present models and simulations of the moving constricted arc burning in metal vapor and in air. And we describe a particular switching application of TMF arc control and explain a scaling law of the contact size with the current interruption capability.

Keywords: constricted arc, vacuum arc, gas arc.

1. Introduction

The method of controlling high-current switching arcs by transverse magnetic fields (TMF) forcing the constricted arc to rotate in a contact system is being applied successfully in the development of vacuum interrupters and gas circuit breakers. Rotating arcs distribute the energetic stress of the arc over the contact surfaces and to prevent overheating of the contacts and, in case of gas arcs, the gas environment. The track and the velocity of the arc strongly influence the switching capacity of the contact system. Detailed optical and magnetic experiments in combination with physical modeling and numerical simulation of the arc provide a profound understanding of the functionality of contact systems with superimposed TMF and thus allow a purposeful improvement of the performance of vacuum and gas switches up to high short-circuit currents.

2. Experiments

2.1. Concept of Arc Control by Tranverse Magnetic Fields

Without any technique of arc control, a switching arc constricts at high currents. The resulting arc diameter is determined by the balance of magnetic force and internal pressure gradient (pinch). The constricted arc is either immobile or, due to existing current loops, moves to the edge of the contact system, where it remains fixed. In both cases, the contact surfaces and an existing gas environment are overheated locally. Local overheating can be prevented by forcing the constricted arc to move across the contacts by applying a Lorentz magnetic force on the arc current.

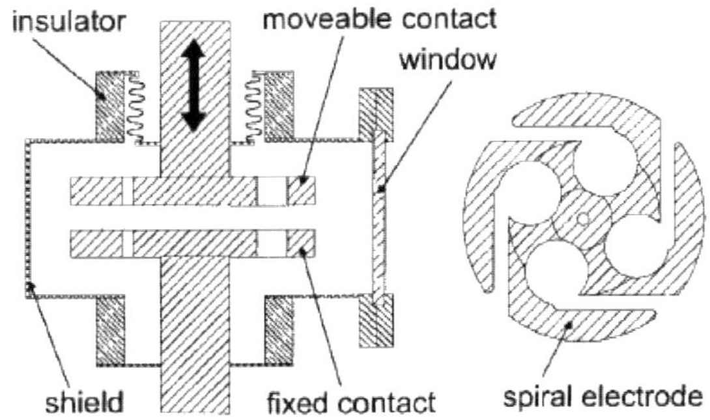

Figure 1. Conventional spiral-type TMF vacuum interrupter contact system.

2.2. Investigation of Vacuum Arcs

We studied experimentally cup-shaped and spiral-type "transversal magnetic field" (TMF) vacuum interrupter contacts with different diameters and arc currents. Fig. 1 (taken from [1]) shows a conventional spiral contact. The constricted metal vapor arc is drawn by separating a pair of 75 mm diameter copper-chromium (CuCr) electrodes to a gap of 8 mm at a velocity of typically 1 m/s. The arc is driven outward and rotates along the spiral arms by the Lorentz magnetic force.

The experiments were performed in a synthetic test circuit using a capacitor bank to generate 50 Hz sine half-cycles of up to 120 kA peak amplitude.

Arc position and arc velocity were measured by side-on observation of the rotating arc with a high-speed movie camera. For this purpose, the sealed-off vacuum tube was equipped with a viewing port window. Details are described in [1]. Simultaneous measurements with a set of several magnetic field probes provided independent information on the arc velocity movement. The arc velocity was recomputed from the modulation of the probe signal reflecting the times when the arc passes the probes.

2.3. Investigation of Gas Arcs

In case of gas arcs, the movement of the arc and the effect of external magnetic fields are important factors influencing switch gears. These features are some of the main reasons to simulate the switching arc to get a better knowledge to improve the geometry of the contact system. We developed a method of arc simulation for medium voltage switch gears, for example switches using "Rotating Arc" contact designs with permanent magnets. To validate the simulation, experiments were done in atmospheric air with a simple linear system of a stationary and a moving copper contact. The maximum contact gap was 22 mm, the experiments have been done with a current of 150 A and 500 A at 50 Hz, respectively. The arc position was measured by high-speed movie cameras. We used two cameras to get an approximate 3D position (one at the front side, one at the top side) of the arc. Simultaneous emission spectroscopy with side-on optical access to the arc provides information on the plasma temperature by fitting Boltzmann-plots to suitable spectral lines of the nitrogen gas species. This allows the validation of the simulated electrical arc conductivity. The third information gained in the experiment was the arc voltage. As boundary condition for the simulation, the movement of the contact and also the r.m.s. current of the experiment were measured.

3. Arc Modeling and Simulation

A moving arc that is not gradually accelerated requires a well-defined balance of driving and "frictional" forces acting on the arc. The arc experiences a permanent loss of momentum (restraining ("friction") force) that is balanced by the permanent acceleration of the plasma ions due to the Lorentz magnetic force (driving force). The computation of the arc velocity requires the solution of the complete mass, momentum, and energy balance of the arc in combination with all driving and frictional forces.

3.1. Analytical Vacuum Arc Models

We used different analytical models [1–3] to interpret the movement of constricted vacuum arcs and deliver expressions for the arc velocity.

The model published in [1] assumes a fixed surface temperature and a fixed total energy flux at the contact surfaces. The assumptions are justified insofar as the evaporation of contact material, necessary for the existence of metal vapor vacuum arcs, is insufficient below the boiling point. Most of the heat into the contact is used for evaporation.

An extended model developed in [3] is based on the energy balance equation similar to the model found in [1], but supplemented by an additional term that considers the momentum loss caused by the loss of neutral vapor leaving continually the arc column by diffusion. It explicitly takes into account the thermal response of the electrodes, assuming a variable surface temperature and a self-consistent heat transport. In particular, evaporation from the contacts and the corresponding cooling is treated quantitatively, using empirical relations from arc voltage and root diameter measurements at anode and cathode. The model [3] results in an explicit equation for the arc velocity,

$$v_A = \frac{\text{const.}\ I^{\frac{23}{16}}\ b_r\ h\ (c\ T_M\ +\ H_M\ +\ H_B)}{\alpha\ U\ I^{\frac{7}{16}}\ -\ \text{const.}'\ (\gamma\ T_M\ -\ T_0)\ \sqrt{v_A}}, \quad (1)$$

v_A = arc velocity, I = current, b_r = specific transversal magnetic flux density, h = contact distance, U = arc voltage, T_M and T_0 = surface temperatures of arc root (T_M) and before arc attachment (T_0), c, H_M, H_B = specific heat, melting and boiling enthalpy of contact material, γ = numerical sensitivity parameter, α = fraction of arc power dissipated to contacts.

3.2. Numerical Vacuum Arc Simulation

In principle, the simulation of a rotating constricted TMF arc has to be solved three-dimensionally (3D) because of the lack of symmetry of relevant contact geometries. Because of the high spatial gradients of the arc variables occurring in the intermediate zones towards electrodes, we chose in a first step a planar 2D simulation using the COMSOL Multiphysics platform.

In the model, the usual two circular contacts are replaced by two parallel straight electrode rails separated by a gap distance. The plasma column is described by a fluid dynamic approach with temperatures for heavy species and electrons. The energy loss by radiation is taken into account by a net emission coefficient. Two attachment zones at cathode and anode act as sheaths that describe energy and mass transfer between plasma column and electrodes. The electrodes are treated as current and heat conducting boundary zones. The TMF is superimposed as background field. The approach is similar to that treated in [4]. We applied the 2D model of the arc moving in the linear rail configuration to a realistic arc plasma that rotates in a circular cup-shaped TMF contact system and returns to the same location several times. For this purpose, the length of the rails was chosen to be equal to the circumference of the circular contact. After each arc transfer across the segment, the arc location is set back to the beginning of the segment, and the plasma and electrode parameters are stored to provide the initial conditions for the next arc transfer. The segment is allowed to cool until the arc returns and transfers additional energy to the electrodes. For test purposes, a gas arc model (SF_6 , ambient pressure = several bars) and a vacuum arc model with simplified electrode boundary conditions were implemented. The simulations were run for currents of several 10 kA, specific TMF amplitudes on the order of 1 T, and an electrode gap of 15 mm.

3.3. Gas Arc Simulation

Also in this case, the simulation of the switching arc has to be done by 3D. Because of the symmetry of the simulation model and the movies of the experiment, it was decided to do the simulation in a 1/4-model with symmetry conditions. The simulation (CFD and electromagnetic) was done in Ansys CFX in combination with the MHD-module. During the opening of the contacts, a switching arc appears between the fixed and the moving copper electrode. The plasma is described by a fluid dynamic approach with temperatures for heavy species and electrons. The radiation is considered by the net emission coefficient. The anode and cathode are included by well known formulations. The simulation involves moving meshes. After a mesh criterion was reached, the simulation was stopped, a new mesh was loaded, and the results were interpolated on the new mesh. After the interpolation, the simulation restarted with the time code of the previous step. This includes also the position and current of the simulation at the last time step. The recorded current of about 150 A and 500 A (rated current) was simulated for a time period of 10 ms. After this time the gap between the electrodes reached 18 mm.

4. Results and Discussion

In this paper, we present particular examples of experimental and theoretical results.

Fig. 2 illustrates the simulated spatial structure of an arc plasma burning in atmospheric air environment and moving between two electrode rails separated by 15 mm. The distributions of arc velocity, current density, and temperature within the arc column 1 ms after arc ignition are plotted for a current of 10 kA and a TMF of 1 T. Pressures of several tens of bars are deduced from the simulation. Due to the action of the TMF, the flow of plasma is bended to the direction of arc propagation, and plasma velocities of the order of 250 m/s are approached in the mid-plane of the gap. When the gap distance is increased, the deviation from a straight column becomes more pronounced.

Fig. 3 shows a comparison of experimental and theoretical propagation velocities of the TMF vacuum arc as a function of the current. The results based on the model of [3] are obtained in a numerical computation. The iteratively calculated values of v_A correlate quite well with the data reduced from experimental tracks. The arc velocity increases almost linearly with the current amplitude and attains values of more than 300 m/s for currents of 120 kA. The results can be used to describe the number of rotations of the arc on a TMF contact and to explain the scaling law of the contact diameter with the current.

Fig. 4 documents a particular scaling law for the current interruption capability of a short-gap spiral-type TMF contact design used in low-voltage high-

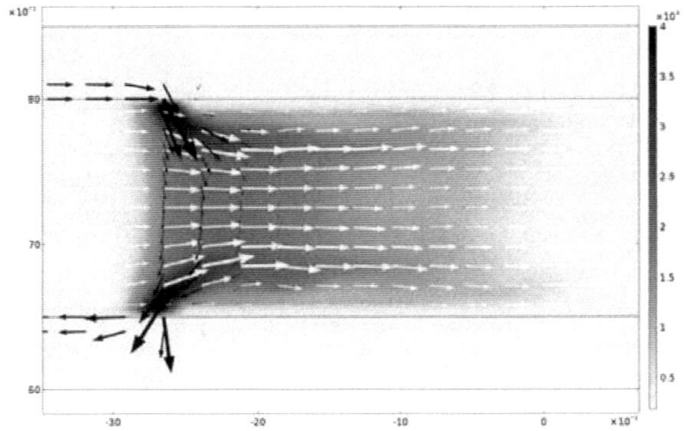

Figure 2. Simulated distribution of arc velocity (white arrows), current density (black arrows), and temperature (colors, legend in K) within a gas arc plasma moving between two electrode rails. Details see text.

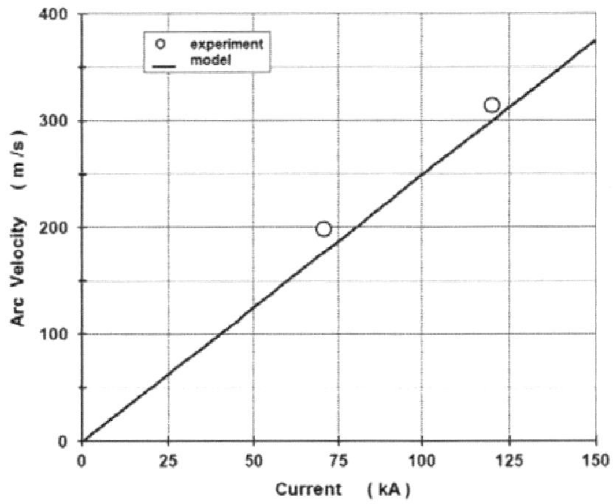

Figure 3. Comparison of experimental and theoretical propagation velocities of TMF vacuum arc as a function of current. Line: Model of [3]. Circles: measured values for CuCr contacts.

current vacuum circuit breakers. The figure shows the dependence of the breaking current of the contacts on their diameter. The results of synthetic tests and of computations based on the model of [3] are in good agreement. The capability linearly increases as function of the contact diameter. This finding can be motivated as follows: The temperature of the contact surface, besides the residence time of the metal vapor in the inter-electrode gap, impacts the performance of the contact. The temperature is determined by the time for one arc circulation on the contact and by the heating time at a fixed arc location. The cooling time, which is given by the ratio of the circumference of the arc rotation and the arc velocity, stays constant if the current varies linearly with the contact diameter. An almost linear behavior is expected and is indeed confirmed by experiment.

Fig. 5 compares the measured total voltage (including fall voltages) and the computed column voltage of the arc burning in atmospheric air. Experimental and simulated temporal evolutions are in satisfying

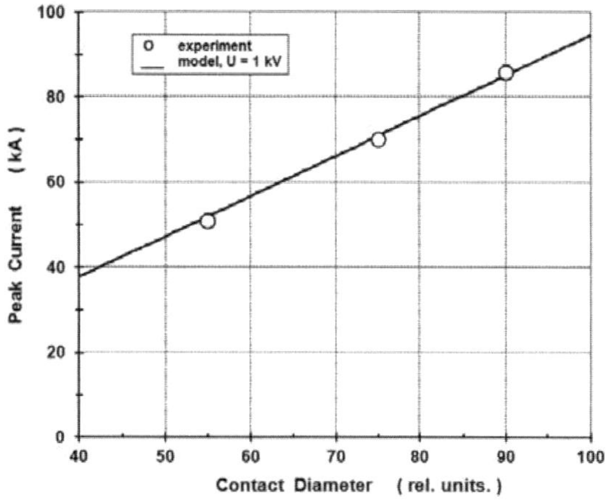

Figure 4. Experimental and theoretical short-circuit current interruption capability of TMF vacuum contacts as a function of contact diameter for the voltage level $U = 1\,kV$. Line: Model of [3]. Circles: measured values for CuCr contacts.

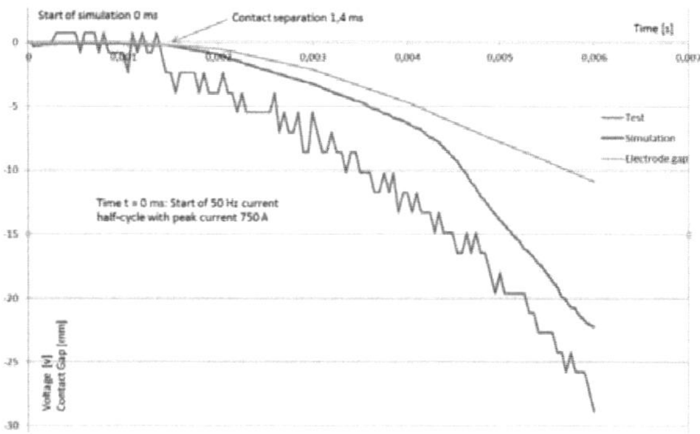

Figure 5. Experimental (blue) and simulated (red) arc voltages of gas arc vs. time. Measured contact gap (green) is input quantity for simulation.

agreement taking into account the measured voltage step at contact separation.

Spiral-type electrodes similar to the vacuum contact described in Fig. 1 were used in [1] for experiments with arcs burning in a low-pressure gas fillings (p = several tens of Pa), such as He, Ne, Ar, H_2, or D_2. The results documented that low-pressure gas arcs propagate in a pseudo-diffuse mode, do not need the metal vapor emission from hot electrode surfaces, and thus are able to propagate without the need of concentrated arc roots. The arcs achieved velocities far exceeding 10 km/s for currents of up to 25 kA. Low-pressure gas arcs move significantly faster than expected for atmospheric-pressure gas arcs and determined for vacuum arcs that move with velocities below 1 km/s.

5. Conclusion

The investigations presented here have been proven to be suitable to determine the velocity of rotating constricted TMF vacuum arcs and thus to predict the switching capability of TMF vacuum interrupter contacts. The applied experimental techniques are mature; more detailed results could be obtained by stereoscopic high-speed videography and tomographic 2D reconstruction of magnetic probe measurements. Though the analytical models describe the arc movement explicitly, relevant model parameters still have to be derived from experiments. The feasibility of a self-consistent planar 2D simulation of the TMF controlled arc moving in a linear electrode rail configuration has been demonstrated. A 2D simulation is suitable to describe the arc motion in a ring-shaped TMF contact and even the jumping over slits occurring in spiral-type TMF electrodes, as has been reported by [4]. Nevertheless, a quantitative optimization of real TMF contacts requires a full 3D treatment. The computational effort necessary to generate results from such simulations remains high; this has been shown by [5] who applied a 3D model to a rail-type electrode configuration.

The investigations show good results for switching arc simulation with the fluid air. Next steps are the simulation of SF_6 arcs and spectroscopic measurements with a high speed camera for transient thermal profiles. Also 3D modeling has to be performed to simulate the effects of external magnetic fields on the switching arc. Furthermore, there will be a change from linear moving contacts to rotatory moving contacts. Application of the 3D method to rotary contacts will deliver more detailed information and a better understanding of physical effects during the switching process, thus enabling a faster development of switch gears in medium voltage systems.

References

[1] J. Teichmann, M. Romheld, and W. Hartmann. Magnetically driven high current switching arcs in vacuum and low pressure gas. *IEEE Transactions on Plasma Science*, 27(4):1021–1025, Aug 1999. doi:10.1109/27.782274.

[2] E. Dullni, E. Schade, and Wenkai Shang. Vacuum arcs driven by cross-magnetic fields (rmf). *IEEE Transactions on Plasma Science*, 31(5):902–908, Oct 2003. doi:10.1109/TPS.2003.818445.

[3] D. W. Branston, W. Haas, W. Hartmann, R. Renz, and N. Wenzel. Thermodynamic model of moving vacuum arcs on contrate contacts. In *22nd ISDEIV*, volume 1, pages 277–280, Matsue, Japan, Sept 2006. doi:10.1109/DEIV.2006.357285.

[4] T. Delachaux, O. Fritz, D. Gentsch, E. Schade, and D. L. Shmelev. Simulation of a high current vacuum arc in a transverse magnetic field. *IEEE Transactions on Plasma Science*, 37(8):1386–1392, Aug 2009. doi:10.1109/TPS.2009.2020403.

[5] O. Fritz, D. Shmelev, K. Hencken, T. Delachaux, and D. Gentsch. Results of 3d numerical simulations of high-current constricted vacuum arcs in a strong magnetic field. In *24th ISDEIV 2010*, pages 359–362, Braunschweig, Germany, Aug 2010. doi:10.1109/DEIV.2010.5625802.

doi:10.14311/ppt.2017.2.177
Plasma Physics and Technology 4(2):177–181, 2017

AN EFFICIENT IMPLEMENTATION OF THE FINITE-VOLUME METHOD FOR THE SOLUTION OF RADIATION TRANSPORT IN CIRCUIT BREAKERS

A. Mazaheri[a,*], J.Y. Trépanier[a], R. Camarero[a], P. Robin-Jouan[b]

[a] *École Polytechnique de Montréal, 2900 boul. Édouard-Montpetit, Campus de l'Université de Montréal, 2500 chemin de Polytechnique, Montréal (Québec), H3T 1J4, Canada*

[b] *GE Grid-Solutions, Villeurbanne,France*

[*] ali.mazaheri@polymtl.ca

Abstract. In this paper, we propose to revisit the method to solve the radiation transport equation in circuit breakers to reduce the computation time. It is based on an explicit approach using a space marching algorithm. The method can further be accelerated using a Cartesian grid and using the axisymmetric assumption. Comparisons performed in terms of accuracy and efficiency between the P1 model, the implicit finite-volume discrete ordinate method and the space-marching finite-volume discrete ordinate method show that the explicit approach is more that an order of magnitude faster than the implicit approach, for the same accuracy.

Keywords: Radiation, Circuit-breaker, Finite-volume method, Space-Marching.

1. Introduction

The numerical modeling and simulation of arc-flow interaction in circuit-breakers are based on the solution of compressible fluid flow, coupled with electrical and magnetic field equations. In circuit breakers, during the high current arcing phase, the high plasma temperature (around 20 kK) results in dominant energy transport by radiation, and intense radiation fluxes reach the nozzle material, causing critical vaporization of the nozzle. Consequently, most recent models require accurate solution of the radiative energy transfer. The numerical modeling of the radiation phenomena is highly complex [1, 2] as the radiation field depends on temperature, gas composition, frequency and direction. Over the years, various radiation models have been proposed, often to provide a compromise between accuracy and computation time. The P1 model [3] has been extensively used because of its relatively low computational cost. However, boundary conditions are difficult to apply for this model and the model fails for transparent media [4]. The most recent modeling involve the solution of the radiative transport equation for a small number of frequency bands using finite-volume discretization (FVM) of the angular directions of propagation [5]. Although it is generally admitted that the FVM method is superior in terms of accuracy over the P1 model, its computational cost often limit its usage for practical applications. In the present paper, we revisit the FVM approach and propose, under few assumptions, an explicit low cost extension of the FVM for circuit-breaker arcing simulation.

2. Governing Equations

The radiative transfer equation (RTE) for a non-scattering medium(eg. circuit breaker) in cylindrical coordinate system is shown in Eq 1:

$$\begin{aligned}\frac{1}{r}\frac{\partial}{\partial r}[\mu r I(\vec{R},\vec{S})] + \frac{1}{r}\frac{\partial}{\partial \varphi_0}[\eta I(\vec{R},\vec{S})] \\ + \frac{\partial}{\partial z}[\xi I(\vec{R},\vec{S})] - \frac{1}{r}\frac{\partial}{\partial \phi}[\eta I(\vec{R},\vec{S})] \\ = -\beta(\vec{R}) I(\vec{R},\vec{S}) + \kappa_a(\vec{R}) I_b(\vec{R})\end{aligned} \tag{1}$$

In this equation, $\vec{S} = \vec{e}_r \sin\theta\, \cos\phi + \vec{e}_{\varphi_0} \sin\theta \sin\phi + \vec{e}_z \cos\theta$, $\mu = \sin\theta\, \cos\phi$, $\eta = \sin\theta\, \sin\phi$, and $\xi = \cos\theta$ are the cosines of the path $\vec{S}$, κ_a is the absorption coefficient. This equation governs the variation in the radiant intensity $I(\vec{R},\vec{S})$ at a specific location $\vec{R}$ through the length ds in the direction of $\vec{S}$ as shown in Fig. 1.

For an axisymmetric configuration the term of $\frac{\partial}{\partial \varphi_0}$ in Eq. 1 is removed and reduces to:

$$\begin{aligned}\frac{1}{r}\frac{\partial}{\partial r}[\mu r I(\vec{R},\vec{S})] + \frac{\partial}{\partial z}[\xi I(\vec{R},\vec{S})] - \frac{1}{r}\frac{\partial}{\partial \phi}[\eta I(\vec{R},\vec{S})] \\ = -\kappa_a(\vec{R}) I(\vec{R},\vec{S}) + \kappa_a I_b(\vec{R})\end{aligned} \tag{2}$$

Assuming a non-scattering medium, the final form of the RTE becomes:

$$\frac{dI(\vec{R},\vec{S})}{ds} = -\kappa_a I(\vec{R},\vec{S}) + \kappa_a I_b(\vec{R}) \tag{3}$$

In this analysis all boundaries are assumed diffuse-gray where the RTE is subject to the following bound-

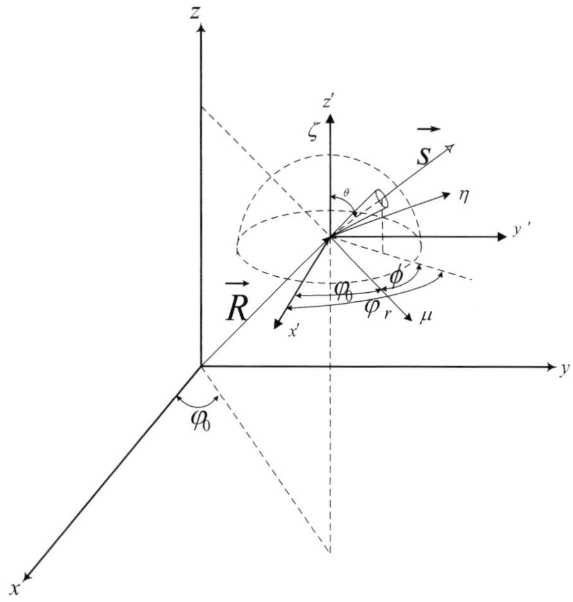

Figure 1. Reference coordinates system for the RTE as given by [6].

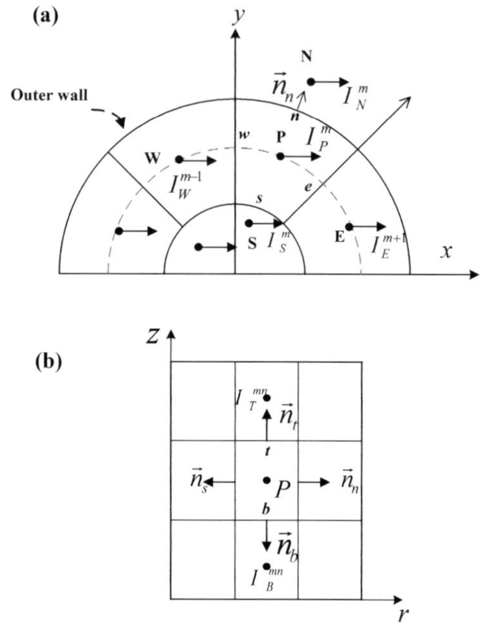

Figure 2. The control volume in a cylindrical enclosure around node P and molecule of nodal points:(a)top view (b)side view.

ary condition for the intensity on the wall I_w:

$$I_w(R_w, \vec{S}) = \varepsilon_w I_b(R_w) + \frac{1-\varepsilon_w}{\pi}\int_{\vec{n}_w.\vec{s}'<0} I(r_w, \vec{s}')d\omega' \quad (4)$$

In a circuit breaker, all boundaries are assumed as black body ($\varepsilon_{wall} = 1$), so for a surface with a specific temperature (T_B) the boundary condition is:

$$I_w(R_w, \vec{S}) = \sigma_B T_B^4/\pi \quad (5)$$

2.1. Finite Volume Solution for the RTE

The finite volume method (FVM) enforces the conservation laws for discrete volumes (computational mesh) to solve the governing equations in fluid dynamics and heat transfer simulation. Eq. 1 conveys the conservation of radiant energy in a specific direction over a differential control volume and within an infinitesimal control angle.

To solve this equation, the computational domain is discretized as shown in Fig. 2 for a cylindrical domain. The intensity is located on the nodal points P, E, W, S, N, T, and B. Each nodal point P is enclosed by six control surfaces identified by e, w, s, n, t, and b. Furthermore, direction is divided into discrete solid angles ω^{mn} that are defined for m_{th} polar and n_{th} azimuthal solid angle in a range of $\phi^{n-1/2}$ to $\phi^{n+1/2}$ and $\theta^{m-1/2}$ to $\theta^{m+1/2}$ as shown in Fig. 3. The integration points are located on the surfaces of the control volume [6]. Eq. 3 is integrated over a control volume, V_P, and a control angle ω^{mn} to obtain the discretized form of the RTE:

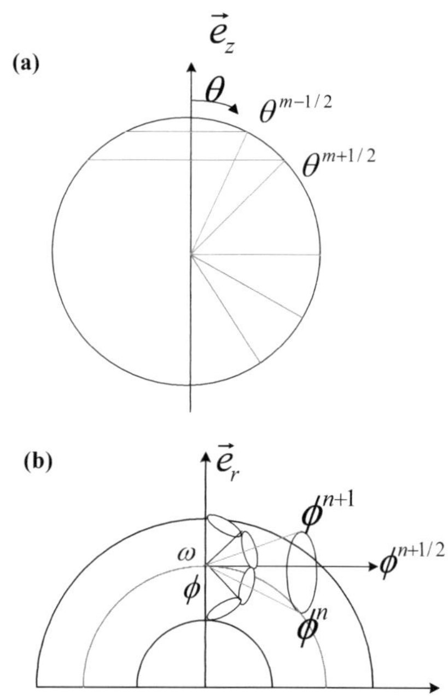

Figure 3. The angular discretization:(a)azimuthal direction (b)polar direction.

$$\sum_{i=n,s,t,b,w}^{6} A_i D_i^{mn} I_u i^{mn} = \left[-\kappa_{a,P} I_P^{mn} + \kappa_{a,P} I_{b,P}\right]\omega^{mn} V_P \quad (6)$$

where A_i is the control surface area and I_b is the black body intensity which is determined with domain temperature and:

$$D_f^l = \int_{\omega^l} \vec{s}.\vec{n_f} d\omega \qquad (7)$$

2.2. Evaluating the Intensity on the Control Surfaces

In a circuit breaker the RTE and the gas dynamics equations are solved on the same grid and due to presence of high gradient in the flow, this grid should be fine enough in order to have accurate results. In the present work, a first order step scheme that is accurate on fine grid is used to evaluate the intensity on the surfaces of the control volume. In each direction, for the surfaces which are located in the upstream of the node P while the intensity on the surface is considered as I_P^{mn} and for the surfaces which are in downstream of the node P the intensity of downstream nodes are:

$$I_{i=n,t,e}^{mn} = I_P^{mn} \qquad (8a)$$

$$I_s^{mn} = I_S^{mn} \qquad (8b)$$

$$I_w^{mn-1/2} = I_W^{mn-1} \qquad (8c)$$

$$I_b^{mn} = I_B^{mn} \qquad (8d)$$

2.3. Final Form of the Equations

Replacing the value of the intensity in Eq. 6, the final form of the equations for the radiant intensity on node P is:

$$a_p^{mn} I_p^{mn} = \sum_{I=N,S,T,B,W} a_I I_I^{mn} + b_P^{mn} \qquad (9)$$

In order to solve this equation explicitly an appropriate marching order is necessary.

2.4. Axisymmetric Extension

The radiant heat transfer is axisymmetric when the intensity does not depend on φ_0 and the intensity is indicated as $I(r, z, \phi, \theta)$. The intensities shown in Fig. 4a are located in the same r, z, and θ. The spatial angular difference $\Delta\varphi_0$ for all points (i=1,2..., 8) is $\pi/4$. The points (i=1,2..., 8) are locate at $\varphi_0 = 7\pi/8, 5\pi/8, 3\pi/8, \ldots$ respectively, and the angle $\phi_r = 0$. Thus, the angle $\phi = \varphi_r - \varphi_0$ is $-7\pi/8, -5\pi/8, -3\pi/8, \ldots$ for the intensities $I_1^1, I_2^1, I_3^1, \ldots$. However, in Fig. 4b the intensities are located at same r, z, and θ and at spatial angle $\varphi_0 = 7\pi/8$, with angular azimuthal angle $\varphi_0 = 0, \pi/4, \pi/2, \ldots$ for the intensities $I_1^2, I_1^2, I_1^2, \ldots$, respectively and the $\phi = 7\pi/8, -5\pi/8, -3\pi/8, \ldots$ ($\Delta\varphi_0 = \Delta\phi$). Therefore, r, z, ϕ, and θ are the same for I_1^m and I_m^1 in Fig. 4a and Fig. 4b. Due to this mapping between intensities in Fig. 4a and Fig. 4b, it is more convenient to solve the domain in Fig. 4a to obtain the intensity for each node.

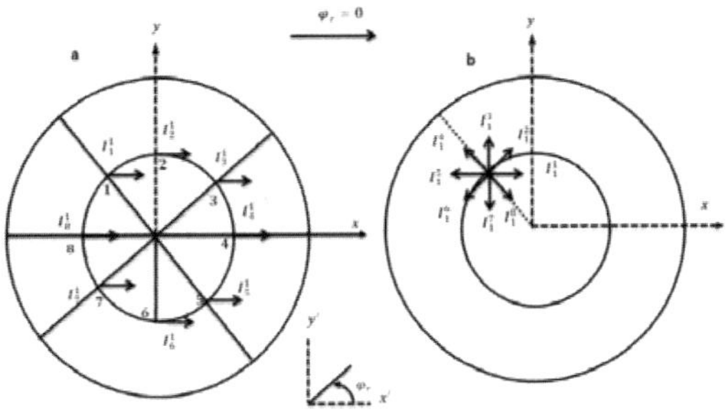

Figure 4. A mapping between the distribution of intensity in cylindrical coordinate system and intensity at specific point in all directions.

2.5. Solution Procedure

A marching order map for nodes is employed to solve the radiant heat transfer within a non-scattering cylindrical medium. For $\theta < \pi/2$, the marching order is from outer wall to the center point for segment 1 in Fig. 2a for each z level from bottom to the top of the cylinder Fig. 2b. For the segments with $\phi_0 > \pi/2$, the marching is from center point to the wall for each z level from bottom of the cylinder to the top of it. The same manner is carried out for all the segments with $\theta > \pi/2$, but the marching order in z direction is from the top to the bottom. The nodal intensity is obtained explicitly for a given medium temperature if the enclosure walls are black.

3. Results and Discussions

3.1. Isothermal cylinder

Figure 5 shows the results of a verification study for a cylindrical geometry filled with isothermal gas at T=1000 K. The cylinder has a $r = 1$ m radius and its length is 2 m. The temperature at the cylindrical wall and at both ends of the cylinder is fixed at $T = 0$ K. The results show the normalised radiative flux on the cylindrical wall for three different values of the optical thickness (defined as $\tau = r\kappa_a$). Two cartesian grids are computed with, respectively, 17×33 and 50×75 grid points. For the two cases, the discretization of angular space is 14×16 directions. Results are compared to the analytical solution from ref. [7]. Globally, the results are excellent and accurate, even on coarse grid.

Figure 6 shows the effect of the angular discretization for the same cylindrical case. One can see that an angular discretization using 10×12 directions provides almost the same result as a 14×16 discretisation. Since the computational time increases linearly with the number of directions, the following results will be computed using a 10×10 angular discretization.

3.2. Semi-industrial test case

Figure 7 illustrates the second geometry used to test the present computational approach. The geometry is representative of elements found in circuit-breakers to illustrate the characteristics of the method. As seen

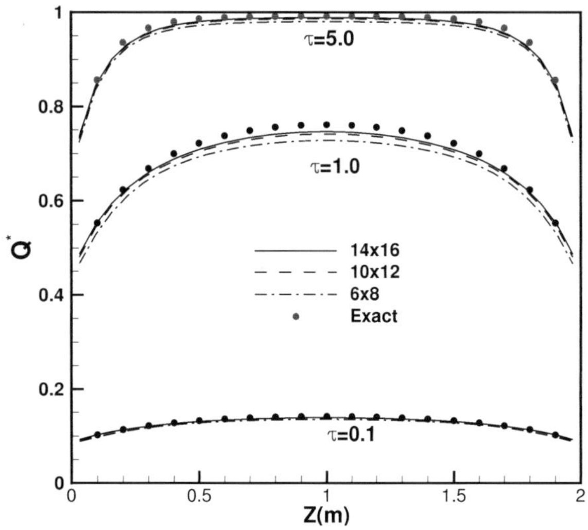

Figure 5. The radiative heat flux on the cylinder wall in various angular grids $N_r \times N_z \times N_\phi \times N_\theta$.

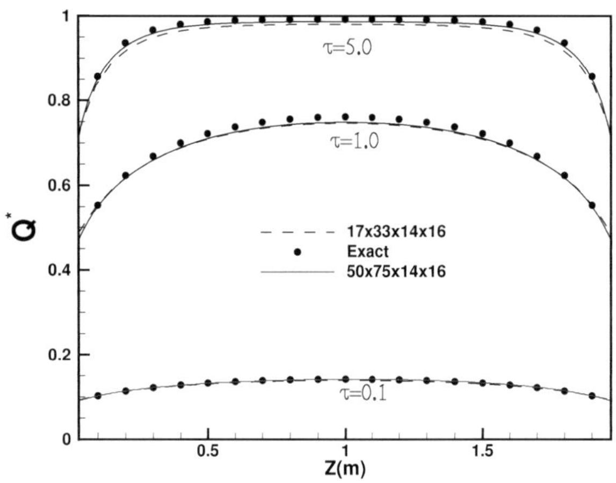

Figure 6. The comparison of the radiative heat flux obtained with numerical simulation on the cylinder wall in various spatial grids $N_r \times N_z \times N_\phi \times N_\theta$.

in Fig. 7, a hot region is placed on the axis with a temperature $T = 1000\,\mathrm{K}$. Two blocking regions are inserted into the domain. The rest of the domain contains a gas at $T = 0\,\mathrm{K}$ with a constant absorption coefficient.

Figures 8, 9 and 10 illustrate a comparison of the normalized radiative heat flux on the upper boundary obtained with various methods. The FVM-MC3 method is described in ref. [5]. It implements an implicit version of the FVM method for the solution of the RTE. The P1-MC3 method is described in ref. [3]. It implements the classical P1 method. Both methods are compared with the present work, for three different values of the absorption coefficient. The FVM-MC3 and the P1-MC3 method are computed on a triangular grid composed of 20780 nodes. The present method use a 20700 nodes Cartesian grid. First, one can observe that the agreement between the two FMV methods is very good. The differences

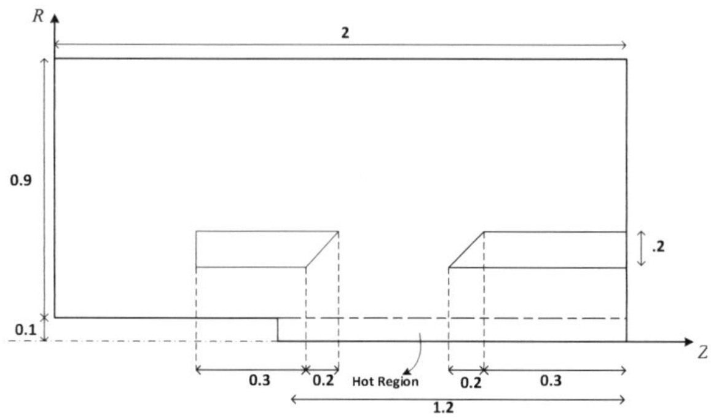

Figure 7. Semi-industrial test case geometry.

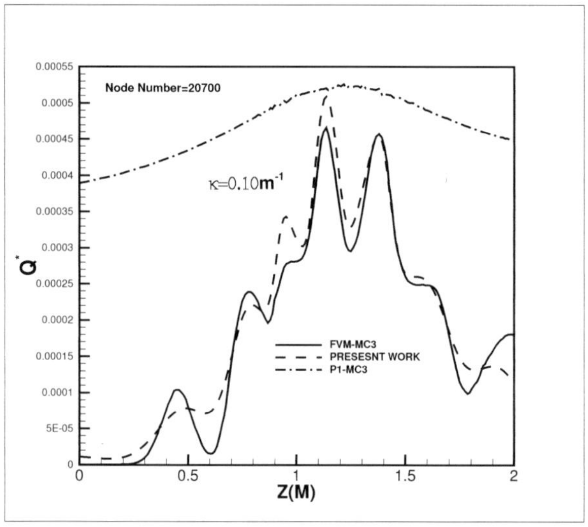

Figure 8. The comparison of the radiative heat flux obtained using current numerical approach and results obtained with MC^3 for $\kappa = 0.1\,m^{-1}$.

can be attributed to the different grids used. Second, the results confirm the poor performance of the P1 method for such cases, especially for low and high coefficient of absorption and for long distance propagation of radiation over obstacle in the domain.

3.3. CPU time

The Table 1 give the CPU time required for the radiation computation for the semi-industrial case. It can be observed that on a 10×10 angular discretization, the present method is more that 50 times faster than the previous FVM-MC3. Also, the computation time is close to the one required by the current implementation of the P1 model. This result confirms the very high potential of the method in reducing computation time for the FVM solution of the RTE.

FVM-MC3	P1-MC3	Present Method
40 sec.	0.3 sec.	0.7 sec.

Table 1. Comparison of CPU time

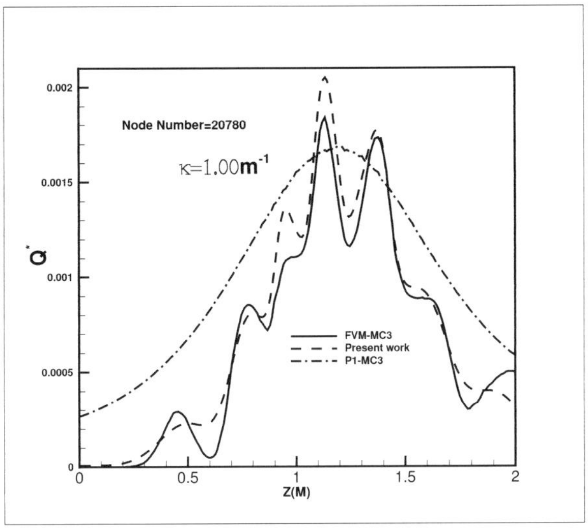

Figure 9. The comparison of the radiative heat flux obtained using current numerical approach and results obtained with MC^3 for $\kappa = 1.0\,m^{-1}$.

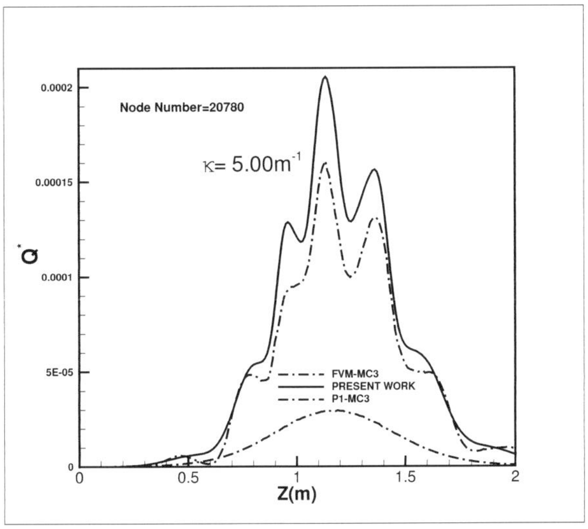

Figure 10. The comparison of the radiative heat flux obtained using current numerical approach and results obtained with MC^3 for $\kappa = 5.0\,m^{-1}$.

4. Conclusion

In the present work, we have described and implemented an explicit version of the FVM for the solution of the RTE in circuit-breakers. The main assumption is that the boundaries of the computational domain can be assumed as black bodies, absorbing all incident radiation. This assumption allows to use a one sweep explicit marching solution procedure of the RTE, reducing the computation time in comparison to implicit methods. The results shows the correctness of the implementation and confirm the CPU saving expected.

Acknowledgements

The authors would like to thanks the NSERC of Canada for the support of this work.

References

[1] H. Nordborg and A.A. Iordanidis. Self-consistent radiation based modelling of electric arcs: I. efficient radiation approximations. *Journal of Physics D: Applied Physics*, 41(13):135205, 2008. doi:10.1088/0022-3727/41/13/135205.

[2] A.A. Iordanidis and C.M. Franck. Self-consistent radiation-based simulation of electric arcs: II. application to gas circuit breakers. *Journal of Physics D: Applied Physics*, 41(13):135206, 2008. doi:10.1088/0022-3727/41/13/135206.

[3] S.D. Eby, J.Y. Trépanier, and X.D. Zhang. Modelling radiative transfer in SF_6 circuit-breaker arcs with the P_1 approximation. *Journal of Physics D: Applied Physics*, 31(13):1578, 1998. doi:10.1088/0022-3727/31/13/012.

[4] M.F. Modest. Chapter 16 - the method of spherical harmonics (P_N-approximation). In M.F. Modest, editor, *Radiative Heat Transfer (Third Edition)*, pages 495–540. Academic Press, Boston, 2013.

[5] M. Melot, J.Y. Trépanier, R. Camarero, and E. Petro. Comparison of numerical models in radiative heat transfer with application to circuit-breaker simulations. *Mathematics and Computers in Simulation*, 82(12):2982 – 2996, 2012. doi:10.1016/j.matcom.2012.07.002.

[6] M.Y. Kim. Assessment of the axisymmetric radiative heat transfer in a cylindrical enclosure with the finite volume method. *International Journal of Heat and Mass Transfer*, 51(21):5144–5153, 2008. doi:10.1016/j.ijheatmasstransfer.2008.03.012.

[7] S. Dua and P. Cheng. Multi-dimensional radiative transfer in non-isothermal cylindrical media with non-isothermal bounding walls. *International Journal of Heat and Mass Transfer*, 18(2):245–259, 1975. doi:10.1016/0017-9310(75)90157-X.

doi:10.14311/ppt.2017.2.182
Plasma Physics and Technology **4**(2):182–185, 2017

DEVELOPMENT OF CURRENT LIMITING DEVICE FOR SHORT AND AUTONOMOUS NETWORKS

K. Vostrov*, V. Frolov, E. Safonov

Department of Electrical Power Engineering and Equipment, Peter the Great Saint-Petersburg Polytechnic University, Polytechnicheskaya 29, 195251 St. Petersburg, Russia

* vostrov_ki@spbstu.ru

Abstract. In this paper, the problem of tripping of high values of short-circuit currents in medium-voltage networks will be considered. Disconnection of short-circuit currents which can reach 100–300 kA in short networks is difficult task for modern switching equipment due to the impossibility of damping the electric arc generated by such currents, and due to strong thermal and electrodynamic effects on elements of the switching device. Solution based on active current limiting reactor will be presented.

Keywords: current limiting, short circuit, arcless commutation, arc extinction.

1. Introduction and problem defining

Short circuits have a chance of occurrence in any network types. To protect network and equipment from the destructive effects of high short circuit current the principle of fast disconnection of the fault place from the power source is commonly used. Among all types of electrical networks we would like here to consider short and autonomous networks and consider the features of short circuit disconnection process in such networks. When we are talking about common long lines and short circuit protection on it, we can find that modern commercially available switching units are able to operate with currents up to 63 kA for 220–550 kV and 27.5 kA for 35 kV [1–4].

In short networks the transmission line impedance is negligibly small, which becomes a reason for the occurrence of high value short-circuit currents. Short circuit current that occurs in short and autonomous network may exceed mentioned limits several times [5, 6] and reach 100–300 kA for example, in field of marine construction. For present day, disabling such high currents is a very difficult task due to high electrodynamic force, acted to the contacts, and high temperatures, caused by the electrical arc burning. In addition, quenching of the electric arc at currents of the order of 100 amperes and above will require a significant complication and increase in the cost of the circuit-breaker design, increasing the mass and dimensions of the arc extinguishing chamber and its cooling system.

One of the effective solutions in the question of short circuit protection in such low-impedance networks is the design and operating of the device which could limit the short-circuit current down to the values which can be switched off by commercially available circuit breakers. In that case it becomes possible to decrease the arc current and use more simple and small arc extinguishing chamber.

The task of limiting of short circuit currents and facilitating of the switching of emergency mode networks of class 110 kV and above is currently trivial. The electrotechnical industry has extensive experience and the theory of designing and creating current-limiting devices for networks of such stress classes. Nevertheless, the permissible continuous current for power lines does not exceed 1 kA. In connection with this, passive current-limiting reactors are the most widely used to date. Passive reactors have relativity simple design and low price and high reliability. Fig. 1 shows the volt-ampere characteristics of different reactor types. The passive reactors usually have linear or close to linear characteristics (shown on figure as curves "1" and "2" respectively)[7].

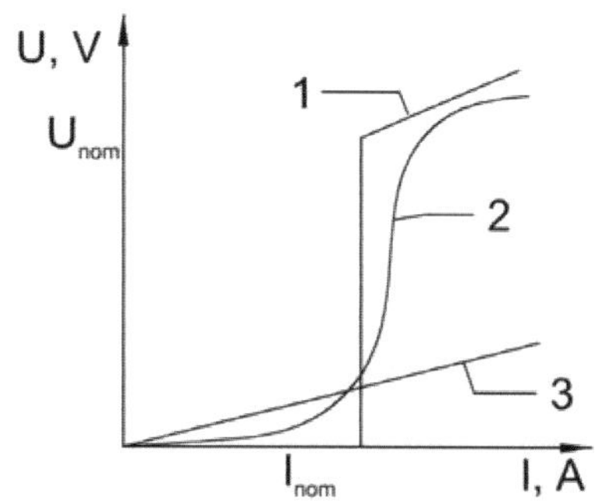

Figure 1. The volt-ampere characteristics of current limiting devices. 1 - ideal characteristics; 2 - non-linear characteristics; 3 - linear characteristics.

Autonomous and short networks, especially designed for special use purposes, operate at voltages not exceeding 35 kV. Accordingly, the rated operating current can reach 30 kA in such networks [5, 6]. Under these conditions it is impossible to use passive current-limiting reactors actively used at voltages of 110–220 kV because of their linear volt-ampere characteristics [8] and high energy losses in normal oper-

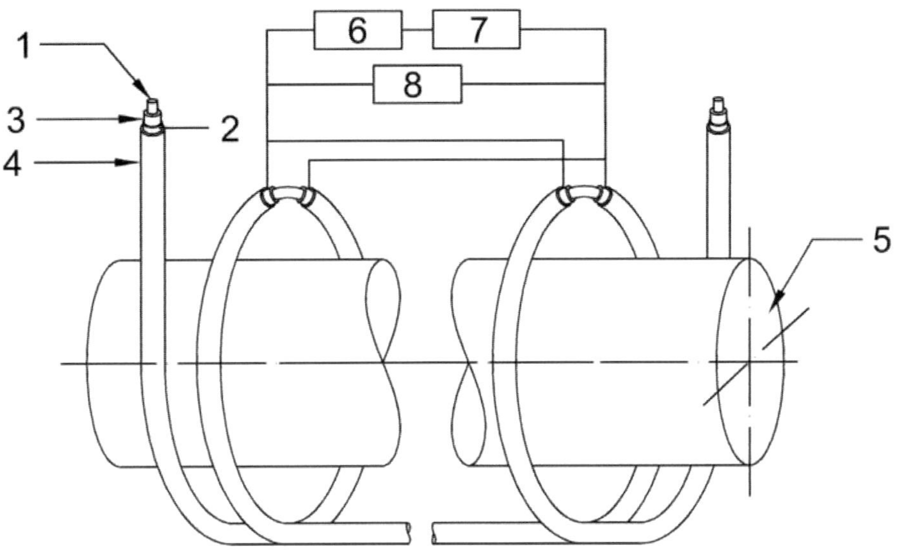

Figure 2. The sketch draw of the coaxial cable transformer-based current limiting device. 1 - cable core; 2 - cable shield; 3 - insulation between the primary and secondary windings; 4 - External cable insulation; 5 - the core; 6 - the battery of capacitors; 7 - active losses in the capacitor branch; 8 - semiconductor switch.

ating mode. In case of high nominal currents, it is necessary to provide very low impedance during the nominal currents range with its sharp increase when the current increases up to short circuit value. The volt-ampere characteristics of such device should have a view of ideal current-limiter curve with zero normal state impedance, which shown on fig. 1 under the label "3" or close to that.

As was mentioned earlier, different types of current-limiters was developed for high voltage application, but due to economical and practical reasons did not go into mass production. Considering the knowledge background and currently invented principles, it was found that the use of active current limiting reactor, instead of passive, can provide a desired volt-ampere characteristics. The working principle and design of active reactor for the class of 110 kV was described in [8, 9]. The principle may be successfully applied for the medium voltage networks and the desired level of current limiting may be reached. The basic idea of the principle will be given in further parts of the paper. The computer simulation model of the active reactor, designed taking into account the specifics of the operation modes of autonomous and high-current networks was created and tested. Simulation results are presented and described in corresponding section.

2. The current limiting device design and simulation

As a current-limiting device for networks with high short-circuit currents, we can use the active current limiting reactor. The reactor is built on the principle

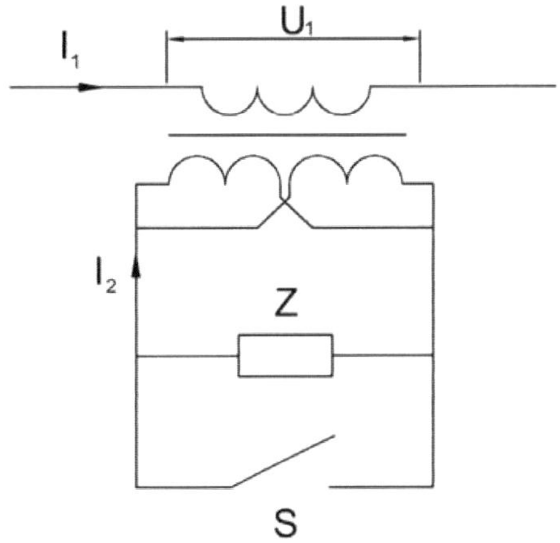

Figure 3. The coaxial cable transformer-based current limiting device circuit diagram.

of coaxial cable transformer [8, 9], made from the conventional coaxial cable (figure 2). The transformer's primary winding is a current-carrying core, while the secondary winding is a shield of coaxial cable, which divided to several sections and joined in parallel. The circuit diagram of the current limiting device is shown on figure 3.

Regardless of the splitting of the secondary winding, the working principle of the current limiting device is following. In normal operation, when the current in the cable is nominal, the current flows through the cable core, which is the primary winding. The secondary winding is short-circuited by a semiconductor switch and does not actually affect the current in normal operation.

As known, the transformer's short-circuit voltage usually does not exceed 10 percent from its nominal voltage [10], thus, we can find the impedance of the system in normal operation state. Total impedance of the transformer in short-circuited state can be found using the equation 1.

$$Z_{\mathrm{nom}} = \frac{U_{\mathrm{n}} \times 0.1}{I_{\mathrm{n}}} \tag{1}$$

Considering, for instance, a 35 kV nominal voltage network, with steady state operating current of 30 kA, we can find that full current limiting reactor impedance in regular operation state will not exceed the value of 0.17 Ω. Such impedance is comparable with the linear portion of the cable.

In the magnetic system of the device there is practically no flow, since the entire magnetic flux is concentrated inside the cable. The secondary winding sections are connected in parallel for obtaining the required transformation ratio. Splitting the screen into sections gives the option to reduce the voltage on the transistor module and makes possible to use only one transistor key, which greatly simplifies the electronic switch construction. A control semiconductor device (for example, based on IGBT - transistors) is connected to the screen clamps and a capacitor bank.

In case of a short circuit, a trip signal based on the current derivative is applied to the semiconductor switch. Thus, a capacitor is turned on in the secondary winding and the total resistance of the cable transformer increases. In dependence on the required level of limited current, different impedance capacitors may be used. The secondary winding impedance Z_{s} can be found as

$$Z_{\mathrm{s}} = \frac{1}{C \times \omega} + R; \tag{2}$$

total impedance of current limiter referred to the primary winding Z_{p}, can be calculated as

$$Z_{\mathrm{p}} = k^2 \times Z_{\mathrm{s}} \tag{3}$$

where ω is angular frequency, k is transformer's turns ratio, R and C are resistance and capacitance of the capacitor bank respectively.

Taking as an example a condenser battery with a capacity of 400 µF and circuit resistance of 0.05 Ω, we obtain the impedance Z_{p} equal to 175 Ω.

In case of smaller capacitance installed, the limited short-circuit current may be much lower than nominal current. In such case the coaxial cable transformer operates at the mode close to open-circuit mode.

The given principle was modelled using Matlab Simulink environment for the case of autonomous network short circuit protection. For the simulation it was chosen the 35 kV low-impedance alternating current network, which consists of the power source, the active load and current limiting device. The circuit breaker in examined system is not modelled since the aim of the research is to limit the current. The nominal operating current of the modelled system is 0.7 kA (1 kA amplitude value), and the unstitched short circuit current is tends to 115 kA. The current limiting device under the simulation was built according to the principle shown on fig. 3 and basically consists of the cable transformer, where secondary winding is divided by 8 sections. The active resistance of the reactor is determined by the length of the cable and the specific active resistance and the inductance parameters was found using the equation 4 as it was described in [8, 11].

$$L = \mu_0 \omega^2 \frac{\pi}{4} \frac{D}{0.44} k_1 k_2 \tag{4}$$

where D is the core diameter, k_1 and k_2 - Kuhler coefficients, which depends on the geometrics [8, 11]. In section 3 the simulation results are shown and described.

3. Results discussion

Figure 4 shows the protected line current vs. time dependence for the process of short circuit current being limited. In the simulation it was assumed that the moment of the short circuit occurs at 0.095 seconds after the start of the observation. Also it was assumed that current sensors and the semiconductor switch control device has the response time of 0.001 s. It means that after the time of 1 ms after the moment of the short circuit, the impedance of the active reactor will be increased. On figure 4 one can see the normal operating mode from 0 to 0.095 s with the current amplitude value of 1 kA; at the period 0.095–0.096 s we can see uncontrolled current raise up to 2.3 kA peak; after the moment of 0.096 s the current becomes limited by the active reactor and the steady-state short circuit current is set at a value of 65 A. At this point the task of the current limiting device is performed and further shutdown is performed by the circuit-breaker.

Presented example was tested for the case of using the cable transformer with ratio k=8. Such winding ratio, especially with high impedance Z_2, allows to reduce the short circuit currents very effective, down to values when the switching arc may be extinguished by itself. In the case where there is no need for strong limitation of the short-circuit currents it is possible to use smaller winding factor transformer and disconnect the short circuit by conventional circuit breaker, operates with common short circuit currents.

Figure 5 presents the current vs. time dependence for the secondary circuit's capacitor current. As we can see, there is no current in capacitor during the normal operation state. When the current limiting

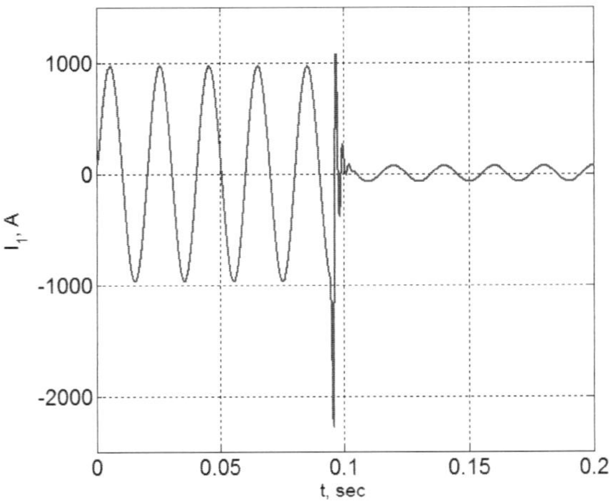

Figure 4. The cable core current waveform at the moment of occurrence of a short circuit and the current-limiting device reacting.

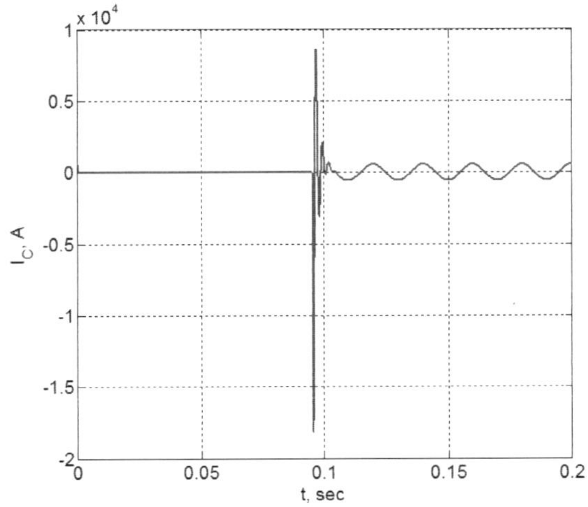

Figure 5. The cable core current waveform at the moment of occurrence of a short circuit and the current-limiting device reacting (secondary circuit's capacitor current).

function is switched on, the steady-state current becomes at the level of 500 A amplitude value, with transient peak value of 18 kA.

4. Conclusion

As we can see from figure 4, the short-circuit current limited by the active current-limiting reactor can be even lower than normal operation current. The degree of decreasing depends on the impedance, which may vary by setting different capacitance placed in secondary winding and different transformer's winding factor. In presented model it was used the 400 µF capacitor, which is commercially available and does not ask for extra large volume to be placed. Depending on the circuit breaker, installed in series with the current limiter, the capacitance and transformation ratio may vary until the value at which the short-circuit current is close to the breaking capacity of the circuit-breaker but does not exceed it. Thus, a compromise between capacitor's size, cable transformer design and short circuit current value can be found. Possibility of such significant current reducing makes it possible to greatly reduce the arc or carry out arcless commutation, which has the great development potential in the field of shipbuilding, submarine construction, oil platforms building, mining industry, can increase reliability and time response for civil and special energy systems. The obtained theoretical and simulation results will be utilized during the design of full-size full-voltage working prototype. The form-factor and size of the device designed for general application can be limited only by the transportation issues and the size of regular truck load. The 35kV device designed for the shock current up to 300 kA can be implemented by the coil radius of about 1 m, the length of which is not going to exceed 4 m.

References

[1] K. Antipov, A. Vostrosablin, and V. Zhukov. On the problem of coordination of levels of short-circuit currents in power systems. *Power Plants*, (4):19–32, 2005.

[2] K. Mozgalev, B. Neklepaev, and A. Shuntov. On stabilization of levels of short-circuit currents in 110 kV and higher networks. *Power Plants*, (12):40–42, 2001.

[3] ABB. Dead tank circuit breaker 245 kv. `http://new.abb.com/high-voltage/AIS/selector/DTB_245`.

[4] Electroshield Samara. Vacuum circuit breakers catalogue. `https://electroshield.ru/catalog/vakuumnie-vykluchateli`.

[5] I. Kalinin. *Development of the theory and calculation methods of operating modes of marine integrated electrical networks.* PhD thesis, Krylov State Research Centre, 2014.

[6] A. Baran, I. Klimkovich, and A. Andrukevich. Necessity of taking into account of electrodynamic forces in electrical installations and apparatus. In *Actual Problems of Energy 2016*. Belarusian National Technical University, 2016. `http://electro.bntu.by/user/SNTK-2016_1.pdf`.

[7] P. Elagin. On the problem of coordination of levels of short-circuit currents in power systems. *News ElectroTechnics*, 30(6), 2004. `http://www.news.elteh.ru/arh/2004/30/03.php`.

[8] A. Sazonov. Transformer-resonant current-limiting device for 110 kV. Master's thesis, Peter the Great Saint-Petersburg Polytechnic University, 2016. `http://elib.spbstu.ru/dl/2/v16-268.pdf/view`.

[9] P. Dunaev. Current-limiting device 110 kV, 400 A with a semiconductor switch. Master's thesis, Peter the Great Saint-Petersburg Polytechnic University, 2015.

[10] I. Ivanov and G. Soloviev. *General electrical engineering and electronics. Part 1. General electrical engineering. Laboratory practice.* First printing. Peter the Great Saint-Petersburg Polytechnic University, 2001.

[11] L. Leites. *Electromagnetic calculations of transformers and reactors.* First printing. Moscow: Enegria, 1981.

doi:10.14311/ppt.2017.2.186
Plasma Physics and Technology **4**(2):186–189, 2017

VACUUM ARC ION FLUX FROM VACUUM INTERRUPTER CONTACT GAP

M. Kurrat

Technische Universitaet Braunschweig, Institute for High Voltage Technology and Electrical Power Systems Schleinitzstrasse 23, 38106 Braunschweig, Germany

m.kurrat@tu-braunschweig.de

Abstract. Dielectric properties of vacuum interrupters are influenced by metal vapour deposition on the inner ceramic surfaces. Therefore these surfaces are partly protected by shield electrodes. The metal vapour originates from the vacuum arc in the contact gap during switching off process. Metal layer thickness on ceramics after switching operations were measured by Gramberg [1] and the negative effect on dielectric performance was demonstrated. In this paper we address the principle transport process of metal vapour in vacuum interrupters. Düning [2] and Beilis [3] performed ion flux measurements from vacuum arcs. We regard experiments with 5 kA effective arc current. There seems to be experimental evidence that the final particle transport onto the ceramic surface within a shield gap can be explained by particle tracing. The measured deposited layer thickness is taken as quantity to validate vapour densities and transport processes in vacuum interrupters.

Keywords: vacuum interrupter, vacuum arc, ion flux, metal vapour, layer deposition, particle tracing.

1. Introduction

The contact system of a vacuum interrupter is surrounded by the main shield made of copper. Breaking currents generates a metal vapour arc between the contacts (fig. 1). This arc emits metal vapour into the vacuum chamber at increasing gas pressure. After cooling the metal vapour condenses on main and end shields, but also on the inner surface of the ceramic insulator. The deposited conductive layers bridge partly the dielectric gaps between the shields (fig. 1, layer length L). The dielectric strength is decreased by shortening the dielectric gap and a flash over across the inner ceramic surface may result by overvoltage incidents (fig. 1).

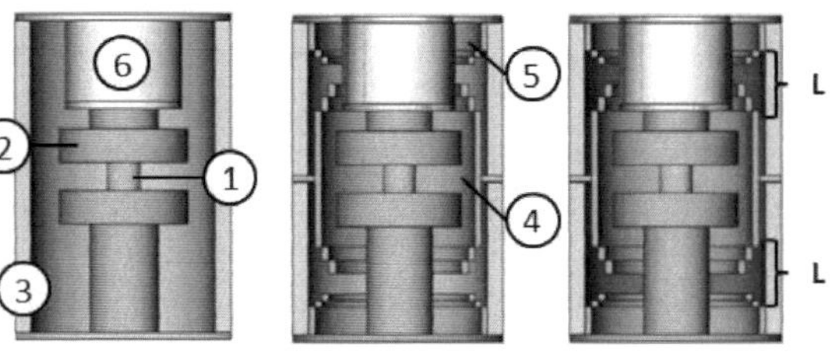

Figure 1. Metal vapour arc (1) between contacts (2) in a ceramic tube (3) without and with main (4) and end (5) condensation shields / metal vapour deposition through inter shield gap (L) / bellow (6).

To investigate the reduction of dielectric strength, vacuum interrupter with simple shield geometry were tested. High numbers of short circuit breaking operations were done in a test circuit. After breaking tests the significant dielectric performance reduction was evaluated by lightning impulse tests [1]. To improve the knowledge about the metal vapour layer deposition process the vacuum interrupters were opened after the tests. The layer position and thickness were measured. Figure 2 shows the layer thickness on the ceramic surface of the inter shield gap after 40 breaking operations at 5 kA.

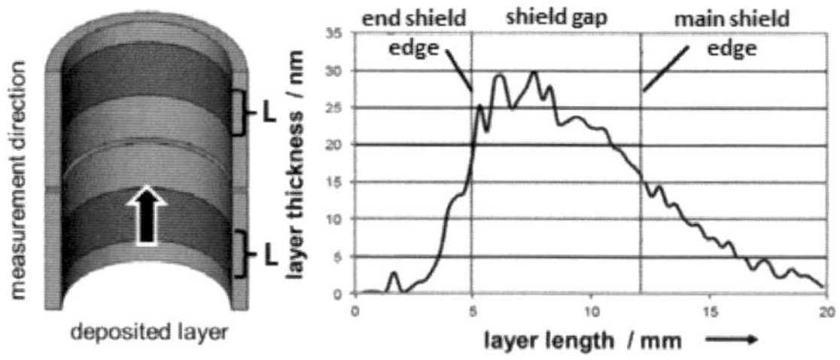

Figure 2. Measurement of deposition layer L thickness on ceramics from end shield to main shield (arrow).

The vertical lines mark the position of the inter electrode gap between position 5 and 12 mm. In between the thickness of the deposition layer reached the maximum value. But the deposition layer was also found behind the shields. The thickness decreases with increasing distance from the inter shield gap.

2. Particles Generation and Transport

The metal vapour flux from the arc between the contacts to the main shield is generated by the arc pressure. But the question arises what kind of transport mechanisms controls the deposition process behind the shields. Is it still a diffusion process and what is about the mean free path length? A stepwise approach for a simple calculation will be proposed in turn below. The particles are generated in the arc and released. Secondly the particles diffuse into the whole volume

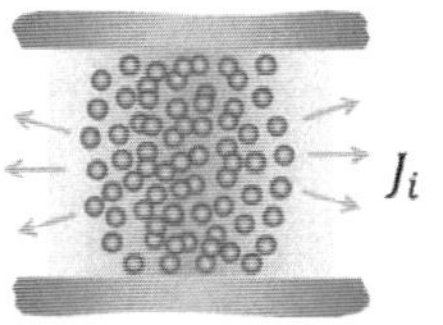

Figure 3. Radial ion current density leaving the thermal arc plasma.

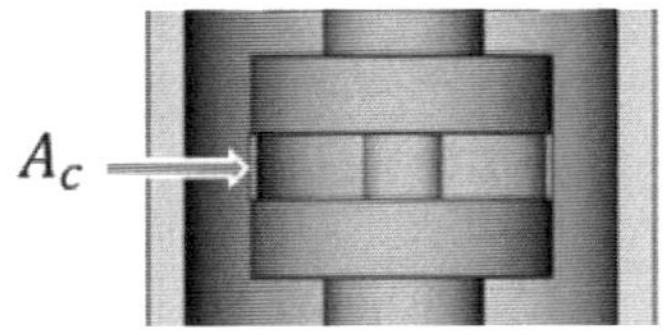

Figure 4. Cylindrical control area A_c for radial ion flux calculation.

creating a gas pressure in the vacuum chamber. The last step is the deposition process between main and end shield.

2.1. Ion Flux from Contact Gap

Rusteberg [4] and Düning [2] measured the ion flux from a diffuse mode vacuum arc with currents up to 10 kA. Ion current densities and ion energies were measured with a retarded field analyzer. For the calculations they assumed thermal arc plasma completely ionized and dominated by ion collisions. In result scattered ions with radial velocity components leave the contact gap with an ion current density J_i (fig. 3).

The calculated radial ion current density leaving the contact gap refers to a cylindrical control area A_c around the contact gap as shown in fig. 4.

Some more quantities can be deduced from the measured quantities [5] using equations (1) and (2). The measured ion energies allow the calculation of the ion velocities v_i. And the measured ion current densities J_i include the information about the ion densities n_i regarding ion charge q.

$$J_i = n_i \, v_i \, q \tag{1}$$

$$W_i = \frac{1}{2} m_i \, v_i{}^2 \tag{2}$$

The calculated ion densities of the radial flux increase with the arc current up to values of $10^{19}\,\mathrm{m}^{-3}$ as shown in fig. 5.

The maximum ion energies were observed at lower short circuit currents. Thermal ion energies at high arc currents yield high velocities about $v_i = 10^4\,\mathrm{m\,s}^{-1}$. High ion densities result in a low mean free path λ_i in the mm range calculated for high arc currents with equation (3).

$$\lambda_i = v_i/(c_0\, n_i), \; c_0 = 8 \cdot 10^{-13}\,\mathrm{m^3 s^{-1}}. \tag{3}$$

2.2. Axial Collisional Transport

The burning arc emits continuously ions from the contact gap into the vacuum chamber through the

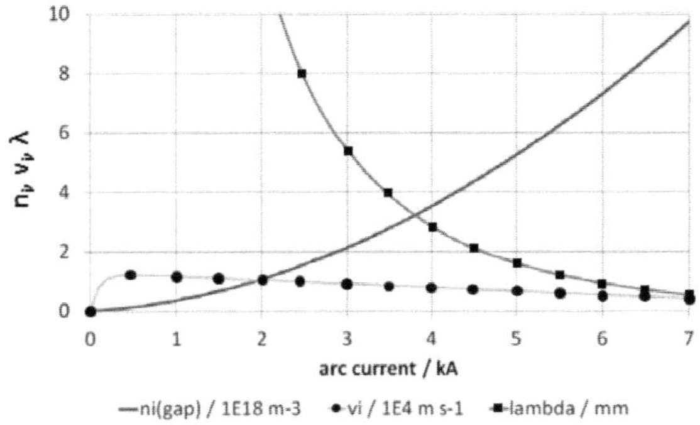

Figure 5. Ion density n_i, ion velocity v_i and mean free path lengths λ_i of the ion flux through the control area calculated from measurements.

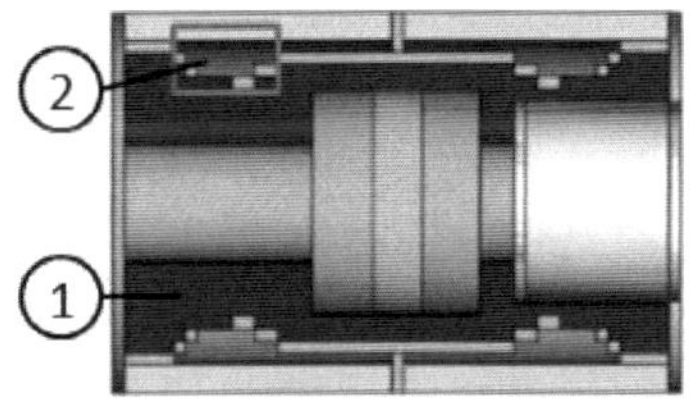

Figure 6. Chamber volume (1) with constant metal vapour particle density and inter shield gap volumes (2, red box).

control area A_c. The total ion number escaping from the arc gap within one time step Δt is given by eq. (4).

$$N_i = n_i(\mathrm{gap}) \cdot v_i(\mathrm{gap}) \cdot A_c \cdot \Delta t \tag{4}$$

To obtain a particle density in the vacuum chamber the total number of ions from the arc gap is divided by the chamber volume V_{chamber} as shown by eq. (5).

$$n_i(\mathrm{chamber}) = N_i / V_{\mathrm{chamber}} \tag{5}$$

The dark blue volume in fig. 6 without electrode gap and inter shield gaps is taken as chamber volume V_{chamber} filled by metal vapour. It is assumed that the initial ion flux is distributed as metal vapour over the chamber volume within one time step of 0.1 ms. The distribution process is not completely understood but some physical aspects are described in turn. The positive ion flux is directed to the main shield. When the ions hit the copper shield, they are neutralized by electrons. Due to the ionization energy further electrons may be liberated from the metal (secondary electron emission) and further atoms are neutralised. The low sticking coefficient about 2 % may be a result of this surface process. In addition the main shield is charged by the ion flux and the positive ions are retarded and do not reach the shield. In result a diffusion process is assumed and electric forces are not regarded.

With the arc current dependence of the measured ion flux shown in fig. 5 the mean density of the metal vapour in the vacuum chamber can be calculated for the sinusoidal arc current. The particle density in the

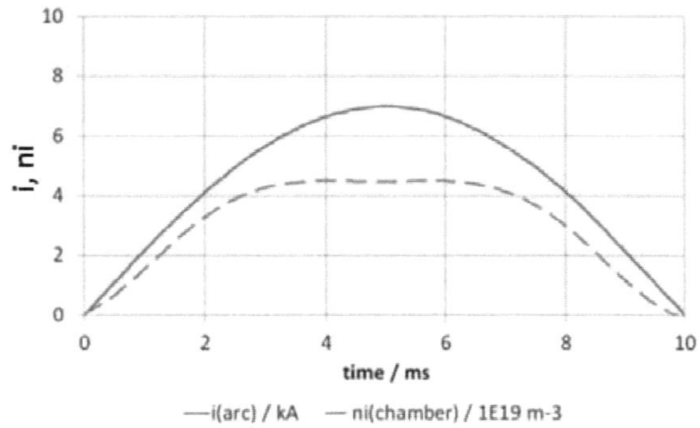

Figure 7. Time curves of arc current and particle densities in the chamber, $\Delta t = 0.1$ ms.

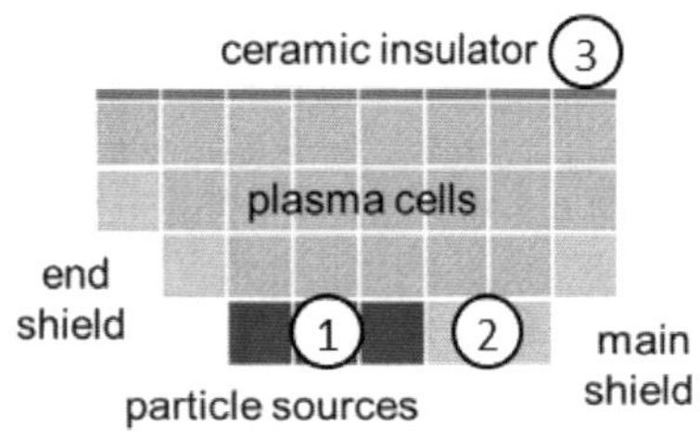

Figure 8. Discretisation of inter shield gap volume with particle source elements (1), shield elements (2), insulator surface elements (3).

vacuum chamber reaches $10^{-19}\,\mathrm{m}^{-3}$ calculated with an integration time step of 0.1 ms. The mean free path has accordingly a length in the mm range. The particle density reduction through sticking to metal surfaces during one time step is neglected. The total particle density in the vacuum chamber is shown in fig. 7 for one half sine of a 5 kA arc current (rms). The high initial ion velocity of $10^4\,\mathrm{m\,s}^{-1}$ supports the metal vapour diffusion through the whole vacuum chamber within one time step.

2.3. Radial Transport in Shield Gap

The last transport path is the gap between the shields. The regarded transport volume is marked with a red box in fig. 6. The metal vapour density in the chamber volume in front of the shield gap is given in fig. 8 as "particle source" within one time step. To calculate the particle number which are collected by the ceramic insulator surfaces a rough two-dimensional discretisation of the inter shield gap volume is applied as shown in fig. 8. Depending on the mean free path two transport processes are possible: Diffusion or crossing the inter shield gap without collisions.

2.3.1. Collisonal Transport

The first transport process test assumes diffusion as predominant. The two dimensional partial differential equation (6) is solved numerically by difference equations with total particle numbers N in cells of

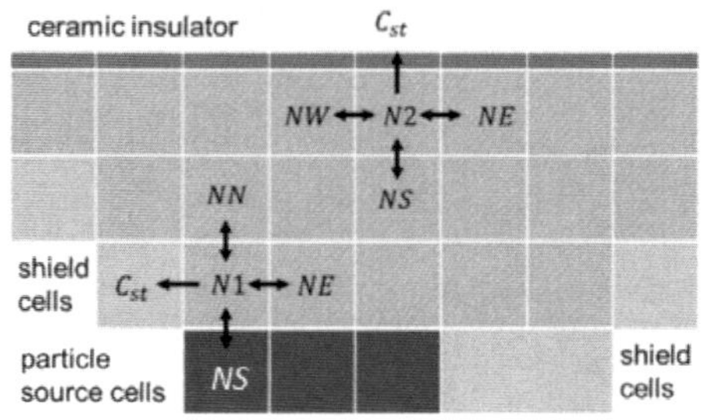

Figure 9. Two dimensional cell model for calculation of particle transport by diffusion.

constant volumes and diffusion coefficient D.

$$\frac{\partial N}{\partial t} = D \cdot \left[\frac{\partial^2 N}{\partial x^2} + \frac{\partial^2 N}{\partial y^2} \right[\qquad (6)$$

The particle densities of the particle source cells are given in equation 5. The particle distribution is calculated within one time step $\Delta t = 0.1$ ms with 5 iterations of 5-points difference equations (compass notation). The shield cells and insulator cells are particle drains and the sticking coefficients C_{st} determine the particle collection of these cells. The particle numbers $N1$ and $N2$ in the example cells in fig. 9 are given in the iteration equations (7), (8).

$$N1 = (NN + NE + NS)/(3 + C_{\mathrm{st}}(\mathrm{shield})) \qquad (7)$$

$$N2 = (NE + NS + NW)/(3 + C_{\mathrm{st}}(\mathrm{ceramic})) \qquad (8)$$

The cell definitions are shown in fig. 9 and they are applied at every volume cell connected to a shield cell or a ceramic cell. Due to the low sticking coefficient C_{st} only a small part of particles is collected from the insulator cells.

2.3.2. Particle Tracing

The second transport process test assumes low particle densities and no particle collisions in the shield gap. Particles from the vacuum chamber volume fly from the source cells i directly to the ceramic insulator cells j. For each insulator cell a receiving coefficient $R_{\mathrm{c}}(i,j)$ is calculated within one time step according to the geometry. fig. 10 shows the particle tracing lines from the centre of the source cells to the receiving insulator cells. The given receiving coefficients $R_{\mathrm{c}}(i,j)$ for the insulator cells show the geometrical dependance examplarily. The particle numbers in each insulator cell j is the sum in equation (9) over all plasma source cells i. But only a small part of particles is adsorbed from the insulator surface due to the low sticking coefficient.

$$N_j(\mathrm{insulator}) = \sum_i R_{\mathrm{c}}(i,j) \cdot N_i(\mathrm{source}) \qquad (9)$$

2.4. Calculation of Layer Thickness

The numbers of metal particles reaching the insulator surface cells are summed up during one half sinewave

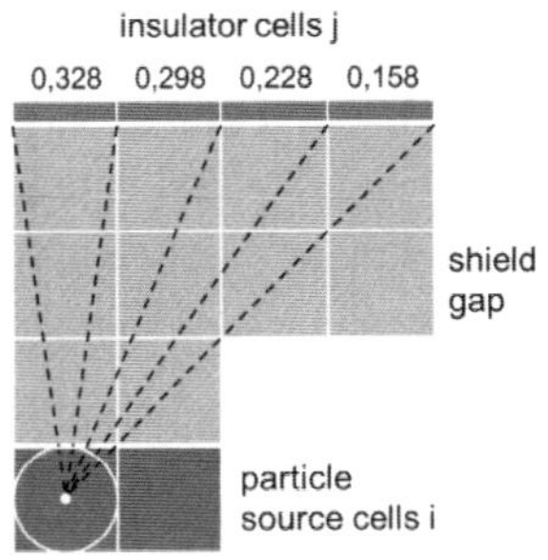

Figure 10. Particle tracing model with receiving coefficients of the red ceramic insulator cells on top.

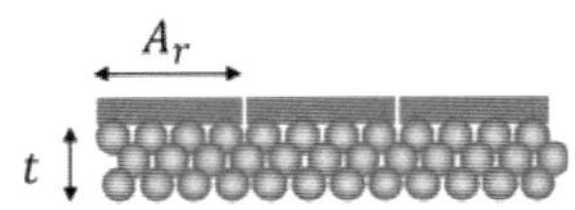

Figure 11. Insulator cell with surface area A_r and metal vapour layer thickness t.

of arc current. A low sticking coefficient of copper on ceramic surface about 2 % is assumed. The layer thickness t is calculated due to the particle number per surface area A_r as shown in fig. 11. The mass of the layer is given by the number of particles and the atomic mass unit $u = 63$. The deposited layer thickness t is calculated by the insulator cell surface area A_r and the specific density $\rho = 8.9\,\mathrm{g\,cm^{-3}}$ of copper as given by equations (10), (11).

$$m = N \cdot u \cdot 1.66 \cdot 10^{-24}\,\mathrm{g} \qquad (10)$$

$$t = m/(\rho \cdot A_r) \qquad (11)$$

3. Results

The diffusion approach assumes a low average free path length and therefore fluid dynamics are applicable. Figure 12 shows the layer thickness after 40 breaking operations at 5 kA arc current. The curve shows the measured thickness values of 30 nm with a steep decrease behind the end shield and a broader one behind the main shield. The calculated layer thickness for the insulator cells are given as bars. The calculated values exhibit a broad distribution of layer thickness behind the shields.

The particle tracing approach assumes a collisional free transport in the shield gap. The bar graph in fig. 13 shows the layer thickness of the insulator cells obtained by free particle flight through the inter shield gap. The bar distribution is quite similar to the measured distribution curve. The layer thickness is influenced by the "optical properties". The distribution exhibits a shadow effect of the end and main shield.

4. Conclusion

A simple model calculation for the deposition of metal vapour is compared to experimental results. The following are found.

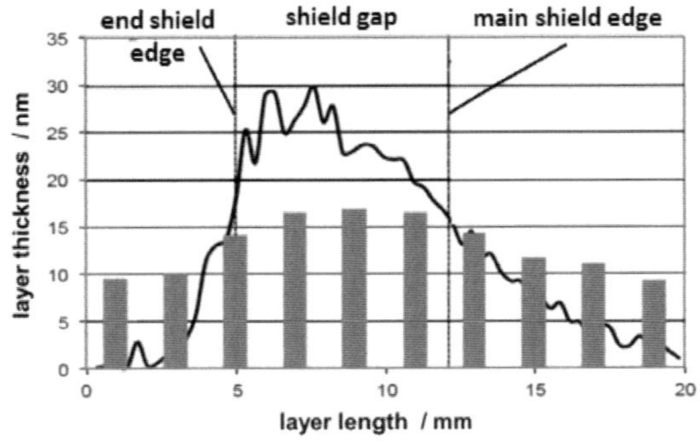

Figure 12. Measured (line) and calculated (bars) vapour layer thickness by diffusion.

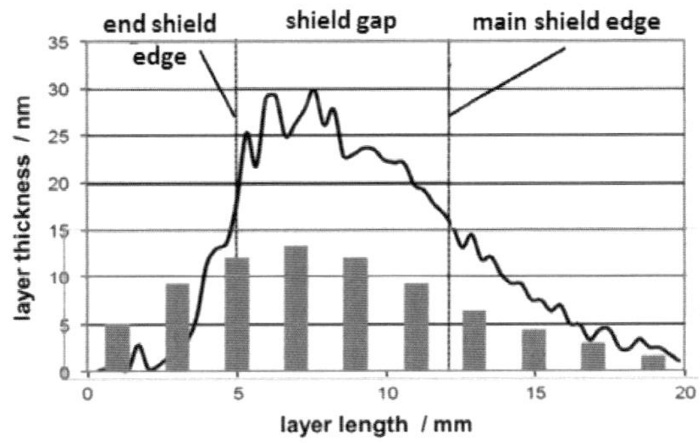

Figure 13. Measured (line) and calculated (bars) vapour layer thickness by particle tracing.

- ☐ The ion flux from vacuum arc is one important source for metal vapour emission into the vacuum chamber.
- ☐ The ions are partly or completely converted into neutrals and are distributed over the vacuum chamber volume by diffusion.
- ☐ It seems to be a free particle flight in the inter shield gap to the insulator surface which shows the measured shadowing effect of the shields.
- ☐ The calculated layer thickness with low sticking coefficient about 2 % agrees quite well with measured values.

References

[1] I. Gramberg et al. Investigations of copper chrome coatings on vacuum circuit breaker ceramics. *IEEE Trans. Plasma Science*, 41(8):2074–2080, 2013. doi:10.1109/TPS.2013.2273260.

[2] G. Düning and M. Lindmayer. Energy of ions in vacuum arcs. In *Int. Symp. Discharges and El. Insul. in Vacuum, ISDEIV*, Xi'an, 2000.

[3] I. Beilis et al. Measurement of ion flux as a function of background gas pressure. *IEEE Trans. Plasma Science*, 35(4):973–979, 2007. doi:10.1109/TPS.2007.896751.

[4] C. Rusteberg. Plasma parameters in vacuum arcs. Phd Thesis (in German), TU Braunschweig, 1995.

[5] M. Kurrat. Plasma diagnostics for vacuum arcs. In *Symp. on Physics of Switching Arc (FSO)*, Nove Mesto, Tschechien, 2009.

doi:10.14311/ppt.2017.2.190
Plasma Physics and Technology **4**(2):190–193, 2017

DETERMINATION OF CR DENSITY IN THE ACTIVE PHASE OF A HIGH-CURRENT VACUUM ARC

S. Gortschakow[a,*], A. Khakpour[a], S. Popov[b], St. Franke[a], R. Methling[a], D. Uhrlandt[a]

[a] *Leibniz Institute for Plasma Science and Technology e.V. (INP Greifswald), Felix-Hausdorff-Str. 2, 17489 Greifswald, Germany*

[b] *Institute of High Current Electronics, Siberian Branch of the Russian Academy of Sciences, Tomsk 634055, Russia*

[*] sergey.gortschakow@inp-greifswald.de

Abstract. Melting and evaporation of the anode surface strongly influence the interruption capability of vacuum circuit breakers, because they lead to injection of atomic vapour into the inter-electrode gap. Determination of the vapour density and its dynamics with respect to different anode phenomena is therefore of great importance. Results of Cr density measurements in a high-current vacuum arc by using broadband absorption spectroscopy are presented. The vapour density of atomic Cr is determined after the formation of anode spots as well as close to the current zero. Cr I resonance lines at 425.43 nm have been used for the analysis. An AC current pulse with maximum value of 7 kA and a frequency of 100 Hz is applied to a vacuum arc between two cylindrical butt electrodes made of CuCr7525 with a diameter of 10 mm. The high-current anode modes are observed by means of high-speed camera imaging. The temporal evolution of the Cr ground state density is presented and discussed.

Keywords: vacuum arc, anode spot, absorption spectroscopy.

1. Introduction

The application of vacuum circuit breakers in medium voltage is still growing. Among other advantages this type of devices has a high grade of environmental compatibility and, therefore, has distinct perspectives for future applications. The great interest in developing vacuum circuit breakers for high-voltage applications leads to continuous basic research in the field of vacuum arcs. It is well known that the high-current anode phenomena have a distinct impact on contact erosion and interrupting capability [1, 2]. Therefore, studies of related processes during the arcing as well as after the arc extinguishing are of great importance for application-driven research.

In the high-current modes the anode emits metal vapour into the arc column which can exist also after current interruption and is a potential reason for failure of vacuum interrupters [1]. Axial and radial behaviour of the arc spectra which consist of different atomic and ionic lines during various high-current anode modes was recently investigated by means of emission spectroscopy [3].

In the case of low optical emission, like e.g. during and after the current zero crossing, another techniques must be used for quantitative analysis. The broadband absorption technique is applied in high-current vacuum arcs to determine the Cr I density within first several milliseconds after current zero in case of anode spot type 2 in [4]. The chromium density after current zero is compared for two types of anode spot in [3].

In the present work a broadband absorption technique is used for diagnostics of high-current vacuum arcs in order to determine the chromium ground state density in the presence of anode spot during the active phase and close to the current zero. AC 100 Hz as an interrupting current with a peak value of about 7 kA is applied.

2. Experimental Setup

The schematic representation of the experimental setup is shown in Fig. 1. The power source which consists of capacitors, inductances, charging and control units was used to generate a sinusoidal alternating current at 100 Hz. Five LC elements are applied in parallel. The capacitance and inductance of each branch amounts to 512 µF and 4700 µH, respectively.

An ultra high vacuum chamber with a basic pressure below 2×10^{-8} mbar is equipped with mounting adapters for different electrodes. In this work, cylindrical butt electrodes made of CuCr7525 alloy with a diameter of 10 mm are used. The upper electrode is fixed and serves as an anode which is connected to the power supply. Whereas the lower electrode which is the cathode can be moved down simulating the opening of a vacuum circuit breaker. The contact opening speed is adjusted to 2 m/s. The chamber is equipped by two optical quartz windows which make it accessible for optical diagnostics. The diameter of the windows of about 20 cm allows for application of several diagnostic techniques simultaneously.

The observation of the arc dynamics and of the anode activity is realized by means of a high-speed

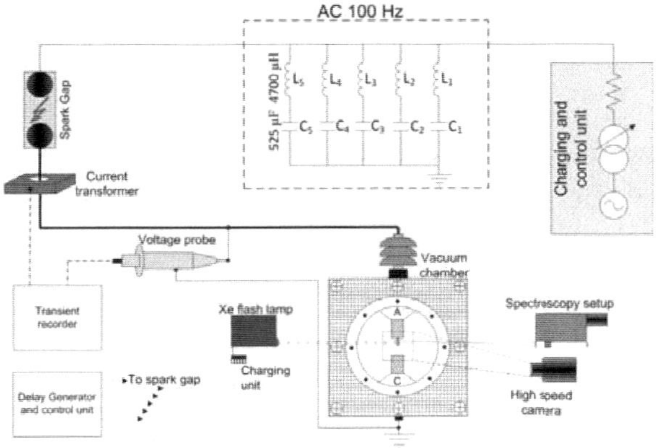

Figure 1. Electrical setup and test object configuration. Five LC elements are coupled in parallel in order to obtain pulses of AC 100 Hz. Spark gap and delay generator are applied for triggering and time control of the experiment.

video camera (IDT MotionPro Y4) with a recording speed of 20 000 fps. A pulsed high-intensity xenon lamp with a pulse width of 1 ms acts as background radiation source [4]. The xenon lamp is aligned on the left side window of the vacuum chamber. Its radiation is directed through the vacuum arc on the optical axis and coupled to the spectrometer entrance slit that is located near the right side window [4]. Making use of a deflecting and a focusing mirror the arc is observed along a line perpendicular to the electrode surfaces, i.e. along the arc axis. The radiation is spectrally dispersed using a spectrograph (Andor Technology Ltd. Shamrock 750) with 0.75 m focal length. To achieve a high spectral resolution, a grating with 1800 lines/mm and an entrance slit of 50 µm width are used. Registration of acquired spectra spatially resolved in one direction is performed by an intensified charge coupled device (iCCD) camera (Andor Technology Ltd. iStar-DH334T-18H) with a 1024×1024 pixels chip. The arc current and voltage are measured by a Pearson current monitor and a voltage probe Tektronix 6015A, respectively. The voltage probe is connected to the external electrical contacts of the vacuum chamber.

3. Theory

Absorption spectroscopy is performed by positioning the xenon lamp as background radiation source behind the test object. One and the same spectrograph is used to capture the radiation of the plasma and the radiation of the background light. For this purpose two optical paths were aligned to the spectrograph slit making use of different mirrors. One optical path is aligned in such a way that the plasma radiation is captured only, while the other optical path allows for the registration of the radiation of both the plasma and the lamp (cf. Fig. 2). Both optical paths are positioned in the same horizontal plane in parallel to the anode surface.

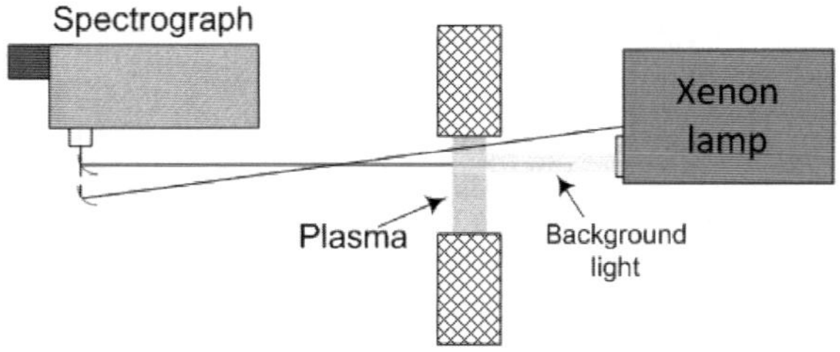

Figure 2. Schematic representation of absorption experiment. Both optical paths are in the same horizontal plane in parallel to the anode surface.

Fig. 3 represents the image of iCCD camera which contains the space-dependent absorption spectra of the background light source and the plasma (Fig. 3a) as well as of the plasma only (Fig. 3b). Both spectra are measured in a wavelength range covered by the spectrometer. The grey colour background corresponds to the light emitted by xenon lamp. As it can be seen from the figure the background light source radiation is less intense than the plasma radiation. Chromium atoms actively participates in the plasma processes becoming excited in collisions with the electrons and emitting the light due to the decay of corresponding excited states. This radiation is added to that of the lamp. After the current zero crossing the density of excited state rapidly decreases and there is no considerable light emission from the plasma. In this case the intensity of the lamp is much higher than that of the plasma leading to dark stripes in absorption spectra [4]. The intensity of the background source is nearly constant in the analysed spectral interval. The three thin vertical lines at 425.43 nm, 427.78 nm, and 428.97 nm correspond to three resonance lines of Cr I. For further evaluation the transmission T has to be calculated by [4]:

$$T = \frac{PL - P}{L} \tag{1}$$

where PL is the radiation of the plasma and the background light source, P means the plasma radiation only and L the background light source only. The background light source spectrum is measured separately and verified to be stable in terms of shot to shot variation. For particle density estimation the results obtained for Cr I 425.43 nm resonance line will be shown. The analysis made in Ref. [4] has clarified that the density obtained with the data for other two lines is very similar. According to the Lambert-Beer law the absorption coefficient k can be obtained from transmittance T by [4]:

$$T(\lambda) = \exp(-k(\lambda)D) \tag{2}$$

where $k(\lambda)$ is the absorption coefficient and D is the effective absorption length which is estimated to be equal to the electrode diameter of 10 mm in the present experiment. The absorption coefficient is related to

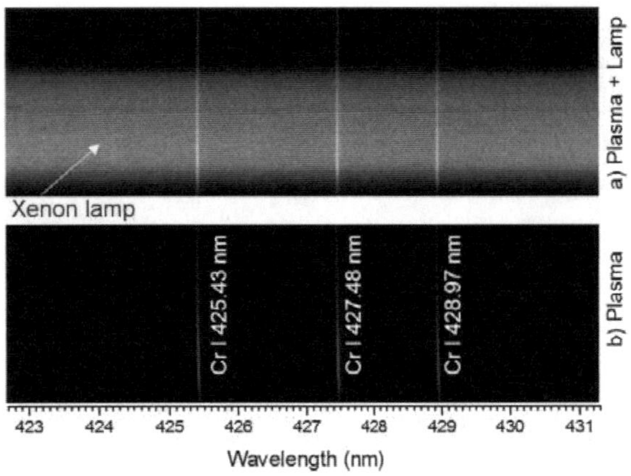

Figure 3. Sample 2D spectrum with three Cr I lines. The vertical dimension of each sub-image is about 3 mm. Anode and cathode are in top and bottom direction outside of the image, respectively.

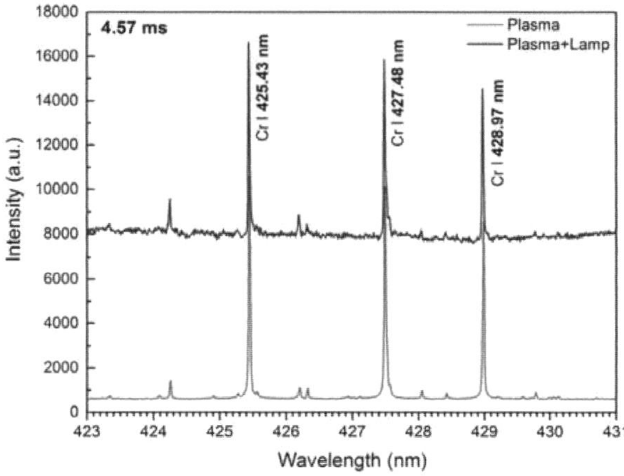

Figure 4. The relative intensities of the plasma emitted light and the background light source integrated along the wavelength.

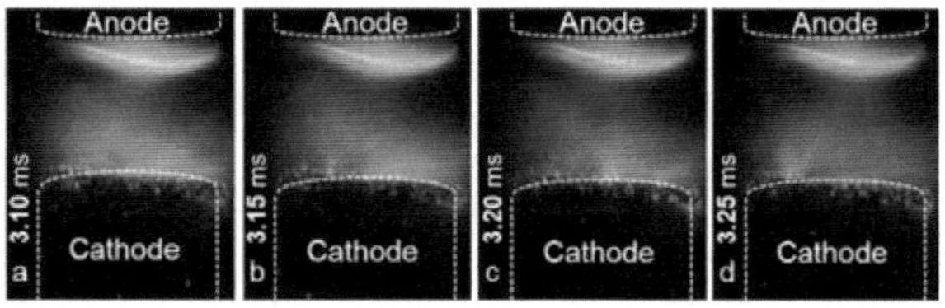

Figure 5. The high-speed camera images during the formation of anode spot type 1. The times are corresponding to the acquisition time of spectrograph.

the lower state population density n_l of the transition by following formula [4]:

$$\int_0^\infty k\,D\,d\lambda = \int_0^\infty \ln\left(\frac{L_\lambda}{PL_\lambda - P_\lambda}\right) d\lambda = \frac{\pi e^2 \lambda_0^2 D n_l f_{\text{lu}}}{4\pi\varepsilon_0 m_e c^2} \tag{3}$$

where ε_0 is the electric permittivity, m_e denotes the electron mass, c is speed of light in vacuum, e is the elementary charge, f_{lu} denotes the oscillator strength, and λ_0 is the centre wavelength of the considered transition.

A spectral interval 424–431 nm is chosen to study the ground state chromium density during active phase of the arc discharge. Spectroscopic constants for the selected lines including wavelength, transition probability (A_{ki}), and oscillator strength (f_{lu}) are presented in [4]. The focus of the present study is set on the chromium lines which have better optical accessibility (visible spectral range) than the copper lines (UV range) and to the comparison with the results presented in [4] for the density development after the current zero.

The relative intensities of the plasma emitted light and the background light source integrated along the wavelength corresponding to Fig. 3 are presented in Fig. 4. The signal is averaged over a horizontal stripe of certain height in axial direction in front of anode in order to improve the signal to noise relation. Such procedure imposes the situation when the axial variation of the density is small within considered region. Taking into account that the neutral species are not subjected to the action of electric field the assumption of an homogeneous density distribution can be applied here. The total absorption within one line is obtained by integration over the spectral line assuming a Lorentz profile [4]. As an example, the calculated ground state chromium densities according to (3) for the instant of 300 µs after current maximum using the data for 425.43 nm line is about $2.1 \times 10^{19}\,\text{m}^{-3}$.

4. Experimental Results

In the present work the Cr I density during the active phase is determined applying the method described in section 3. The appearance and extinguishing of the anode spot type I can be detected optically by high-speed camera observation of the anode surface.

The high-speed camera images during the development of anode spot type 1 are presented in Fig. 5. In the case of the anode spot type 1 a considerable anode activity with usually one large bright spot can be obtained on corresponding electrode. The averaged arc voltage is increased compared to the case without the anode spot.

The corresponding current and voltage traces together with an example of the acquisition time window of the spectrograph are presented in Fig. 6. The maximum current is about 7 kA. The xenon lamp as a background radiation source started earliest at about 2.1 ms. After 500 µs the intensity of lamp is almost constant, the iCCD connected to the spectrograph is triggered to record an image with acquisition time of 100 µs. Notice that different shots at different time instants during formation of anode spot type 1 are performed. A stable operation of the setup is, therefore, necessary to obtain the reliable data. This has been

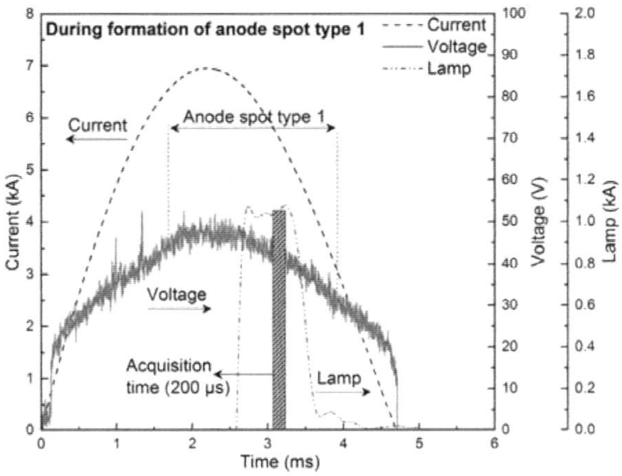

Figure 6. The current and voltage traces together with acquisition time of spectrograph corresponding to Fig. 5. Only anode spot type 1 is observed. The maximum current is about 7 kA.

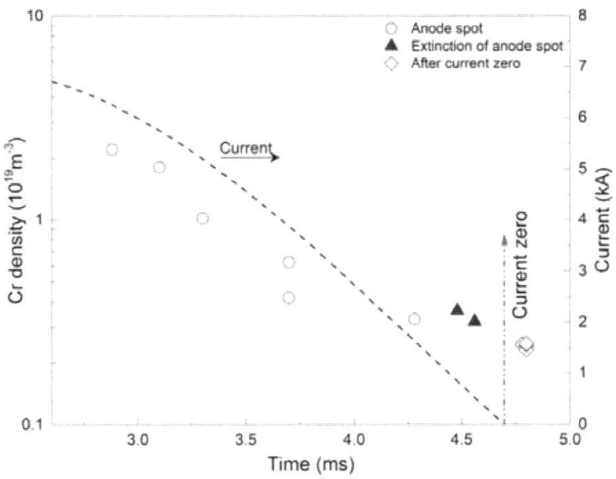

Figure 7. The Cr density during anode spot type 1 based on the method presented in section 3.

obtained by the electrode preconditioning through several shots at the moderate currents (below 5 kA).

The ground state chromium density during the anode spot type 1, after extinction of the anode spot, at the instant of current zero crossing and after the current zero from 2.8 ms to 4.8 ms is presented in Fig. 7. The Cr I density in the time interval between 2.8 and 3.2 ms in which the current decreases from about 5.5 kA to about 4 kA, respectively is about $2.1 \times 10^{19}\,\mathrm{m}^{-3}$ and $1 \times 10^{19}\,\mathrm{m}^{-3}$. When the arc current decreases, the ground state chromium density also decreases. After extinction of the anode spot of type 1, the Cr I density is about $3 \times 10^{18}\,\mathrm{m}^{-3}$. The density of the ground state chromium decreases shortly after the instant of current zero crossing (about 100 µs) to the value of about $2.6 \times 10^{18}\,\mathrm{m}^{-3}$. These values are in a reasonable agreement with the results presented in [4] $1.5 \pm 0.5 \times 10^{18}\,\mathrm{m}^{-3}$.The uncertainty in the absolute values of the density is similar to that discussed in ref. [4] and remains less than 15 %.

5. Summary and Outlook

In the present work chromium density during the active phase with pronounced high-current anode modes and after the extinction of the anode spot as well as after the current zero crossing is determined by using a broadband absorption spectroscopy technique. The Cr I resonance line at 425.43 nm is used for estimation of the ground state density. Chromium density determined during the discharge operation with the anode spot type 1 at 2 ms before current zero is about 10 times higher than that after the current zero crossing. The determination of the copper density during and after the high-current anode modes and the anode plume is considered as a future work. The general behaviour of the copper component is expected to be very similar to that of chromium. The differences are expected in the absolute values of particle densities.

References

[1] E. Schade and E. Dullni. Recovery of breakdown strength of a vacuum interrupter after extinction of high current. *IEEE Trans. Dielectr. Elect. Insul.*, 9(2):207–215, 2002. doi:10.1109/94.993737.

[2] E. Schade and E. Dullni. Investigation of high-current interruption of vacuum circuit breakers. *IEEE Trans. Dielectr. Elect. Insul.*, 28(4):607–620, 1993. doi:10.1109/14.231543.

[3] A. Khakpour, St. Franke, R. Methling, D. Uhrlandt, S. Gortschakow, S. Popov, A. Batrakov, and K.D. Weltmann. Optical and electrical investigation of transition from anode spot type 1 to anode spot type 2. *IEEE Trans. Plasma Sci.*, 45(8):2126 – 2134, 2017. doi:10.1109/TPS.2017.2690572.

[4] A. Khakpour, S. Popov, St. Franke, R. Kozakov, R. Methling, D. Uhrlandt, and S. Gortschakow. Determination of Cr density after current zero in a high-current vacuum arc considering anode plume. *IEEE Trans. Plasma Sci.*, 45(8):2108 – 2114, 2017. doi:10.1109/TPS.2017.2681898.

doi:10.14311/ppt.2017.2.194
Plasma Physics and Technology 4(2):194–197, 2017

INTERRUPTION OF WEAKLY COOLED ARCS IN AIR AND AIRPLUS

N. Ranjan*, J. Carstensen, S. Scheel

ABB Switzerland Ltd, Corporate Research, Segelhofstrasse 1, 5405 Baden-Dättwil, Switzerland

* nitesh.ranjan@ch.abb.com

Abstract. Switching of low current arcs in free burning or weakly cooled conditions is mainly determined by the thermal properties of the gas. Products with such switching conditions are widely found in secondary distribution medium voltage (MV) gas insulated switchgears (GIS). In this study, we compare the current interruption capability of synthetic air and AirPlus^{TM}, i.e. a mixture of synthetic air with $C_5F_{10}O$ fluoroketone (C5-FK). We focus on thermal interruption performance of the gases. AirPlus mixture corresponds to -25°C condensation temperature of C5-FK. An arc is drawn between the contacts and cooled by blowing cold gas from a tank. Blowing pressure required for current interruption is compared. Within the measurement accuracy, the current interruption performance of both gases is similar. Chemical analysis of the AirPlus mixture after 69 shots was performed using Gas Chromatography Mass Spectroscopy (GCMS) and it shows very little decrease in the concentration of C5-FK.

Keywords: weakly cooled arcs, load break switches, C5-FK, thermal interruption.

1. Introduction

Load break switches (LBS) are often used in medium voltage (MV) secondary distribution for interrupting low currents typically in the range of few hundreds of Ampere. The arc in the LBS is either free burning or very weakly blown. In these situations, current interruption is mainly determined by the intrinsic physical properties of the arcing medium. SF_6 is currently used in LBSs owing to its extremely good switching and dielectric properties.

SF_6 has a high global warming potential (GWP) and this might lead to regulations in future to minimize its usage in specific applications, e.g. European Union (EU) has targeted to replace SF_6 specifically from MV secondary distribution GIS [1]. Recently, the search for an alternative gas with low GWP value has intensified for high and medium voltage applications. It is challenging to find a gas which has as good switching, dielectric, and thermal properties as that of SF_6. In last years, different families of molecule like perfluoroketones (PFK) [2], perfluoronitriles (PFN) [3] have emerged as a promising candidate for an alternative gas. These gases have a relatively high boiling point and so they must be used in mixture with other gases (like air or CO_2) for low temperature applications (−5°C or below). Different groups have reported on the switching [4, 5] and insulation [6, 7] properties of these gases for HV [3, 8] and MV [9–11] applications. In this article, we compare the current interruption capabilities of AirPlus, i.e. a mixture of $C_5F_{10}O$ fluroketone (C5-FK) with air. Current and voltage waveforms were based on the IEC norm for active load test duty of 24 kV/630 A rating of LBS [12]. We specifically evaluate the thermal interruption capability of AirPlus mixture and compare it with air. Thermal interruption phase is dominated by current ($\frac{di}{dt}$) and voltage ($\frac{du}{dt}$) stresses in a small time window (few 10 µs) around current zero (CZ). We did not explore the performance of the gas in the presence of the transient recovery voltage (TRV) peak or the recovery voltage which appears later. For our experiments, the arc is weakly cooled, i.e. the flow is sub-sonic. After the experiments were completed, gas samples were taken and analysed for the formation of decomposition product and decrease in the C5-FK concentration.

2. Experimental method

The test object (TO) has been designed in a way that several parameters (blowing pressure, travel, working gas) can be varied independently. It is a metal enclosed housing of 250 litre volume which can be filled to different pressures. The housing encloses a model LBS and a drive to move the contacts. The model LBS consists of a plug-tulip contact made of tungsten-copper and a nozzle made from transparent Poly-methyl-methacrylate (PMMA). A simple pneumatic cylinder is used as a mechanical drive to move the contacts. The stroke was adjusted to 85 mm and the average speed was in the range of 3–5 m/s. The metal housing also has a quartz window to gain direct optical access to the arcing zone. A high speed camera with a maximum frame rate of 14000 fps is used to capture videos of the arcing zone during current interruption.

2.1. Test circuit

During the experiment, an arc is drawn between the contacts by moving them apart. Such an arcing event is referred to as "shot" in the rest of this article. During a typical shot, the plasma created during the high current phase starts to cool down as CZ is approached. If

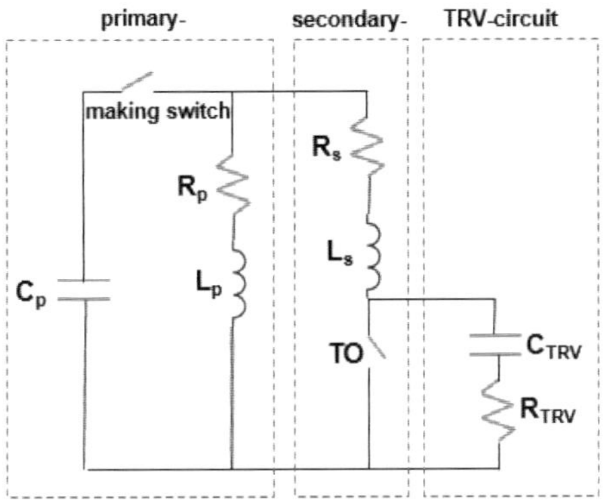

Figure 1. Electrical circuit.

the plasma is not sufficiently cooled and stays conductive, the current continues after CZ. This is referred to as "thermal failure". Both the current and voltage stresses at CZ are decisive for successful thermal interruption. The circuit was built accordingly to measure the thermal interruption performance of the gas.

The electrical circuit consists of three parallel branches as shown in Figure 1. The primary circuit is optimized for low damping by using coils with low ohmic losses (large conductor cross section) and tuned to a frequency of 50 Hz. The TO is located in the secondary circuit, where the current is limited by a high inductance L_s. Since $L_s > L_p$, only a small fraction of the current is bypassed through the secondary circuit so that the damping rate is mostly determined by the primary circuit, and there is only a minor change in the power frequency due to the influence of the secondary circuit. The rate of rise of recovery voltage (RRRV) is adjusted by a capacitor and a resistor parallel to the TO. The nominal values of the circuit elements are compiled in Table 1. The circuit impedance $Z_{\mathrm{circuit}} = \left.\frac{du/dt}{di/dt}\right|_{\mathrm{CZ}}$ seen by the test object is mostly determined by the resistor R_{TRV}, and Z_{circuit} was adjusted to 250 Ω by choosing an appropriate R_{TRV}. During the experiment, if the current through the TO changes, the RRRV is changed as well, since the Z is fixed ($\frac{du}{dt} = Z\frac{di}{dt}$). The value of Z for common ratings of LBS is shown in Table 2. The circuit components were adjusted so that at a charging voltage of ≈ 3 kV of C_p the desired current ($I_{\mathrm{rms}} \approx 630$ A) is flowing through the model LBS. Lower charging voltages would have led to arc circuit interaction around CZ, while the highest achievable charging voltage was limited by the voltage rating of the capacitors C_p to 4 kV. The TRV peak was typically in the range of 3–4 kV depending on the charging voltage.

Circuit element	Values
C_p	7 mF, charged to 4 kV
L_p, R_p	1.66 mH, 15.5 mΩ
L_s, R_s	7.4 mH, 0.1–1 Ω
C_{TRV}, R_{TRV}	2 µF, 310 Ω
f	≈ 50 Hz
Z_{circuit}	≈ 250 Ω

Table 1. Circuit elements corresponding to the schematic shown in Figure 1.

Rated voltage kV	$\mathrm{RRRV}_{\mathrm{CZ}}$ V/µs	Z Ω
12	51.5	184
24	70	250
36	86	308

Table 2. RRRV and effective impedance at current zero according to the IEC mainly active load type test duty for a current of 630 A and different rated voltages [12].

2.2. Gas handling system

The TO was filled to 1.3 bar, a pressure typical for LBS. In some LBS, the puffer mechanism is used to create a gas flow which cools the arc [11]. We wanted to control the pressure required to cool the arc independently from the stroke and so a closed gas flow system was built around the TO. The schematic of the gas handling system is shown in Figure 2. At the beginning of each experiment, the high pressure tank (V_{HP}) is charged with gas to high pressures by the compressor. The pressure in V_{HP} is highest amongst all volume elements as it controls the pressures in the tank and TO. Two precision pressure regulators are directly connected to V_{HP} and are used to maintain a constant pressure in the tank (p_{tank}) and TO (p_{TO}). In this way, the pressure in the TO is kept constant at 1.3 bar and the pressure in the tank is independently and precisely controlled. During any experiment the valve was opened for a certain amount of time and the gas flew from the tank to the TO to cool the arc. This system was used for both the series of experiments related to air and AirPlus.

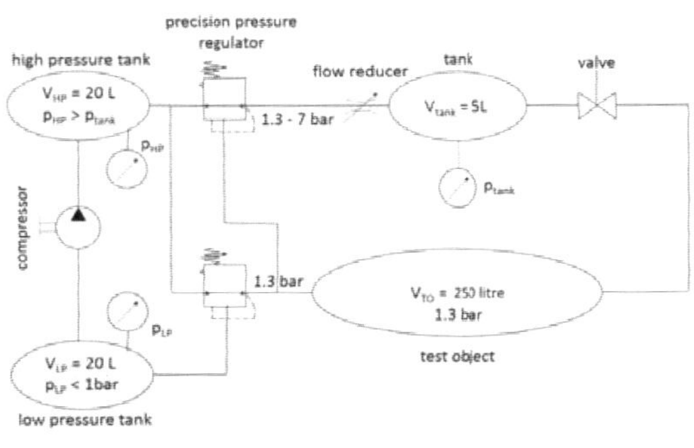

Figure 2. Sketch of the gas handling system.

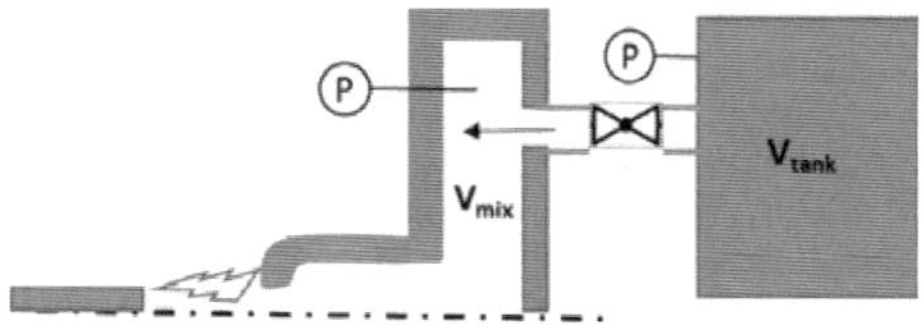

Figure 3. Sketch of gas flow path from tank to arc during any shot. Ⓟ denotes pressure sensor mounted to the tank and mixing volume.

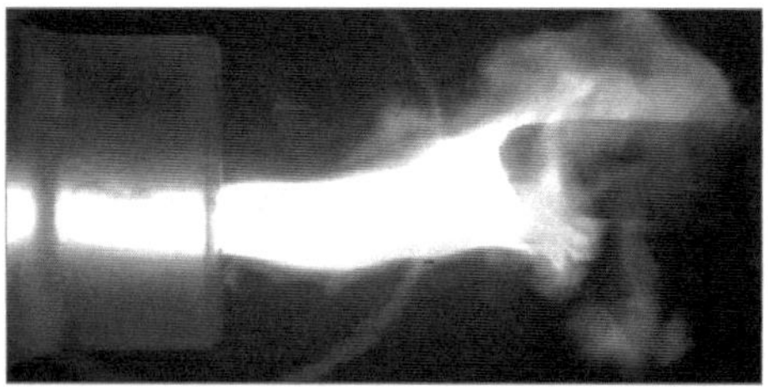

Figure 4. Image from high speed camera.

2.3. Experimental procedure

During any shot, the arc was drawn between the contacts and cooled by blowing the gas from the tank. The gas flows from the tank to the "mixing volume" inside the TO and then through the tulip into the arcing zone. The schematic of the blowing mechanism is shown in Figure 3. The mixing volume ($V_{\text{mix}} \approx 0.5$ litre) was equipped with a differential pressure sensor to measure the pressure p_{mix}. Since the pressure in the TO is fixed to 1.3 bar, measurement of the pressure inside the mixing volume directly gives the pressure drop across the tulip Δp. During any shot, the blowing was started first and a flow was established before an arc was drawn between the contacts. The current was injected just before contact separation so that the arcing time is in the range of t_{arc}= 8–10 ms. Interruption was considered successful, when it took place at the first CZ after the contact separation, otherwise it was considered as a failure. The high speed camera was used to make video during the entire arcing time. Figure 4 shows an image of the arc drawn between the contacts. Figure 5 shows current through the TO and the arc voltage for a typical shot (resulting in interruption).

We used the standard Up&Down statistical method [13] to determine the current (I_{50}) at which holds and fails occur with 50 % probability for a fixed set of parameters (blowing pressure, Z, travel, speed).

We used a mixture of ≈ 8 % C5-FK and air. This mixture remains fully gaseous down to temperature of −25 °C. The total filling pressure was 1.3 bar. For the case of air, synthetic air was filled up to 1.3 bar inside the TO.

3. Results and dicussion

3.1. Interruption performance

Figure 6 shows the result from the two experimental test series. I_{50} is plotted against the blowing over-

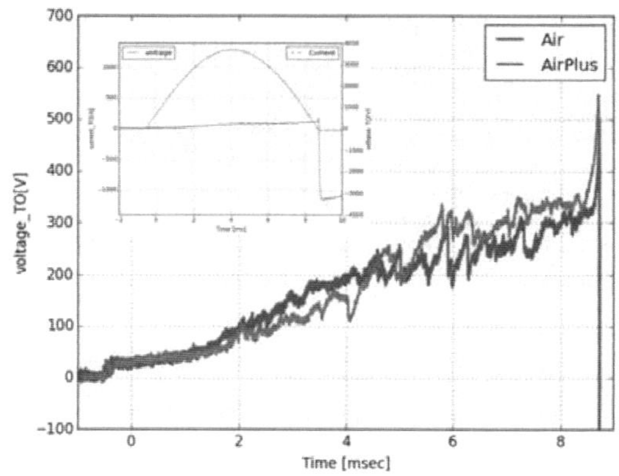

Figure 5. Arc voltage of AirPlus and air. Inset shows the current and the TRV-peak.

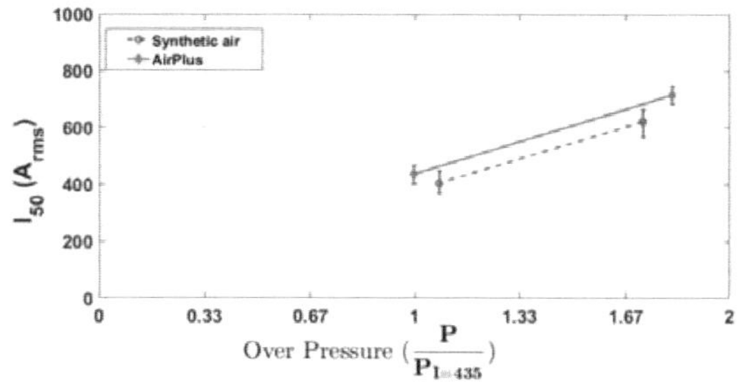

Figure 6. I_{50} against the normalized pressure drop across the tulip.

pressure. The X-axis shows the pressure required to interrupt a current normalized by the pressure required to interrupt 435 A of current for AirPlus. From the plot, we can conclude that within the measurement accuracy, the interruption performance of both gases are similar. Addition of C5-FK has no significant influence and the thermal interruption performance of the mixture is dominated by air which contribute to about 92 % of the mixture. Figure 5 shows the arc voltage from a measurement of air and AirPlus for the same current and blowing over-pressure. From the plot one sees that there is no significant difference in the arc voltage of the two gases. The inset in the figure shows the current and the full TRV peak.

The energy injected into the gas volume can be estimated by the following procedure. Taking an average arc voltage of 200 V from Figure 5 and 630 A current and 10ms of arcing time leads to an approximate value of 87 kJ of energy for all the 69 shots. But the actual energy input should be higher than the approximate value because we have performed up-down statistics on current and many of the shots have higher than 630 A of current flowing. For example Figure 5 shows about 920 A of current. Besides this, about 50 % of the shots resulted in fails and for these shots, the arcing time varied between 20–50 ms. Hence the total energy input can be estimated to be 3–4 times higher

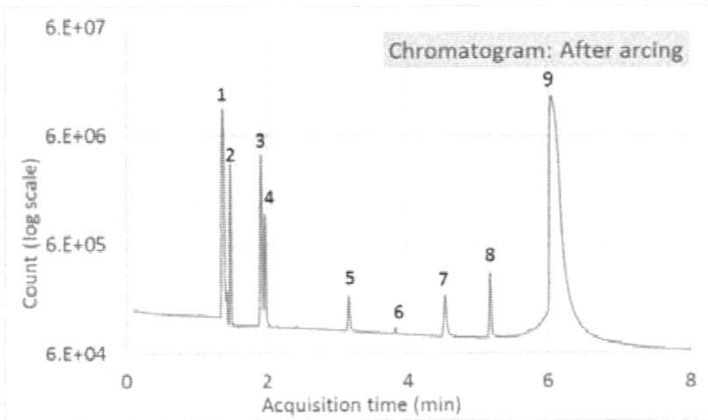

Figure 7. Chromatogram of gas sample taken after the arcing experiment.

Nr.	Chemical name	Formula
1	Nitrogen/Oxygen	N_2, O_2
2	Tetrafluoromethane	CF_4
3	Carbondioxide	CO_2
4	Hexafluoroethane	C_2F_6
5	Octafluoropropane	C_3F_8
6	Hexafluoropropene	C_3F_6
7	Decafluorobutane	C_4F_{10}
8	2H-Heptafluoropropane	C_3HF_7
9	C5-FK	$C_5F_{10}O$

Table 3. Molecules found after the test.

than the approximate value. Taking an integral of $I \times V$ for all shots we get 341 kJ which is close to the approximate calculation.

3.2. Decomposition products

Experiments were performed without any adsorbent inside the TO to get the full composition of the by-products. For experiments with AirPlus, gas samples were taken before and after the test and given for analysis using GCMS [14]. The chromatogram of the sample after testing is shown in Figure 7. The peak in retention time corresponds to different species present in the gas sample. The molecules corresponding to these peaks are shown in Table 3. These created molecules are also found in other investigations related to switching in C5-FK mixture [5]. All of these compounds have LC50 value (4 hour exposure to rats) greater than 2500 ppmv and hence they are not toxic according to GHS regulation [15]. Fourier transform infrared spectroscopic measurements were not performed and therefore the presence of CO or CF_2O cannot be excluded. Comparison of the GCMS results from gas samples taken before and after the experiments show very little drop in the concentration of C5-FK. Pressure was measured in all the volume elements during the experiment and negligible pressure rise below 20 mbar was observed.

4. Conclusions

In this report we compared the low current thermal interruption performance of AirPlus with air in weakly blown conditions. The dielectric performance of the mixture is not compared. The results show that AirPlus has similar thermal interruption performance as compared to air. Chemical analysis of the gas samples with GCMS shows no toxic by-products and very little decrease in the composition of C5-FK. There was negligible pressure rise after the experiment.

References

[1] Regulation (EU) no 517/2014 of the European parliament and the of council of 16 april 2014 on fluorinated greenhouse gases and repealing regulation (EC) no 842/2006.

[2] J.D. Mantilla et al. Investigation of the insulation performance of a new gas mixture with extremely low GWP. In *Electrical Insulation Conference*, Philadelphia, USA, 2014. doi:10.1109/EIC.2014.6869432.

[3] Y. Kieffel et al. Alternative gas to SF_6 for use in high-voltage switchgears: g^3. In *CIRED*, Lyon, France, 2015. Paper 0230.

[4] P.C. Stoller et al. Mixtures of CO_2 and $C_5F_{10}O$ perfluoroketone for high voltage applications. *submitted to IEEE Trans. Dielectrics and Electrical Insulation*, 2017.

[5] B. Radisavljevic et al. Switching performance of alternative gaseous mixtures in high-voltage circuit breakers. In *submitted to 20th Int. Symp. on High Voltage Engineering*, Buenos Aires, Argentina, 2017.

[6] C. Preve et al. Validation method and comparison of SF_6 alternative gases. In *Cigre Session*, Paris, France, 2016.

[7] P. Simka et al. Dielectric strength of C5 perfluoroketone. In *19th Int. Symp. on High-Voltage Engineering*, Pilsen, Czech Republic, 2015.

[8] D. Tehlar et al. Ketone based alternative insulation medium in a 170 kV pilot installation. In *Cigre Colloquium*, Nagoya, Japan, 2015.

[9] M. Hyrenbach et al. Alternative gas insulation in medium-voltage switchgear. In *CIRED*, Lyon, France, 2015. Paper 0587.

[10] M. Saxegaard et al. Dielectric properties of gases suitable for secondary medium voltage switchgear. In *CIRED*, Lyon, France, 2015. Paper 0926.

[11] M. Saxegaard et al. Low current interruption in SF_6-alternatives. In *CIRED*, Glasgow, UK, 2017. Paper 0614.

[12] High-voltage switchgear and controlgear part 103: Switches for rated voltages above 1 kV up to and including 52 kV. 62271-103. IEC International Standard, 2011.

[13] D.J. Dixon and A.M. Mood. A method for obtaining and analyzing sensitivity data. *J. of the Am. Stat. Association*, 43(241):109–126, 1948. doi:10.2307/2280071.

[14] P. Simka et al. Decomposition of alternative gaseous insulation under partial discharge. In *submitted to 20th Int. Symp. on High Voltage Engineering*, Buenos Aires, Argentina, 2017.

[15] Globally harmonized system of classification and labelling of chemicals (Sixth revised ed.), New York and Geneva: United Nations, 2015.

doi:10.14311/ppt.2017.2.198
Plasma Physics and Technology 4(2):198–201, 2017

MODELING OF PERFLUOROCARBONS DECOMPOSITION IN NITROGEN THERMAL PLASMA

S.W. Chau[a,*], S.H. Chen[b]

[a] *Department of Engineering Science and Ocean Engineering, National Taiwan University, 1, Sec. 4, Roosevelt Rd., 10617 Taipei, Taiwan, ROC*

[b] *Physics Division, Institute of Nuclear Energy Research, 1000 Wenhua Rd., Longtan, 32546 Taoyuan, Taiwan, ROC*

[*] chausw@ntu.edu.tw

Abstract. This paper develops an axisymmetric flow model to predict the flow field inside a direct-current plasma torch reactor, where the magneto-hydrodynamic equations, including the continuity, momentum, energy and current continuity equations as well as turbulence transport equations are solved with a finite volume discretization in a segregated manner. The thermodynamic and transport coefficients of thermal plasma are obtained from the condition of local thermal equilibrium, whereas a kinetics model is employed to consider the production and destruction of species due to the chemical reactions in a decomposition process. Species transport phenomena arising in the decomposition process are described by the transport equations for different species. The perfluorocarbons decomposition characteristics in a well-type, non-transferred nitrogen torch reactor are analyzed by the proposed numerical model to disclose the influence of working parameters on the decomposition efficiency.

Keywords: perfluorocarbons decomposition, nitrogen, thermal plasma, direct-current, plasma torch.

1. Introduction

Compared with other methods to remove hazardous gases arising in modern industrial processing, thermal plasma delivers an advantage of removal efficiency, processing flexibility, operation safety and installation cost that attract the attention of recent researchers [1, 2]. Perfluorocarbon (PFC), such as CF_4, C_2F_6, and NF_3, can be first reduced into an atomic or molecular form of its chemical components in the high-temperature environment created by thermal plasma flow. A further introduction of water into the reacting process can enhance the production of HF and CO_2, which can be easily treated in the following processes. This paper is first to develops an axisymmetric flow model to predict the thermal plasma inside a direct-current torch reactor operating with nitrogen. The perfluorocarbons decomposition characteristics in a well-type, non-transferred plasma torch reactor are then analyzed by the proposed numerical model to disclose the influence of working parameters on the decomposition efficiency.

2. Thermal plasma model

The following assumptions are adopted for a steady and axisymmetric thermal plasma flow: The plasma is optically thin, the plasma is assumed in local thermal equilibrium, the gravitational effect and the external magnetic field is ignored. The magneto-hydrodynamic equations consisting of the continuity equation, momentum equation, energy equation, and current continuity equation are given as follows:

Continuity equation

$$\frac{\partial(\rho u)}{\partial z} + \frac{1}{r}\frac{\partial(\rho r v)}{\partial r} = 0 \tag{1}$$

Momentum equation (axial-direction)

$$\frac{\partial(\rho u^2)}{\partial z} + \frac{1}{r}\frac{\partial(\rho r u v)}{\partial r} = -\frac{\partial p}{\partial z} + 2\frac{\partial}{\partial z}[\mu(\frac{\partial u}{\partial z})] + \frac{1}{r}\frac{\partial}{\partial r}[r\mu(\frac{\partial u}{\partial r} + \frac{\partial v}{\partial z})] + j_r B_\theta \tag{2}$$

Momentum equation (radial-direction)

$$\frac{\partial(\rho u v)}{\partial z} + \frac{1}{r}\frac{\partial(\rho r v^2)}{\partial r} = -\frac{\partial p}{\partial y} + \frac{2}{r}\frac{\partial}{\partial r}[\mu r(\frac{\partial v}{\partial r})] + \frac{\partial}{\partial z}[\mu(\frac{\partial u}{\partial r} + \frac{\partial v}{\partial z})] - \mu(\frac{2v}{r^2}) + \frac{\rho w^2}{r} - j_z B_\theta \tag{3}$$

Momentum equation (azimuthal-direction)

$$\frac{\partial(\rho u w)}{\partial z} + \frac{1}{r}\frac{\partial(\rho r v w)}{\partial r} = \frac{\partial}{\partial z}[\mu(\frac{\partial w}{\partial z})] + \frac{1}{r^2}\frac{\partial}{\partial r}[\mu r^3 \frac{\partial}{\partial r}(\frac{w}{r})] - \frac{\rho v w}{r} \tag{4}$$

Energy equation

$$\frac{\partial(\rho u h)}{\partial z} + \frac{1}{r}\frac{\partial(\rho r v h)}{\partial r} = u\frac{\partial p}{\partial z} + v\frac{\partial p}{\partial r} + \frac{\partial}{\partial z}(k\frac{\partial T}{\partial z}) + \frac{1}{r}\frac{\partial}{\partial r}(rk\frac{\partial T}{\partial r}) + \frac{j_z^2 + j_r^2}{\sigma} - S_R + \frac{5}{2}\frac{k_B}{e}(j_z\frac{\partial T}{\partial z} + j_r\frac{\partial T}{\partial r}) + \mu|\dot{\gamma}|^2 \tag{5}$$

Turbulence model

$$\frac{\partial(\rho u K)}{\partial z}+\frac{1}{r}\frac{\partial(\rho r v K)}{\partial r}=G-\rho\epsilon+ \frac{\partial}{\partial z}[(\mu_m+\frac{\mu_t}{\sigma_K})\frac{\partial K}{\partial z}]+\frac{1}{r}\frac{\partial}{\partial r}[(\mu_m+\frac{\mu_t}{\sigma_K})\frac{\partial K}{\partial r}] \quad (6)$$

$$\frac{\partial(\rho u \epsilon)}{\partial z}+\frac{1}{r}\frac{\partial(\rho r v \epsilon)}{\partial r}=c_{\epsilon 1}G\frac{\epsilon}{K}-c_{\epsilon 2}\rho\frac{\epsilon^2}{K}+ \frac{\partial}{\partial z}[(\mu_m+\frac{\mu_t}{\sigma_\epsilon})\frac{\partial \epsilon}{\partial z}]+\frac{1}{r}\frac{\partial}{\partial r}[(\mu_m+\frac{\mu_t}{\sigma_\epsilon})\frac{\partial \epsilon}{\partial r}] \quad (7)$$

Current continuity equation

$$\frac{\partial}{\partial z}(\sigma\frac{\partial\phi}{\partial z})+\frac{1}{r}\frac{\partial}{\partial r}(\sigma r\frac{\partial \phi}{\partial r})=0 \quad (8)$$

where ρ denotes the fluid density, u the velocity component in the axial direction (z), v the velocity component in the radial direction (r), w the velocity component in the azimuthal direction (θ), p the pressure, μ the apparent viscosity ($= \mu_m + \mu_t$), μ_m the fluid viscosity, j_z the axial component of current density, j_r the radial component of current density, B_θ the self-induced magnetic field in the radial direction, h the enthalphy, S_R the radiation loss, k the thermal conductivity, T the temperature, k_B the Boltzmann constant, e the electric charge, K the turbulent kinetic energy, ϵ the dissipation rate of K, G the energy production due to turbulence [3], μ_t the turbulent viscosity ($= c_\mu \rho K^2/\epsilon$), ϕ the electric potential, σ the electric conductivity, (c_μ, σ_K, σ_ϵ, $c_{\epsilon 1}$, $c_{\epsilon 2}$) the constants of turbulence model [4] and $|\dot\gamma|$ the magnitude of shear rate. The current density j_z and j_r are determined by the Ohm's law:

$$j_z=-\sigma\frac{\partial\phi}{\partial z},\qquad j_r=-\sigma\frac{\partial\phi}{\partial r} \quad (9)$$

The self-induced magnetic field B_θ is calculated according to the Biot-Savart law:

$$B_\theta=\frac{\mu_0}{r}\int_0^r j_z\,\xi\,\mathrm{d}\xi \quad (10)$$

3. Kinetic model

The transport equation to describe the neutral species arising in the destruction process is given as follows:

$$\nabla\cdot(n_i\mathbf{U})+\nabla\cdot\mathbf{\Gamma}_i=R_{i,p}+R_{i,d} \quad (11)$$

where n_i denotes the density of the ith species, $\mathbf{U}$ the flow velocity, $\mathbf{\Gamma}_i$ the flux of the ith species, $R_{i,p}$ the production of the ith species, $R_{i,d}$ the destruction of the ith species. The decomposition mechanism of fluorinated compounds disclosed in [1] together with the kinetic model for water plasma suggested in [5] is adopted in this paper to describe the perfluorocarbons decomposition in nitrogen thermal plasma. The following table summarizes the decomposition mechanisms for CF_4 considered in this study.

No.	Reactions
1	$CH_2F_2 + H \rightarrow CH_2F + HF$
2	$CF_4 + H \rightarrow CF_3 + HF$
3	$CF_4 \rightarrow CF_3 + F$
4	$CH_2F + F \rightarrow CHF + HF$
5	$CF_3 + O \rightarrow COF_2 + F$
6	$CF + O \rightarrow CO + F$
7	$CF_2 + O \rightarrow CO + 2F$
8	$CF_3 + H \rightarrow CF_2 + HF$
9	$CF_2 + H \rightarrow CF + HF$
10	$CF + H \rightarrow CH + F$
11	$CF_2 + F \rightarrow CF_3 + M$
12	$CF_3 + F \rightarrow CF_4 + M$
13	$CF_2 + H_2 \rightarrow CH_2F_2$
14	$CF_3 + H_2 \rightarrow CHF_3 + H$
15	$CF_2 + O_2 \rightarrow COF_2 + O$
16	$COF_2 + H \rightarrow COF + HF$
17	$COF_2 + O \rightarrow CO_2 + F_2$
18	$COF + O \rightarrow CO_2 + F$
19	$COF + H \rightarrow CO + HF$
20	$COF + F \rightarrow CO + F_2$

Table 1. Decomposition mechanisms for CF_4.

4. Numerical method

A finite volume method [6] is employed to discretize the governing differential equations on a Cartesian grid and then a integral form of governing equation over an arbitrary control volume V is obtained:

$$\int_S \rho\varphi(\mathbf{U}\cdot\mathbf{n})\,\mathrm{d}S=\int_S(\mathbf{T}_\varphi\cdot\mathbf{n})\,\mathrm{d}S+\int_V \rho f_\varphi\,\mathrm{d}V \quad (12)$$

where the control volume V is defined by the control surface S with an outer normal vector $\mathbf{n}$, φ the generalized variable, $\mathbf{U}$ the velocity vector, $\mathbf{T}_\varphi$ the generalized surface force, f_φ the generalized source term. Using the value of neighboring nodes to approximate each term appearing in the governing equations, a linearized equation is obtained for every governing equation. A SIMPLE-like algorithm is employed to solve the coupling of dependent variables iteratively. For the current density distribution at cathode, an exponential profile is assumed as follows [7]:

$$j_c(\xi)=J_{\max}\,\mathrm{e}^{-b\xi} \quad (13)$$

where $J_{\max}$ denotes the maximum current density, ξ the decreasing direction of current density and the constant b is determined by the system current I and the area of hot spot A:

$$I=2\pi\int_A j_c(\xi)\,\mathrm{d}A \quad (14)$$

5. Boundary conditions

In the numerical modeling, the DC torch is simplified as a hollow cylinder with length of 89 mm and radius of 12 mm, Fig. 1. One end of the hollow cylinder

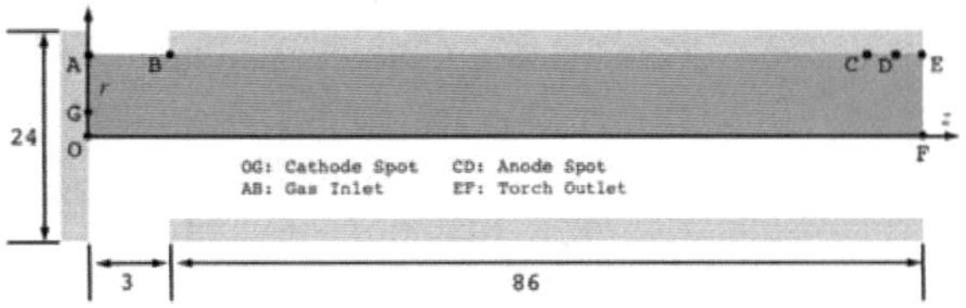

Figure 1. Computational domain and the location of electrodes.

is closed, while the other end is open for outgoing plasma. The inlet boundary conditions are specified at the annular gas inlet adjacent to the torch bottom. The flow rate of nitrogen is 180 slpm and a swirler design additionally yields a tangential velocity of 20 m/s at the gas inlet. CF_4 and H_2O are supplied together with the working gas (nitrogen) in a concentration of 10^4 ppm and $2{\cdot}10^4$ ppm, respectively. The working current of the simulated case is 145 A, which results in a measured voltage drop of 194 V between two electrodes. The location of cathode is at the center of torch bottom due to the installment of a cylindrical cathode. Anode's position is then fixed by the measured voltage obtained in the experiment. The arc root sizes of both electrodes are assumed 3 mm. About fifty thousand cells are used to discretize the computational domain, where the grid points close to arc root and gas inlet, as well as the torch wall, are clustered to resolve the steep gradient of field variables, such as velocity, temperature and current density.

6. Numerical results

Figure 2 depicts the plasma temperature and velocity profile at the torch outlet, where the predicted voltage drop is 191 V. The exiting plasma flow is accelerated by the self-induced magnetic field to deliver a blunt axial velocity profile. Except in the region close to torch wall ($r > 10$ mm), the nitrogen plasma can reach the speed range above 45 m/s along with a maximum value of 72 m/s at the centerline. Due to the installation of swirler at the gas inlet, a strong rotational flow in introduced into the hollow torch. The peak of tangential velocity is predicted for about 39 m/s located at about one third of the torch radius (4 mm). The position of the maximum tangential speed clearly defines the size of plasma core, which owns a relatively high velocity as well as temperature when compared with its outer region. Inside the core region, the plasma jet can attain a flow velocity over 60 m/s, whereas the temperature profile is between 3300 K and 7200 K. Figure 3 shows the plasma characteristics along the centerline. Because cathode is installed at the torch bottom, high current density and temperature as well as significant axial velocity is expected in the vicinity of the torch bottom. The spike of current density is predicted to be 62 A/mm^2, which gives a maximum temperature of 26000 K along the centerline. The axial velocity along the torch axis strongly oscillates as the plasma moves away from the torch bottom. The flow quickly soars from -50 m/s to 290 m/s near the torch end, and substantially loses its velocity magnitude at approximately one-third of the torch length. With the help of nontrivial current density existing in the middle part of plasma torch, the plasma jet obtains an obvious velocity increase and exits from the torch at 90 m/s. Figure 4 illustrates the plasma characteristics inside the plasma torch. The right part of the figure disclose the velocity distribution, where a nontrivial vortex motion with the center located about 25 mm in front of the cathode is observed in the neighborhood of cathode. The temperature field as well as the current density vector is given in the left part of the figure. Significant current density emerges in the high-temperature core along the centerline.

Figure 5 gives the radial density profile of various species stemming from the decomposition mechanism of CF_4, where low density products are neglected in the profile plot. Water is relatively abundant (between 200 mole/cm^3 and 300 mole/cm^3) and smoothly distributed in the outer region of torch outlet (r>5 mm), whereas it rapidly decline with the decrease of radial distance. Similar behavior is observed for CF_4, and at the same radial position the density of CF_4 is about one tenth of the water density. CF_3 as well as CH_2F shows a very low concentration in the plasma core, whereas an exponential grow of density is expected outside the hot plasma core. In contrast, the predicted density of fluorine atom varies in the range between 7 mole/cm^3 and 20 mole/cm^3 in the plasma core but a strong fall of concentration is predicted outside the high-temperature zone. The product group, i.e., HF, CO, and CHF, also deliver a hill-shape profile along the radial direction, where the peak density approximately occurs near the radial boundary of the plasma core. CHF_3 is the only material given in the profile plot, which exhibits a consistent tendency of growing concentration with r. The efficiency of CF_4 decomposition η in the investigated torch arrangement is 97%, where η is defined as the ratio of the mass flux at the gas inlet to the mass flux at the torch outlet

$$\eta = \frac{\dot{m}_{CF_4}|_{\text{inlet}}}{\dot{m}_{CF_4}|_{\text{outlet}}} \tag{15}$$

7. Conclusions

This paper proposes a numerical model to simulate the perfluorocarbons decomposition within a well-type, non-transferred nitrogen torch. An operation condition of 180 slpm, 145 A and 193 V is predicted using nitrogen as background gas, where water vapor and Tetrafluoromethane are initially with a concentration of 10^4 ppm and $2{\cdot}10^4$ ppm, respectively, at the gas inlet. The numerical simulation predicts the major products stemming from the decomposition mechanism of CF_4 are H_2O, F, HF, CO, CHF , CH_2F, CF_3,

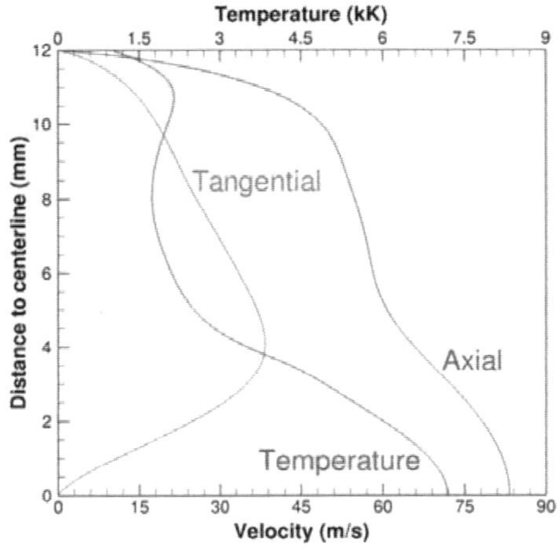

Figure 2. Plasma temperature and velocity profile at the torch outlet.

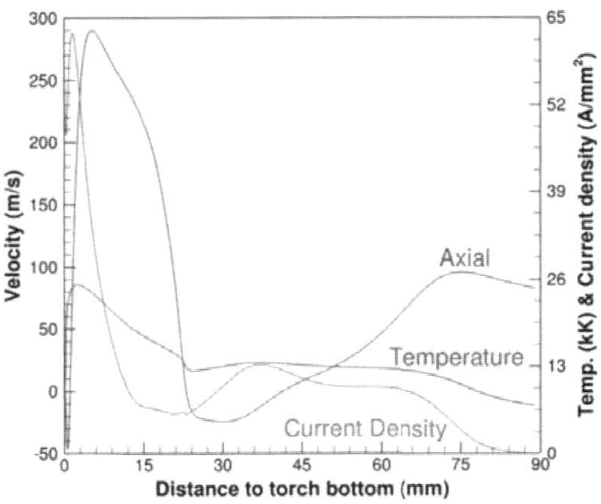

Figure 3. Plasma characteristics along the centerline.

CF_4 and CHF_3. The efficiency of CF_4 decomposition via a nitrogen plasma torch under the studied operation condition is estimated about 97 %.

Acknowledgements

This study was supported by the Institute of Nuclear Energy Research, ROC (NL1050877).

References

[1] S.H. Narengerile and T. Watanabe. Decomposition mechanism of fluorinated compounds in water plasmas generated under atmospheric pressure. *Plasma Chem. Plasma Process*, 30(6):813–829, 2010. doi:10.1007/s11090-010-9259-y.

[2] M.S. Lim, S.C. Kim, and Y.N. Chun. Decomposition of PFC gas using a water jet plasma. *J. Mech. Sci. Tech.*, 25(7):1845–1852, 2011. doi:10.1007/s12206-011-0422-z.

[3] S.W. Chau, S.Y. Lu, and P.J. Wang. Modeling of axis-symmetric steam plasma flow in a non-transferred torch. *Comp. Phys. Comm.*, 182(1):152–154, 2011. doi:10.1016/j.cpc.2010.08.011.

[4] B.E. Launder and D.B. Spalding. The numerical computation of turbulent flows. *Comp. Methods in Appl. Mech. Eng.*, 3(2):269–289, 1974. doi:0.1016/0045-7825(74)90029-2.

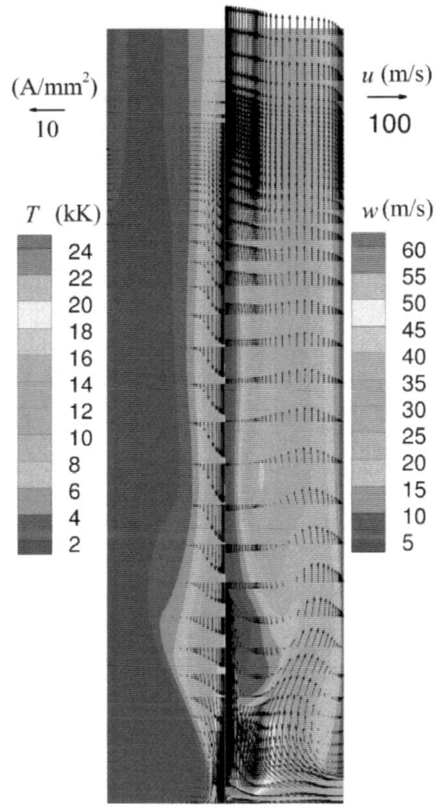

Figure 4. Plasma characteristics inside the torch.

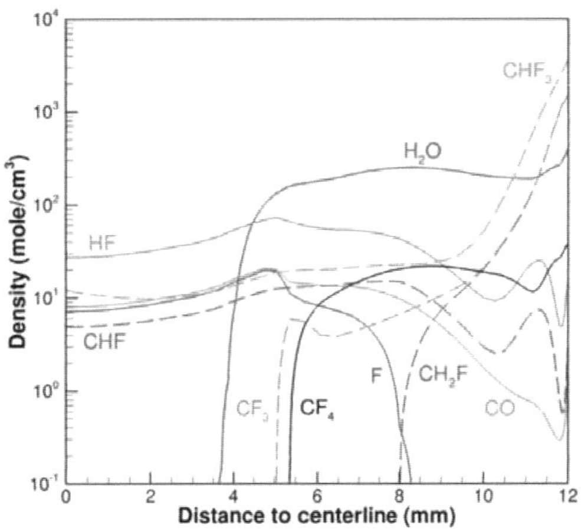

Figure 5. Density profile at the torch outlet.

[5] S.W. Chau, C.M. Tai, and S.H. Chen. Nonequilibrium modeling of steam plasma in a nontransferred direct-current torch. *IEEE Tran. Plasma Sci.*, 42(12):3797–3808, 2014. doi:10.1109/TPS.2014.2365479.

[6] S.W. Chau. *PhD Disseration.* Universität Hamburg, 1997.

[7] K.C. Hsu, K. Etemadi, and E. Pfender. Study of the free-burning high-intensity argon arc. *J. Appl. Phys.*, 54(3):1293–1301, 1983. doi:10.1063/1.332195.

doi:10.14311/ppt.2017.1.1
Plasma Physics and Technology **4**(1):1–4, 2017

CONTACT AND NOZZLE WEAR FROM 100 INTERRUPTIONS FOR A PUFFER-TYPE AIR LOAD BREAK SWITCH

N. Støa-Aanensen*, E. Jonsson, M. Runde

SINTEF Energy Research, 7465, Trondheim, Norway

* nina.stoa-aanensen@sintef.no

Abstract. One type test requirement for medium voltage load break switches is to interrupt 100 consecutive "mainly active loads". A puffer-type switch with axial-blown arc has been tested according to the 630 A/24 kV ratings. The nozzle and contact wear were measured regularly to investigate design requirements and the impact from nozzle wear on gas flow. The contact wear is only moderate, while the nozzle wear causes a decrease in pressure build-up, which in turn may influence the interruption performance.

Keywords: load break switch, switchgear, medium voltage, electric arc, puffer switch, nozzle wear.

1. Introduction

In the recent years, substational efforts have been made to develop SF_6-free switchgear to reduce the use of greenhouse gases in the electrical industry [1]–[4]. One of the most challenging aspects when designing and producing a compact SF_6-free load break switch (LBS) is to make it cost-efficient. Since both currents and voltage stresses are moderate compared to fault current interruptions and higher voltage ratings, the difficulty of making a medium voltage (MV) LBS is not necessarily the current interruption itself, but to do it at fairly low gas pressure, in a compact and small device, and to a low manufacturing cost.

The various parts of a switch are subjected to repeated stresses during their life-time, both mechanical and from the current and arc heating. Consequently, they need to be made of durable and often expensive materials. Often, cost reduction comes with a lower performance, calling for trade-offs. Some examples are reducing the use of silver plating versus the increased heat losses, or decreasing the sizes of the expensive arcing contacts versus wear of these by the electric arc.

This paper reports a study on the arcing contact and nozzle wear from 100 consecutive interruption operations with a single-phase version of the 24 kV/630 A "mainly active load" test duty as described by IEC [5]. A puffer-type setup with axial cooling flow and air as the interrupting medium is used.

There are several issues related to the wear of a puffer switch during testing. The arcing contacts are expensive and should not be made larger than necessary. Thus, it is important to observe their mass losses and any change in diameter. Moreover, the nozzle wear could change the outlet area from the puffer volume, and consequently the cooling air flow itself. These topics are covered and discussed in this paper.

2. Setup and procedures

The experimental work is carried out with a test circuit that is directly powered by the MV grid. The circuit setup is described in detail elsewhere [6]–[7]. The current to be interrupted is approximately 630 A. The transient recovery voltage (TRV) can be described by a rate of rise of recovery voltage (RRRV) of approximately 80 V/μs and a first voltage peak of 7.5 kV. This corresponds to the type tests for a 24 kV LBS. Air is used as the interrupting gas, and the starting pressure is atmospheric.

The puffer test switch is shown in Figure 1. A stationary contact set including main contact and the pin arcing contact is seen to the right in the figure. The puffer volume is created with a cylinder, a stationary back wall and a moving puffer disc. The moving main and arcing ring contacts are attached to the puffer disc. Before testing, the puffer disc is positioned to the right end of the puffer cylinder, and the main contact is closed. During a current interruption, the puffer disc and moving contacts are pulled away from the stationary contacts, the puffer volume decreases and the pressure rises. The compressed air vents out from the puffer volume through ten holes in the middle of the puffer disc and further out through the arcing ring contact after contact separation and to the arcing zone. A cylindrical polytetrafluoroethylene (PTFE) nozzle is attached on the outside of the arcing ring contact to guide the air flow onto the arc, and to prevent flashover from the arcing contact to the main contact set. The total contact movement is around 90 mm, where 24 mm is before arcing contact separation.

The diameter of the ring and pin arcing contacts is 7 mm. The nozzle has an inner diameter of 7.3 mm at the start of the experiment, and the nozzle channel is 12 mm long. A pre-compressed spring is released by a trigger signal to start the movement of the ring contact and the compression of the puffer volume. To vary how far the contact has moved before current zero (CZ),

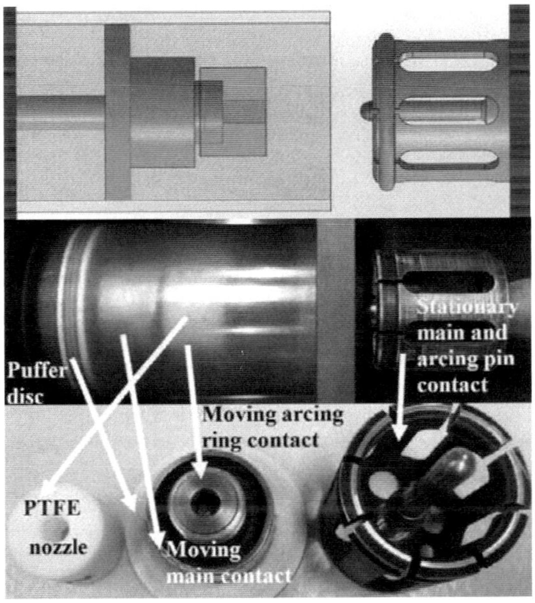

Figure 1. The puffer test switch and its main components. Arcing contacts are made of copper tungsten, while the main contacts are of brass.

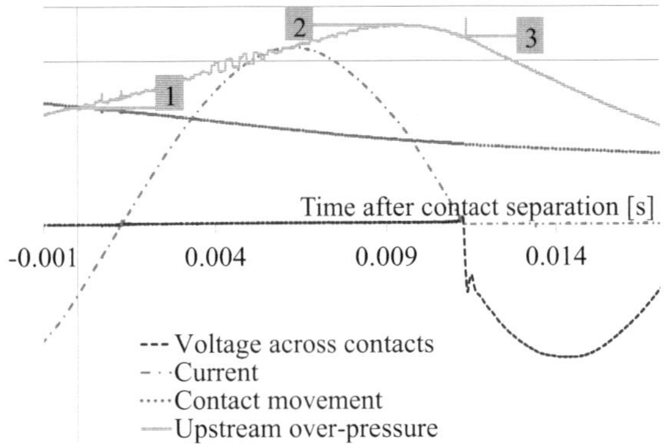

Figure 2. Current interruption test with measurements of contact position, voltage and current across the contact gap, and the over-pressure in the puffer volume. Pressure at contact separation (1), the peak pressure (2), and the pressure at interruption (3).

the triggering set point is changed throughout the test series. The spring or the amount of pre-compression is not changed, but friction from the puffer disc or between the arcing contacts may slightly alter the contact velocity from test to test. The pressure build-up itself also affects the movement; the higher the over-pressure in the puffer volume, the larger force must be applied to keep the movement at a constant velocity.

In order to keep track of the nozzle and contact wear during the 100 interruptions, the test switch is disassembled after every 10 tests. The nozzle and contact diameters and masses are measured. Moreover, the cold flow (no arc) pressure build-up as a function of contact movement is recorded to see to what extent the wear changes the pressure profiles and behaviour of the puffer device.

3. Experimental results and discussion

Figure 2 shows typical measurements from a current interruption test. The first CZ was a failed interruption (occurred with the pin contact still inside the nozzle), while the second attempt was successful. All interruption tests were successful at first attempt outside the nozzle. A few even interrupted with the pin contact still inside the nozzle channel.

The current was in the range 336 A to 349 A. The contact velocity around current interruption was 2–3 m/s, depending on the pressure build-up and the contact gap. An LBS in normal operation cannot control at what contact position the CZ crossings occur, and the switch should be able to interrupt the current within the first two CZs, independent of when the contacts separate. In this test series, the triggering time was varied, and the interruptions occurred at contact gaps from 1 to 35 mm.

3.1. Nozzle and contact wear

The arcing contacts had no notable change in their diameters during the 100 tests. Their weight reduction was 0.04–0.05 g, or less than 0.6% of the total mass of the pin contact tip and below 0.3% of the ring contact. The 100 consecutive interruptions of the "mainly active load" test duty is one of the toughest requirements for the arcing contacts, and often determines the minimum dimensions of the expensive copper-tungsten contacts. In this test, the diameter was 7 mm, which does not seem to be a limiting factor when designing a durable switch for load currents of 630 A. The main contacts must be dimensioned based on the rated current amplitude and the maximum permissible temperature during normal operation.

The measured inner diameter of the nozzle is listed in Table 1. Both side-to-side and top-to-bottom were measured, that is, at a 90° angle. Although there is some asymmetry, the nozzle inner diameter increased by approximately 1 mm, or 14%, during 100 interruption tests. The total mass loss was 0.5 g, or 2.2%.

The nozzle wear can affect the current interruption capability in several ways. First, a larger nozzle-to-contact diameter ratio requires an increased over-pressure for successful interruption [8]. Second, an increase in the nozzle diameter changes the outlet area when the pin contact is still inside the nozzle, and could thus change the pressure build-up in the puffer volume.

3.2. Pressure build-up

Figure 3 shows the cold flow measurements of pressure and contact position carried out before and during

Test no.	Side-to-side [mm]	Top-to-bottom [mm]
0	7.31	7.31
10	7.39	7.40
20	7.58	7.60
30	7.65	7.66
40	7.76	7.69
50	7.87	7.86
60	7.96	7.96
70	8.07	8.08
80	8.15	8.13
90	8.26	8.25
100	8.32	8.37
Difference	1.01	1.06

Table 1. Measured nozzle inner diameters.

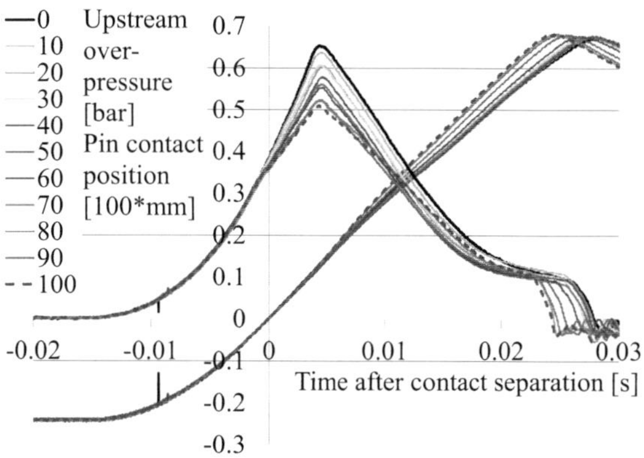

Figure 3. Cold-flow measurements of puffer volume over-pressure and contact movement from before start of test series (0) to after 100 tests (100).

the test series. The black lines represent the measurements before the interruption tests (labelled "0"), the red lines represent the measurements after 90 tests, etc. There are two lines per cold flow test, namely the upstream over-pressure (starts at 0 bar) and the pin contact position relative to contact separation (starts at -24 mm).

The movement and pressure build-up before the contacts separate do not change much during the 100 tests. In this period, the pin contact is blocking the only outlet, and the pressure rise is consistent with adiabatic compression of the puffer volume. As long as the pin and ring contact do not change size during the test series, this pressure build-up is expected to be the same each time.

The variations start after contact separation. The newer (and narrower) the nozzle, the higher upstream over-pressure is achieved in the puffer volume. The maximum over-pressure is around 0.65 bar before the test series, and about 0.5 bar after 100 interruptions. It is important that also the 100th interruption test builds up high enough over-pressure to interrupt.

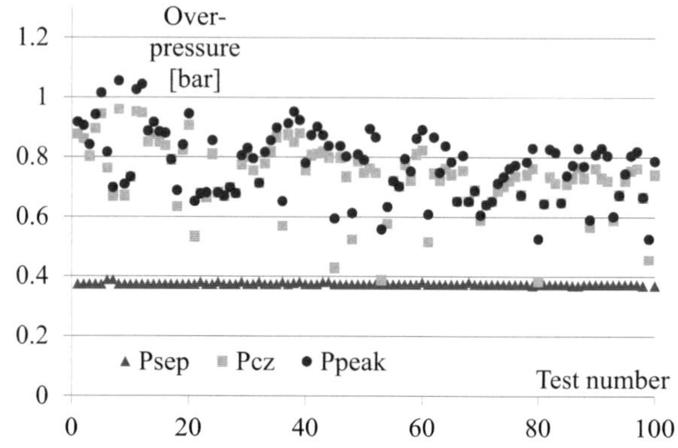

Figure 4. Over-pressure in puffer volume at contact separation (Psep), at CZ and current interruption (Pcz), and the maximum over-pressure (Ppeak).

The trend is similar for the pressure measurements with arcing. Figure 4 shows the pressures at contact separation, current interruption and the peak pressure for all tests. Whereas the pressure at contact separation is constant, and similar to the cold-flow cases, the peak pressure and pressure at current interruption decrease during the course of the tests. Furthermore, the pressure is in general higher than in the cold flow cases, due to clogging of the outlets by the arc. The maximum over-pressure in the cold flow cases was not above 0.65 bar. Now, the over-pressure exceeds 1 bar, depending on the contact position at current interruption.

With the switch design used in this work, 100 "mainly active load" test duties are sufficitent to significantly alter the PTFE nozzle inner diameter and consequently the cooling gas flow. Obviously, a switch must be designed by taking nozzle wear and its influence on the gas flow into consideration.

3.3. Arcing energy

As the triggering time and the contact gap were varied throughout the test series, so was the arcing energy per test. The arcing energy for each test, together with the cumulative arcing energy are shown in Figure 5. The arcing energy per test varied from as low as 53 J (a test where the current was interrupted right after contact separation) and up to 835 J. In total, the arcing energy of the test series was 31.7 kJ. The total nozzle mass loss was 0.5 g.

In [9], thermogravimetric analyses (TGA) were carried out to measure the energy needed to evaporate PTFE by slow heating of the material. In addition, arcing experiments were carried out in the vicinity of PTFE plates, and the PTFE mass losses were compared to the energy dissipated in the arc. The energy required to evaporate or "remove" 1 g PTFE was found to be 1485 kJ/g and 87.1 kJ/g, respectively, in these two cases.

If it is assumed that all the energy dissipated by the arc during the 100 interruption tests is absorbed

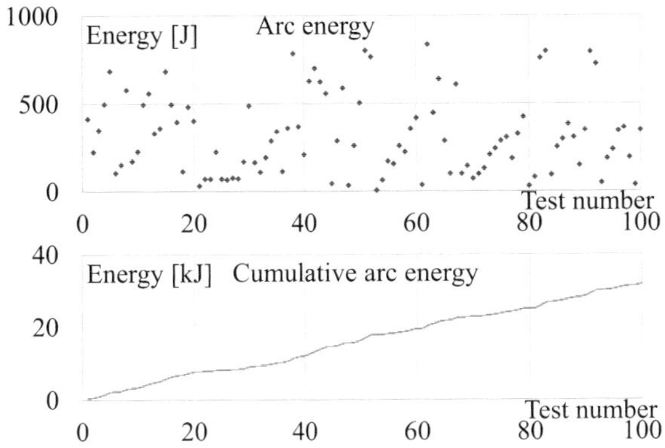

Figure 5. The arcing energy of each interruption test, and the cumulative arc energy for the entire test series.

by the PTFE nozzle, it corresponds to 63.4 kJ/g. In reality, the arc energy also goes into heating of the arcing contacts and the surrounding air. An energy per evaporated mass of PTFE of 63.4 kJ/g is well below the TGA value, but is comparable to the arcing tests in [9]. Here, the tests were carried out with pure copper contacts and not copper tungsten contacts, which means that a significant part of the arcing energy probably went into evaporation of the arcing contacts and not only to the surrounding PTFE plates.

Both current interruption experiments result in significantly lower energies required to ablate the PTFE nozzle material than the case where the PTFE is heated uniformly until evaporated. This indicates that the stresses during arcing are clearly a dynamic process where e.g., local heating of the nozzle can cause small pieces or fragments of solid PTFE to break off. Thus, the TGA value is obviously not usable for estimating nozzle wear from arcing.

4. Conclusions

A test series of 100 consecutive current interruptions in a 24 kV/630 A "mainly active load" test duty have been carried out for a puffer-type LBS design. The contact and nozzle wear was monitored, as well as the development of the pressure build-up during the test series. The main findings are:

- All interruptions were successful at first or second CZ (that is, first CZ occuring with the pin contact outside the nozzle channel).
- The wear of the 7 mm diameter arcing contacts is not critical. Thus, a contact set of this size is believed to be acceptable for an MV LBS design with respect to wear during load current interruptions.
- The nozzle inner diameter increased from approximately 7.3 to 8.3 mm, or 14%. An increase in the nozzle-to-contact diameter ratio is not beneficial with respect to interruption capability, and could perhaps be a limiting design factor.
- The maximum pressure obtained in the puffer volume during both arcing and cold-flow measurements decreases with the number of interruptions.
- The pressure build-up is substantially higher when an arc is present. The maximum pressure in a cold-flow case is typically 0.6 bar over-pressure, while the maximum pressure for tests with arc can reach more than 1.0 bar over-pressure.
- A total of 31.7 kJ was dissipated by the arc during the 100 interruptions. Assuming that all the energy goes into nozzle ablation, this corresponds to a rate of 63.4 kJ/g. This value is significantly lower than found with TGA on this material.

Acknowledgements

This work is supported by the Norwegian Research Council.

References

[1] H. Okubo, and A. Beroual, Recent trend and future perspectives in electrical insulation techniques in relation to sulfur hexafluoride (SF_6) substitutes for high voltage electric power equipment, *IEEE Electrical Insulation Magazine*, vol. 27, no. 2, pp. 34–42, 2011. doi:10.1109/MEI.2011.5739421

[2] P. C. Stoller, M. Seeger, A. A. Iordanidis, and G. V. Naidis, CO_2 as an arc interruption medium in gas circuit breakers, *IEEE Trans. Plasma Science*, vol. 41, no. 8, pp. 2359–2369, 2013. doi:10.1109/TPS.2013.2259183

[3] N. Ranjan, J. Carstensen, P. Stoller, M. Schwinne, D. Over, and T. Votteler, Switching performance of weakly cooled arcs, *Plasma Physics and Technology*, vol. 2, no. 2, pp. 191–194, 2015.

[4] Y. Kieffel, T. Irwin, P. Ponchon, and J. Owens, Green gas to replace SF_6 in electrical grids, *IEEE Power and Energy Magaizine*, vol. 14, no. 2, pp. 32–39, 2016. doi:10.1109/MPE.2016.2542645

[5] *High-voltage switchgear and controlgear - Part 103: Switches for rated voltage above 1 kV up to and including 52 kV*, IEC International Standard no. 62 271-103, ed 1.0, 2011.

[6] E. Jonsson and M. Runde, Medium voltage laboratory for load break switch development, the *Proc. Int. Conf. Power Systems Transients*, Vancouver, 2013.

[7] N. S. Aanensen, E. Jonsson, and M. Runde, Air Flow Investigation for a Medium Voltage Load Break Switch, *IEEE Trans. Power Del.*, vol. 30, no. 1, pp. 299–306, 2015. doi:10.1109/TPWRD.2014.2334360

[8] N. S. Støa-Aanensen, M. Runde, E. Jonsson , A. D. Teigset, Empirical Relationships Between Air Load Break Switch Parameters and Interrupting Performance, *IEEE Trans. Power Del.*, vol. 31, no. 1, pp. 278–285, 2016. doi:10.1109/TPWRD.2015.2435804

[9] E. Jonsson, M. Runde, G. Dominguez, A. Friberg, and E. Johansson, Comparative study of arc-quenching capabilities of different ablation materials, *IEEE Trans. Power Del.*, vol. 28, no. 4, pp. 2065–2070, 2013. doi:10.1109/TPWRD.2012.2227834

INVESTIGATION OF PULSE GLOW DISCHARGE EXCITATION MODES IN N_2 IN A HOLLOW CYLINDRICAL COMBINED CATHODE

S. Bordusau*, S. Madveika, A. Bozhko, O. Tsikhan

Belarusian State University of Informatics and Radioelectronics, P.Brovki 6, 220013 Minsk, Belarus

* bordusov@bsuir.by

Abstract. Electric discharge excitation modes and N_2 pressure range for a stable formation and maintaining of pulse glow discharge plasma in a hollow combined cathode have been investigated. It has been found that both of these parameters are influenced by the constructive peculiarities of the cathode. It has been established experimentally that the value of N_2 breakdown voltage for the studied construction of the cathode within the range of 50–250 Pa does not exceed -700 V.

Keywords: hollow cathode, glow discharge, electric pulses, vacuum.

1. Introduction

The application of ion-plasma treatment is characterized by high technological opportunities to change physico-chemical and mechanical properties of constructional materials parts' surfaces because of their complex treatment: obtaining optimal structure, phase composition and the degree of phase doping [1].

Concerning a number of technological problems the use of glow discharge with hollow cathode effect (HCE) for materials heating up in vacuum (including melting) is of interest [2], [3], [4]. It is connected with the fact that a glow discharge with HCE possesses a number of specific peculiarities such as a possibility of significant plasma density increase at a low average power level [5], high temperature in plasma-forming zone because of high power input [6], etc.

The studies of glow discharge plasma-forming with HCE modes are presented in a number of publications. But because of specific construction peculiarities of such plasma discharge systems it is impossible to define in advance the optimal modes of plasma-forming for a certain construction and dimensions of hollow cathode. Besides, the modes of plasma-forming are greatly influenced by such factors as the kind of plasma-forming gas, the frequency and the form of electrical signals exciting the plasma, the form and the location of the anode, the cathode's spatial position, etc.

Therefore, for unconventional hollow cathode constructions the choice of electrical plasma-forming modes and the definition of gas pressure ranges that provide a stable exciting and maintaining of plasma are performed experimentally.

2. Experimental part

The investigation was carried out with the use of a gas discharge system in the form of a hollow cylindrical electrode-cathode (Figure 1) made of stainless steel. The internal diameter of the electrode-cathode

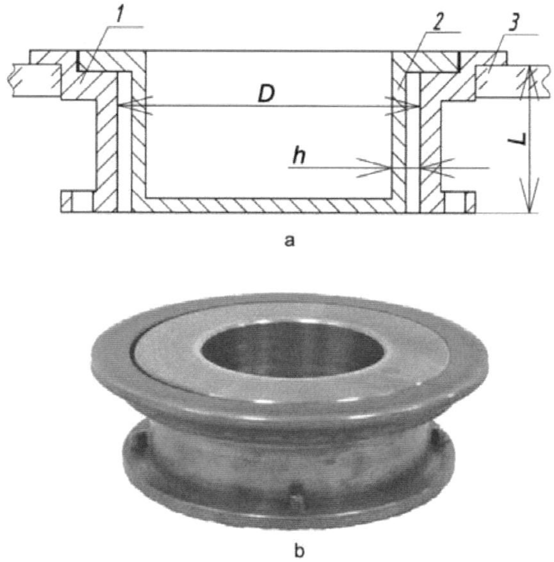

Figure 1. Construction (a) and external view (b) of the hollow cylindrical electrode-cathode

body was $D = 60$ mm, the height of the body was $L = 40$ mm.

The electrode-cathode consists of a round body 1 which can contain replaceable insertions 2 of various construction. The cathode is held by a quartz plate 3 located on the top end of the vertical quartz cylinder standing on the metal base of the vacuum chamber. Such construction of the electrode-cathode holder provides separation of plasma region (the discharge area is inside the quartz cylinder) from the rest volume of the installation vacuum chamber. A current lead is connected to the body of the electrode-cathode placed inside the quartz cylinder. The voltage is applied to the electrode through it.

The glow discharge in the electrode-cathode is excited in the space h between the internal body surface

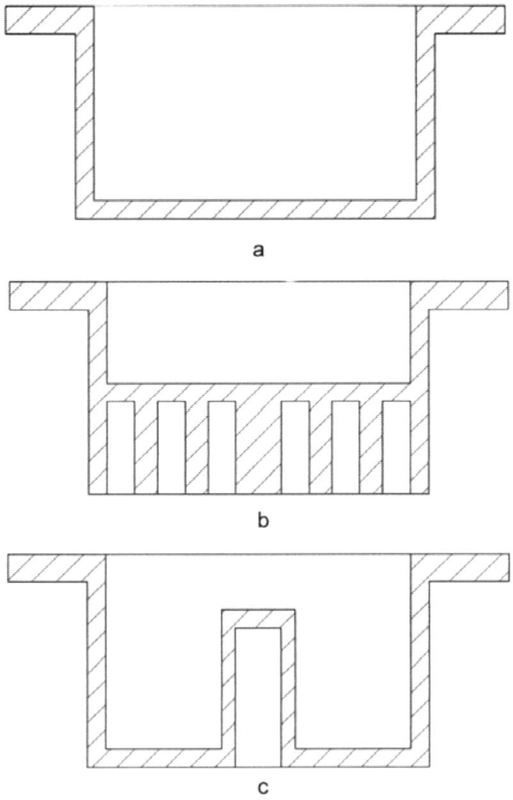

Figure 2. Forms of replaceable insertions of the electrode-cathode; a — solid cylindrical; b — with ring cavities in the bottom part; c — with a hollow protrusion

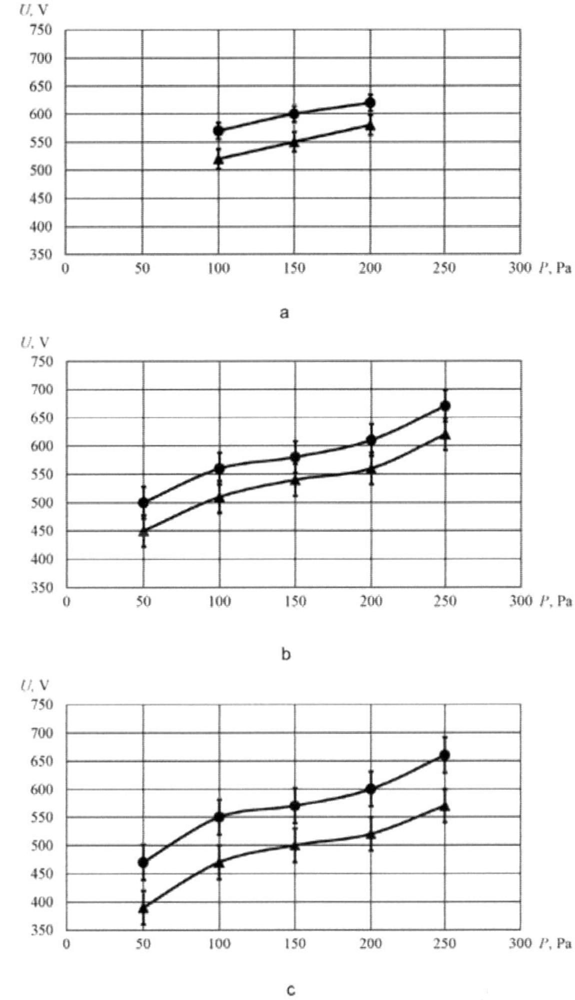

Figure 3. Excitation and extinction voltage of the discharge with HCE in N_2 for solid cylindrical insertions at $h = 2$ mm (a), $h = 3$ mm (b) and $h = 5$ mm (c):
• — discharge exciting voltage
▲ — discharge extinction voltage

and the external surface of the replaceable insertion. Replaceable insertions of various construction were used during experiments: cylindrical of various diameter (Figure 2a), with ring cavities in the bottom part (Figure 2b), with a hollow protrusion (Figure 2c).

To excite the discharge with HCE the pulse voltage of negative polarity with the rectangular pulse repetition frequency $f = 47$ kHz and pulse ratio $S = 4$ was applied to the cathode.

High purity nitrogen N_2 was used as plasma-forming gas.

The moments of discharge excitation and extinction were defined according to the indications of the plasma optical luminescence sensor connected to an oscillograph.

Before the experiments, the vacuum chamber was pumped out to $P = 1$–2 Pa extreme residual pressure.

During the experiments the ranges of plasma-forming process regulation modes were: the amplitude of generator voltage pulses changed from -300 V to -1000 V; the pressure of N_2 in the vacuum chamber varied within the interval of 30–600 Pa.

3. The results and discussion

The experimental data on the conditions of discharge excitation and extinction for solid cylindrical insertions at various distance h from the insertion's external surface to the electrode-cathode's body internal surface are given in Figure 3.

The presented dependencies show that lower values of discharge with HCE excitation voltage for investigated N_2 pressure ranges are provided at $h = 5$ mm distance between plasma-forming surfaces.

With the decrease of this distance the discharge excitation voltage increases and the N_2 pressure range of discharge with HCE excitation decreases. It may be caused by the increasing probability of plasma particles losses on the cathode surface and the decrease of near to cathode area size in the plasma-forming region.

The value of discharge excitation voltage in the range of plasma-forming gas pressure monotonously increases with the growth of pressure.

N_2 pressure range at which the stable excitation of

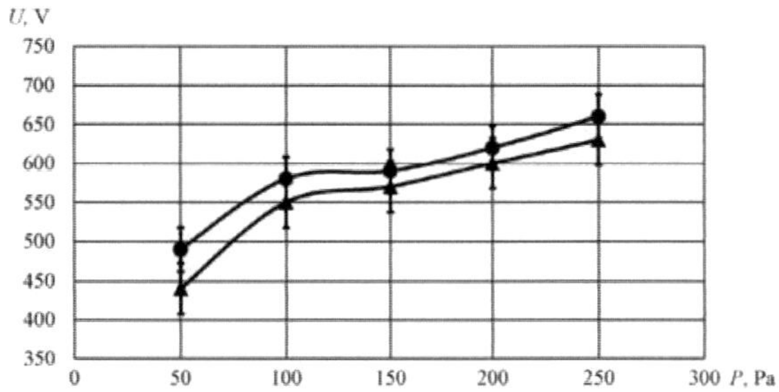

Figure 4. Voltage of discharge with HCE excitation and extinction in N_2 for an insertion with a hollow protrusion:
- *— discharge exciting voltage*
▲ *— discharge extinction voltage*

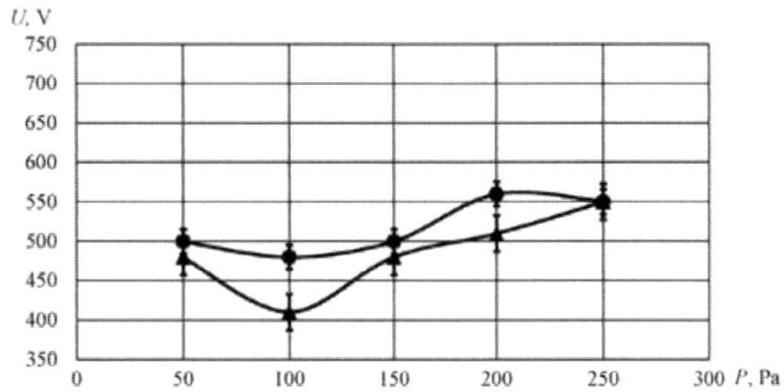

Figure 5. Voltage of discharge with HCE excitation and extinction in N_2 for an insertion with ring cavities in the bottom part:
- *— discharge exciting voltage*
▲ *— discharge extinction voltage*

the discharge with HCE takes place for $h = 5\,\mathrm{mm}$ and $h = 3\,\mathrm{mm}$ was 50–250 Pa. For $h = 2\,\mathrm{mm}$ the range narrowed to 100–200 Pa.

Data analysis of the discharge excitation and extinction values shows that the distance between the walls $h = 5\,\mathrm{mm}$ provides better conditions not only for a breakdown but for maintaining a discharge as well. In this case the discharge disappears at voltage values lower than for $h = 2\,\mathrm{mm}$ and $h = 3\,\mathrm{mm}$.

The presence of a cylindrical protrusion with the internal diameter $D = 5\,\mathrm{mm}$ in the end part of a replaceable insertion (Figure 2c) has not essentially influenced the modes of exciting the discharge with HCE (Figure 4).

The presence of ring cavities in the bottom part of the insertion (Figure 2b) led to lower values of discharge with HCE excitation voltage in the whole range of plasma-forming pressure (Figure 5).

Note, the discharge with HCE appeared in the cavities as well.

The discharge in cavities at various gas pressure values took place in different ways.

At the pressure about 50 and 250 Pa the discharge appeared only in the outside cavity. The appearance of the discharge in the outside and central cavities was observed at the pressure 100 and 200 Pa. At the pressure about 150 Pa the discharge occurred in all cavities.

Therefore, the performed experiments showed that the exciting of a pulse glow discharge with HCE in the investigated construction of the hollow combined cylindrical cathode in nitrogen at an average vacuum takes place at breakdown voltage values not exceeding -700 V. Consequently, there is no need to use generators with a higher amplitude value of electric pulses.

Note, the exciting voltage of the glow discharge with HCE may be lowered by modifying the electrode-cathode's construction.

4. Conclusions

Electrical modes have been studied and N_2 pressure ranges for a stable formation and maintaining of the pulse glow discharge with HCE plasma in a hollow combined cylindrical cathode have been established.

It has been shown that cathode's constructive peculiarities influence both the electrical modes of discharge formation and the extent of N_2 pressure range in which plasma is formed and maintained.

It has been established experimentally that the values of nitrogen breakdown voltage for the investigated electrode-cathode construction in the pressure range of 50–250 Pa does not exceed -700 V.

Acknowledgements

This research has been supported by the Belarusian State Program of Scientific Research "Science of physical materials, new materials and technologies" assignment 4.1.01.

References

[1] Berin E.V., Koval N.N., and Seydman L.A. *Plasma chemical-thermal treatment of the surface of steel parts.* Moscow: Technosphere, 2012.

[2] Janosi. S., Kolozsvary. Z., and Kis. A. Controlled hollow cathode effect: New possibilities for heating low-pressure furnaces. *Metal Science and Heat Treatment*, 46(7-8):310, 2004.

[3] McDonald M. S., Gallimore A. D., and Goebel D. M. Improved heater design for high-temperature hollow cathodes. *Review of Scientific Instruments*, 88(3):026104, 2017.

[4] Cherednichenko V.S. and Yudin B.I. *Vacuum plasma electric furnaces.* Krasnoyarsk: SFU, 2011.

[5] Asselin D. J. *Characterization of the Near-Plume Region of a Low-Current Hollow Cathode.* Worcester Polytechnic Institute, 2011.

[6] Gushenets V. I., Bugaev A. S., Oks E. M., Schanin P. M., and Goncharov A. A. Self-heated hollow cathode discharge system for charged particle sources and plasma generators. *Review of Scientific Instruments*, 81(6):02B305, 2010.

RECENT TRENDS IN DEVELOPMENT OF HIGH VOLTAGE CIRCUIT BREAKERS WITH SF_6 ALTERNATIVE GASES

M. Seeger[a,*], R. Smeets[b], J. Yan[c], H. Ito[d], M. Claessens[a], E. Dullni[e], L. Falkingham[f], C. M. Franck[g], F. Gentils[h], W. Hartmann[i], Y. Kieffel[j], S. Jia[k], G. Jones[c], J. Mantilla[a], S. Pawar[l], M. Rabie[g], P. Robin-Jouan[j], H. Schellekens[h], J. Spencer[c], T. Uchii[m], X. Li[k], S. Yanabu[k]

[a] *ABB Switzerland Ltd, Corporate Research Center, Segelhofstrasse 1K, CH-5405 Baden-Dättwil, Switzerland*
[b] *DNV GL, Utrechtseweg 310, 6800 ET, Arnhem, Netherlands*
[c] *University of Liverpool, Department of Electrical Engineering and Electronics, Liverpool, L69 3GJ, UK*
[d] *Mitsubishi Electric Corp., Tokyo Building 2-7-3, Marunouchi, Chiyoda-ku, Tokyo 100-8310, Japan*
[e] *ABB AG, Bahnstrasse 39-4, D-40832 Ratingen, Germany*
[f] *Vacuum Interrupters Limited, Sir Frank Whittle Business Centre, Great Central Way, Rugby CV21 3XH, UK*
[g] *ETH Zürich, Rämistrasse 101, 8092 Zürich, Switzerland*
[h] *Schneider Electric, 35, rue Joseph Monier, 92500 Rueil-Malmaison, France*
[i] *Siemens AG, Corp.Techn., Otto-Hahn-Ring 6, 81739 Munich, Germany*
[j] *GE Grid Solutions, 51 Espl. du Général de Gaulle, 92900 Puteaux, France*
[k] *Xi'an Jiaotong University, 28 Xianning West Road, Beilin, Xi'an, Shaanxi, China*
[l] *Crompton Greaves, Dr. Annie Besant Road, Worli, Mumbai – 40001, India*
[m] *Toshiba Corp., Power and Industrial Systems R&D Center, 8 Shinsugita-cho, Kanagawa 235-8523, Japan*
[*] martin.seeger@ch.abb.com

Abstract. The available knowledge of state-of-the-art of SF_6 alternative gases in switching applications was collected and evaluated in an initiative of the Current Zero Club [1] together with CIGRE. The present contribution summarizes the main results of this activity and will also include the latest trends. The main properties and switching performance of new gases are compared to SF_6. The most promising new gases are at the moment perfluoroketones and perfluoronitriles. Due to the high boiling point of these gases, in HV applications mixtures with CO_2 are used. For MV insulation perfluoroketones are mixed with air, but also other combinations might be possible. The dielectric and switching performance of the mixtures, with mixing ratios that allow sufficiently low operating temperatures, is reported to be only slightly below SF_6. Minor design changes or de-rating of switchgear are therefore necessary. Differences between the gas mixtures are mainly in the boiling point and the GWP.

Keywords: SF_6 alternative gases, CO_2, Circuit Breaker.

1. Introduction

SF_6 is widely used in electric power transmission and distribution systems, as for example in gas insulated switchgear (GIS), circuit breakers (CB) and medium voltage (MV) load break switches. It combines unique electrical insulation and arc interruption capability [2]. However, it is also a very strong greenhouse gas with a global warming potential (GWP) of about 23500 over a time horizon of 100 years, e.g. [3] and its use is regulated and restrictions are discussed. Therefore, search for alternative gases for use in power applications has been ongoing since about two decades ago e.g. [4, 5]. The state of the art of SF_6 alternative gases for switching applications was recently addressed in an initiative of the current zero club (CZC) [1] in collaboration with CIGRE. A survey was done collecting all the available recent literature on the topic. The result was presented and discussed at a joint workshop at the CIGRE session 2016. The present paper gives the main results of this survey. Since vacuum switching technology is a separate ongoing activity [6], it will be left out in the present review.

2. Alternative gases

The intensification of search for alternative gases started about two decades ago [4, 5] after the Kyoto protocol was agreed in 1997 and further increased in the last 10 years (e.g. [7–15]). Important requirements for alternative gases were identified as: Low GWP, zero ozone depletion (ODP) potential, low toxicity, non-flammability, high dielectric strength, high arc quenching and heat dissipation capability, stability and material compatibility and availability on market.

From various studies of gases of natural origin, CO_2

	CAS number	Boiling point °C	GWP	ODP	Flammability	Toxicity LC50 (4h) ppmv	Toxicity TWA[1)] ppmv	Dielectric strength at 0.1 MPa	Ref
SF_6	2551–62–4	-64 [2)]	23500	0	No	-	1000	1	[7, 16]
CO_2	124–38–9	-78.5 [2)]	1	0	No	>300000	5000	≈0.3	[4, 5, 17]
C5–PFK	756–12–7	26.5	<1	0	No	≈20000	225	≈2	[13, 16, 18]
C4–PFN	42532–60–5	-4.7	2100	0	No	12000 . . . 15000	65	≈2	[7, 16, 19–21]

1) The occupational exposure limit is given by a time-weighted-average (TWA), 8-hr

2) Sublimation point

Table 1. Properties of pure gases compared to SF_6.

turned out to be the most promising arc quenching gas, e.g. [8, 11], possibly enhanced in performance by some additives [12], like e.g. O_2 or CF_4. However, as was shown, the switching and dielectric performances of CO_2 are both below those of SF_6, e.g. [11, 17]. Other interesting gases were identified to be fluorinated gases like CF_3I, hydrofluoroolefins (HFO1234ze and HFO1234yf), perfluoroketones (e.g. C5F10O), perfluoronitriles (C_4F_7N), fluoroethers (HFE245cb2), fluorooxiranes and hydrochlorofluoroolefins (HCFO1233zd), e.g. [7, 13–16, 22]. Taking all the requirements into account, the most promising candidates at present appeared to be the C5 perfluoroketone (CF3C(O)CF(CF3)2 or C5–PFK) [18] and the iso-C4 perfluoronitrile ((CF3)2-CF-CN or C4–PFN) [19]. The dielectric performance of pure gases scales with the boiling point, i.e. gases with high dielectric strength usually also have a high boiling point, see e.g. [10]. For C5–PFK and C4–PFN, the boiling points at 0.1 MPa are 26.5 °C and −4.7 °C, respectively. Thus, for application in switchgear, where a sufficiently low boiling point is needed for low temperature requirements, an admixture of a buffer gas is needed. CO_2 is selected for this role in HV due to its good arc quenching capability, e.g. [9, 11, 12]. In MV application air is also reported as the buffer gas in combination with C5–PFK for insulation purposes [23–25]. The concentration of C5–PFK and C4–PFN, and by this the performance of the mixtures, will depend on the minimum operating temperature requirement of the switchgear. An additional alternative approach is proposed to use air for insulation and vacuum CB (VCB) for switching [26, 27].

3. Properties of pure gases and mixtures

The properties of the selected alternative gases with reference to SF_6 are shown in table 1. The GWP for the various gases are different: the C4–PFN has a much higher GWP than CO_2 or C5–PFK that are both around 1. All the gases of interest are not flammable, have no ODP and are non-toxic according to technical and safety data sheets available from the chemical manufacturer [18–20, 28, 29]. The dielectric strength of pure C4–PFN and C5–PFK is nearly twice that of SF_6. CO_2 has a dielectric withstand comparable to air [4, 17], i.e. significantly below that of SF_6. The properties of gases and mixtures when used in switchgear are shown in table 2. The concentration of admixtures of C4–PFN and C5–PFK with the buffer gas is given in the second column and is typically below 13 % (molar concentration). Note that for the use of C5–PFK in CO_2 additionally an oxygen admixture is reported, since the presence of oxygen reduces the generation of harmful by-products like CO and solid by-products such as soot [30]. Due to a reduced dielectric withstand of the mixtures compared to SF_6 (column 6) at the same pressure the minimum operating pressure needs to be increased to about 0.7 . . . 0.8 MPa for C5–PFK and C4–PFN when using CO_2 as the buffer gas for HV application, see column 3 in table 2. For Air/C5–PFK mixtures in MV application 0.13 MPa can be kept and the dielectric withstand of SF_6 is approached. The high dielectric withstand of mixtures with relatively low admixture ratios of C4–PFN or C5–PFK can be explained by a synergy effect [7, 30, 31], i.e. a non-linear increase of the dielectric strength with the admixture ratio, as it is known for SF_6/N_2 mixtures [32]. The GWP of mixtures with C5–PFK is negligible, at the cost of a higher minimum operating temperature. Low temperature applications of e.g. −25 °C for HV can be covered by pure CO_2 or CO_2+C4–PFN mixtures. This is at the cost of significantly reduced dielectric withstand in case of pure CO_2 or significantly higher GWP in case of C4–PFN mixtures. Due to strong dilution, the toxicity of the mixtures is well below that of the pure substances, see e.g. [7, 33].

4. Switching performance of alternative gases

Preliminary information on the switching performance of pure CO_2 and CO_2 mixtures is collected in table 3. The performance of SF_6 is given for comparison. With an enhanced operating pressure compared to SF_6 the cold dielectric strength, which is e.g. a measure of the performance in capacitive switching, can reach that of SF_6.

In the scanned literature, only qualitative statements on the switching performance of C4–PFN and C5–PFK mixtures could be found. For CO_2 a few quantitative comparisons exist. Very roughly, for pure CO_2 at an increased filling pressure of about 1 MPa, about 2/3 of the dielectric and thermal interruption

	C_{ad} 1)	p_{min} 2) MPa	T_{min} 3) °C	GWP	D.S. 4)	Toxicity LC50 ppmv	Ref
SF_6	-	0.43 ... 0.6	−41 ... −31	23500	0.86 ... 1	-	
CO_2	-	0.6 ... 1	≤ -48 6)	1	0.4 ... 0.7	> 3e5	[8, 11, 12]
CO_2/C5–PFK/O_2 (HV)	≈ 6/12	0.7	−5 ... +5	1	≈ 0.86	> 2e5	[13, 22, 27, 30]
CO_2/C4–PFN (HV)	≈ 4 ... 6	0.67 ... 0.88	−25 ... −10	327 ... 690	0.87 ... 0.96	> 1e5	[7, 15, 21, 33, 34]
Air/C5–PFK (MV)	≈ 7 ... 13	0.13	−25 ... −15	0.6	≈ 0.85 5)	1e5	[16, 24, 25]
N2/C4–PFN (MV)	≈ 20 ... 40	0.13	−25 ... −20	1300 ... 1800	0.9 ... 1.2	> 2.5e4	[15]

1) Concentration of admixture is in mole % referred to the gas mixture

2) Typical lock out pressure range

3) Minimum operating temperature for p_{min}

4) Dielectric strength compared to SF_6 at 0.55 MPa. For the scaling of SF_6 breakdown field E_d with pressure correction in the form of $E_d = 84p^{0.71}$ was used [32]

5) Compared to SF_6 at 0.13 MPa, measurements were for a mixture at −15 °C

6) Calculations with Refprop: `https://www.nist.gov/srd/refprop`

Table 2. Properties/performances of pure gases and mixtures in MV and HV switchgear applications.

performance of SF_6 might be expected. With the admixture of O_2 to CO_2 in the mixing ratio range up to 30 %, an increase of the thermal interruption performance [12] and also a slight increase in dielectric strength (e.g. [35]) is expected. With the admixture of C4–PFN and C5–PFK into CO_2 the dielectric performance can be close to SF_6. The short-line fault (SLF) switching performance for the mixtures of CO_2/O_2/C5–PFK is reported to be 20 % below that of SF_6 [30]. For an adapted CB with CO_2/C4–PFN a similar SLF performance to that of SF_6 is stated, e.g. [7]. There are, however, also direct comparisons of pure CO_2 with CO_2/C4–PFN and CO_2/C5–PFK mixtures using identical geometry and pressure, which show similar thermal interruption performance of CO_2 with and without admixtures [27]. IEC test duties L90 (SLF) and T100 (100 % terminal fault) with the new mixtures are passed with some design modifications [36] or certain de-rating [30], suggesting that the switching performance of the new mixtures is not significantly lower than that of SF_6. This has also been shown to be valid for the bus transfer switching duty of disconnector switches, e.g. [36, 37]. It is expected that dedicated design improvements can still increase the switching performance in the future.

An important point is the toxicity of the gas after arcing. C5–PFK and C4–PFN are complex molecules which start to decompose above approximately 650 °C in case of C4–PFN, e.g. [32]. After decomposition C5–PFK and C4–PFN molecules do not recombine to their original structure, but form smaller molecules. A decomposition rate of 0.5 Moles/MJ under high current switching is reported for CO_2/O_2/C5–PFK mixtures [30]. For partial discharges decomposition rates of more than one order of magnitude lower are observed for this mixture [38]. No quantitative information is given so far on the decomposition rates of C4–PFN. Note that this decomposition involving the new gases is not comparable with the decomposition of SF_6 because the latter only occurs due to chemical reactions with ablated contact and nozzle material. The decomposition involving the new gases is not seen as a problem over lifetime, but concentrations in the equipment need to be monitored or regularly checked, in a way similar to SF_6 [39]. Most toxic decomposition products for HV, i.e. mixtures with CO_2, are CO and HF, e.g. [30, 32]. The arced mixtures are regarded to have similar or lower toxicity as arced SF_6. It is recommended, therefore, to treat this in a way similar to arced SF_6. It must, however, be noted that the above statement is made only based on the limited knowledge available on the toxicity of the new gases. Formation of critical by-products under repetitive switching in a small volume is discussed in [16]. Considerable more experience seems to be needed on the post arcing toxicity of the potential SF_6 substitute gases. Additional reported issues are: material compatibility [22, 32] (e.g. effects on sealings and grease), gas tightness and gas handling procedures. Therefore, it should not be expected that existing HV equipment can be filled with the new gases without design or material changes. Internal arc tests were done with all mixtures and no critical issues are reported, e.g. [7, 22, 24]. Heat dissipation of the mixtures is slightly inferior to SF_6 [7, 22], i.e. moderate de-rating or design changes might be necessary with respect to the current carrying capability. At present, field experience is gained with CO_2 live-tank CB [40], being started some years ago. A CO_2 filled CB is also commercially available [41]. With the C5–PFK mixtures for HV and MV pilot installations have been in operation successfully since 2015 in Switzerland [22, 39] and Germany [42]. Pilot installations with the CO_2/C4–PFN mixture are planned in several European countries [7], such as a 145 kV indoor GIS in Switzerland, 245 kV outdoor Current Transformers in Germany and outdoor 420 GIL in UK and Scotland [7, 34, 36].

5. Conclusions and outlook

Published information on alternative gases for SF_6 in switching applications has been reviewed. In their present state, these investigations have just started and are by far not as extensive as for SF_6. The presently available manufacturer information on prop-

	Operating pressure MPa	Dielectric strength	SLF performance compared to SF_6 1)	Dielectric recovery speed	Ref
SF_6	0.6	1	1	1	
CO_2	0.8...1	0.5...0.7	0.5...0.83	≥ 0.5	[8, 11, 12, 27]
CO_2+C5–PFK/O_2	0.7...0.8	close to SF_6	0.8...0.87	close to SF_6	[22, 27]
CO_2/C4–PFN	0.67...0.82	close to SF_6	0.83...(1) 2)	close to SF_6	[7, 27, 32]

1) At same pressure build up

2) Same performance as SF_6 is stated but it is not clear if this was under same condition

Table 3. Switching performance of gases and mixtures compared to SF_6 at increased operating pressures in HV applications

erties shows that new gases (e.g. C5–PFK and C4–PFN) are available, which can compete with, but may not fully reach the performance of, SF_6 when used in mixture with CO_2 as the buffer gas. Main differences are in the insulation and interruption performances and boiling point with the latter defining the minimum operating temperature specified for the switchgear. The lowest operating temperatures (e.g. −50 °C) can be reached with CO_2. However, CO_2 seems to have an overall lower interruption performance, especially in dielectric interruption and withstand, than gas mixtures containing C4–PFN or C5–PFK. The advantage of CO_2/C5–PFK mixture compared with CO_2/C4–PFN mixture is the negligible GWP of about 1 compared to 427...600 of the latter. The advantage of CO_2/C4–PFN compared to CO_2/C5–PFK is the lower minimum operating temperature of about −25 °C compared to about −5 °C of the latter. Since research and development of these new SF_6 alternatives has just started, design improvements can be expected in the future. Exhaustive studies on decomposition products after current switching and their level of toxicity are still required, as it was performed in the past for SF_6, in different operating conditions. Probably from all different alternatives, a convergence to a single solution can be expected on the longer term. For sure, much more investigations and experimental validations have to be carried out.

References

[1] Current Zero Club (CZC). `http://www.currentzeroclub.org/`.

[2] P. Glaubitz, S. Stangherlin, JM Biasse, F. Meyer, M. Dallet, M. Pruefert, R. Kurte, T. Saida, K. Uehara, P. Prieur, H. Ito, E. Kynast, A. Janssen, R. Smeets, and Dufournet D. CIGRE position paper on the application of SF_6 in transmission and distribution networks. *Electra*, 34(274):34–39, 2014.

[3] T. Stocker et al. Climate change 2013: The physical science basis, contribution of working group I to the fifth assessment report of the intergovernmental panel on climate change, 2013. `https://www.ipcc.ch/rephort/ar5/wg1/`.

[4] L. Niemeyer. A systematic search for insulation gases and their environmental evaluation. In *Gaseous Dielectrics VIII*, pages 459–464. Springer Science & Business Media, 1998.

[5] L. Christophorou, J. K. Olthoff, and D. S. Green. Gases for electrical insulation and arc interruption: Possible present and future alternatives to pure SF_6. US Department of Commerce, Technology Administration, National Institute of Standards and Technology, 1997.

[6] R. Smeets, L. Falkingham, M. Backman, T. Betz, I. Cotton, R. Doche, L. Drews, M. Eiselt, M. Glinkowski, D. Johnson, S. Hambarde, V. Hinrichsen, N. Inoue, H. Ito, K. Kagawa, S. Kulkarni, Y.-G. Kim, R.K. Smith, Z. Liu, W. Pepper, P. Meyer, S. de Morais, F. Richter, J.-S. Ryu, N. Kubota, H. Saito, S. Shoval, L. Violleau, S. Yanabu, and Y. You. Cigre 589: The impact of the application of vacuum switchgear at transmission voltages. WG A3.27, 2014.

[7] Y. Kieffel, T. Irwin, P. Ponchon, and J. Owens. Green gas to replace SF_6 in electrical grids. *IEEE Power and Energy Magazine*, 14(2):32–39, 2016.

[8] T. Uchii, Y. Hoshina, T. Mori, H. Kawano, T. Nakamoto, and M. Toyoda. Fundamental research on SF_6–free gas insulated switchgear adopting CO_2 gas and its mixtures. In *Proceedings of International Symposium on EcoTopia Science ISETSO7*, 2007.

[9] T. Ueno, K. Mochiduki, H. Mizoguchi, S. Yanabu, S. Okabe, and S. Yuasa. Evaluation of SLF interruption capability on various gases. In *XV International Conference on Gas Discharges and their Applications*, Toulouse, 2004.

[10] M. Rabie and C. M. Franck. Predicting the electric strength of proposed SF_6 replacement gases by means of density functional theory. In *18th International Symposium on High Voltage Engineering*, Seoul, Korea, 2013.

[11] P. C. Stoller, M. Seeger, A. A. Iordanidis, and G. V. Naidis. CO_2 as an arc interruption medium in gas circuit breakers. *IEEE Trans. Plasma Science*, 41(8):2359–2369, 2013.

[12] T. Uchii, A. Majima, T. Koshizuka, and H. Kawano. Thermal interruption capabilities of CO_2 gas and CO_2–based gas mixtures. In *Proceedings of the XVIII International Conference on Gas Discharges and Their Applications*, Greifswald, Germany, 2010.

[13] J. D. Mantilla, N. Gariboldi, S. Grob, and M. Claessens. Investigation of the insulation performance of a new gas mixture with extremely low GWP. In *Electrical Insulation Conference (EIC)*, Philadelphia, USA, 2014.

[14] M. Taki, D. Maekawa, H. Odaka, Mizoguchi H., and S. Yanabu. Interruption capability of CF_3I gas as a substitution candidate for SF_6 gas. *IEEE Transactions*

on Dielectrics and Electrical Insulation, 14(2):341–346, 2007.

[15] Y. Kieffel, A. Girodet, F. Biquez, Ph. Ponchon, J. Owens, M. Costello, M. Bulinski, R. van San, and K. Werner. SF_6 alternative development for high voltage switchgears. In *Cigre*, 2014. Cigre Paper D1-305, Paris.

[16] C. Preve, R. Maladen, D. Piccoz, and J.-M. Biasse. Validation method for SF_6 alternative gas. In *Cigre*, 2016.

[17] K. Juhre and E. Kynast. High pressure N_2, N_2/CO_2 and CO_2 gas insulation in comparison to SF_6 in GIS applications. In *14th International Symposium on High Voltage Engineering (ISH)*, pages 1–6, 2005. Paper C-01.

[18] 3M™ Novec™ 5110 dielectric fluid. Technical Data Sheet, 2015.

[19] 3M™ Novec™ 4710 dielectric fluid. Technical Data Sheet, 2015.

[20] J.G. Owens. Greenhouse gas emission reductions through use of a sustainable alternative to SF_6. In *IEEE Electrical Insulation Conference (EIC)*, 2016.

[21] H. E. Nechmi, A. Beroual, A. Girodet, and P. Vinson. Fluoronitriles/CO_2 gas mixture as promising substitute to SF_6 for insulation in high voltage applications. *IEEE Transactions on Dielectrics and Electrical Insulation*, 23(5), 2016.

[22] D. Tehlar, T. Diggelmann, P. Müller, R. Buehler, N. Ranjan, and C. Doiron. Ketone based alternative insulation medium in a 170 kV pilot installation. Cigre Colloquium, Nagoya, Japan, 2015.

[23] M. Saxegaard, M. Kristoffersen, P. Stoller, M. Seeger, M. Hyrenbach, and H. Landsverk. Dielectric properties of gases suitable for secondary MV switchgear. CIRED, 2015. Paper 0926.

[24] M. Hyrenbach, T. Hintzen, P. Müller, and J. Owens. Alternative gas insulation in medium voltage switchgear. CIRED, 2015.

[25] M. Mann, F.-J. Koerber, R. Uhl, and P. Mueller. Ein Beitrag zur Evaluierung von alternativen Isoliergasen in Gasisolierten Hochspannungs–Schaltanlagen. In *Hochspannungs–Schaltanlagen: Anwendungen, Betrieb und Erfahrungen*, Darmstadt, 2015.

[26] N. Presser. Advanced insulation and switching concepts for next generation high voltage substations. In *Cigre*, 2016. B3–108.

[27] A. Kosse. Development of CB with SF_6 alternatives. In *Cigre*, 2016. Presentation at Workshop of Current Zero Club with CIGRE SC A3 on Switching in Alternative Gases.

[28] 3M™ Novec™ 5110 dielectric fluid. Safety Data Sheet, 2014. 31-6591-7 (UK), rev. date 24/09/2014.

[29] 3M™ Novec™ 4710 dielectric fluid. Safety Data Sheet, 2016. 33-6330-6 (UK), rev. date 19/08/2016.

[30] J. D. Mantilla, M. Claessens, and M. Kriegel. Environmentally friendly perfluoroketones-based mixture as switching medium in high voltage circuit breakers. In *Cigre*, 2016. A3-348.

[31] P. Simka and N. Ranjan. Dielectric strength of C5 perfluoroketone. In *Proceedings of 19th International Symposium on High Voltage Engineering*, Pilsen, Czech Republic, 2015.

[32] H. Hama, S. Okabe, M. Muhr, T. Britten, K. Juhre, C. Neumann, U. Prucker, and A. Sabot. Insulation co-ordination related to internal insulation of gas insulated systems with SF_6 and N_2/SF_6 gas mixtures under AC conditions. In *Cigre*, 2008. WG C4.302.

[33] K. Pohlink, Y. Kieffel, J. Owens, Meyer F., F. Biquez, Ph. Ponchon, and R. van San. Characteristics of fluoronitrile/CO_2 mixture–an aternative to SF_6. In *Cigre*, 2016. D1-204.

[34] E. Laruelle, A. Ficheux, Y. Kieffel, and M. Waldron. Reduction of greenhouse gases in GIS pilot project in UK. In *Cigre*, 2016. Paper C3-304.

[35] H. Zhao, X. Li, S. Jia, and A. B. Murphy. Prediction of the critical reduced electric field strength for carbon dioxide and its mixtures with 50 % O_2 and 50 % H_2 from boltzmann analysis for gas temperatures up to 3500 K at atmospheric pressure. *Journal of Physics D: Applied Physics*, 47(32):325203, 2014.

[36] D. Gautschi. Application of a fluoronitrile gas in GIS and GIL as an environmental friendly alternative to SF_6. In *Cigre*, 2016. B3-106.

[37] Y. Kieffel and F. Biquez. SF_6 alternative development for high voltage switchgears. In *IEEE Electrical Insulation Conference (EIC)*, Seattle, Washington, USA, 2015.

[38] T. Hammer. Decomposition of low GWP gaseous dielectrics caused by partial discharges. In *21st Int Conference on Gas Discharges and Their Applications*, Nagoya, Japan, 2016.

[39] P. Müller. Praxiserfahrungen der Ersten 170-kV-GIS mit Alternativem Isolationsmedium Basierend auf Ketonen. In *GIS Anwenderforum*, Darmstadt, Germany, 2016.

[40] P. Söderstöm, J. Lidholm, and U. Åkesson. Suitability evaluation of improved high voltage circuit breaker design with drastically reduced environmental impact. In *Cigre*, 2012.

[41] ABB. Live tank circuit breaker – LTA 72.5 kV. `http://new.abb.com/high-voltage/AIS/selector/lta`.

[42] M. Engel. Einsatz von ökoeffizientem Isoliergas in einer Mittelspannungsschaltanlage. In *GIS Anwenderforum*, Darmstadt, Germany, 2016.

doi:10.14311/ppt.2017.1.13
Plasma Physics and Technology **4**(1):13–16, 2017

THE INFLUENCE OF MICROWAVE CF_4 PLASMA ACTIVATION ON THE CHARACTERISTICS OF REACTIVE ION ETCHING OF MONO-SI

S. Bordusau*, S. Madveika, M. Lushakova, N. Kovalchuk

Belarusian State University of Informatics and Radioelectronics, P.Brovki 6, 220013 Minsk, Belarus
* bordusov@bsuir.by

Abstract. The experiments have shown that microwave preliminary ionization of the plasma-forming gas increases the rate of the reactive ion etching of monocrystalline silicon (mono-Si) by four and more times in comparison with the etching process without it. It is established that the mode of plasma activation (the power of microwave discharge) and the value of plasma-forming gas pressure significantly affect the characteristics of mono-Si surface micro-roughness obtained in the result of etching.

Keywords: microwave discharge, activation, etching, roughness, surface.

1. Introduction

One of the main characteristics of material's plasma etching is the processing rate [1]. This parameter becomes of utmost importance during plasma etching of microelectromechanical systems [2], deep etching of integrated circuits elements [3], making sensors using planar technologies methods [4], etc. Note, the necessity of plasma etching process acceleration is connected not only with the requirement to improve the efficiency of technological equipment but with the need to reduce the duration of plasma component influence (primarily of charged particles) on the surface layers of the material and the elements of integrated circuits structures [5]. Direct increase of power applied to the electric discharge plasma is reasonable within certain values. The exceeding of this level can lead to such negative effects as discharge instability, significant heating of technological discharge systems construction elements, uncontrolled heating of treated materials' surface, etc.

Because of this, the investigation and development of such method of non-thermal intensification of the materials' etching process as preliminary etching gas activation in the off-site plasma source becomes interesting. Microwave plasmatron has been chosen as such source. It is determined by the fact that microwave discharge plasma has a number of specific peculiarities that make it rather attractive for the purpose of materials plasma treatment [6],[7].

2. Experimental part

Figure 1 shows schematically the construction of a vacuum gas discharge technological system used for investigation of the reactive ion etching of monocrystalline silicon (mono-Si) wafers.

The preliminary plasma activation of etching gas (CF_4) was performed in a microwave plasmatron 1. It is developed on the base of the rectangular waveguide 6 bent into a ring with holes on the internal surface for microwave energy 7 feed into the internal region 5.

Quartz discharge chamber 2 was placed along the waveguide's axis in which microwave discharge 4 had been excited in vacuum. From one side, the quartz chamber is closed with vacuum-tight cover 3 in the plane of the microwave applicator 5 end face, and from the other side it enters technological chamber 8.

Mono-Si wafer 11 under treatment was placed in vacuum chamber 8 out of the microwave plasma-formation region on the wafer-holder electrode 12. Between grounded electrode 9 and potential electrode 12 of the planar discharge E-type system a glowing low-frequency discharge 10 was excited in which the reactive ion etching process was performed.

The experiments were carried out with the wafers of 76 mm in diameter. The level of microwave power (at 2.45 GHz frequency) was within the range of $P_{MW} = 250–650$ W. CF_4 was used as working gas.

The value of mono-Si etching rate was defined by measuring the depth of the etched hole, dividing it by the time of the wafers' treatment. The local etching of the holes on the surface of wafers was provided with the masking coating which was removed after the treatment. The measuring of the etched holes depth was performed according to the interferometric method by the MII-4 interferometer.

The analysis of mono-Si morphology and surface microrelief after the reactive ion etching was performed by the atomic-force microscopy method. The surface roughness parameter Rz (ten-point mean roughness) was measured using the standard method.

The experimental data were obtained at fixed low frequency generator output power.

The pressure value of CF_4 and output microwave magnetron power served as operation parameters of the treatment process.

3. The results and discussion

It has been experimentally established that microwave plasma activation of the gas used for reactive ion etching allows to accelerate the etching process by

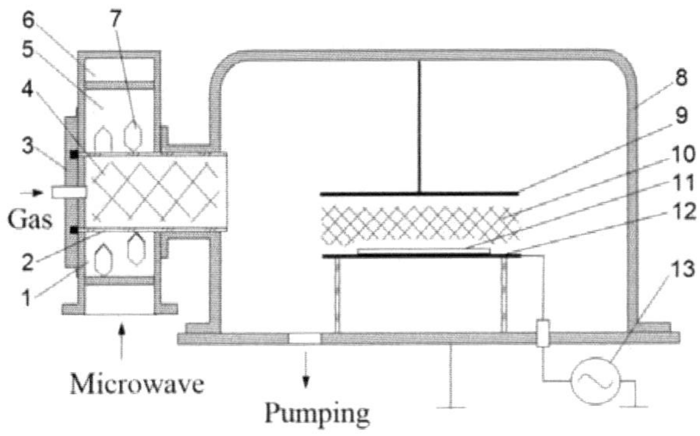

Figure 1. Diagram of the experimental plasma discharge system:
1 — microwave plasmatron; 2 — cylindrical quartz chamber; 3 — vacuum-tight flange cover; 4 — microwave discharge zone; 5 — internal region; 6 — rectangular waveguide bent into a ring; 7 — connection holes; 8 — vacuum chamber; 9 — grounded electrode; 10 — glow discharge region; 11 — wafer; 12 — wafer-holder electrode; 13 — low frequency generator

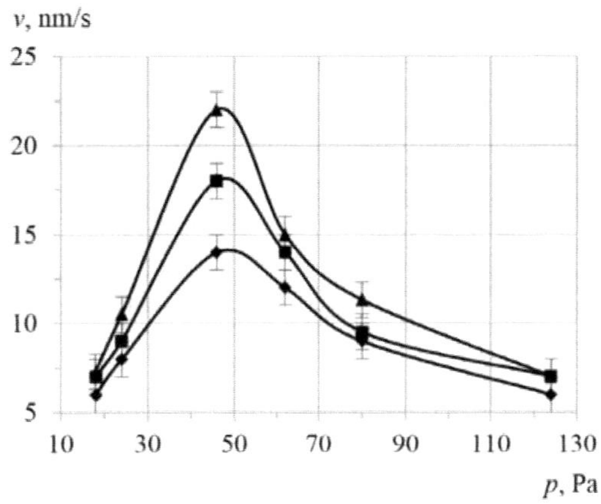

Figure 2. Dependence of the rate values of reactive ion etching of mono-Si in CF_4 *on the values of pressure in the vacuum chamber at various microwave power of etching gas plasma activation:*
♦ — $P_{MW} = 250\,W$; ■ — $P_{MW} = 500\,W$;
▲ — $P_{MW} = 650\,W$

four and more times in comparison with the etching process in E-type discharge without activation.

Fig. 2 shows the dependence of reactive ion etching rate of mono-Si in CF_4 on the value of pressure in the vacuum chamber at various microwave power of working gas plasma activation. Within the range of 10–133 Pa pressure these dependencies have extreme character. Maximum rate of mono-Si wafers etching in CF_4 is reached at the pressure of 40–50 Pa in the vacuum chamber.

Fig. 3 shows the dependence of hole bottom surface roughness values Rz after reactive ion etching of mono-Si in CF_4 on the value of pressure in the vacuum chamber. It has an extreme form similar to the data of Fig.2.

Fig. 4 shows the dependence of the rate values of reactive ion etching of mono-Si in CF_4 on the value of microwave plasma activation power of the etching gas. The presented data show that the maximum effect of microwave plasma activation is observed at the gas pressure of $p = 46$ Pa, at which the rate of mono-Si reactive ion etching is maximum.

For other pressure values in the chamber (23 and 80 Pa) the influence of gas plasma activation microwave power on the etching rate is lower.

Fig. 5 shows the results of investigation of the hole bottom surface morphology after the process of reactive ion etching of mono-Si in CF_4 in the region of microwave discharge plasma afterglow at various pressure values. The roughness value of the treated samples in the range of the investigated pressure values (20–80 Pa) varies from $Rz = 7.75$ nm to $Rz = 31.5$ nm and correlates with the value of etching rate.

The experiments have shown that in the area of lower pressure values the surface relief uniformity is higher than at 40–50 Pa (Fig. 5) but the rate values

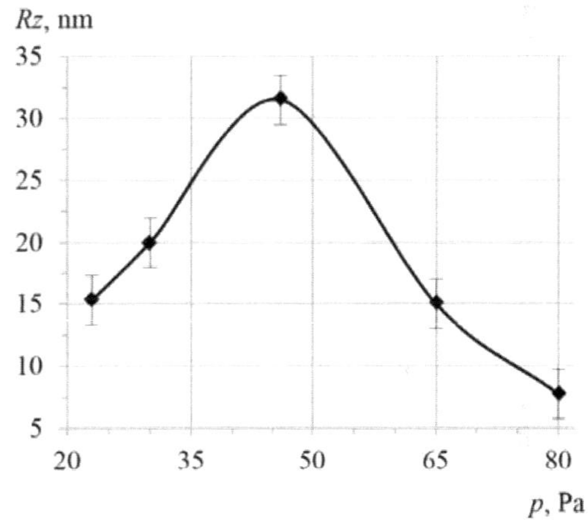

Figure 3. Dependence of hole bottom surface roughness values Rz after reactive ion etching of mono-Si in CF_4 *on the values of pressure in the vacuum chamber at* $P_{MW} = 250\,W$ *microwave discharge power*

of mono-Si etching are lower. In the area of higher pressure values, the wafer surface morphology has a more ordered structure in comparison with the area of lower pressure values (Fig. 5c). The mono-Si wafer surface microrelief at $p = 46$ Pa pressure in the vacuum chamber has a more disordered character (Fig. 5b).

Analysing the character of mono-Si surface microrelief after reactive ion etching in CF_4 at various pressure values it is possible to assume that 40–50 Pa pressure provides better plasma-formation conditions and the silicon wafer surface is bombarded by ions most intensively.

The analysis of experimental data also shows that in case of CF_4 microwave plasma activation presence the etched surface of mono-Si holes bottom has a more

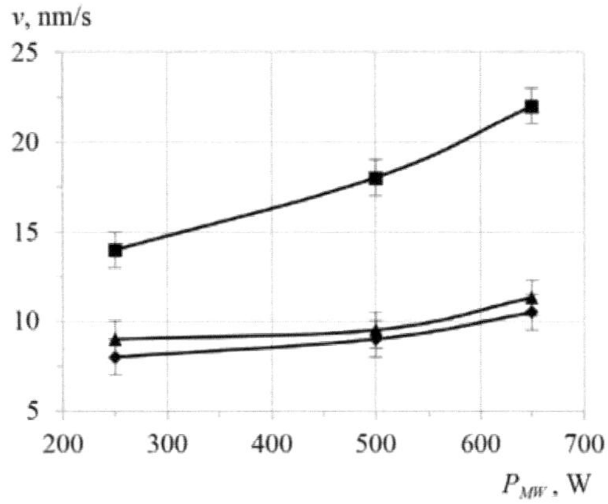

Figure 4. Dependence of the rate values of mono-Si reactive ion etching in CF_4 *on the values of microwave gas plasma activation power:*
♦ — $p = 23\,Pa$; ■ — $p = 46\,Pa$; ▲ — $p = 80\,Pa$

developed periodic structure, the roughness is more even and regular in form than in case of reactive ion etching without it.

4. Conclusions

The results of the performed experiments show the significant effect of preliminary microwave plasma activation of CF_4 on the rate of reactive ion etching process of mono-Si as well as on the character of treated surface microrelief. Choosing the mode of pre-ionisation (the value of microwave discharge power) and the value of plasma-forming gas pressure, it is possible to control the character of surface profile during the process of the reactive ion etching of mono-Si. Note, the character of the surface profile roughness after the reactive ion etching with microwave pre-ionisation may be more ordered and uniform than without it.

Acknowledgements

This research has been supported by the Belarusian State Program of Scientific Research "Science of physical materials, new materials and technologies", assignment 3.5.02.

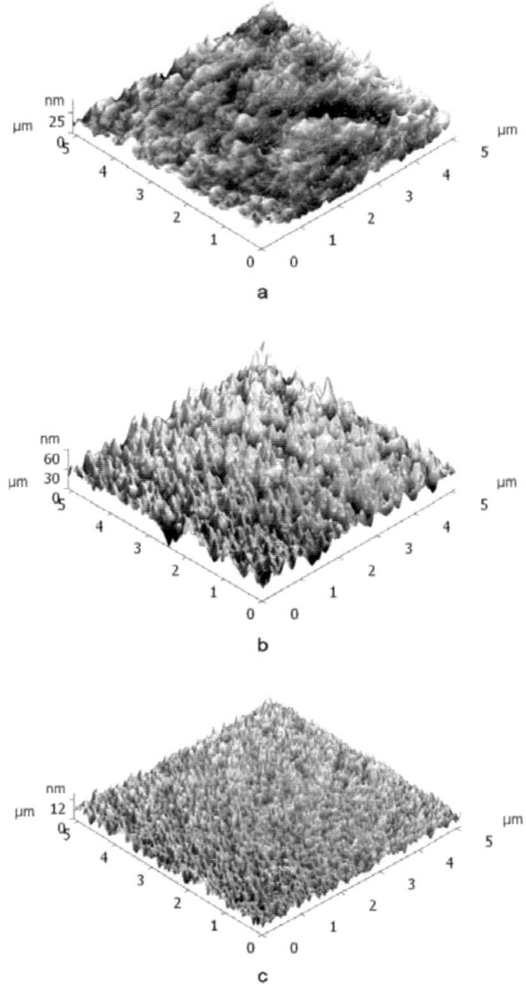

Figure 5. Hole bottom surface microrelief after reactive ion etching of mono-Si in CF_4 *in the conditions of gas microwave plasma activation at various values of pressure in the vacuum chamber:*
a — $p = 23\,Pa$; $Rz = 15.32\,nm$; *max height 32.4 nm;*
b — $p = 46\,Pa$; $Rz = 31.48\,nm$; *max height 62.8 nm;*
c — $p = 80\,Pa$; $Rz = 7.75\,nm$; *max height 15.8 nm;*

References

[1] Nojiri Kazuo. *Dry Etching Technology for Semiconductors.* Springer International Publishing, 2015.

[2] Bliznetsov V., Lin Hua Mao, Zhang Yue Jia, and Johnson D. Deep SiO_2 etching with Al and AlN masks for MEMS devices. *Journal of Micromechanics and Microengineering*, 25(8):087002, 2015. doi:10.1088/0960-1317/25/8/087002.

[3] Porter D. and Berfield T. Die separation and rupture strength for deep reactive ion etched silicon wafers. *Journal of Micromechanics and Microengineering*, 23(8):085020, 2013. doi:10.1088/0960-1317/23/8/085020.

[4] Narducci M., Yu-Chia L., Fang W., and Tsai J. CMOS MEMS capacitive absolute pressure sensor. *Journal of Micromechanics and Microengineering*, 23(5):055007, 2013. doi:10.1088/0960-1317/23/5/055007.

[5] Fonash S. Plasma processing damage in etching and deposition. *IBM J.RES. DEVELOP.*, 43(12):103–106, 1999.

[6] Bordusov S. *Microwave plasma technologies in the production of electronic devices.* Minsk: Bestprint, 2002.

[7] Lebedev Yu. Microwave discharges at low pressures and peculiarities of the processes in strongly non-uniform plasma. *Plasma Sources Science and Technology*, 24(5):053001, 2015. doi:10.1088/0963-0252/24/5/053001.

doi:10.14311/ppt.2017.1.16
Plasma Physics and Technology **4**(1):16–19, 2017

Investigation of Vacuum Arc Anode Temperatures of Cu–Cr and Pure Cu Contacts

St. Franke[a,*], R. Methling[a], S. Gortschakow[a], M. Abplanalp[b], R.-P. Sütterlin[b], T. Delachaux[b], K. O. Menzel[b]

[a] *INP Greifswald, Felix-Hausdorff-Str. 2, 17489 Greifswald, Germany*
[b] *ABB Corporate Research, ABB Switzerland, Segelhofstrasse 1K, 5405 Baden-Dättwil, Switzerland*
[*] steffen.franke@inp-greifswald.de

Abstract. The present contribution reports on investigations of electrode temperatures for pure Cu electrodes and Cu–Cr electrodes of different diameters exposed to vacuum arcs with sinusoidal currents of 5-15 kA and an axial magnetic field up to 180 mT. It is found that surface temperatures of pure Cu electrodes are significantly lower than for Cu–Cr electrodes of the same diameter. This must be explained by different thermal properties of both materials. Reducing the diameter of Cu–Cr electrodes it is found that surface temperatures increase, but moreover it is shown that the enthalpy stored in the electrode bulk material may effect electrode temperatures on timescales much longer than the current pulse width, particularly if there is no effective heat dissipation after current zero.

Keywords: vacuum arc, anode temperature, thermography.

1. Introduction

Surface temperatures of vacuum arc anodes are of dedicated interest in the research of vacuum arc interrupters, because they affect the switching capability of circuit breakers due to the temperature dependent evaporation of electrode material. Recently investigations on experimental determination of anode temperatures have been reported by means of a two-temperature pyrometer [1]. However, measurements where restricted to situations after current zero. In Dullni & Schade 1993 [2] extensive investigations have been presented including calculation of the melt depth for different electrodes. A comparable approach was followed by Niwa et al. 2000 [3], which also considered the melt depth of electrodes. This contribution presents vacuum arc anode surface temperatures before and after current zero and discusses selected effects observed for different electrode materials and geometries.

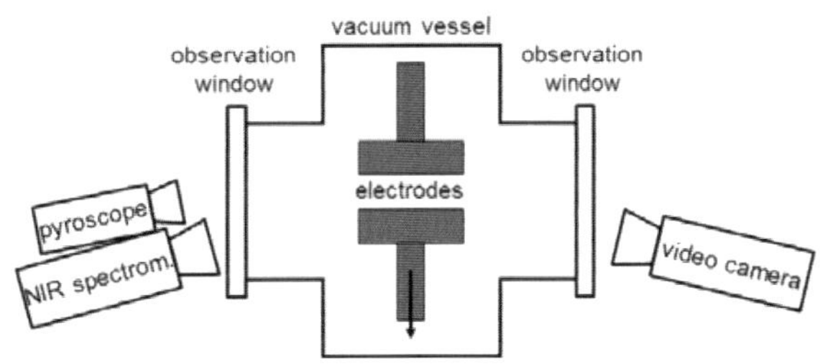

Figure 1. Experimental setup for investigation of vacuum arc anode temperatures (upper electrode). Three different optical detectors applied through observation windows: high-speed video camera, pyrometer and NIR spectrometer.

2. Experimental methods

Three different types of electrodes have been investigated, namely pure Cu butt electrodes with 38 mm diameter, Cu–Cr50 butt electrodes with 38 mm diameter, and Cu–Cr50 pin electrodes with 10 mm diameter enclosed in Macor caps to restrict the arc attachment to the front surface. The upper electrode was fixed and the lower electrodes was drawn downwards by a pneumatic actuator up to an electrode distance of around 10 mm. The distance from the electrode center to the observation windows was around 190 mm, and the observation window diameter was about 150 mm. The electrodes where exposed to sinusoidal currents of 5 – 15 kA. External axial magnetic fields up to 180 mT have been applied to affect the constriction of the arc. Hence, effects of different magnetic fields on anode temperatures could be investigated as well as effects of electrode material and electrode diameter.

Details of the experimental methods are described elsewhere [4]. Nevertheless, the experimental procedure is summarized here. Three different detectors have been used to determine anode surface temperatures, namely high-speed video camera, pyrometer and near-infrared (NIR) spectrometer as represented in Fig. 1.

To reduce effects by metallization of observation windows, the observation window used for high-speed video imaging usually was displaced by a vacuum extender to a total distance from the electrode of about 500 mm. The angle of inspection was therefore about 9° from the horizontal. For pyrometer and IR spectrometer an angle of inspection of about 15° was realized due to the shorter distance to the electrodes. In case of Cu–Cr pin electrodes additional protective glasses have been introduced between the electrodes and the corresponding observation window, which could be exchanged without opening the vacuum vessel in case of significant window blackening by evaporated electrode material.

Near infrared spectroscopy was carried out by means of a fiber optics spectrometer in the wavelength range from 1000 to 1600 nm with a temporal resolution of 1 – 2 ms (Hamamatsu TG series C11482GA). The fiber optic spectrometer was focused by an imaging optics to the center of the anode surface. Due to the small angle of inspection the effective area was elliptic with an approximate dimension of (0.4×1.6) mm. The spectrometer has been calibrated by means of a tungsten strip lamp (OSRAM Wi 17/G) in units of spectral radiance. Electrode temperatures after current zero were obtained by fitting Planck functions to spectral radiances of the radiating surfaces. For proper signals with high signal-to-noise ratio surface emissivities could be obtained as well.

The pyrometer (Kleiber 274 C LWL - High Speed Pyrometer) was focused to the anode surface by a long distance microscope (Questar LaVision QM 100). Pyrometric measurements were performed in the spectral range around $2\,\mu$m with high temporal resolution ($< 100\,\mu$s) but also without spatial resolution. Calibration was performed making use of the tungsten strip lamp at different currents and the calibrated NIR spectrometer. Application of the pyrometer to other materials than tungsten (like Cu–Cr) requires consideration of the emissivity of the material under investigation, and application to the active phase of the arc requires consideration of plasma background radiation. For these reasons absolute calibration was lost. However, relative calibration was maintained assuming temperature independence of emissivities. Finally, it was found that the pyrometer signal was dominated by thermal radiation after current maximum and that a characteristic feature in the temporal signal could be identified with a phase transition of the Cu–Cr material. Pyrometer signals of 650 mV therefore corresponded to the temperature of the eutectic point at 2020 K[5]. Therefore absolute temperatures could be given with the pyrometer.

Two-dimensional spatio-temporal thermography could be realized by a high speed-video camera (Integrated design Tools Motion Pro Y4, 10 000 frames per second) applying a Metal Interference Filter (MIF) at 890 nm with a full width at half maximum of 10 nm. It was proven that there is no spectral line emission in this spectral range by a compact spectrometer (Avantes VIS-spectrometer, AvaSpec-ULS2048). Relative calibration was performed by calculating the Planck radiance transmitted by the MIF for different temperatures. Absolute calibration was matched to a pyrometer signal after current maximum (around 6 ms). After removing spatial distortion 2D temperature maps could be obtained.

Before presentation of results and discussion a general remark has to be given. Occasionally it is referred

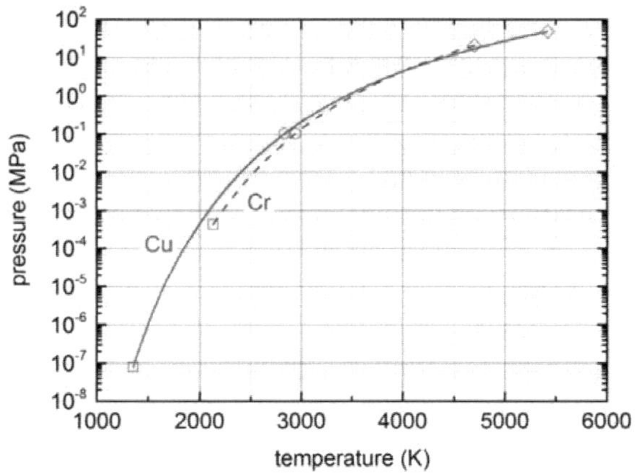

Figure 2. Vapor pressures of copper and chromium. Open squares mark the triple point, open circles the boiling point at standard pressure, and open diamonds the critical temperature.

to boiling temperatures of copper and chromium in the context of vacuum arc anodes. However, boiling temperatures depend on the ambient pressure. According to Clausius-Clapeyron the vapor pressure curve can be calculated, if the boiling temperature T_1 at $p_1 = 0.101325$ MPa standard pressure is given.

$$\ln\left(\frac{p_2}{p_1}\right) = \frac{\Delta H_{m,v}}{R} \times \left(\frac{1}{T_2} - \frac{1}{T_1}\right) \qquad (1)$$

In this equation R is the universal gas constant and $\Delta H_{m,v}$ is the molar enthalpy of evaporation, which is assumed to be a constant here. In Fig. 2 the vapor pressure curves for pure Cu and Cr are given[6] obtained from eqn. (1) making use of the boiling point at atmospheric pressure (2835 K for pure Cu, 2944 K for pure Cr). The vapor pressure curve is limited by the critical point at higher temperatures and the triple point at lower temperatures. The melting temperature is assumed to be independent on the pressure (1358 K for pure Cu, 2130 K for pure Cr). Boiling occurs in an equilibrium situation, if the vapor pressure of the liquid is equal to the ambient pressure. In the context of vacuum arcs it is therefore not suggested to use the term boiling, as neither the equilibrium situation nor the ambient pressure are specifically given. For instance in front of the anode pressures around 0.003 MPa are estimated in [7], and simulations reveal pressures up to 0.04 MPa [8]. However, within the scope of this contribution the vapor pressure curve in Fig. 2 demonstrates that up to atmospheric pressures the vapor pressure for Cu is significantly higher than for Cr.

3. Results and discussion

From NIR spectra temperatures are deduced by fitting Planck functions to the curves as exemplified in Fig. 3 for Cu–Cr pin electrodes.

$$L_\lambda(T) = \theta\,\varepsilon\,B_\lambda(T) \qquad (2)$$

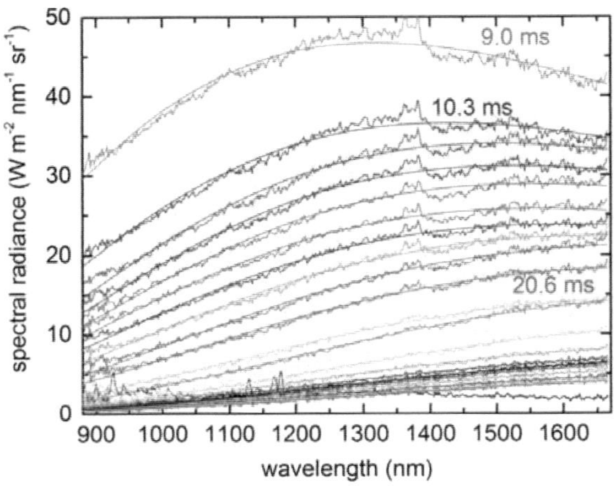

Figure 3. Planck fits to spectral radiances of Cu–Cr pin electrodes after current zero.

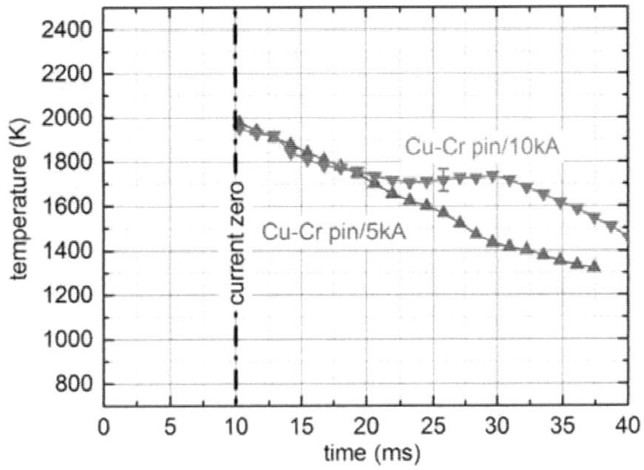

Figure 4. Anode surface temperatures for Cu–Cr pin electrodes after current zero subjected to sinusoidal currents with 5 kA and 10 kA peak currents, respectively.

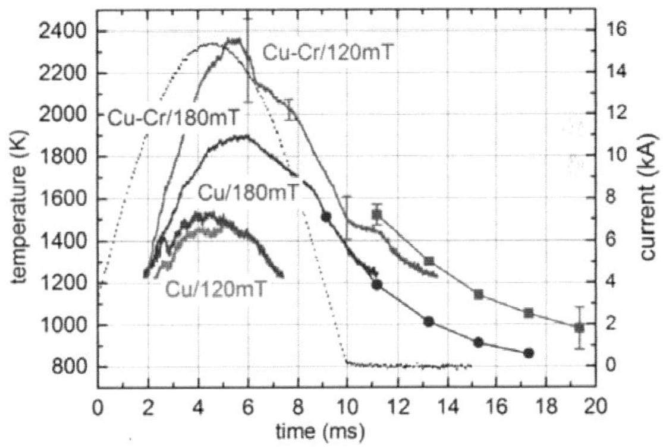

Figure 5. Anode surface temperatures for Cu–Cr and pure Cu butt electrodes. Peak current is 15 kA in all cases. The axial magnetic field varies between 120 and 180 mT. Black dotted line: current. Solid lines: temperatures obtained from pyrometer. Solid lines with symbols: temperatures obtained from NIR spectrometer.

Here $B_\lambda(T)$ denotes temperature and wavelength dependent Planck function, ε emissivity of anode surface, and θ transmission of observation windows. For a vastly long time (up to 30 ms after current zero) Planck fits could be performed. Due to strong evaporation of electrode material during the active phase of the arc there is a significant reduction of observation windows transmission which is considered as time independent after current zero. The spectral dependence of transmission has been checked by independent measurements. Thermal radiance therefore is affected by both, surface emissivity and transmission, as given in the equation above. Due to the limited optical access no pyrometer data could be recorded.

Surface temperatures are given in Fig. 4. The statistical uncertainty is assumed to be ± 50 K. Around current zero surface temperatures are close to the eutectic temperature of Cu–Cr at 2020 K[5]. For higher currents applied to the electrodes 10 ms after current zero there is a delay in the temporal temperature decay. The most reasonable explanation for this feature is to assume that the melt depth is significantly deeper for 10 kA peak currents (see also [2]). Therefore the stored enthalpy in the electrode bulk material is higher and after an effective cooling period, immediately after current zero, excess heat is fed towards the electrode surface slowing down the temperature decay. This effect is potentially enhanced by the Macor cap which decreases heat dissipation of the electrodes towards the lateral surface of the cylindric pin electrode.

A similar effect has been obtained for Cu–Cr butt electrodes as well. In Fig. 5 surface temperatures in the active phase and after current zero are given. The black dotted line represents an example of the sine-like current reaching current zero after 10 ms. Solid lines show surface temperatures obtained from pyrometer and lines with symbols denote temperatures obtained from NIR spectrometer after current zero. For Cu–Cr/120 mT one can see a plateau-like phase directly after current zero at temperatures between 1400 and 1600 K. This kind of plateau has been found for several conditions, preferably at higher currents and lower magnetic fields, always after current zero but at different temperatures. The most reasonable explanation is again that due to the constricted arc attachment and local heating the melt depth was increased in these cases and excess heat in the bulk material slowed down the temperature decrease of the anode surface.

As expected higher magnetic fields avoid constriction of the arc attachment and hence lead to considerably lower surface temperatures. The agreement with temperatures from NIR spectroscopy are within experimental uncertainties, which are given in the figure. Lowest uncertainty is given around the eutectic point with a temperature of 2020 K marking a phase transition which is identified repeatedly with a pyrometer signal of 650 mV and by a clear structure in almost all temporal courses of the surface temperature.

Surface temperature for pure Cu butt electrodes are much lower. There is almost no difference between the surface temperatures for the two different magnetic fields. This corresponds to high-speed videos which show a diffuse arc attachment for both reported cases. Reasons for the lower Cu temperatures are the higher thermal conductivity of Cu compared to Cu–Cr, lower melting temperature and hence stronger evaporation in liquid phase, which leads to additional cooling. No NIR spectra could be recorded for pure Cu electrodes after current zero due to low thermal emission.

One should keep in mind that only temperatures in the center of the electrodes are given up to now, due to the diagnostics discussed above. High-speed thermography is much more sophisticated, but 2D surface temperature distributions could be estimated from high-speed videos as given in Fig. 6. This figure demonstrates that surface temperatures of Cu electrodes are not only significantly lower than for Cu–Cr but furthermore Cu electrodes are much more homogeneously heated. In contrast the lower part of Fig. 6 shows an inhomogeneous distribution of surface temperatures for a Cu–Cr electrode. This evidently demonstrates the necessity of 2D surface temperatures in case of constricted arc attachments to observe inhomogeneous temperature distributions.

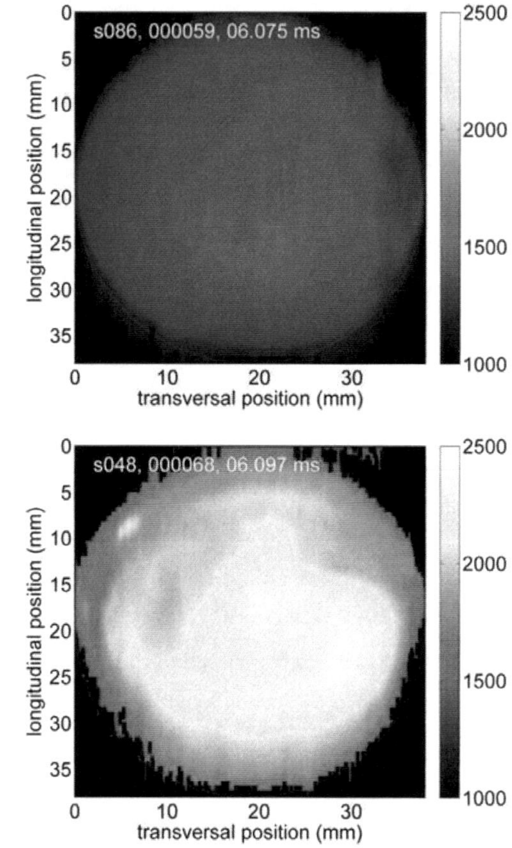

Figure 6. 2D anode surface temperature distribution around current zero at 15 kA, 120 mT. Upper subfigure: pure Cu butt electrode. Lower subfigure: Cu–Cr butt electrode.

4. Summary

Several methods for vacuum arc anode surface temperatures have been applied to different electrode materials and geometries. Vapor pressure curves propose massive evaporation of anode material, if the melting point is exceeded significantly. Furthermore it is argued that enthalpy stored in the bulk material of electrodes may affect surface temperatures, if they are heated at elevated levels in the active arc phase. This is partly in contradiction to previous publications which identified plateau-like phases of the temperature evolution by a phase transition [1, 3]. However, the present findings do not suggest a similar conclusion. Finally it is shown that 2D temperature distributions seem to be advantageous in case of constricted arc attachments, because they allow to conclude on the homogeneity of temperature distributions.

References

[1] Z. X. Wang, H. Ma, G. W. Kong, Z. Y. Liu, Y. S. Geng, and J. H. Wang. Decay Modes of Anode Surface Temperature After Current Zero in Vacuum Arcs - Part I: Experimental Study. *IEEE Trans. Plasma Sci.*, 42(5):1464–1473, 2014. doi:10.1109/TPS.2014.2316131.

[2] E. Dullni and E. Schade. Investigation of High-Current Interruption of Vacuum Circuit-Breakers. *IEEE Trans. Elec. Insu.*, 28(4):607–620, 1993. doi:10.1109/14.231543.

[3] Y. Niwa, J. Sato, K. Yokokura, T. Kusano, E. Kaneko, I. Ohshima, and S. Yanabu. The Effect of Contact Material on Temperature and Melting of Anode Surface in the Vacuum Interrupter. In *Proc. 19th Int. Symp. on Disch. and El. Insul. in Vac.*, pages 524–527, Xi'an, China, 2000. doi:10.1109/DEIV.2000.879042.

[4] R. Methling, St. Franke, S. Gortschakow, M. Abplanalp, R.-P. Sütterlin, T. Delachaux, and K.O. Menzel. Comparison of methods of electrode temperature determination in high-current vacuum arcs. *submitted to IEEE Trans. Plasma Sci.*, 2016.

[5] R. Müller. Arc-melted Cu-Cr alloys as contact materials for vacuum interrupters. *Siemens Forsch. Entw.*, 17(3):105, 1988.

[6] Dan Linder and Dean Linder. Chromium. URL: http://www.knowledgedoor.com/2/elements_handbook/chromium.html.

[7] R. Methling, S. Gorchakov, M. V. Lisnyak, S. Franke, A. Khakpour, S. A. Popov, A. V. Batrakov, D. Uhrlandt, and K. D. Weltmann. Spectroscopic Investigation of a Cu-Cr Vacuum Arc. *IEEE Trans. Plas. Sci.*, 43(8):2303–2309, 2015. doi:10.1109/TPS.2015.2443856.

[8] Y. Tian, Z. Wang, Y. Jiang, H. Ma, Z. Liu, Y. Geng, J. Wang, K. Nordlund, and F. Djurabekova. Modelling of crater formation on anode surface by high-current vacuum arcs. *J. Appl. Phys.*, 120(18):183302, 2016. doi:10.1063/1.4967528.

doi:10.14311/ppt.2017.1.20
Plasma Physics and Technology **4**(1):20–23, 2017

UTILISATION OF THE HIGH SPEED CAMERA FOR THE PIN-HOLE DISCHARGE DIAGNOSTICS

Z. Kozáková*, F. Krčma, L. Dřímalková, L. Hlochová

Faculty of Chemistry, Brno University of Technology, Purkyňnova 118, 612 00 Brno, Czech Republic

* kozakova@fch.vut.cz

Abstract. The high speed camera was utilised for plasma diagnostics of the DC pin-hole discharge in electrolyte solutions. Two discharge modes were determined. Plasma channels were observed either in the bubble or outside the bubble in the bulk solution, which confirms both thermal and electron theory of the discharge ignition in liquid. In the diaphragm discharge, plasma streamers were better visible on the cathode side of the dielectric barrier because they formed significantly longer channels.

Keywords: discharge in liquid, pin-hole discharge, high speed camera, plasma diagnostics.

1. Introduction

The problem of discharges in liquids has been in an intensive focus of researches for the last three decades, only. Although several applications such as water treatment [1], surface treatment [2], or plasma sterilization[3] have been developed during the underwater research, exact mechanisms of the discharge creation and related processes still desire a detail study and description.

Non-thermal plasma of underwater discharges can be generated by various electrode configurations and voltage regimes. The most studied configuration is a point-to-plate electrode geometry[4] where DC pulsed high voltage up to tens of kV is applied. Next configurations suitable for the discharge creation under the water surface are coaxial[5] and pin-hole systems[6][7][8]. Besides the DC pulsed high voltage, also high frequency, microwave or DC non-pulsed voltage regimes can be used for the discharge generation.

The pin-hole discharge can be distinguished as a diaphragm or capillary configuration. In the case of the diaphragm system, high voltage electrodes are separated by a thin dielectric barrier which thickness is approximately the same size as the diameter of an orifice connecting both reactor parts[9]. On the other hand, the capillary system uses a thick dielectric barrier which thickness is remarkably bigger than the orifice diameter[6]. In both cases, the discharge can be generated in the orifice using either DC or AC high voltage. The liquid medium is an electrolyte solution.

Plasma diagnostics by the ICCD camera has an important force among other diagnostic methods utilised for the characterisation of the discharge in liquids. Due to the still indefinite mechanism of the discharge ignition and propagation in the liquid phase, the usage of the high speed camera enables a proper capture of different processes taking place before, within and after the discharge breakdown moment in the system. Together with an employment of electric measurements, the ICCD imaging can reveal processes such as the bubble formation, the discharge breakdown in the bubble, the propagation of plasma streamers in and outside the bubble, the bubble cavitation, etc.

Following pages demonstrate a complex image study of the pin-hole discharge in the electrolyte solution. It is focused on the determination of particular processes taking place in the plasma reactor during the discharge ignition, and on the confirmation of the discharge breakdown theory (electron or thermal theory[10]). Further, a comparison of the diaphragm and capillary discharge is enabled due to the utilisation of the high speed camera records.

2. Experimental setup

Plasma diagnostics of the pin-hole discharge was carried out in plasma reactor shown in Figure 1. Its total volume was 100 ml. Two high voltage electrodes made of platinum were separated by a wall made of POM-C (Polyoxymethylene (Copolymer), thickness of 1 mm). There was an orifice (10 mm in diameter) in the centre of the wall into which a dielectric barrier made of ceramics Shapal-M^{TM} (thickness variable from 0.3 to 2 mm) was installed. One central pin-hole with the diameter of 0.2–0.6 mm was connecting both electrode parts.

High voltage was supplied from the DC self-pulsing source (high voltage amplitude up to 4 kV, power up to 300 W). Water solutions containing NaCl electrolyte with initial conductivity of 275 µS cm^{-1} were used in presented experiments.

An intensified CCD camera (ICCD) Andor iStar with an objective Cosmicar/Pentax TV lens 50 mm (1:2.8) was used for visual records of bubbles and plasma streamers. The microsecond time resolution was used for these measurements. The ICCD camera was synchronized with the discharge operation by current evaluation determined by a four-channel oscilloscope LeCroy LT374L with a high voltage probe ELDITEST GE 3830 30 kV (attenuation 1:1000).

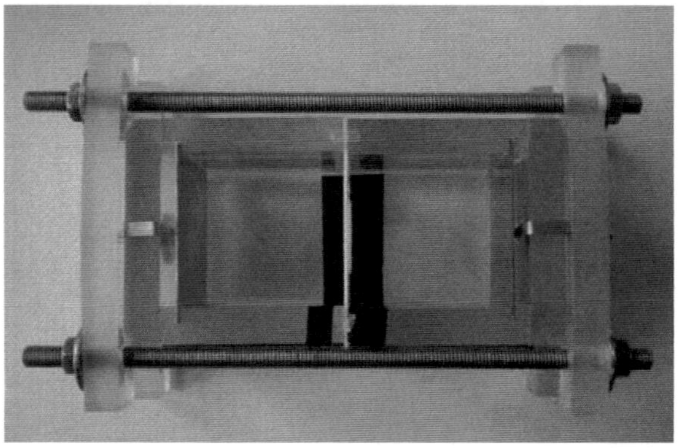

Figure 1. Plasma reactor used for the image diagnostics of the pin-hole discharge.

3. Results

The utilisation of the ICCD high speed camera had enabled the detail characterization of the discharge ignition, its regular operation as well as the identification of related processes such as the bubble formation. The moment representing the discharge breakdown in the pin-hole was confirmed both by time resolved current-voltage characteristics and light emission records by optical emission spectroscopy[9][11][12]. Camera records represented by photographs of a particular discharge moment had been completed by time resolved electric characteristics of discharge current and camera triggering. The ICCD camera was switched on automatically when the current signal rapidly increased which indicated the discharge ignition (see Figure 2).

The camera images were taken vertically to the dielectric barrier separating the anode and cathode parts of the plasma reactor. In the photograph, the dielectric barrier is the white central stripe from the left to the right (its thickness is approximately 1 mm). There is an orifice in the barrier onto which a particular diaphragm or capillary is installed. The diaphragm or capillary is not visible in the photograph because it is hidden within the barrier. The pin-hole connecting both electrode parts is approximately in the middle of the picture. The anode is set in the upper part while the cathode lies in the bottom part of the photographs.

Increasing the applied voltage on the electrodes, tiny microbubbles appear in the pin-hole due to the liquid heating. Microbubbles are dragged from the pin-hole to the bulk solution. When the applied voltage exceeds the breakdown value (over 1200 V), the discharge is ignited in the pin-hole. According to the image records, the discharge is created primary in the bubble of the evaporated liquid. This fact confirms the thermal theory of the discharge ignition in liquids.

Formation of the bubble is associated with a substantial increase of system resistance. Subsequently, different potentials formed on the inner and outer surface of the bubble lead to the breakdown of the liquid vapour inside the bubble. Simultaneous decrease of system resistance induces an intensive current increase which is represented by a significant peak on the current characteristics. The camera triggering was adjusted to this current enhancement, as it seen in the Figure 2.

Further increase of the applied voltage ensures the stable discharge operation. However, two discharge modes have been detected using the high speed camera images. The first one is represented by the further discharge propagation in the bubble (Figure 3), according to the thermal theory. The photograph clearly shows plasma streamers in the bubble on the cathode side of the dielectric barrier.

The second mode of the pin-hole discharge operation is represented by further propagation of plasma streamers outside the bubble (Figure 4). We suppose that this discharge mode corresponds to the electron theory. When plasma streamers primary created in the vapour phase reach the bubble inner surface, ionisation of liquid molecules starts outside the bubble, and plasma channels can propagate further into the bulk solution. If the applied voltage is high enough, the discharge operation is possible even though the bubble explodes.

Based on obtained images we have assumed that both breakdown theories fit for the discharge ignition in liquids. Primary, the discharge ignition starts in the bubble of the liquid vapour due to the lower ionisation potential needed for the breakdown. This phase corresponds to the thermal (bubble) theory. On the other hand, further propagation of plasma channels into the liquid is possible outside the bubble as well. This fact reflects on the electron theory of electron avalanches. Thus mechanisms of the pin-hole discharge operation are a combination of both theories. This fact is valid in systems where a slow voltage increase is used for the discharge ignition. Moreover, this phenomenon was also observed when gas bubbles were additionally introduced into the system through the HV electrode[13]. In the case of pulsed nanosecond discharges, a proceeding of the thermal theory is impossible due to a lack of time necessary for the liquid evaporation. In such systems, the discharge operates directly by streamers in the liquid phase.

3.1. Comparison of the diaphragm and capillary discharge

This result section compares the shape of plasma streamers created in the diaphragm and capillary configuration. The diaphragm configuration means that the dielectric barrier parameters are approximately equal, i.e. the barrier thickness of 0.3 mm and the pin-hole diameter of 0.3 mm. In the capillary configuration, the barrier thickness is substantially higher than the pin-hole diameter, i.e. the thickness of 1.5 mm and the pin-hole diameter of 0.3 mm.

The camera imaging of the discharge created in the pin-hole was adjusted according the same conditions

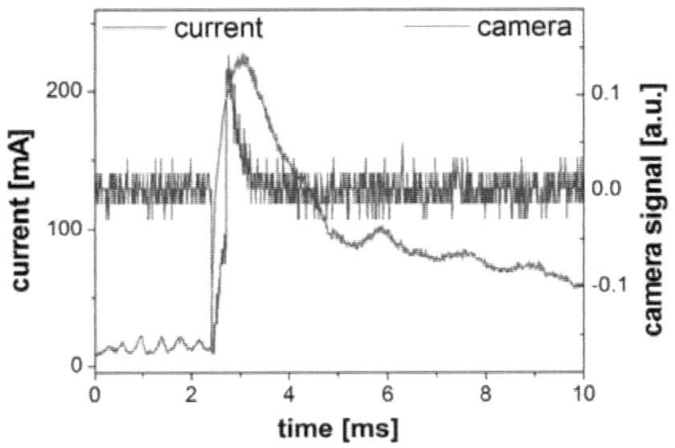

Figure 2. Time resolved record of discharge current and camera signal during the discharge breakdown in the diaphragm.

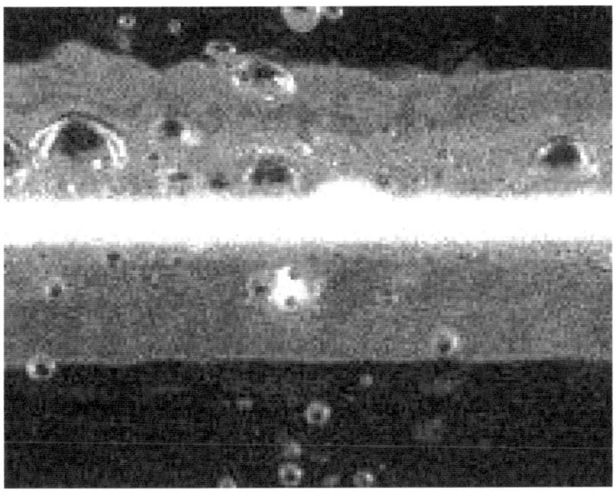

Figure 3. Discharge creation in the bubble on the cathode side of the dielectric barrier; mean applied voltage of 1540 V (diaphragm configuration: ceramic barrier thickness of 0.3 mm, central pin-hole with diameter of 0.3 mm).

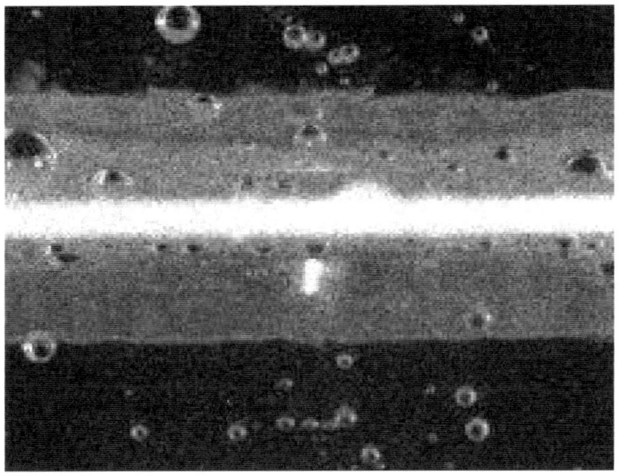

Figure 4. Plasma streamers propagating outside the bubble from the pin-hole towards the cathode; mean applied voltage of 1540 V (diaphragm configuration: ceramic barrier thickness of 0.3 mm, central pin-hole with diameter of 0.3 mm).

as in the previously mentioned experiments. The camera triggering was set at the current increasing phase. Both discharge modes (i.e. plasma streamers propagation in the bubble or outside the bubble) has been recorded in all tested configurations. Images obtained in the capillary configuration are presented in Figures 5 and 6. Figure 5 represents plasma streamers formed in the bubble while Figure 6 shows plasma streamers propagating outside the bubble into the bulk solution.

In the diaphragm configuration (Figures 3 and 4), plasma channels are well visible only on one side of the dielectric barrier where the cathode is installed. This phenomenon indicates a different shape of plasma streamers in the DC voltage regime. According to the theoretical assumption[14], longer plasma channels are created on the cathode side while shorter plasma channels on a spherical shape are formed on the anode side. Presented images confirm this assumption for the diaphragm discharge because longer streamers are observed just on the cathode side.

In the capillary configuration (Figures 5 and 6), plasma streamers are visible on both sides of the dielectric barrier. However, the shape of the channels seems different similarly to the diaphragm configuration. Recorded plasma streamers are more intensive on the cathode side of the barrier. Moreover, there is an interesting difference of the discharge creation in the bubble or in the liquid, only. When a bubble is formed in the pin-hole and the discharge is ignited in it, plasma channels fill almost the whole space of the bubble. This effect is more obvious in the case of the thicker barrier (1.5 mm). Therefore, the discharge seems more intensive in the bubble mode because energy is mostly dissipated inside the bubble, i.e. in the gas phase. Another effect of the capillary configuration is that both bubbles and plasma channels are dragged from the pin-hole (capillary) towards the bulk solution. Especially in the case of the thicker barrier (1.5 mm), the bubble on the cathode side is significantly enhanced, and it forms a conical shape. It corresponds to the effect of pumping which was already observed in the capillary configuration[6].

4. Conclusions

The high speed camera was utilised for plasma diagnostics of the DC pin-hole discharge in electrolyte solutions. This device has been assumed as an important tool for a detailed imaging of plasma streamers and bubble formation. Based on the camera records together with the time resolved characteristics of electric parameters, particular processes before, within and after the discharge breakdown have been detected. According to detailed photographs, two discharge modes have been determined. Plasma channels were observed either in the bubble or outside the bubble in the bulk

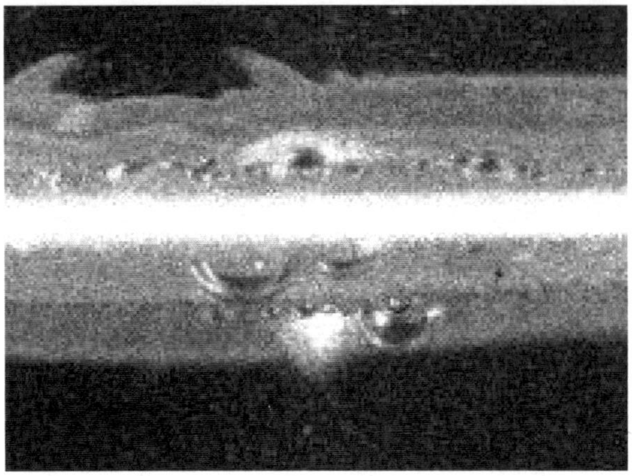

Figure 5. Discharge creation in the bubble; mean applied voltage of 2020 V (capillary configuration: ceramic barrier thickness of 1.5 mm, central pin-hole with diameter of 0.3 mm).

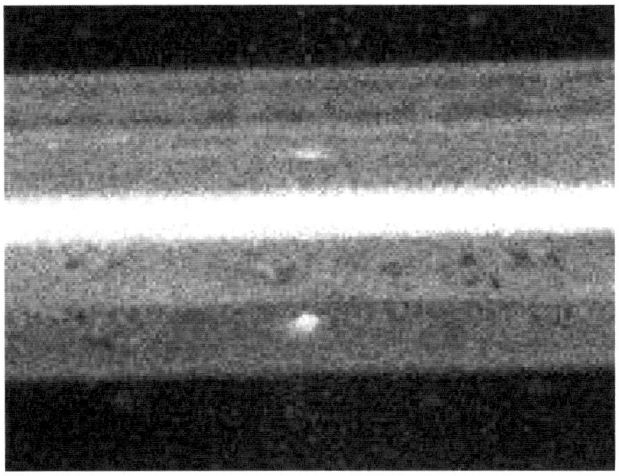

Figure 6. Discharge creation outside the bubble; mean applied voltage of 1920 V (capillary configuration: ceramic barrier thickness of 1.5 mm, central pin-hole with diameter of 0.3 mm).

solution. These results confirm both theories of the discharge ignition in liquid, thermal and electron theory.

The ICCD imaging also helped to compare the discharge creation in the diaphragm and capillary configuration. In the case of the diaphragm discharge, plasma streamers were better visible on the cathode side of the dielectric barrier because they formed significantly longer channels. In the case of the capillary, plasma channels propagate further from the pin-hole into the solution and thus they were well visible on both sides of the dielectric barrier.

Acknowledgements

This work was supported by the Czech Ministry of Education, Youth and Sports, project No. LD14014.

References

[1] P. Lukeš et al. Degradation of phenol by underwater pulsed corona discharge in combination with tio2 photocatalysis. *Res. Chem. Intermed.*, 31(4-6):285–294, 2005. doi:10.1163/1568567053956734.

[2] M. Klíma et al. Hf plasma pencil and dc diaphragm discharge in liquids - diagnostics and applications. *Czech J. Phys.*, 56:B1051–B1056, 2006. doi:10.1007/s10582-006-0325-x.

[3] K.R. Stalder et al. Electrosurgical plasmas. *J. Phys. D: Appl. Phys.*, 38:1728–1738, 2005. doi:10.1088/0022-3727/38/11/014.

[4] M.J. Kirkpatrick and B.R. Locke. Hydrogen, oxygen, and hydrogen peroxide formation in aqueous phase pulsed corona electrical discharge. *Ind. Eng. Chem. Res.*, 44(12):4243–4248, 2005. doi:10.1021/ie048807d.

[5] P. Šunka et al. Potential applications of pulse electrical discharges in water. *Acta Physica Slovaca*, 54(2):135–145, 2004.

[6] F. De Baerdemaeker et al. Characteristics of ac capillary discharge produced in electrically conductive water solution. *Plasma Source Sci. Technol.*, 16:341–354, 2007. doi:10.1088/0963-0252/16/2/018.

[7] M. Sato et al. Decoloration of dyes in aqueous solution by pulsed discharge plasma in water through the pinhole. *Transaction of the institute of fluid-flow machinery*, (107):95–100, 2000.

[8] Z. Stará et al. Organic dye decomposition by dc diaphragm discharge in water: Effect of solution properties on dye removal. *Desalination*, 239(1-3):283–294, 2009. doi:10.1016/j.desal.2008.03.025.

[9] F. Krčma et al. Diaphragm discharge in liquids: Fundamentals and applications. *J. Phys.- Conf. Series*, 207:012010(6 pages), 2010. doi:10.1088/1742-6596/207/1/012010.

[10] R.P. Joshi et al. Electrical network-based time-dependent model of electrical breakdown in water. *J. Appl. Phys.*, 92(10):6245–6251, 2002. doi:10.1063/1.1515105.

[11] Z. Kozáková et al. Diagnostics of electrical discharges in electrolytes: Influence of electrode and diaphragm configuration. *Book of Contributed Papers: 18th Symposium on Application of Plasma Processes and Workshop on Plasmas as a Planetary Atmospheres Mimics*, pages 83–87, 2011.

[12] L. Hlavatá et al. Influence of electrode configuration on dc diaphragm discharge breakdown in electrolyte solution. *J. Phys.- Conf. Series*, 768:012007–1(4 pages), 2016. doi:10.1088/1742-6596/768/1/012018.

[13] A. Nikiforov et al. The influence of water vapor content on electrical and spectral properties of an atmospheric pressure plasma jet. *Plasma Sources Sci. Technol.*, 20:015014, 2011. doi:10.1088/0963-0252/20/1/015014.

[14] I.P. Kuzhekin. Corona electrodeless discharge in water. *Proceedings of 9th International Symposium on High Voltage Engineering*, pages 8073–1–8073–3, 1995.

doi:10.14311/ppt.2017.1.24
Plasma Physics and Technology 4(1):24–27, 2017

EXPERIMENTAL INVESTIGATIONS ON ELECTRICAL PLASMA CONDUCTIVITY IN A MODEL SPARK GAP FOR SURGE CURRENTS

T. Runge*, T. H. Kopp, M. Kurrat

Technische Universität Braunschweig, Institute for High Voltage Technology and Electrical Power Systems, 38106 Braunschweig, Germany

* t.runge@tu-braunschweig.de

Abstract. In this experimental investigation the electrical conductivity of plasma is measured during surge current using potential probes. The measurements were carried out in a narrow gap arrangement based on spark gap technology. In order to investigate the electrical conductivity during surge this model is tested using 8/20 µs surge currents according to the IEC 62475. The measured behaviour of the electrical conductivity during surge and the uncertainty of these measurements are discussed.

Keywords: arc, electrical conductivity, plasma, spark gap, surge current, surge protective device.

1. Introduction

In case of current limiting switching devices in inductive circuits the voltage drop of the arc has to be higher than the the supply voltage. The arc voltage can be calculated as the sum of the electrode sheath and the plasma voltage, which can be calculated using the electrical conductivity σ, current I, length l and cross section A.

$$U_{\text{Sup}} < U_{\text{Arc}} = U_{\text{Electrode sheath}} + \frac{I * l}{A * \sigma} \qquad (1)$$

In order to design switching devices it is important to know how to influence the arc voltage. Thus following influences are possible in order to increase the arc voltage:

- Increasing of the electrode sheath voltage by using deion arc splitters
- Extending of the length
- Decreasing of the cross section
- Decreasing of the electrical conductivity

The electrode sheath voltage, length and cross section is mostly determined by the geometry and influenced easily, whereas the electrical conductivity cannot be influenced directly. In order to identify different influence possibilities the electrical conductivity has to be determined firstly.

Two ways are possible to identify the conductivity: Firstly a simulation of the device and secondly a measurement. To guarantee high quality of the results it is best to do both ways with a comparison of the results. In this experimental investigation a spark gap is tested with surge currents according to surge protective devices. The applicability of a compact measuring method is verified in pre-investigations and used to determine the electrical conductivity behaviour of plasma. Further a statement about the uncertainty of the measurement is given.

2. Experimental Set-Up

The measurements are carried out in a spark gap model using surge currents described in subsection 2.1. The electrical conductivity is measured using two potential probes introduced in subsection 2.2. In order to verify the applicability of the method some pre-investigations are presented in subsection 2.3.

2.1. Experimental Model

The spark gap model consists of a box-shaped narrow gap which is formed by two tungsten copper electrodes and surrounding chamber wall made of Polyoxymethylen. Additional two outlet ducts are mounted at the sides of the gap (Figure 1). The dimensions are: side lengths of the spark gap: 10 mm, height: 1 mm and length of the outlet duct: 20 mm. A surface discharge device is used to ignite the spark gap connected through a varistor. More details are given in [1].

As surge current the 8/20 µs impulse is used according to the IEC 62475 with amplitudes of 23 kA. The behaviour of the spark gap during impulse can be diverted into different phases (see also Plasma Propagation in section 2.3) [2]: Ignition of the spark gap, commutation of the current from varistor to plasma, high current phase and undershoot.

2.2. Measuring Method

The aim of this investigation is to measure the electrical conductivity of the plasma. One method is to use the current density and electrical field of the plasma, which is the voltage drop of the plasma divided by the length:

$$\sigma = \frac{j}{E} = \frac{j * l}{\Delta U} = \frac{I * l}{A * \Delta U} \qquad (2)$$

The current density can be calculated by using the measured current and the cross section of the spark gap. Also the length is given by the geometry whereas the voltage drop of the plasma is unknown in most

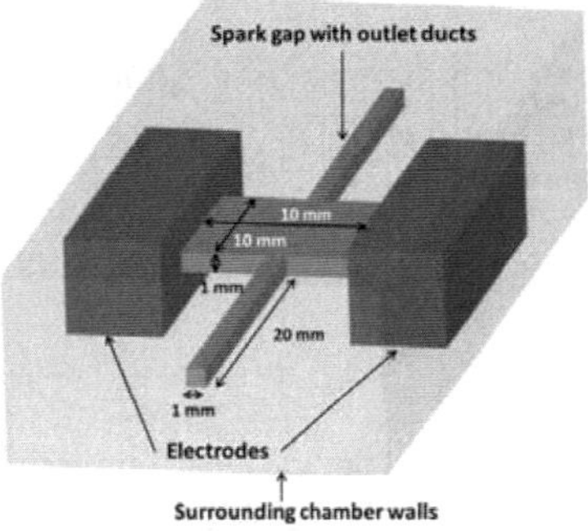

Figure 1. 3-D model of the spark gap model without SDD.

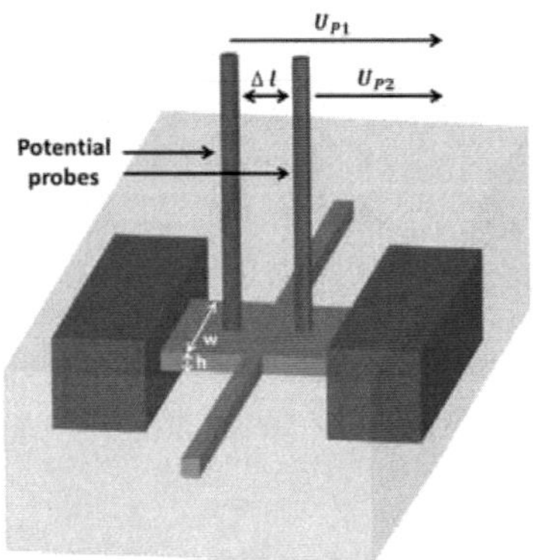

Figure 2. Measuring set-up for electrical conductivity.

cases, caused by the voltage drop of the electrode sheath. Two methods can be used to identify the voltage drop of the plasma: Firstly the voltage of the arc can be measured. In order to determine the plasma voltage the measured voltage is subtracted by the voltage drop of the electrode sheath which can be estimated based on data like in [3]. Secondly a measuring method can be used, which eliminates the voltage drop of the electrode sheath. In this paper, the second method is presented. In order to eliminate the voltage drop potential probes are used, which are already investigated like in [4]. The concept of the presented method is to use two voltages with the same voltage drop like in [5]. Thus two identical potential probes are located in the plasma with floating potential. The voltage drop between both potential probes and a reference electrode is measured (U_{P1} and U_{P2} in Figure 2).

With the assumption that the voltage drops of the electrode sheaths of both measured voltages are similar it can be subtracted by calculating the difference between them. Thus the electrical conductivity can be estimated using (3).

$$\sigma = \frac{I * \Delta l}{A * (U_{P1} - U_{P2})} \tag{3}$$

But due to the ablation of the chamber wall a thin cold gas layer is located at the chamber wall which height is about 20 % of the spark gap height [6]. This layer does not make a contribution to the current flow. Thus about 80 % is filled with plasma in which current flow occurs. Therefore the spark gap height (h) multiplied with the width (w) and 80 % has to be used to calculate the conductivity (4).

$$\sigma = \frac{I * \Delta l}{0.8 * w * h * (U_{\mathrm{P1}} - U_{\mathrm{P2}})} \tag{4}$$

The inserting of the probes in the plasma caused electrical and thermic influences which disturbs the plasma properties. Thus in order to minimise these influences following requirements for the used probes have to be fulfilled [5, 7, 8]: Firstly the ablation of the probes has to be minimised to reduce the thermic influence. Secondly cylindrical probes with small diameter have to be used to reduce the influence on the plasma flux. And thirdly the current through the probes has to be negligible compared with the current in the plasma. These requirements are fulfilled by using cylindrical probes made of pure wolfram with a diameter of 0.4 mm and voltage probes with high resistance.

2.3. Pre-Investigations

Before the method is applied at the spark gap three questions must be attempted: The influence of the probes on the plasma, the inducted voltages in the measurement and the propagation of the plasma during surge.

Influence of the potential probes on plasma

A huge challenge of measurement is that the measurement itself is not allowed to influence the result excessively. Thus in this case the electrical conductivity has to be the same with and without measurement setup. In order to approve this, the voltage of the arc (plasma with arc root voltage) is measured and compared with and without measurement setup (Figure 3). The comparison reveals a voltage within the standard deviation in both cases especially in the high current phase. Thus it can be assumed that the electrical conductivity is not influenced by the measuring setup.

Induced Voltage

Another challenge to deal with is the voltage measurement. Due to the high $\mathrm{d}i/\mathrm{d}t$ values caused by the surge current induced voltages occur in the measurement. These voltages are measured for U_{P1} and U_{P2}. In this case the spark gap is filled with copper, thus a short circuit occurs and the measured voltages have to be near zero due to the negligible resistance of copper. In Figure 4 the measured induced voltages are presented, which is maximal about 110 V in U_{P1} and thus not negligible. Due to the predefined current the induced voltages are similar during the cases of short circuit and plasma. Using the same geometry setup the induced voltages will

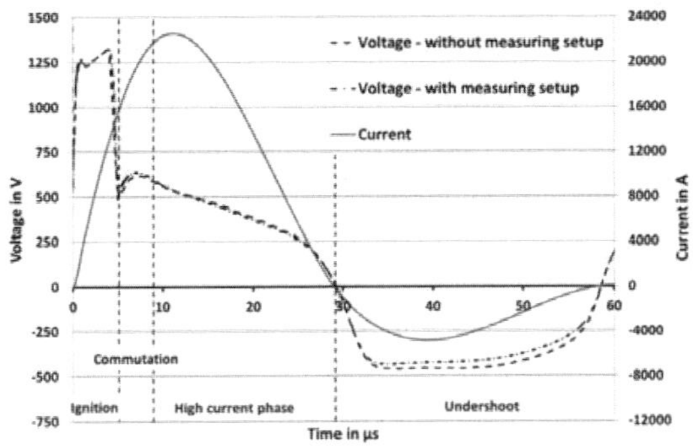

Figure 3. Influence of the measuring setup on plasma.

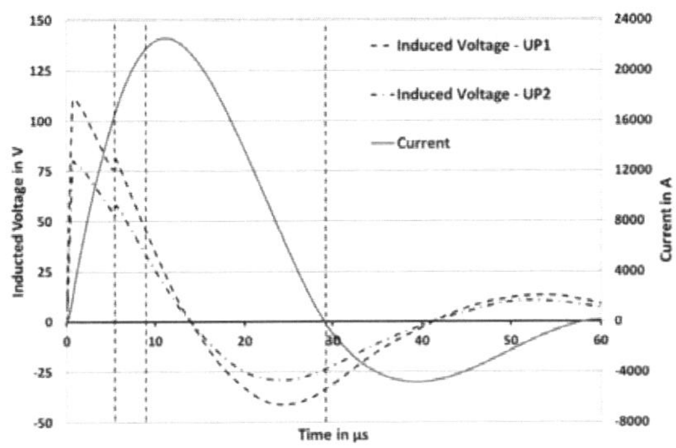

Figure 4. Induced voltages during surge current.

remain equal and can be subtracted in the calculation.

Plasma Propagation

In order to calculate the conductivity the current density respectively the cross section of the current has to be known. It can be assumed as the plasma propagation. In [9] it is shown that plasma during surge is spreading. Using a much smaller volume than [9] it is assumed that in this case the plasma fills out the spark gap completely. In order to verify this assumption the plasma propagation in the spark gap was recorded using a high speed camera and neutral density filters as shown in [10] (Figure 5):

During ignition phase, the spark gap ignites at the SDD at the right electrode and the plasma spreads towards the left electrode. The SDD is connected by a varistor, thus in this phase the current flows through a varistor until the plasma reaches the opposite electrode and a breakdown of the spark gap occurs. Afterwards the current commutates from the varistor to the spark gap. In high current and undershoot phases current flows through the plasma which fills the chamber nearly homogenous and completely. Further due to high pressure, the plasma steams out through the outlet ducts. Thus the assumption of homogenous plasma propagation in the spark gap is verified.

3. Uncertainty of Measurement

For the interpretation of the measurement results a statement about the relative uncertainty is necessary.

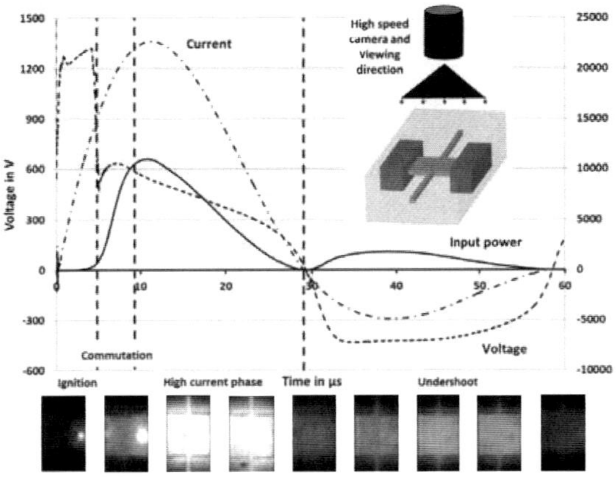

Figure 5. Plasma propagation during surge current.

Thus an estimation of the uncertainty u using the Guide to the Expression of Uncertainty in Measurement (GUM) is presented using the data at maximum electrical conductivity.

Due to the calculation of the electrical conductivity using different values the uncertainty is a geometrical addition of the uncertainties of these values (5). Thus uncertainty of current and voltage measurements have to be determined together with the uncertainty in the distance of the probes and cross section of current.

$$u_{\sigma,r} = \sqrt[2]{(\frac{u_I}{\bar{I}})^2 + (\frac{u_{\Delta U}}{\bar{\Delta U}})^2 + (\frac{u_{\Delta l}}{\bar{\Delta l}})^2 + (\frac{u_A}{\bar{A}})^2} \quad (5)$$

The uncertainty of the current respectively of the voltage is the sum of the variation with different measurements and the uncertainty of the measurement setup. For both (current and voltage measurement) the uncertainty of the measurement setup can be assumed to 2.5 %. Thus for the total uncertainty of the current this calculation reveals 2.52 % and 2.8 % for the voltage including the influence of variation in the inducted voltage at confidence interval of 68 %.

To determine the uncertainty of the distance of probes and cross section of current another method has to be used. In these cases it is assumed that the probability of these values is a Gaussian distribution. Further a most probably value is estimated with boundaries in whose area the real value is located with a probability of 99 %. Thus the standard deviation can be calculated which is used as the uncertainty of these values.

The probe distance should be 4 mm due to the design drawing. Thus this is the most probable value. Caused by tolerances in the manufacturing and slightly movements during surge the mentioned boundaries are estimated to be 3.9 mm and 4.1 mm, same procedure for the cross section: the most probable value is estimated to be 8 mm^2 (see 2B). The boundaries are estimated to be about ± 20 % of this value and thus about 6.5 mm^2 and 9.5 mm^2 (Figure 6). This yields into a relative uncertainty of 0.97 % for the

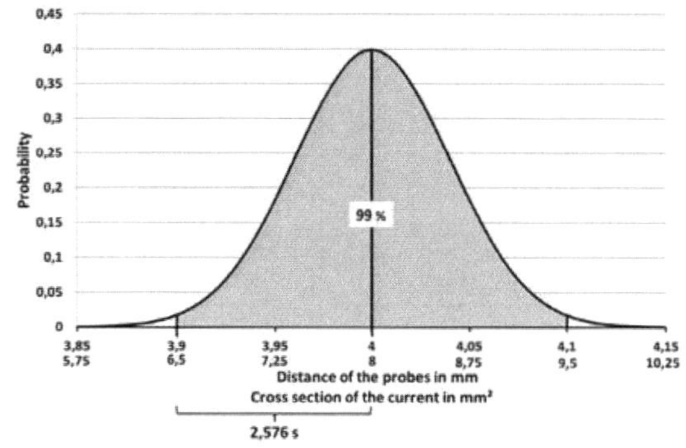

Figure 6. Gaussian distribution of the probe distance and cross section.

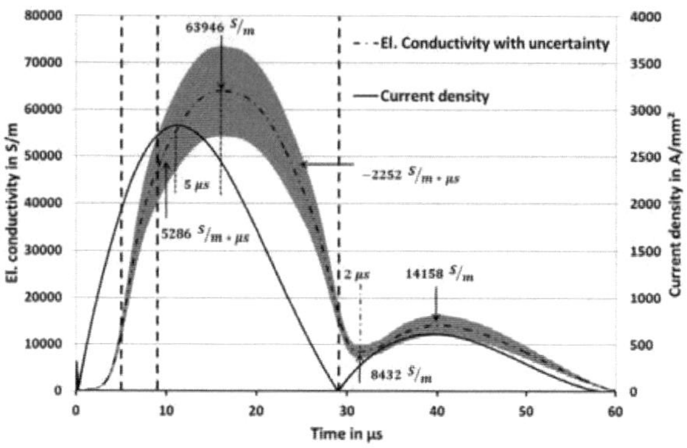

Figure 7. Behaviour of electrical conductivity during surge current.

probe distance and 7.28 % for the cross section (68 % confidence interval).

The combination of these different uncertainties in (5) yields into a relative uncertainty of the electrical conductivity of 15 % with an increased confidence interval to 90 %. This uncertainty is mostly determined by the uncertainty of the cross section. For example a reduction of the boundaries in the crosssection to ± 10 % (instead of ± 20 %) yields into a relative uncertainty of the electrical conductivity of 10 %.

4. Results

The electrical conductivity is shown in Figure 7 together with the current density. Due to the influence of the ignition using a SDD, commutation behaviour of current and unspecified propagation of the plasma in the ignition and commutation phase the electrical conductivity is an assumption during both phases. Afterwards the electrical conductivity is calculated using the measurement data.

The presented slopes are calculated as the linear increasing respectively decreasing between the 60 % and 80 % value of the maximum. The conductivity has a slope of about 5300 S per m and µs at the rising edge respectively 2250 S per m and µs at falling edge and a maximum of 64000 S per m. A comparison of the conductivity and current density reveals a similar behaviour with a time delay between the maxima of about 5 µs.

5. Conclusion and Outlook

A method to determine the electrical conductivity of plasma in spark gaps is applied at surge currents using two potential probes. The most important conclusions can be summarised as follows:

Firstly the investigations on influence of the probes on plasma, induced voltages and plasma propagation yield into the applicability of the method. Secondly the uncertainty of the result is estimated to 15 % at a confidence interval of 90 %. And thirdly the electrical conductivity during surge current was determined.

This method can be used in order to characterize the electrical conductivity in dependence of different boundary conditions like the current density and dimensions of the spark gap.

Furthermore measurements with follow current will help to get a more detailed knowledge about transition behaviour between surge and follow current like described in [10] in order to determine the current limiting behaviour.

References

[1] T. Runge, T. H. Kopp, and M. Kurrat. Experimental investigation on plasma pressure in an narrow gap for short time currents. In *Int. Conf. on Electrica Contacts*, Edinburgh, Great Britain, 2016.

[2] B. Schottel and M. Kurrat. Experimental investigation of short-time arc-wall interaction. In *Int. Conf. on Gas Disch. and Their Appl.*, Bejing, China, 2012.

[3] M. S. Benilov and A. Marotta. A model of the cathode region of atmospheric pressure arcs. *J. Phys. D: Appl. Phys.*, 28(9):1869–1882, 1995. doi:10.1088/0022-3727/28/9/015.

[4] H. M. Mott-Smith and I. Langmuir. The theory of collectors in gaseous discharges. *Phys. Rev.*, 28(4):727–763, 1926. doi:10.1103/PhysRev.28.727.

[5] G. R. Jones. *High pressure arcs in industrial devices.* Cambridge University Press, 1988.

[6] E. Z. Ibrahim. The ablation dominated polymethylmethacrylat arc. *J. Phys. D: Appl. Phys.*, 13(11):2045–2065, 1980. doi:10.1088/0022-3727/13/11/015.

[7] V. Bergmann. Untersuchungen zur anwendbarkeit elektrischer sondenmessungen im hochstromlichtbogen. PhD Thesis, 1987.

[8] G. Hertz and R. Rompe. *Einfürhung in die Plasmaphysik und ihre technischen Anwendungen.* Akademie Verlag, 1986.

[9] B. Schottel, T. H. Kopp, T. Runge, and M. Kurrat. Experimental investigations of short-time plasma propagation recorded by high speed camera. In *Int. Conf. on Gas Disch. and their Appl.*, Orleans, France, 2014.

[10] T. H. Kopp, T. Runge, and M. Kurrat. Investigation of arc interruption behavior of a model spark gap after surge current. In *Int. Conf. on Electric Contacts*, Edinburgh, Great Britain, 2016.

doi:10.14311/ppt.2017.1.28
Plasma Physics and Technology **4**(1):28–31, 2017

PLASMA ASSISTED GENERATION OF MICRO- AND NANOPARTICLES

A. Veklich[a,*], A. Lebid[a], T. Tmenova[a], V. Boretskij[a], Y. Cressault[b], F. Valensi[b], K. Lopatko[c], Y. Aftandilyants[c]

[a] *Taras Shevchenko National University of Kyiv, 64/13, Volodymyrska str., Kyiv, Ukraine*

[b] *Université de Toulouse; UPS, INPT; LAPLACE (Laboratoire Plasma et Conversion d'Energie); 118 route de Narbonne, F-31062 Toulouse Cedex 9, France*

[c] *National University of Life and Environmental Sciences of Ukraine, Kiev, Ukraine*

* van@univ.kiev.ua

Abstract. In this research, the peculiarities of micro- and nanoparticles generation are considered. Two techniques of micro- and nanoparticles' formation using electric arc and underwater discharge plasma sources are proposed. Molybdenum oxide crystals were deposited on side surface of the bottom electrode (anode) of the free-burning discharge between metallic molybdenum electrodes. Friable layer of MoO_3, which consists of variously oriented transparent prisms and platelets (up to few hundreds of µm in size), was formed by vapor deposition around the electrode. In the second technique, plasma of the underwater electric spark discharges between metal granules was used to obtain stable colloidal solutions with nanoparticles of 20–100 nm sizes.

Keywords: molybdenum trioxide, plasma evaporation, underwater spark discharge, colloidal liquids.

1. Introduction

Plasma treatment of materials attracts strong interest due to extreme conditions of processes, which can't be achieved in other way. Particularly, plasma technology can be used for fabrication of materials structured on micro- and nanoscale [1, 2]. Application of such materials can significantly improve existing industrial and biological technologies. For example, molybdenum trioxide crystals can be used in solar cells [3], lithium-ion batteries [4, 5], chemical catalysis [6] and gaseous sensors [7]. At the same time, solutions of metallic nanoparticles are already being used for purposes of biological and food applications [8, 9]. Fundamental difference between the colloidal and conventional forms of mineral nutrients is that the first are non-toxic, bioavailable and by virtue of the crystalline structure of the dispersed phase are the most suitable for their use on biological objects. Such metal colloids exhibit high efficacy and have already found their application in the cultivation of vegetable production under conditions of low or disturbed mineral nutrition [8, 9]. Nowadays, plasma assisted technologies are very effective for micro- and nanomaterials generation. Cheapness of such technologies has made them attractive for solution of different scientific and applied problems.

This work deals with two alternative plasma assisted techniques of micro- and nanoparticles generation. Technique of micro structured molybdenum trioxide crystal fabrication in electric arc plasma source and alternative technique of metal nanoparticles generation in underwater spark discharge is described. Main processes stages and material properties are discussed.

2. Techniques of micro- and nanoparticles generation

2.1. Micro-structured molybdenum trioxide crystal fabrication in electric arc plasma source

The vertically oriented free-burning arc was ignited in air between the end surfaces of metallic molybdenum non-cooled electrodes (See Fig.1). The diameter of the rod electrodes was 6 mm, the discharge gap was 8 mm and DC current was 3.5 A. Molybdenum oxide crystals were deposited on the side surface of anode [10]. It must be noted, that the formation of crystals takes place 3-5 mm below the end surface of electrode.

2.2. Underwater spark discharge for production of nanoparticles solution

Colloidal solutions of metal particles are obtained by realization of volumetric electric spark destruction of metal granules. This method lies in the simultaneous formation of spark channels in contacts between metal granules immersed in a liquid. Electric energy was supplied by specially developed pulse generator [11]. As a result of spark erosion, part of metal granules evaporates and, while being tempered into a liquid, forms fine dispersion fraction of spark-erosive particles. The plasma of underwater electric spark discharges between metal granules was used to obtain stable colloidal solutions with nanoparticles of 20-100 nm sizes. The experimental values of voltage were up to 300 V, current - up to 600 A and the capacity of the discharge circuit - up to 1300 µF.

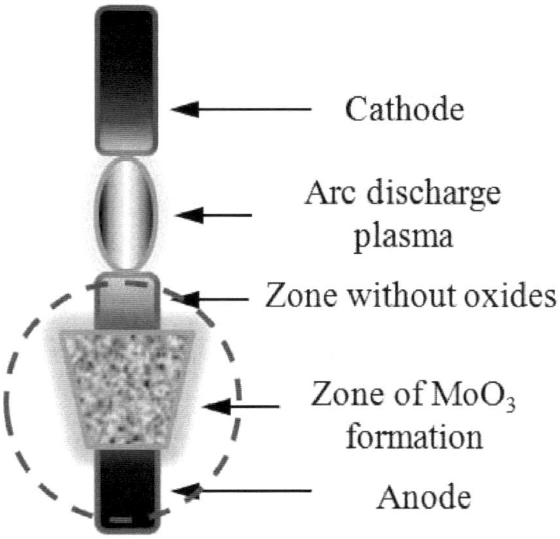

Figure 1. Electrodes arrangement for fabrication of MoO_3 crystals

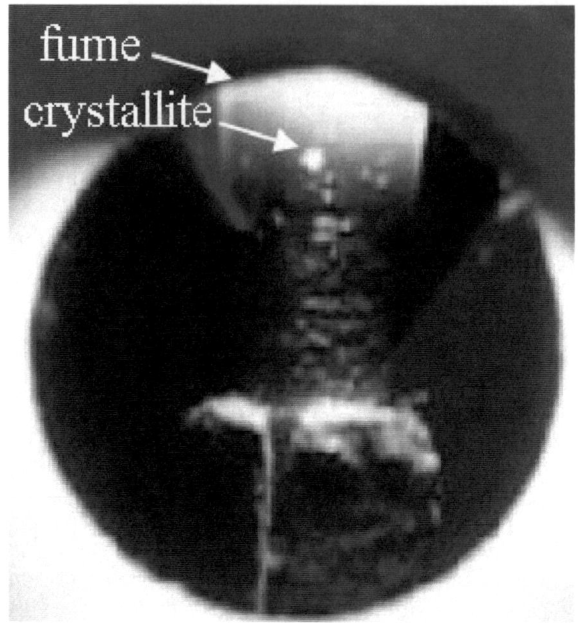

Figure 2. Formation of MoO_3 crystallites.

3. Results and discussions

3.1. Micro-structured MoO_3

The process of crystals formation during arc discharge can be divided into specific sequential stages. Immediately after arc ignition there was no evaporation due to relatively low initial temperature of the anode surface.

Few seconds after, a white fume was observed around the electrode (See Fig.2). This stage can be explained by oxidation of metallic molybdenum and volatilization of oxides as the electrode temperature increases. The oxidation of metallic molybdenum surface and volatilization of the oxide layer were observed during heating in the furnace. The MoO_3 crystallites appeared on the electrode surface during the next process phase (See Fig.2). Crystals start growing from white fume evaporations, which are transported by convectional air flow. Probably, initial crystallization starts on surface defects or on greyish-black particles. Friable layer around electrode, which consists of variously oriented transparent prisms and platelets, was formed by vapor deposition.

It was mentioned in [12] that oxide layer completely evaporates from molybdenum surface at temperature above 1150° C. Therefore, this assumption explains the absence of crystals near the end surface of electrode wherein surface temperature is obviously higher.

Re-evaporation of deposited crystals is absent because of two reasons. The first one is low thermal conductivity between electrode and crystallites due to their irregular orientation. The crystallites are weakly connected to the electrode surface but have numerous connections with each other. The second reason is cooling of crystallites by convectional flows.

Chemical composition and structure of produced crystals were determined by X-rays diffraction method (XRD). Obtained crystals were detached from electrode surface and milled before the investigations. X-ray diffraction study indicates that resulting crystals consist of orthorhombic α-MoO_3 phase. Only small amount of monoclinic β-MoO_3 phase was detected.

Optical microscopy studies indicate formation of feather-like structures (Fig.3). Every "feather" consists of closely packed parallel needles, which are unfinished structures of directional crystal growth. These needles are resembled to microrods observed in [4], but in contrast, these microrods have random orientation. Probably, attachment of building material on these pins takes place during the vapor deposition process and supports further translation of crystal structure.

3.2. Peculiarities of metal nanoparticles generation

The study of energy dissipation channels in underwater discharges can clarify the most effective mechanisms of nanoparticles' generation. Therefore, it is essential to determine the excitation temperature of metal atoms in discharge plasma.

Optical emission spectroscopy (OES) methods were used for diagnostics of the underwater electric spark discharge plasma between metal granules. Plasma emission was registered by the SDH-IV spectrometer with a 4-position manually switchable diffraction gratings turret. The measurements were performed in all four channels of spectrometer, so the total registered spectral range was 200-1200 nm. Toshiba TCD 1304 AP linear image sensor was used as CCD detector. Spectral sensitivity in every spectral range of the SDH-IV spectrometer was determined and taken into account.

Experiments with granules of Cu, Fe, Mn and Mo were performed. Emission spectra of corresponding

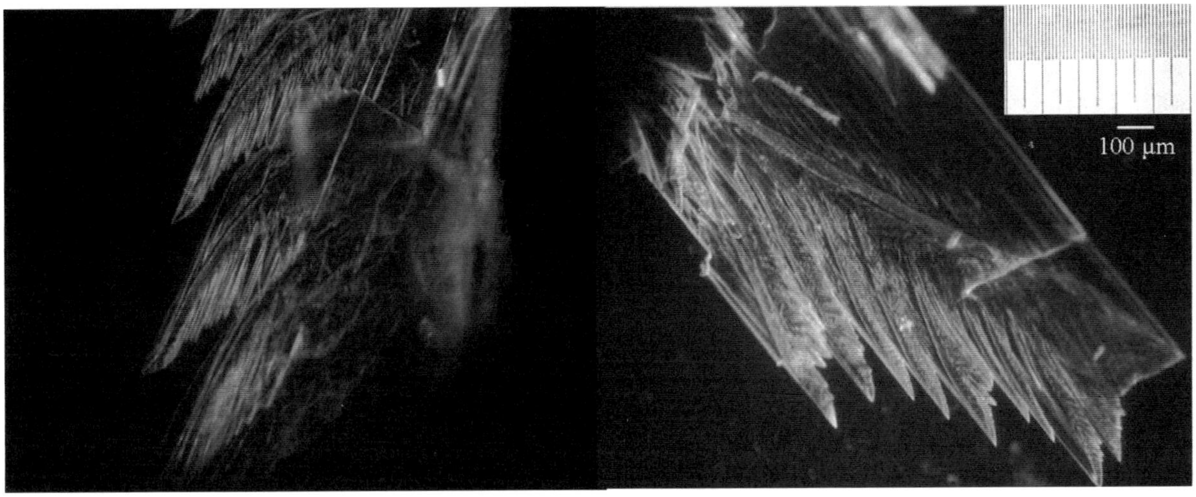

Figure 3. Images of α-MoO_3 crystallite from different view angles

elements were obtained using OES techniques. As an example, the emission spectrum of the discharge between manganese granules is shown in Fig.4 together with the relative emission of manganese lines from NIST database [13]. One can see that the spectrum contains mainly metal emissions lines (Mn I and Mn II). For discharges between copper granules [11] we also observed oxygen triplet (λ=777 nm), and hydrogen Balmer H_α and H_β lines, which are typical for emission spectrum of the electric discharge plasma in water [14].

The excitation temperature of manganese ions in plasma was determined by Boltzmann plot technique using Mn II spectral lines 345.94, 347.32, 349.55, 432.61, 434.37 and 530.11 nm under assumption of Boltzmann distribution of population of ion's energy levels. Spectroscopic data for these lines were used from [13]. Fig.5 shows the Boltzmann plot for emission of Mn II lines derived from spectrum presented in Fig.4. The slope of the line in this Boltzmann plot gives the value of the excitation temperature of 9000 K. Similar estimation of the excitation temperature in discharge between copper granules showed the values of 12000±2000 K [11]. Spectral lines of Cu I 456.1, 510.5, 515.3, 521.8, 570.0, 578.2, 793.3 and 809.2 nm were used in this case. Spectroscopic data for these lines were used from [15].

4. Conclusions

Self-organized vapor-deposition process of MoO_3 crystals formation takes place in the proposed plasma source. The investigation of crystal formation zone allows to confirm the following process mechanisms: molybdenum surface oxidation, evaporation of oxide layer, vapour transportation by convectional air flow and crystal growth.

Studies of plasma of the underwater electric spark discharge with granules of Cu, Fe, Mn and Mo were carried out. Detailed investigation of such discharge plasma between manganese and copper granules was performed. The emission spectra of plasma and its excitation temperature were obtained using OES techniques. The obtained results show that values of the excitation temperatures of 12000 K and 9000 K were obtained in underwater discharge plasma between copper and manganese, respectively.

References

[1] P. Badica. Preparation through the vapor transport and growth mechanism of the first-order hierarchical structures of MoO_3 belts on siliminite fiber. *Crystal growth and design*, 7(4):794–801, 2007. doi:10.1021/cg060893s.

[2] A Laviña, J.A. Aznårez, and C. Ortiz. Electron microscopy study of some molybdenum oxide crystals. *Journal of Crystal Growth*, 48:100–106, 1980. doi:10.1016/0022-0248(80)90198-0.

[3] D. Mariotti, T. Belmonte, J. Benedikt, T. Velusamy, G. Jain, and V. Švrček. Low-temperature atmospheric pressure plasma processes for "green" third generation photovoltaics. *Plasma Processes and Polymers*, 13:70–90, 2016. doi:10.1002/ppap.201500187.

[4] W. Li, F. Cheng, Z. Tao, and J. Chen. Vapor-transportation preparation and reversible lithium intercalation/deintercalation of α-MoO_3 microrods. *J. Phys. Chem. B*, 110:119–124, 2006. doi:10.1021/jp0553784.

[5] D. Mariotti, H. Lindstrom, A. Chandra Bose, and K. Ostrikov. Monoclinic α-MoO_3 nanosheets produced by atmospheric microplasma: application to lithium-ion batteries. *Nanotechnology*, 19:49530, 2008. doi:10.1088/0957-4484/19/49/495302.

[6] K. T. Queeney and C. M. Friend. Site-selective surface reactions: hydrocarbon oxidation processes on oxidized Mo(110). *J. Phys. Chem. B*, 104:409–415, 2000. doi:0.1021/jp991994m.

[7] E Comini, G Faglia, G Sberveglieri, C Cantlini, M Passacantando, S Santucci, Y Li, W Wlodarski, and W Qu. Carbon monoxide response of molybdenum oxide thin films deposited by different techniques.

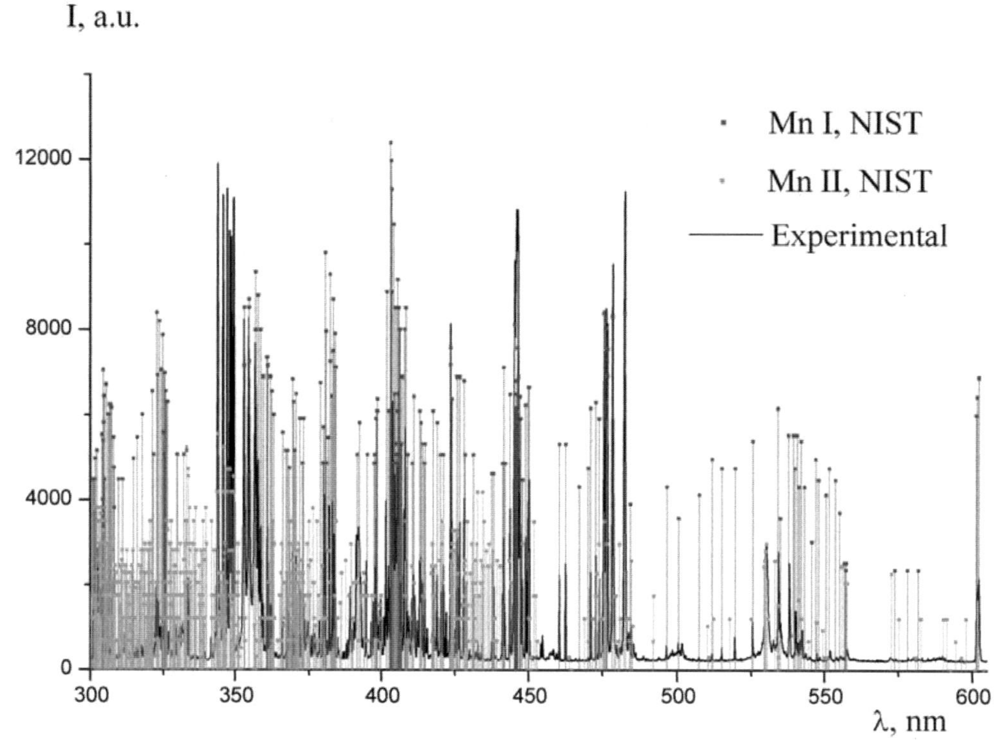

Figure 4. Emission spectrum of plasma of underwater electric spark discharge between granules of Mn.

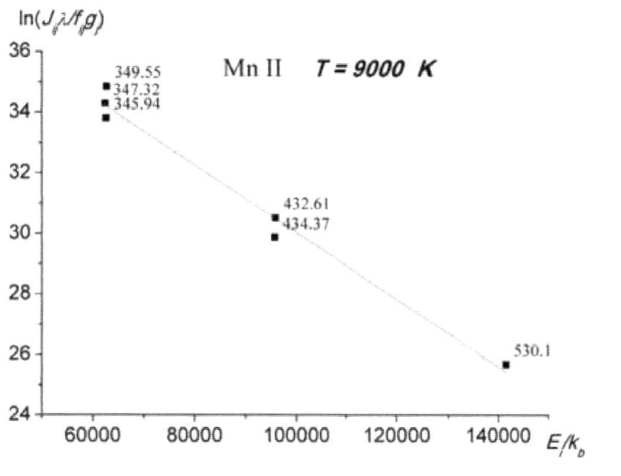

Figure 5. Boltzmann plot for calculation of excitation temperature of the underwater spark discharge plasma between Mn granules.

Sensors and Actuators B, 68:168–174, 2008. doi:10.1016/S0925-4005(00)00484-6.

[8] N. Taran, O. Gonchar, and Lopatko K. The effect of colloidal solution of molybdenum nanoparticles on the microbial composition in rhizosphere of cicer arietinum l. *Nanoscale Res Lett.*, 9:289, 2014. doi:10.1186/1556-276X-9-289.

[9] D. Ehrhardt and W. Frommer. New technologies for 21st century plant science. *The Plant Cell*, 24(2):374–394, 2012. doi:10.1105/tpc.111.093302.

[10] A. Lebid, V. Boretskij, S. Savenok, A. Veklich, and O. Andreev. Thermal plasma source for processing of MoO_3 crystals. *J. Phys.: Conf. Ser.*, 550:012027, 2014. doi:10.1088/1742-6596/550/1/012027.

[11] T. Tmenova, A. Veklich, V. Boretskij, Y. Cressault, F. Valensi, K. Lopatko, and Y. Aftandilyants. Optical emission spectroscopy of plasma of underwater electric spark discharges between metal granules. *Problems of Atomic Science and Technology. Series: Plasma Physics*, 107(23):132–135, 2017.

[12] Vahldiek F.W. Growth and microstructure of molybdenum oxide. *Journal of the Less-Common Metals*, 16:351–359, 1968. doi:10.1016/0022-5088(68)90132-X.

[13] A. Kramida, Yu. Ralchenko, J. Reader, and and NIST ASD Team. NIST Atomic Spectra Database (ver. 5.3), [Online]. Available: http://physics.nist.gov/asd [2017, May 1]. National Institute of Standards and Technology, Gaithersburg, MD., 2015.

[14] H. Li, J. Kang, and K. Urashima. Comparison between the mechanism of liquid plasma discharge process in water and organic solution. *J. Inst. Electrostat. Jpn.*, 37(1):22–27, 2013.

[15] I. Babich, V. Boretskij, A. Veklich, and R. Semenyshyn. Spectroscopic data and stark broadening of cu i and ag i spectral lines: Selection and analysis. *Advances in Space Research*, 54:1254–1263, 2014. doi:10.1016/j.asr.2013.10.034.

doi:10.14311/ppt.2017.1.32
Plasma Physics and Technology **4**(1):32–35, 2017

CHANGING THE SURFACE PROPERTIES OF ABS PLASTIC BY PLASMA

S. Chlupová, J. Kelar, P. Slavíček*

Department of Physical Electronics, Masaryk University, Kotlářská 2, 602 00 Brno, Czech Republic

* ps94@sci.muni.cz

Abstract. The research deals with plasma treatment of acrylonitrile butadiene styrene plastic (ABS). The plastic was treated with Diffuse Coplanar Surface Barrier Discharge (DCSBD) and Gliding arc at atmospheric pressure. ABS was chosen because of its low price, wide use in industry. Samples were analyzed with contact angle measurement and the surface energy was determined as well. The results show that only a few seconds of plasma treatment can cause relatively large wettability change.

Keywords: plasma, surface treatment, DCSBD discharge, ABS plastic, contact angle, surface energy.

1. Introduction

In the last few decades acrylonitrile butadiene styrene (ABS) has attracted a lot of attention for its use in industry; well known are for example productions of protective helmets, small kitchen tools, toys such as dolls, cars and LEGO. The main reasons why ABS plastics are so favorite nowadays are its chemical structure, weathering, aging resistance and thermal properties. Thanks to the chemical composition, ABS belongs to the group of polystyrenes with high impact strength, it is also effectively stabilized against aging or fluctuations in temperature and humidity. Even though the ABS plastic was described chemically and we know its physical properties, it has not been described after plasma treatment very well.

In recent years, the nonthermal dielectric barrier discharge (DBD) at atmospheric pressure has been used extensively for surface treatment of polymers [1], [2]. Plasma treatment is enviromentally friendly in comparison with chemical treatment [3]. Using this type of discharge has a lot of advantages, no vacuum is needed and samples can be treated at atmospheric pressure with easy setup, so it is a low-cost treatment method. We can use different gases, such as He, Ar, N_2. Polymers surface modification have been studied recently because of an improvement of physical properties [4]. However, polymers are appropriate for use in industry because of their low density, flexibility etc., their surface properties do not go hand in hand with demands regarding biocompatibility, wettability, adhesion and friction. This is the main reason we examined ABS; an appropriate surface modification can raise the use of this plastic. After plasma treatment, free surface energy is higher and wettability is better. In presented work we used two atmospheric pressure plasma sources, Gliding arc and Dielectric Coplanar Surface Barrier Discharge (DCSBD). This two sources were used due to their low cost because no vacuum is needed. DCSBD is a special type of DBD with a parallel arrangement of electrodes, which are located in the ceramic plate (serving as a part of dielectric barrier) and dielectric cooling oil [5], [6]. This special type of non-isothermal plasma has been developed for wood, textiles, fibres, polymers treatment etc. [7], [8]. The purpose of this contribution is to show changes in wettability and free surgace energy after the plasma treatment of ABS plastic.

2. Experimental part

Two types of treatments have been used for experiments. The first treatment was carried out with DCSBD and the second one with Gliding arc. These two methods were chosen for their industry applications and small costs.

2.1. DCSBD

DCSBD is an abbreviation for Diffuse Coplanar Surface Barrier Discharge. It is a special type of dielectric barrier discharge operating at atmospheric pressure and in many different gases including air. The scheme of coplanar arrangement of electrodes can be seen at Figure 1. The main purpose of this plasma source is plasma treatment of planar samples in the solid state. This type of barrier discharge has been developed by Cernak's group at the Department of Physical Electronics at Masaryk university and at Comenius University in Bratislava. This plasma source consists of parallel metallic electrodes embedded in ceramic and cooled by a cooling oil, see Figure 2. The cooling oil is necessary because of collisions between the charged particles, which heat the device. The ceramic can be heated by dielectric losses too. In addition to that, cooling oil is cooled by a water cooling system to the temperature about 16–20 °C. The frequency range of supply voltage is 15–50 kHz, the input power is in the range of 50–700 W, the voltage is up to 20 kV peak to peak and amplitude of discharge current is about 0.4 A . The advantage of DCSBD configuration is a large discharge area so big samples can be treated as well.

The effective area of plasma produced by the DCSBD was 8 × 20 cm. The plasma thickness is about 0.3 mm. With the use of DCSBD adhesion, wettability etc can be changed. This type of discharge

Figure 1. The coplanar arrangement of the electrodes

is used for changing material properties very often, for example if we want to change hydrophobic material to hydrophillic one. In presented work, power of 400 W and frequency of 15 kHz were used.

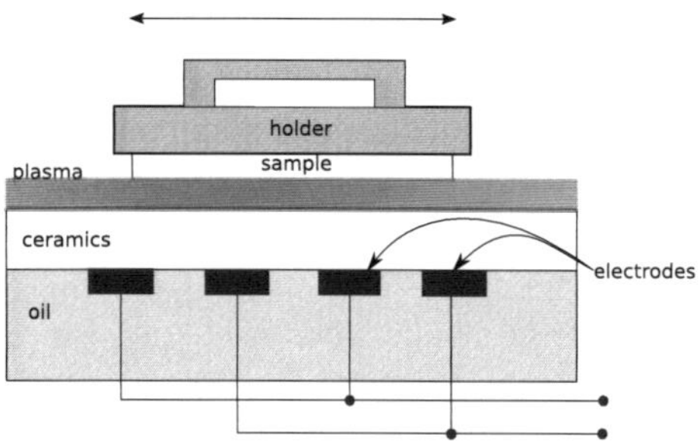

Figure 2. The scheme of the plasma treatment of the sample by the DCSBD

2.2. Gliding arc

The second type of plasma source used was Gliding arc. It is a commercial plasma source generating non-isothermal plasma. A discharge is excited between two corner electrodes made from different materials. A discharge is formed at the narrowest point, then it expands, slides along the electrodes and dissapears. New discharge emerges immediately at the same initial place. The scheme of sample plasma treatment by Gliding arc can be seen at Figure 3.

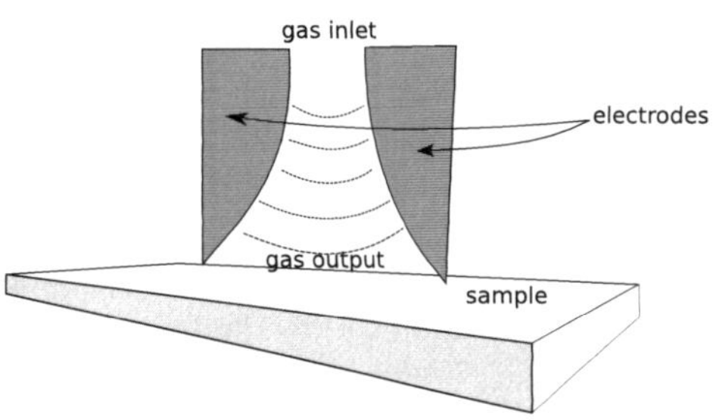

Figure 3. The scheme of the plasma treatment of the sample by the Gliding arc

The discharge journey is given by geometry of the electrodes, airflow and voltage. This plasma source operates at frequency of 40 kHz with the power approx. 500 W. The generator of high voltage produces voltage up to 10 kV. The treatment width is approx. 50–60 mm. Gliding arc is not stable in time thanks to the plasma filaments which are moving constantly, and as the result of the fact that produced plasma is not as homogeneous as DBD.

2.3. Roughness

For the better comparison of these plasma sources we measured the change of roughness before and after plasma treatment. This physical property was measured using laser confocal microscopy to get an R_a coefficient, which is the arithmetic average of the absolute values of the profile height deviations from the mean line, recorded within the evaluation length. Simply put, R_a is the average of a set of individual measurements of the surfaces peaks and valleys. R_a can be calculated as

$$R_a = \frac{1}{L} \int_0^L |Z(x)|\, \mathrm{d}x \tag{1}$$

L is the evaluation lenght and $Z(x)$ is the profile function.

2.4. Experimental procedure

Acrylonitrile butadiene styrene (ABS) was used for experiments . The ABS that was used is used primarily for edges of tables when it is ironed on the edges. This type of ABS is sold in long rolls so it was necessary to warm the plastic to make it straight and then make 5 cm x 2 cm samples. The experimental procedure starts with mechanically cleaning the sample with the mixture of cyclohexane and isopropylalcohol. After ten minutes when sample was dried it was plasma treated. The best way how to treat the sample on the DCSBD is to paste it on the holder as it can be seen in the Figure 2. The holder was specially made for this kind of plasma treatment on the 3D printer because we needed to keep the plasma layer thickness about 0.3 mm. On Gliding arc, the sample was pasted on the stand which is a part of a device delivered by company Electronic Diener. Samples were treated for 1 second, 3 seconds, 5 seconds, 10 seconds and 30 seconds. After this the sample was taken to a See system to analyze the contact angle and the free surface energy. the See system is a device developed by Advex Instruments, a spin-off company of Masaryk

University. The software provided can calculate the free surface energy on the basis of the most often used models. Ten drops of deionized water, ethylene glycol and diiodomethane were used for measurement. Used drop volume was 1 µl.

3. Results

3.1. Changing of surface energy

The dependence of water contact angle on aging time after plasma treatment of ABS plastic by both the atmospheric plasma sources is shown in Figures 4–7. Contact angle and surface energy changes were measured after plasma treatment with five different exposure time and with several different aging times, that means the time interval after the plasma treatment, and before contact angle measurement when samples were stored in the Petri dish in the laboratory, to avoid the change of temperature or humidity.

As it can be seen in the graphs, the exposure time, the aging time and the type of plasma source have a big influence on the samples hydrophilicity. To obtain the lowest contact angle, meaning the highest hydrophilicity of the surface, a longer exposure time was required. The plasma treatment of the ABS caused a big increase of the surface wettability. The water contact angle of the reference is 94.7 degrees, which means that the contact angle was greatly reduced by over 40 degrees. The ABS after plasma treatment was very hydrophilic.

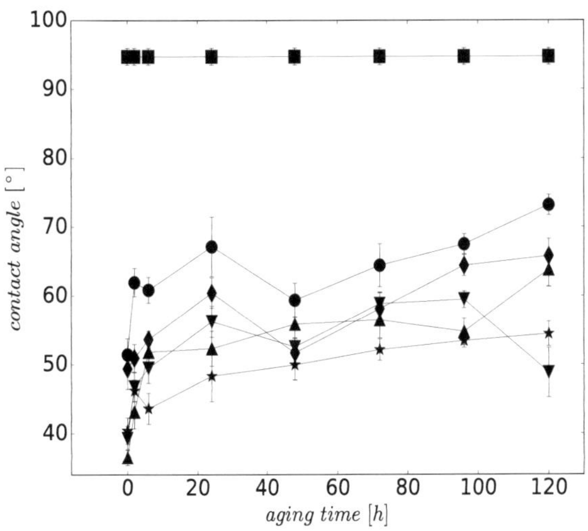

Figure 4. The water contact angle for ABS treated on the DCSBD with different exposure times, (●) 1 s, (◆) 3 s, (▲) 5 s, (▼) 10 s, (★) 30 s, (■) reference

As for the Gliding arc, we have better results in keeping the water contact angle low for a long time. After 5 days, the sample treated on DCSBD have the water contact angle over 50 degrees but for the Gliding arc we get the water contact angle of 45–50 degrees. We have the same results for the ethylene glycol. After 5 days, the sample treated by the Gliding arc had contact angle about 20 degrees, which is lower value than on DCSBD. In the case of the diiodomethane, this difference is not that big because the contact angle after 5 days is about 45–50 degrees for both plasma sources.

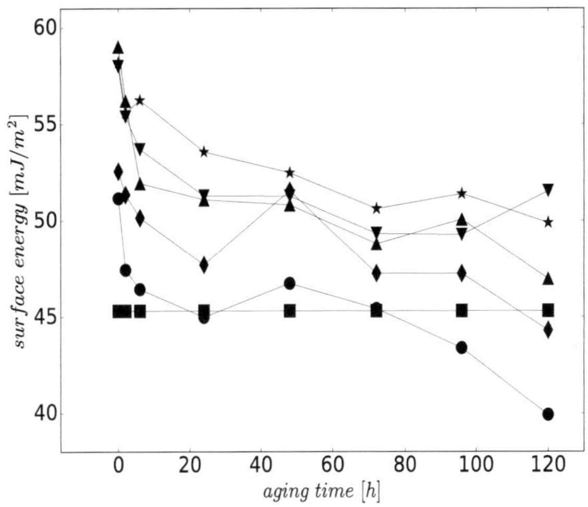

Figure 5. The surface energy for ABS treated on the DCSBD with different exposure times, (●) 1 s, (◆) 3 s, (▲) 5 s, (▼) 10 s, (★) 30 s, (■) reference

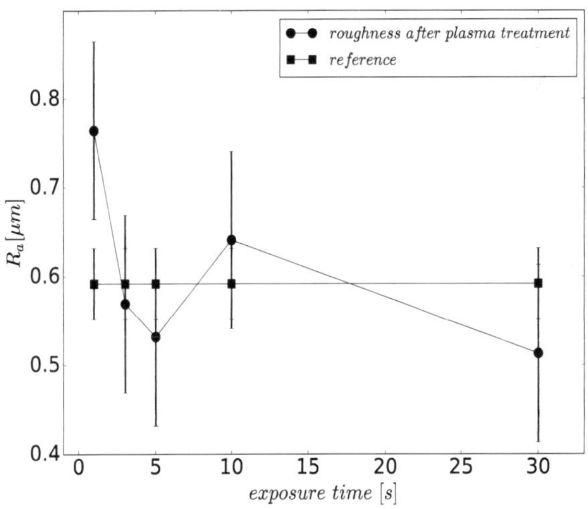

Figure 6. The roughness after the plasma treatment on the DCSBD

3.2. Roughness after plasma treatment

The roughness was measured by the confocal microscope Olympus LEXT OLS4000. Coefficient R_a was calculated from area 646 x 646 µm. Results of roughness after plasma treatment are shown in Figures 6 and 9. Roughness results are the same for both plasma sources. We found with the confocal microscopy that we cannot change the sample roughness very significantly using DCSBD or Gliding arc.

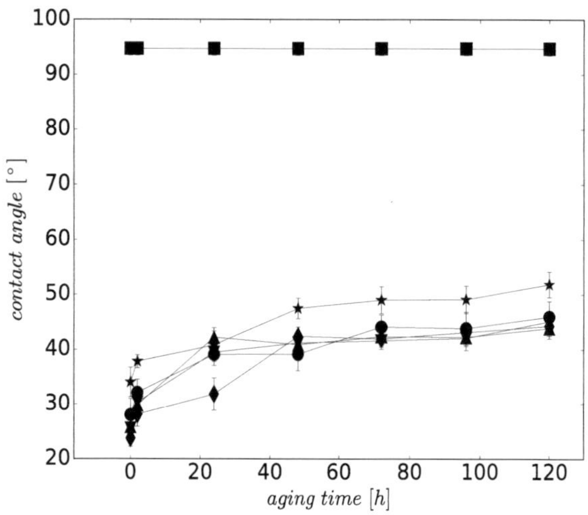

Figure 7. The water contact angle for ABS treated on the Gliding arc with different exposure times, (●) 1 s, (◆) 3 s, (▲) 5 s, (▼) 10 s, (★) 30 s, (■) reference

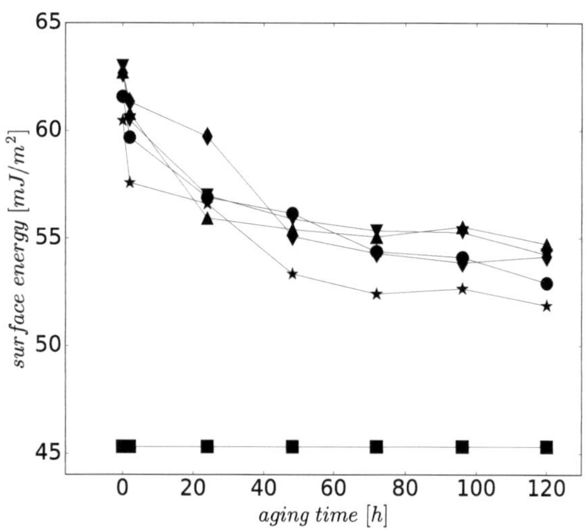

Figure 8. The surface energy for ABS treated on the Gliding arc with different exposure times, (●) 1 s, (◆) 3 s, (▲) 5 s, (▼) 10 s, (★) 30 s, (■) reference

4. Conclusions

In conclusion for this study we can say that the plasma modification in the ambient air is very suitable for increasing the surface energy and decreasing the contact angle, so the surface has better wettability. The main result of this research is the comparison between two plasma sources. Both of them can change surface properties very well. Even though the samples treated by the Gliding arc have lower contact angle, differences are not very significant but the surface energy results show that the Gliding arc is a better choice for changing the surface properties of ABS. The roughness did not change enough to attract our attention any further.

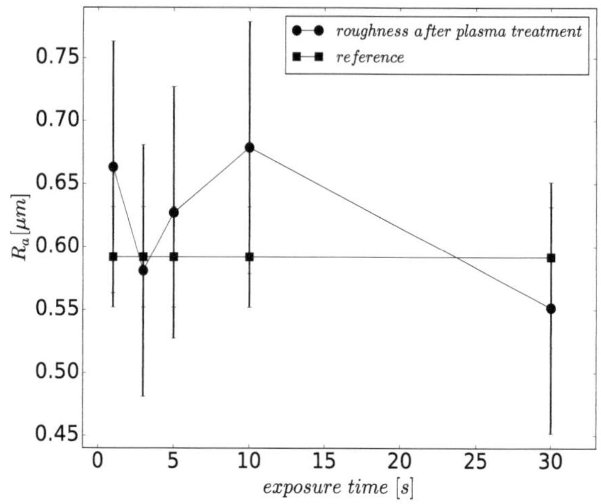

Figure 9. The roughness after the plasma treatment on the Gliding arc

Acknowledgements

This research has been supported by the project LO1411 (NPU I) funded by Ministry of Education Youth and Sports of Czech Republic.

References

[1] D. Pappas. Status and potential of atmospheric plasma processing of materials. *J. Vac. Sci. Technol, A*, 29(2):20801–20817, 2011. doi:10.1116/1.3559547.

[2] H.E. Wagner, R. Brandenburg, K.V. Kozlov, A. Sonnenfeld, P. Michel, and J.F. Behnke. The barrier discharge: Basic properties and applications to surface treatment. *Vacuum*, 71(3):417–436, 2003. doi:10.1016/S0042-207X(02)00765-0.

[3] W. Cui, W. Liu, T. Wang, and Ch. Ma. The application of atmospheric pressure dielectric barrier discharge plasma on the cleaning of photovoltaic panels. *IEEE Transactions on Plasma Science*, 45(2):328–335, 2017. doi:10.1109/TPS.2017.2647991.

[4] Z. Fang, X. Wang, T. Shao, and Ch. Zhang. Influence of oxygen content on argon/oxygen dielectric barrier discharge plasma treatment of polyethylene terephthalate film. *IEEE Transactions on Plasma Science*, 45(2):310–317, 2017. doi:10.1109/TPS.2016.2633063.

[5] V.I. Gibalov and G.J. Pietsch. Dynamics of dielectric barrier discharges in different arrangements. *Plasma Sources Science and Technology*, 21(2):024010, 2012. doi:10.1088/0963-0252/21/2/024010.

[6] M. Simor, J. Rahel, P. Vojtek, M. Cernak, and A. Brablec. Atmospheric pressure diffuse coplanar surface discharge for surface treatments. *Applied Physics Letters*, 81(15):2716–2718, 2002. doi:10.1063/1.1513185.

[7] V. Prysiazhnyi and M. Cernak. Air plasma treatment of copper sheets using diffuse coplanar surface barrier discharge. *Thin solid films*, 520(21):6561–6565, 2012. doi:10.1016/j.tsf.2012.06.069.

[8] T. Homola et al. Plasma treatment of glass surfaces using diffuse coplanar surface barrier discharge in ambient air. *Plasma Chemistry and Plasma Processing*, 33(5):881–894, 2013. doi:10.1007/s11090-013-9467-3.

doi:10.14311/ppt.2017.1.36
Plasma Physics and Technology **4**(1):36–39, 2017

PROPERTIES OF VACUUM ARC INFLUENCED BY ELECTRODE DIAMETER AND MATERIAL IN TMF CONTACT BASED ON FORCED CURRENT ZERO

Huo Wenlei[a,*], Wu Jianwen[a], Deng Yun[a], Jin Xinchen[a], Zhu Liying[b]

[a] *School of Automation Science and Electrical, Beihang University, Beijing 100191, China*

[b] *Institute of Spacecraft System Engineering CAST, Beijing 100094, China*

[*] by1303142@buaa.edu.cn

Abstract. With the increase in electrical equipment in More/All Electric Aircraft, 270V dc power supply systems will be needed. One method for DC interruption is forced current zero (FCZ). Based on FCZ technology with transverse-magnetic-field (TMF) contact, the spiral-type contacts are designed. Experiments with different currents are carried out with contact diameters being 30 mm, 40 mm, and 50 mm, and arcing surface materials Cu-W80 alloy and Cu-Cr50 alloy respectively. It is indicated by the experimental results that breaking capacity of vacuum interrupter and vacuum arc appearance are closely related to the electrode diameter and material. For the same size of electrode diameter, the breaking capacity in Cu-Cr50 is better than that in Cu-W80. With increasing electrode diameter, arc column expansion velocity and diameter increase gradually. Breaking capacity is increasing with larger contact diameter.

Keywords: vacuum arc, forced current zero, transverse magnetic field, breaking capacity, electrode diameter and material.

1. Introduction

Forced current zero (FCZ) is a common method used to interrupt DC currents. This method for DC interruption forces the current to zero by injecting a high-frequency current of opposite polarity [1]. An oscillating current is generated by the resonance of a precharged capacitor and an inductor, and by superimposing this oscillating current over the current in the main circuit, the current is forced to zero. Then the breaker in the main circuit can interrupt the DC current successfully [2]. Thus, the burning arc in the vacuum interrupter (VI) can be extinguished by the FCZ. The experimental analysis of this switching method has been studied frequently [3].

The VI technology with very fast recovery precesses after current zero enables high current interruption with large current gradients ($\mathrm{d}i/\mathrm{d}t$). Thus, it was considered to be suitable for DC current interruption [4]. A successful interruption is significantly influenced by the transient characteristics of the current-commutation process, such as the rates of current and voltage at zero-crossing ($|\mathrm{d}i/\mathrm{d}t|$ and $|\mathrm{d}v/\mathrm{d}t|$). To reduce the size and cost of the forced-commutation branch, a high frequency is usually selected for the commutation branch. However, this condition is unfavorable for the recovery of dielectric strength as well as for the interruption capability of the VI after current zero. Glinkowski mainly researched the impact of the current zero-crossing time $\mathrm{d}i/\mathrm{d}t$ of the arc breaking [5]. In case of high frequency interruption, reignition issues is becoming more likely. There are more publications on the high-frequency currents, which can lead to multiple reignition and interruptions because of the reduced dielectric strength of the vacuum gap, mostly as a result of the small contact gaps.

Electrode and material are important fields in vacuum arc research. Niwa [6] and Sugita [7] investigated the effect of various contact materials on interruption phenomena by measuring the temperature of the anode surface. Liu [8] found the interruption capacity of vacuum interrupters with the slot-type AMF electrodes is related to the ratio of the anode diameter to the gap length. Lindmayer focused on the contact material for high-frequency vacuum arc experimental phenomenon [9]. In [10], properties of intermediate-frequency vacuum arc Influenced by electrode diameter and material in axial magnetic field were researched.

In this paper, the application of forced current zero method to DC circuit breakers for aviation 270 V power system was investigated. Experiments of different currents lower than 5 kA were carried out in interrupters with contact electrode materials being Cu-W80 alloy and Cu-Cr50 alloy at different forced commutation frequency. Besides, the breaking performances were also analysed with the diameters being 30 mm, 40 mm and 50 mm respectively.

2. Experimental setup

Four different commercial VI with spiral-type TMF electrodes without shield are used in this paper. The structural parameters are listed in Table 1. The air pressure inside the VI is 5×10^{-5} Pa. In order to reduce the contact resistance on the condition of ensuring

Contact type	A	B	C	D
Diameter	40 mm	40 mm	30 mm	50 mm
Contact material	CuCr25	CuCr25	CuCr25	CuCr25
Arcing material	CuW80	CuCr50	CuCr50	CuCr50
Contact resistance	12 µΩ	13 µΩ	17 µΩ	12 µΩ

Table 1. Structural parameters of electrodes in VI.

C_1 (µF)	L_1 (µH)	R_1 (mΩ)	f (kHz)
1620	3.4	22	2.1
1080	0.95	14	4.8
250	0.95	14	10.3

Table 2. Parameters of forced commutation branch.

C_0	L_0	C_2	R_2	C_3	R_3
2 F	14 µH	50 mF	2.5 Ω	0.1 µF	0.1 Ω

Table 3. Experimental system parameters for DC forced interruption.

breaking performance, the materials of contact surface and arcing surface are chosen differently. The arcing surface is weld to the base of electrode. When the contact material is Cu-Cr25, the contact resistance is between 12 µΩ to 17 µΩ. The arcing material of electrodes are Cu-W80 alloy and Cu-Cr50 alloy in type A and type B respectively. The contact gap is between 2–2.5 mm.

A direct circuit system is used to study the DC forced commutation interruption, as shown in Figure 1. The instantaneous short-circuit current is supplied from the capacitor C_0 with a capacitance value of 2 F, by triggering the bidirectional thyristor VT_0 controlled by the measure and control system. The main current ratings are obtained by adjusting R_0 which is varied from 55 mΩ to 500 mΩ. As the contact separation time is uncertain, the time of arc ignition is also uncertain; thus, an ignition circuit branch is designed to ignite the arc before the main current is injected. When contact separation occurs, an arc is formed between the anode and the cathode in the VI. At this time, the VI current is small as long as the arc could be maintained, until the contacts are fully open. This current is provided by VT_2, C_2, and R_2. The branch of R_3 and C_3 is designed as a frequency modulation circuit. The forced commutation branch includes VT_1, C_1, and R_1. The different forced commutation frequencies (2.1 kHz–10.3 kHz) are obtained by changing the parameters of forced commutation branch as are shown in Table 2. The varistors are used to absorb the energy stored in inductance when the circuit is interrupted.

The arc current is measured using a Rogowski Coil. A Phantom v7.3 high-speed video camera and a PC are used to record the arc appearances and the arc moving processes with a camera sampling rate of 72000 frames/s and exposure time of 1 µs. The parameters of the experimental system for DC forced interruption are listed in Table 3.

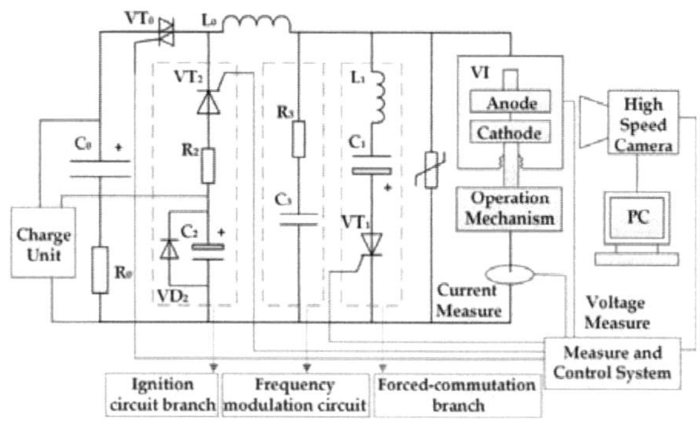

Figure 1. Experimental system for DC forced commutation interruption.

3. Results and analysis

3.1. Interruption characteristics

Waveforms of arc voltage for vacuum interrupters at about 3.3 kA with the same diameter (40 mm) but various forced commutation frequency are shown in Figure 2. The precharged voltages of commutating capacitor in forced commutation branch are 250 V, 224 V and 406 V, corresponding to 2.1 kHz, 4.8 kHz and 10.3 kHz. The waveform of arc voltage at 2.1 kHz is obtained with the branch of varistor, unlike the other two waveforms.

In order to increase the forced commutation frequency, the values of inductance and capacitance in the forced commutation branch need to be reduced. The low frequency is increased by reduction of the commutating capacitor. Thus, the voltage recovery time and interruption time is shorten. However, the over-voltage is much higher with high forced commutation frequency when the current is fixed. Besides, the varistor branch is set to absorb the energy. In order to reduce the over peak voltage, the capacitance in frequency modulation circuit should be larger. When the interruption current is about 3.3 kA, the negative voltages are −195 V, −110 V and −295 V with the increasing of commutation frequency. Comparing the peak of negative voltage in the different tests, it seems that

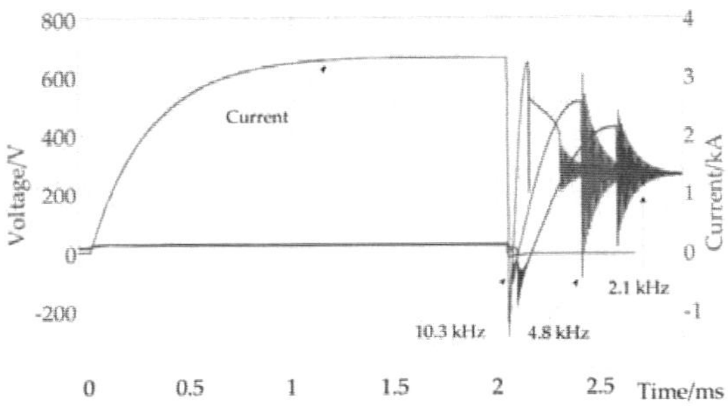

Figure 2. Waveforms of arc voltage for vacuum interrupters. (interruption current is about 3.3 kA, electrode diameter is 40 mm, the arcing time is 2 ms, the forced commutation frequencies are 2.1 kHz, 4.8 kHz and 10.3 kHz)

the peak value is related to the value of precharged voltage in forced commutation branch. The higher precharged voltage, the larger the negative peak value is.

3.2. Properties of vacuum arc influenced by electrode materials

Utilizing the experimental system described above, interruption experiments were carried out with currents of approximately 5 kA, using the type A and type B spiral-type TMF contacts. The typical vacuum arc appearance is shown in Figure 3 for 2 ms arcing time. The arcing materials are Cu-W80 (Figure 3 (a)) and Cu-Cr50 (Figure 3 (b)).

In both two types of contacts, at t=37.5 µs, a constricted arc developed and the arc in the contact anchored at the ignited point. At this early expansion stage, there were few or no cathode spots outside the central column. Then the cathode spots began to move around the arc column. The arcs moved, increasing the cross-sectional area and new cathode spots were formed constantly, which moved to the edges of the contacts. By the time t=2 ms with type A contact, the cathode spots had occupied almost the entire cathode surface, and they did not expand further. The cathode spots in the diffuse vacuum arc was also discussed in [11], and the TMF contact behaved as butt contact at the distance less than 3 mm. When the arcing time was 2063.9 µs, forced commutation was triggered, and the arc current began to decrease, flowing into the forced-commutation branch. With the decreasing of the arc current, the brightness of the arc and particles were reduced, and the metal vapour began to diffuse between the contacts. It was obvious that large amounts of metal vapour particles were filled the contact gap with the Cu-W80 contact.

The melting point and boiling point of W were much higher than that of Cr, which made the contact temperature higher for thermionic emission. The contact may be heated locally by arc under high current. For the locally high temperature, the electrons would be emitted intensively by the cathode and more metal

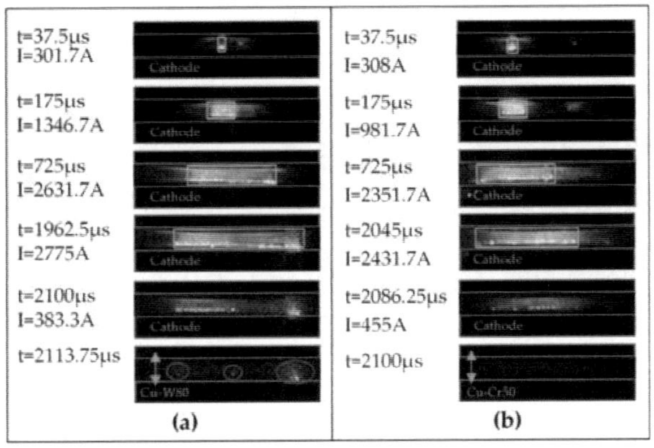

Figure 3. Typical vacuum arc appearance in (a) type A and (b) type B TMF contacts with 2 ms arcing time.

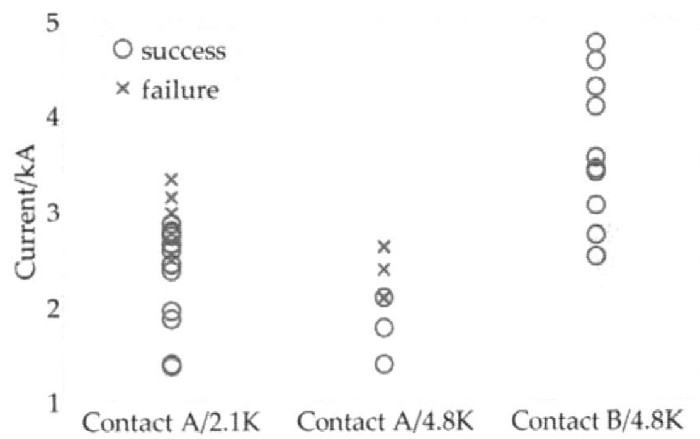

Figure 4. Interruption results of type A and type B vacuum interrupters with 2 ms arcing times at different forced commutation frequencies.

vapours were generated and filled the gap, which was red circle marked in Figure 3 (a). With the increasing the particles, the probability of reignition increased after the current zero. The particle density grew rapidly and the conductivity of arc area was very good when the arc filled the contact gap, which caused a failure to interrupt at first current zero.

Interruption results and breaking capacity of type A and type B vacuum interrupters were shown in Figure 4 with 2 ms arcing times at different forced commutation frequencies. when the current was 2.1 kA and forced commutation frequency was 4.8 kHz the reignition occurred at first current zero, while under the same condition the breaking capacity was larger than 5 kA in Cu-Cr50 contact. Meanwhile, with increase of the forced commutation frequency, the breaking capacity was much lower in Cu-W80 contact.

3.3. Breaking performance with different electrode diameters

In order to analyse the influence of electrode diameters on the breaking capacity, experiments were conducted with type B, type C, and type D contacts for arcing times of 2 ms and 2.3 mm gap length. The forced commutation frequency is 10.3 kHz. The experimental results of the breaking performance are shown in

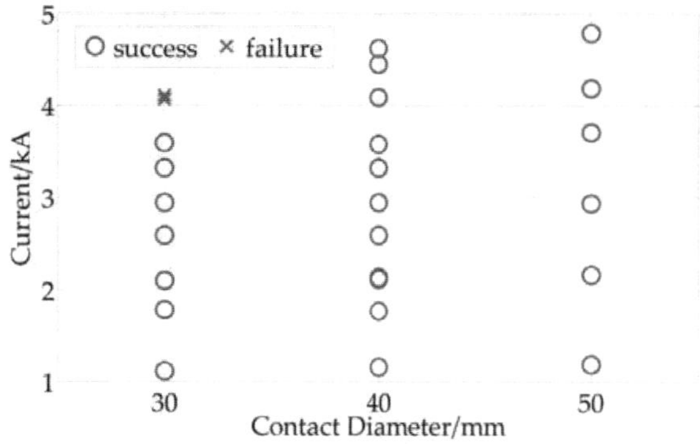

Figure 5. Breaking experiment results of different electrode diameters with 2 ms arcing times at 10.3 kHz forced commutation frequencies.

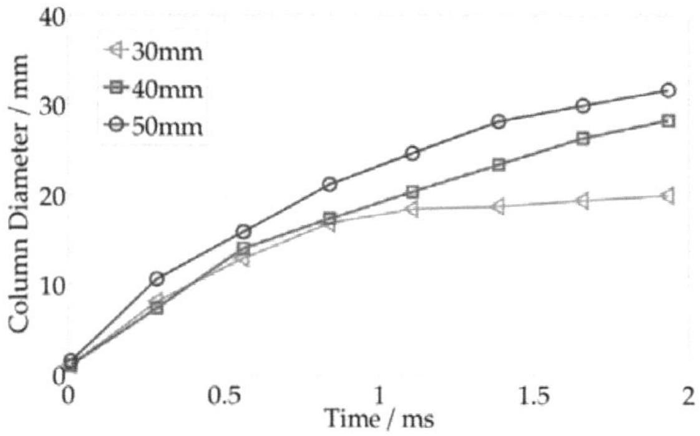

Figure 6. Diameters of diffuse columns at 3.3 kA with different electrode diameters.

Figure 5.

It can be concluded that, as the electrode diameter increases, the breaking capacity of VI is greatly improved. When the electrode diameter is 30 mm, the reignition at first current zero occurs, and the maximum breaking ability is about 4 kA. The diameters of the diffuse columns at 3.3 kA with different electrode diameters are shown in Figure 6. In the early stages of column diffusion, the arc expands rapidly. The larger the electrode diameter, the faster the expansion will be. With increase in the diffusion diameter, the arc diffusion velocity will decrease gradually. After approximately 1.4 ms, the arc tends to expand slowly. The maximum column diameter is proportional to the electrode diameter.

4. Conclusions

Based on method of FCZ, the properties and breaking performances of spiral-type TMF contacts were investigated for 270 V DC application and the results were summarized below:

The negative peak voltage is not only determined by the commutation frequency but also related to the value of precharged voltage in forced commutation branch. The higher precharged voltage, the larger the negative peak value is.

For the locally high temperature, the electrons are emitted intensively by the cathode and more metal vapours are generated and filled the gap in the Cu-W80 contact. Thus, the breaking capacity in Cu-W80 electrode is much lower than that in Cu-Cr50 electrode. With increase of the forced commutation frequency, the breaking capacity is much lower in Cu-W80 contact.

The larger the electrode diameter, the higher speed and the lager maximum column diameter of the expansion will be. As the electrode diameter increased, the breaking capacity of VI is improved.

Acknowledgements

This study was co-supported by the National Natural Science Foundation of China (No. 51677002), the National Natural Science Foundation of China (No. 51407008)

References

[1] A. Klajn et al. Analysis of a forced extinction of a diffuse vacuum arc for digital modelling purposes. In *20th ISDEIV*, 2002. doi:10.1109/ISDEIV.2002.1027396.

[2] Y. Zhang et al. Experimental investigation on hvdc vacuum circuit breaker based on artificial current zero. In *27th ISDEIV*, volume 2, pages 1–4, Sept 2016. doi:10.1109/DEIV.2016.7763995.

[3] L. Bin et al. Study on dynamic characteristic in force interrupted dc vacuum arc. *Trans. on Plasma Sci.*, 42(10):2382–2383, 2014. doi:10.1109/TPS.2014.2334639.

[4] R. Piovan et al. Vacuum breaker for high dc current: Experimental performances and operational limits. *Trans. on Plasma Sci.*, 37(1):229–235, Jan 2009. doi:10.1109/TPS.2008.2005671.

[5] M. Glinkowski et al. Some interruption criteria for short high-frequency vacuum arcs. *Trans. on Plasma Sci.*, 17(5):741–743, 1989. doi:10.1109/27.41189.

[6] Y. Niwa et al. The effect of contact material on temperature and melting of anode surface in the vacuum interrupter. In *19th ISDEIV*, volume 2, pages 524–527 vol.2, 2000. doi:10.1109/DEIV.2000.879042.

[7] M. Sugita et al. Interruption phenomena for various contact materials in vacuum. *IEEE Trans. on Plasma Sci.*, 37(8):1469–1476, Aug 2009. doi:10.1109/TPS.2009.2019654.

[8] Z. Liu et al. Comparison of vacuum arc behaviors between axial-magnetic-field contacts. *IEEE Trans. on Plasma Sci.*, 36(1):200–207, Feb 2008. doi:10.1109/TPS.2007.914184.

[9] M. Lindmayer et al. The influence of circuit parameters and contact materials on the reignition of high-frequency vacuum arcs. *Trans. on Comp., Hybr., and Manufact. Technol.*, 13(1):69–73, Mar 1990. doi:10.1109/33.52851.

[10] J. Yuan et al. Properties of intermediate-frequency vacuum arc influenced by electrode diameter and material in axial magnetic field. In *3rd ICEPE*, pages 85–89, Oct 2015. doi:10.1109/ICEPE-ST.2015.7368343.

[11] P. Slade. The vacuum interrupter contact. *IEEE Trans. on Comp., Hybr., and Manufact. Technol.*, 7(1):25–32, Mar 1984. doi:10.1109/TCHMT.1984.1136331.

doi:10.14311/ppt.2017.1.40
Plasma Physics and Technology **4**(1):40–43, 2017

CALCULATION OF ARC POWER LOSSES IN THE SIMPLIFIED MODEL OF INTENSIVELY BLASTED ELECTRIC ARC

J. Senk*, I. Laznickova, I. Jakubova

Faculty of Electrical Engineering and Communication, Brno University of Technology, Technická 12, 616 00 Brno, Czech Republic

* senk@feec.vutbr.cz

Abstract. In previous versions of the simplified model of intensively blasted electric arc burning in argon in the arc heater's anode channel, the authors used the constant total power loss coefficient for estimation of arc power losses in all anode channel individual parts. Using this approach, the model with relatively low computational complexity has led to very good agreement between the total computed and experimentally obtained values, but when the computed and measured power losses of individual anode channel segments have been compared, considerable differences have been revealed. In the modified model, theoretically computed net emission coefficient of argon is used in the energy equation to express the arc power losses. This way, satisfactory accordance is achieved between not only the total, but also partial measured and computed values. Exemplary results are given in figures and tables and discussed.

Keywords: electric arc, model, experiment, power loss, net emission coefficient.

1. Introduction

The simplified model of intensively blasted electric arc burning in a cylindrical anode channel of the arc heater has been formulated [1] to be used for evaluating measured data and for getting an idea of the arc's behavior inside the experimental device. A modular-type arc heater has been designed [2] to investigate both its technological applications (e.g. for decomposition of stable harmful compounds) and its operational characteristics including also basic parameters of the arc inside, or the plasma jet at the output. Rather extent experience has been collected in the field of the design and operation, as well as in the field of modelling. Although the basics of the model remain unchanged, some presumptions or procedures have been modified in accordance with the results of numerous experiments and their analyses, and new versions of the model are created [3].

2. Estimation of the arc power loss using net emission coefficient

As described in details in [3], in previous versions of the arc model, only the (relative) differences of the measured and computed values of the total voltage and the total power loss of the arc have been evaluated as a condition for termination of the computation. Although such an approach is completely sufficient for estimation of output parameters, substantial deviations have been revealed between the measured and computed power losses of the arc in individual sections of the segmented anode channel. In other words, obviously the previous version of the model is not able to express power loss of the electric arc truly in all the parts of the arc heater's anode channel [3].

To explain this phenomenon, a short description of the used arrangement of the experimental device is useful. The arc heater consists of several separately cooled and electrically and thermally insulated parts, namely of the cathode (subscript $_{\mathrm{cat}}$), the input part of the anode channel including the cathode shell (subscript $_{\mathrm{in}}$), the main part of the anode channel (subscript $_{\mathrm{ch}}$), and the grounded part of the anode channel (the anode itself, subscript $_{\mathrm{an}}$). The arc power loss $P_{\mathrm{loss\_xx}}$ in each of these parts is determined separately. The electric arc (subscript $_{\mathrm{A}}$) is burning between the cathode tip (the arc beginning at the axial coordinate $z = 0$) and the end of the anode ($z = z_{\mathrm{L}}$) inside the cylindrical anode channel.

In the older version of the model the calculation was stopped in the ith step if the minimum of the following function was reached

$$ {}^{i}\delta_{\mathrm{tot0}} = \left\{ \left[1 - \frac{{}^{i}U_{\mathrm{A\_c}}(z_{\mathrm{L}}) I_{\mathrm{m}} + P_{\mathrm{loss\_cat\_m}} + {}^{i} P_{\mathrm{loss\_as\_c}}}{U_{\mathrm{m}} I_{\mathrm{m}}} \right]^2 + \left[1 - \frac{{}^{i}P_{\mathrm{loss\_tot\_c}}}{P_{\mathrm{loss\_tot\_m}}} \right]^2 \right\}^{\frac{1}{2}} \quad (1) $$

Here, subscript $_{\mathrm{m}}$ stands for measured, $_{\mathrm{c}}$ for calculated, $_{\mathrm{tot}}$ for total, subscript $_{\mathrm{as}}$ means anode spot, superscript ^{i} indicates the number of the computational cycle.

As described in [3], the model has been modified to be able to evaluate the power losses of the individual parts of the anode channel (input, channel, and anode) separately, because these power losses are measured separately and thus, more detailed insight on the behavior of the arc could be obtained. Related to this modification, also the objective function has been changed to the following

$$
{}^{i}\delta_{\mathrm{tot1}} = \left\{ \left[1 - \frac{{}^{i}U_{\mathrm{A\_c}}(z_{\mathrm{L}})I_{\mathrm{m}} + P_{\mathrm{loss\_cat\_m}} +^{i} P_{\mathrm{loss\_as\_c}}}{U_{\mathrm{m}}I_{\mathrm{m}}} \right]^2 + \left[1 - \frac{{}^{i}P_{\mathrm{loss\_in\_c}}}{P_{\mathrm{loss\_in\_m}}} \right]^2 + \left[1 - \frac{{}^{i}P_{\mathrm{loss\_ch\_c}}}{P_{\mathrm{loss\_ch\_m}}} \right]^2 \right\}^{\frac{1}{2}} \quad (2)
$$

As demonstrated in [3] on several experiments differing in the input power between approx. 5 and 25 kW, the value of δ_{tot1} (2) has increased significantly in comparison to usual values of δ_{tot0} and reached unacceptable levels, surprisingly under medium input powers mainly (the worst case: $U_{\mathrm{m}}I_{\mathrm{m}} = 18.4\,\mathrm{kW}$, $\delta_{\mathrm{tot1}} = 0.35$).

Deeper analysis of individual terms of (2) has clearly revealed the source of the discrepancy. While the relative error of voltage (i.e. the first fraction in (2)) was lower than 0.03, the second and third fractions, i.e. relative errors of the power losses of the input part and the anode channel, were several times higher and even of opposite sign to each other. Consequently, if only the total power loss was checked, this discrepancy remained hidden. Obviously, the way of evaluating the arc power loss in the model needs a revision.

In all the previous versions of the model the same procedure was used for estimating the arc's power loss. The power loss of the arc in each step in the computational mesh was computed as a portion of the electric input power using the power-loss coefficient p_{loss}. This coefficient was determined from the integral measured values as follows

$$
{}^{i}p_{\mathrm{loss}} = \frac{P_{\mathrm{loss\_tot\_m}} - P_{\mathrm{loss\_cat\_m}} -^{(i-1)} P_{\mathrm{loss\_as\_c}}}{{}^{i}U_{\mathrm{A\_c}}(z_{\mathrm{L}})I_{\mathrm{m}}} \quad (3)
$$

In the ith computational cycle, the measured total power loss $P_{\mathrm{loss\_tot\_m}}$ was decreased by the measured power loss of the cathode $P_{\mathrm{loss\_cat\_m}}$ and the computed power loss of the anode spot $P_{\mathrm{loss\_as\_c}}$, and divided by the computed arc input power $U_{\mathrm{A\_c}}(z_{\mathrm{L}})I_{\mathrm{m}}$. Then this coefficient was applied to determine the power loss of the arc in each step in the computational mesh, thus the same ratio between the arc's input power and its power loss was applied over the total length of the arc, in all the parts of the arc heater's anode channel. This simple approach evidently works well if the output parameters of the device are of main interest. Unfortunately, it fails if the internal distribution is studied in individual parts of the anode channel where significantly different conditions rule. Probably, it is especially the input part of the anode channel which differs from the others as the arc is yet to arise there and its radius and temperature change steeply.

Based on the above given observation and consideration, the computation of the arc's power loss has been modified. Previous experiments and calculations have shown that the arc power loss is due to arc radiation and the gas temperature near the channel wall is equal to the input temperature of cold gas. The experiments discussed here have been carried out with the arc burning in pure argon. The net emission coefficient of argon under normal pressure and in the temperature range up to 30 000 K has been computed by Dr. Petr Kloc [4] and is utilized for estimation of irradiated power from the arc as follows

$$
{}^{i}\Delta P_{\mathrm{loss\_A}}(z_j) = \frac{\pi^2}{2} \left\{ \epsilon\left[{}^{i}T_{\mathrm{A}}(z_{j-1}),{}^{i} r_{\mathrm{A}}(z_{j-1})\right] + \epsilon\left[{}^{i}T_{\mathrm{A}}(z_j),{}^{i} r_{\mathrm{A}}(z_j)\right] \right\} \left[{}^{i}r_{\mathrm{A}}(z_{j-1}) +^{i} r_{\mathrm{A}}(z_j)\right] \sqrt{\Delta z^2 + r_0^2 \left[\left(\frac{z_j}{r_0}\right)^{\frac{1}{n_{\mathrm{r}}}} - \left(\frac{z_{j-1}}{r_0}\right)^{\frac{1}{n_{\mathrm{r}}}} \right]^2}\, \Delta z \quad (4)
$$

Here, ${}^{i}\Delta P_{\mathrm{loss\_A}}$ is the computed power loss of the arc in the ith computational cycle and at the step between z_j and z_{j-1}, ϵ is the net emission coefficient of argon for the given temperature and radius, r_{A} is the arc column radius, r_0 is the radius at the very beginning, at the cathode. Obviously, rather simple computation is applied to test the effect of using the net emission coefficient for calculation of the irradiated power. Averaged values of net emission coefficient are used and the power irradiated through the lateral surface of each elementary segment Δz is taken to be absorbed by the channel wall. Even so the influence on the computed power loss of individual parts of the anode channel has been supposed to be substantial because the net emission coefficient exhibits significant dependence both on the radius and temperature. Summation of power loss increments (4) over the input part, the channel and the anode gives their computed power losses as demonstrated in the following section.

Furthermore, computations have shown that an achievable minimum of (2) is distinctly influenced also by the current density j_0 on the cathode. This parameter defines the radius r_0, and thus the development of the arc radius r_{A} along the axial coordinate z [5]

$$
r_{\mathrm{A}}(z) = r_0[1 + (\frac{z}{r_0})^{\frac{1}{n_{\mathrm{r}}}}] \quad (5)
$$

Previously, $j_0 = 10^8\,\mathrm{A\,m^{-2}}$ was used, but now our further computations have indicated that several times higher values could lead to lower values of δ_{tot1}. Higher values of current density on the tungsten cathode are given e.g. in [6], [7] thus our observation does not contradict experimental results.

After the modifications explained above, such a set of three parameters $[n_r; j_0; T_{\mathrm{A}}(s)]$ has been sought for during the computations, which is giving the minimum value of δ_{tot1} (2). The modified procedure has been tested for the worst-case experimental results where the previous version has led to unacceptable differences.

3. Results and discussion

As shown in [3] and mentioned above, when the older version of the model has been used, the worst agreement between the measured and computed power losses of the input part and of the main part of the anode channel have been found for medium input power. In this section, the modified version of the model is intentionally tested for this worst-case example. In the following, a comparison of the dependences computed by the older version and the modified version of the model is given and discussed.

The analyzed experimental data have been measured on the arc heater whose geometry is defined by the following parameters: the anode channel radius $r = 8\,\mathrm{mm}$, the total anode channel length $z_{\mathrm{L}} = 109\,\mathrm{mm}$, the input part length $l_{\mathrm{in}} = 22\,\mathrm{mm}$, the anode channel length $l_{\mathrm{ch}} = 60\,\mathrm{mm}$, the grounded anode length $l_{\mathrm{an}} = 27\,\mathrm{mm}$. The arc has been blasted by the argon flow rate of $22.5\,\mathrm{g\,s^{-1}}$. The total measured voltage $U_{\mathrm{m}} = 113.4\,\mathrm{V}$, the arc current $I_{\mathrm{m}} = 162.3\,\mathrm{A}$, and the measured power losses of the anode channel's input part, the main part and of the grounded anode are 885.3 W, 1453.7 W and 1553.9 W, respectively.

Table 1 compares the obtained results of the older version of the model (the last row) and four best results obtained by the new version. Obviously, a significant improvement has been achieved comparing δ_{tot0} (1) in the last row with arbitrary one of δ_{tot1} (2) in the first four rows.

Mutual comparison of the results of the new version of the model in the first four rows shows almost negligible differences in the achieved δ_{tot1}. The final choice of the most suitable variant of them has been made with respect to the computed axial dependence of the arc temperature $T_{\mathrm{A}}(z)$ in the close vicinity of the beginning (Figure 1). The variant no. 3 [2.55; $2.67 \times 10^8\,\mathrm{Am^{-2}}$; 12700 K] exhibits the smoothest increase without oscillations. This variant is selected for the following discussion and comparison. Nevertheless, it is worth noting that the computed temperature dependences of all four variants differ only near the beginning and almost merge at about 4 mm from the beginning.

In the following figures, the results obtained by the older version of the model (var. no. 5 in Tab. 1) are denoted with the abbreviation pl (as obtained using the total power loss coefficient p_{loss}) and plotted in dashed lines, while the results obtained by the new version of the model (var. no. 3 in Tab. 1) are denoted with the abbreviation nec (as obtained using the net emission coefficient) and plotted in solid lines.

The left side of Figure 2 shows the computed axial development of the arc temperature and radius.

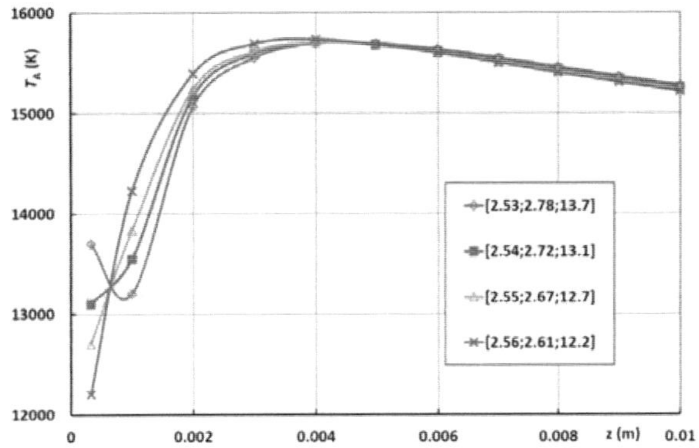

Figure 1. Detailed comparison of the computed arc temperature axial dependence $T_{\mathrm{A}}(z)$ near the beginning of the anode channel, for 4 sets of [n_{r}; $j_0 \times 10^{-8}$; $T_{\mathrm{A}}(s) \times 10^{-3}$] near the lowest value of δ_{tot1}.

Obviously,using the constant total power loss coefficient for all the parts of the arc suppresses the differences between individual parts and thus results in lower temperatures with rather flat axial dependence. On the contrary, the net emission coefficient, which is strongly dependent on radius and temperature, emphasizes the differences and thus results in distinctly higher temperatures with steeper axial dependence. Of course, with the radii, an opposite influence is observed: higher temperatures are accompanied with lower radii and the shape of the dependence $r_{\mathrm{A}}(z)$ is preserved as defined by (5). The results of the new version of the model meet our preliminary assumptions and also the expected behavior of the intensively blasted electric arc.

In the right side of Figure 2 axial dependences of the electric field intensity and the arc potential are given. The arc voltage computed by the new version of the model (var. no. 3 in Tab. 1) is higher than that obtained from the older version (var. no. 5 in Tab. 1). Unfortunately, as can be seen in Tab. 1, it is too high. The relative error of the arc voltage δU for all the four variants (var. no. 1 to 4) is positive and significantly prevails the relative errors of the power losses of the anode channel's input and main parts. Mutual relation between voltage and power-loss relative errors is opposite with the model using the net emission coefficient compared to the model using the constant total power-loss coefficient. While the older version of the model reaches very good agreement between the computed and measured voltages and fails in power losses computation, the modified version achieves very good conformity in power losses but the relative error of voltage worsens. In all the four variants no. 1 to 4 the voltage relative error is almost 0.09, which is about five times worse than in var. no. 5, and represents almost the whole of δ_{tot1}. This issue will be further investigated to reveal its cause and to find an acceptable explanation.

Var. No.	n_r (-)	$j_0 \times 10^{-8}$ ($\mathrm{A\,m^{-2}}$)	$T_A(s)$ (K)	s (mm)	$T_A(z_1)$ (K)	$\delta_{in} \times 10^2$ (-)	$\delta_{ch} \times 10^2$ (-)	$\delta_U \times 10^2$ (-)	$\delta_{tot} \times 10^2$ (-)	loss
1	2.53	2.78	13 700	0.453	13 203	−0.58	−0.35	8.92	8.94	nec
2	2.54	2.72	13 100	0.404	13 552	−0.67	−0.58	8.92	8.96	nec
3	2.55	2.67	12 700	0.372	13 835	−0.63	−0.41	8.95	8.98	nec
4	2.56	2.61	12 200	0.332	14 223	−0.30	−0.81	8.98	9.02	nec
5	2.76	1.00	9 400	0.346	10 111	23.86	−17.72	−1.86	34.83	pl

Table 1. Comparison of several selected variants.

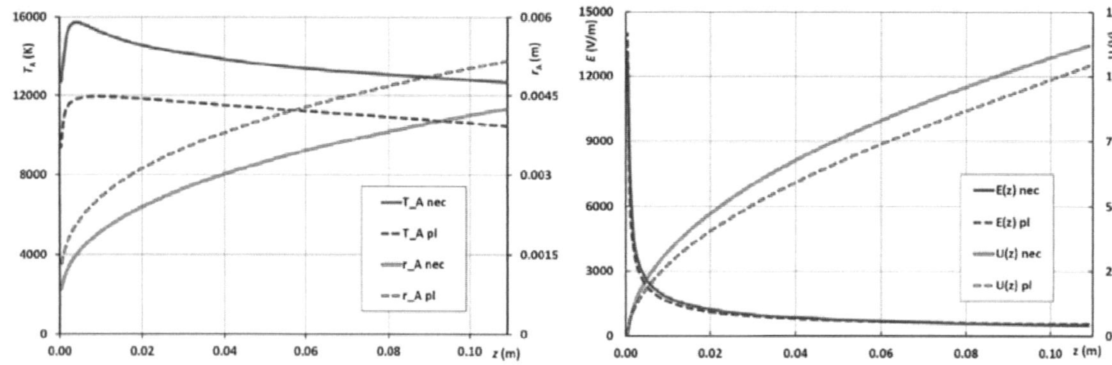

Figure 2. Comparison of axial dependences of the arc temperature $T_A(z)$ and the arc radius $r_A(z)$ (left) and of the electric field intensity $E(z)$ and the arc potential $U(z)$ (right) computed by the arc model using two different methods of computation of the arc power loss: nec - net emission coefficient of argon (var. no. 3), pl - constant power loss ratio (var. no. 5).

4. Conclusions and further work

The paper introduces the modified version of the simplified model of the intensively blasted electric arc burning in argon inside the cylindrical anode channel of the arc heater. As previous versions failed in computation of power losses of individual parts of the anode channel, the new method of estimation the arc's power loss has been applied using the theoretically calculated values of net emission coefficient of argon. Simultaneously, also the initial current density has been included into the set of state variables $[n_r; j_0; T_A(s)]$ instead of being kept constant. The modified model has been tested on the experimental data which previously have exhibited the worst agreement of computed and measured values. Significant improvement has been achieved in comparison with the results of the older version of the model. The obtained dependences meet the preliminary expectations about the behavior of intensively blasted electric arc. Altogether with a significant decrease of relative errors of power losses, the relative error of voltage increased several times in the new model. The explanation could lie in still neglected phenomena and certainly will be sought for during further tests of the modified model with other sets of experimental data.

Acknowledgements

Authors gratefully acknowledge financial support from the Ministry of Education, Youth and Sports of the Czech Republic under NPU I programme (project No. LO1210) and from the Czech Science Foundation under project No. GA 15-14829S. The authors thank Dr. Petr Kloc for computation of net emission coefficient of argon.

References

[1] J. Senk, I. Jakubova, and I. Laznickova. Analysis of intensively blasted electric arc burning in the arc heater's anode channel. *Acta Polytechnica*, 56(5):395–401, 2016. doi:10.14311/AP.2016.56.0395.

[2] J. Senk, I. Jakubova, and I. Laznickova. Remarks on the design of high-temperature devices with electric arc. In *Proc. of the 15th Internat. Sci. Conf. Electric Power Engineering*. FEEC Brno University of Technology, 2014.

[3] J. Senk, I. Laznickova, and I. Jakubova. Power loss distribution along the arc heater with intensively blasted electric arc. In *Proc. of the 18th Internat. Sci. Conf. Electric Power Engineering*. To be published, 2017.

[4] P. Kloc, V. Aubrecht, M. Bartlova, and O. Coufal. Radiation transfer in air and air-Cu plasmas for two temperature profiles. *J Phys D: Appl Phys*, 48(5):1–13, 2015. doi:10.1088/0022-3727/48/5/055208.

[5] Herman Schlichting. *Grenzschicht Theorie*. Verlag G. Braun, Karlsruhe, Germany, 1965.

[6] A. Bauer. Zur theorie des kathodenfalls in lichtbogen. *Z. Physik*, 138:35–55, 1954. doi:10.1007/BF01365533.

[7] A. A. Belevtsev et al. Experimental study of the near-electrode plasma-tungsten cathode system in high-current athmospheric-pressure nitrogen arcs. *High Temperature*, 51(5):583–593, 2013. doi:10.1134/S0018151X13050027.

doi:10.14311/ppt.2017.1.44
Plasma Physics and Technology **4**(1):44–47, 2017

OPTICAL DIAGNOSTICS OF SWITCHING ARCS NEAR CURRENT-ZERO: SPECKLE IMAGING AND INTERFEROMETRY

P. C. STOLLER*, J. CARSTENSEN, B. GALLETTI, C. B. DOIRON, A. SOKOLOV

ABB Switzerland Ltd., Segelhofstrasse 1K, 5405 Baden-Daettwil, Switzerland

* patrick.stoller@ch.abb.com

Abstract. Optical diagnostics can be used to obtain spatially resolved measurements of the density, temperature, conductivity, and electron density of circuit breaker arcs embedded in transonic flows; these can be used to validate the results of simulations, the accuracy of which can currently be assessed in only a limited way. We compare speckle imaging and an interferometric approach. Both use a pulsed nanosecond laser. The speckle imaging setup does not require a reference beam, but only yields information about the gradient of the refractive index. Its accuracy is sensitive to the alignment of the optical components. Interferometry directly yields high resolution images of the index of refraction, from which the density can be calculated using the Gladstone-Dale relation. By using two laser beams, interferometry provides spatially resolved information about the electron density. Such measurements are a significant step towards more accurate CFD models.

Keywords: circuit breaker, current-zero, interferometry, imaging, density, temperature.

1. Introduction

High voltage circuit breakers are used to switch the currents that flow under normal conditions in different sections of the power grid on and off. They must also be able to interrupt short circuit currents that arise when a fault occurs; these currents can be an order of magnitude higher than the typical load currents that flow during normal operation. The current is interrupted by drawing an arc between two contacts. Nozzles surrounding these contacts are used to guide a transsonic flow that cools the arc and interrupts it at a current-zero crossing of the alternating current. The design of more effective and efficient circuit breakers depends on a detailed understanding of the arc physics relevant during the roughly 100 µs before and after the zero-crossing. Information about the arc and the gas flow used to cool it is typically obtained indirectly via measurements of the arc voltage, the current, and the pressure (at specific locations in the arc zone of the circuit breaker). These integral measurements do not give sufficient insight into the structure of the arc to allow a detailed understanding of the processes involved in interrupting it. Such information can be obtained from computational fluid dynamics (CFD) simulations [1], but such models often lack experimental validation (aside from comparison with the measured current, voltage, and pressure). For this reason, several spatially resolved optical diagnostic methods based on well-established techniques that measure the index of refraction (or its gradient) were recently developed or adapted to image circuit breaker arcs near current zero [2–5]. The advantages and disadvantages of interferometric techniques compared to other optical methods (such as emission spectroscopy) are discussed in detail in those studies. Here we seek to briefly outline and compare two of these approaches, speckle imaging and interferometry. Section 2 describes the experimental approaches used. Results of measurements with the two techniques are presented in Section 3, while Section 4 discusses the differences between the two techniques.

2. Experimental techniques

The speckle and interferometry experimental setups are described in detail in [3] and [4], respectively. Here we only briefly present the two techniques; a schematic of the speckle imaging setup is provided in Figure 1; the experimental setup for interferometry is given in Figure 2. In both cases a 532 nm CrystaLaser CRL-QL532-100-YG pulsed, Q-switched laser with a nominal pulse duration of 12 ns was used. To provide an additional wavelength for the interferometry measurements to measure directly the electron density, a 671 nm CNI AO-V-671 pulsed, Q-switched laser with a nominal pulse duration of 16 ns was employed. The laser beam(s) were passed through a beam expander to increase the beam diameter to match the size of the region imaged. In the case of the speckle setup, the arc zone was imaged onto a glass diffuser using a relay lens. A scientific camera (Phantom V7100 or ProSilica GT4905) was focused a defined displacement distance behind the glass diffuser. Shifts in the speckle pattern produced by scattering from the diffuser and observed on the camera (with respect to a reference image in the absence of an arc and/or gas flow) were used to determine the line-of-sight integrated gradient in the refractive index at each point. The approach used to calculate the density and temperature from such measurements is given in [3]. For the interferometry measurements, a reference beam was added to the experimental setup and the probe beam and the reference beam were re-combined at a slight angle using a beam-splitter and recorded by a ProSilica GT4905 or GT2300 camera. The resulting interference patterns were analyzed using Fourier techniques to obtain the line-of-sight integrated refractive index (again making use of a reference image). The procedure employed is detailed in [4].

It is important to note that the circuit breaker test devices used to obtain speckle and interferometric images differed considerably. The test device used in the former tests, described in detail in [3] is shown in Figure 3. Briefly, this test device consisted of two fixed contacts (a hollow contact and a solid contact) and a nozzle system used to guide the flow of gas (synthetic air in the case of the tests discussed in this work) from a high pressure tank through the arc zone. Optical measurements could be performed in the stagnation region (through a window in the nozzle) and at the nozzle exit. This test device provided more direct optical access to the arc (at the nozzle exit), but had the disadvantage that the arc was not well-stabilized by the gas flow (which accelerated out of an intermediate volume between the arc zone and the high pressure tank). This led to an arc that moved within the arc zone and did not always remain parallel to the axis of the test device. As discussed in detail in [3] the instability and asymmetry in the arc sometimes prevented reconstruction of the index of refraction profile from the measured line-integrated deflection angles (corresponding to the line integrated refractive index). This is because a three-dimensional reconstruction based on a single projection is only possible for an object with defined symmetry, such as axi-symmetry (temporal and spatial fluctuations can also wash out features, but this effect is negligible for single, nanosecond laser pulses). One approach to overcome this problem (using almost the same circuit breaker test device) is described in [5]. In that work, multiple beams were used to probe the index of refraction distribution, permitting, in principle, a three-dimensional reconstruction from the line integrated measurements even in the absence of axi-symmetry. However, it likely that at least eight beams are required to achieve a reasonable reconstruction.

The test device used in the interferometry experiments is illustrated in Figure 4. It was designed specifically to ensure a stable, axi-symmetric arc by allowing the gas (argon in the case of the tests discussed in this work) that cools and interrupts the arc to accelerate from rest in the arc zone. The three-dimensional index of refraction and density distribution in this arc and surrounding gas flow can be re-constructed from a single projection. This test device is described in detail in [4].

3. Results

The speckle technique was used to image the index of refraction distribution in the stagnation region of the test device described in the previous section. The density was determined from the measured index of

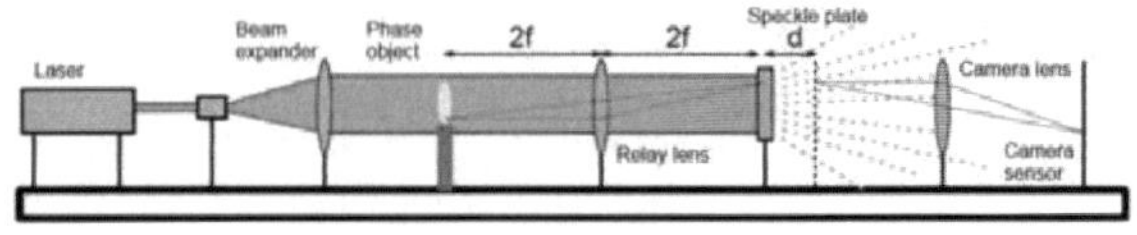

Figure 1. Illustration of the optical setup used for speckle imaging.

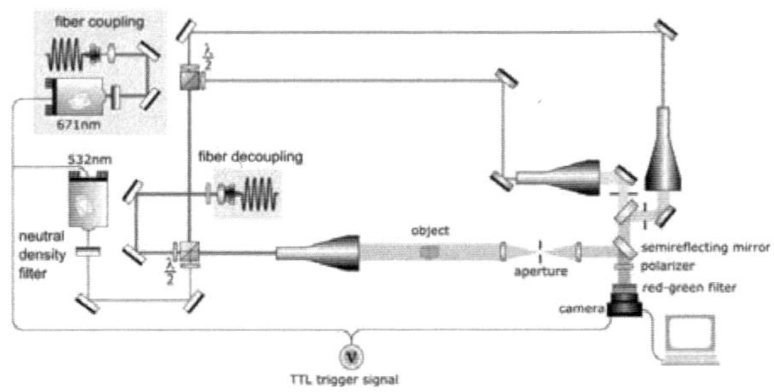

Figure 2. Sketch of the optical setup used for interferometry.

refraction using the approach presented in [3]. Assuming that the pressure distribution is the same as the one obtained from measurements in the absence of an arc (an assumption explained in [3]), the temperature can also be calculated. These results are presented in Figure 5 for illustration; additional results are given in [3]. Figure 5 shows the results for six different measurements and illustrates the scatter (twice the standard deviation). This scatter does not represent the uncertainty in the measurement; instead, it demonstrates that there are fluctuations in the arc from test shot to test shot. The largest variation can be seen in the temperature at the arc boundary; the gradients are very steep, and a large change in temperature corresponds to only a small difference in the refractive index.

Similar measurements were performed with the interferometric technique, although it should be noted that the test device, the gas, and other aspects of the experiment differed. An example of such a measurement for a 100 A quasi-direct-current arc and a stagnation pressure of 1560 kPa are shown in Figure 6. The measurements were also performed in the stagnation point, but in the test device that was optimized to achieve a stable, axi-symmetric arc (Figure 4). The variation in the arc density profile observed from shot to shot with the test device shown in 3 cannot be seen in the measurements made on the stable arc; the results are consistent from test shot to test shot.

The interferometric technique was carried out with two laser wavelengths, permitting the measurement of the electron density in addition to the neutral particle density. Two laser wavelengths allow the measurement of the electron density because the contribution of the electrons to the index of refraction is strongly wavelength dependent, while the contribution of the heavy particles is not. The details of this approach are described in detail in [2, 4]. An example of a

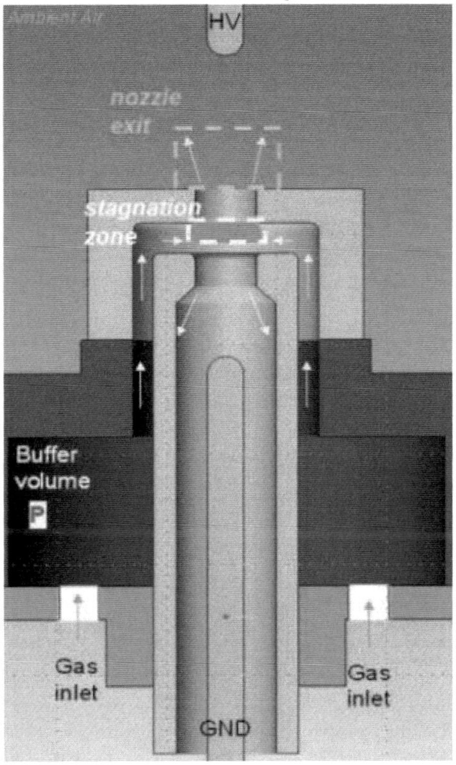

Figure 3. Sketch of the circuit breaker test device used for speckle imaging.

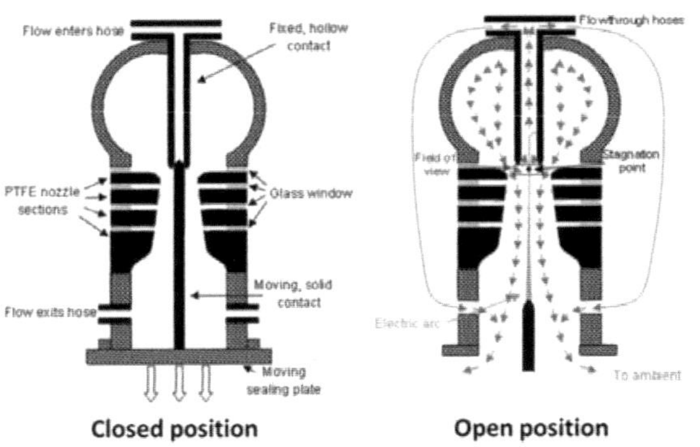

Figure 4. Sketch of the circuit breaker test device used for interferometric imaging.

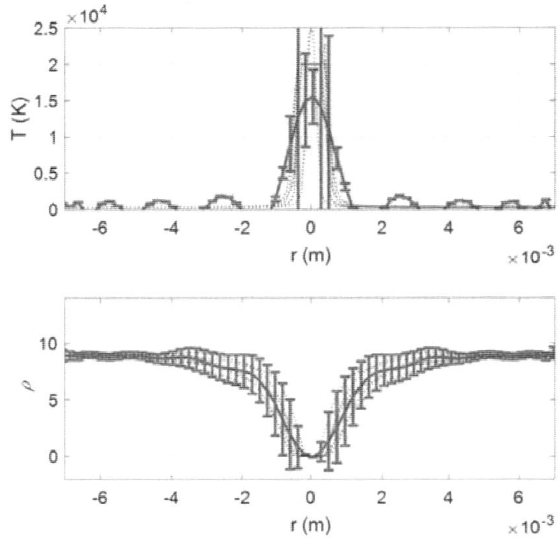

Figure 5. Speckle measurements of the temperature and density in the stagnation zone of the test device shown in Figure 3 illustrating the variation in the measurement. The stagnation pressure was 750 kPa and the current was 12 A. The density is given in units of kg/m^3.

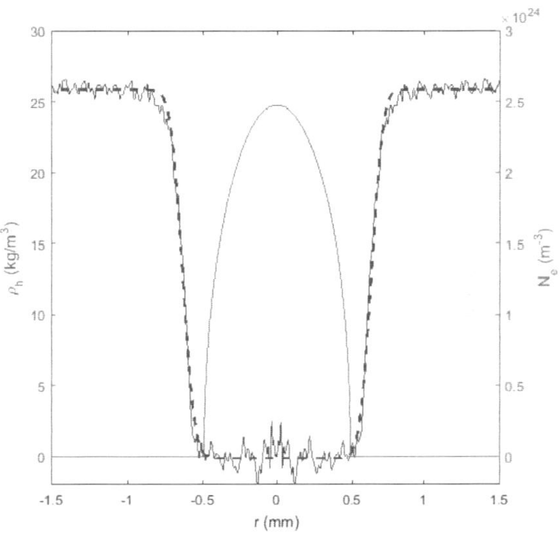

Figure 6. Interferometry measurement of the density and the electron density distribution in the stagnation zone of the test device shown in Figure 4. The stagnation pressure was 1560 kPa and the current was 100 A. Note that noise in the density measurement amplified by Abel inversion results in unphysical dips below zero in the arc core. These unphysical dips are removed by performing a fit to an analytical function [4].

measurement of the electron density in the stagnation point of the arc zone is given in Figure 6. Assuming local thermodynamic equilibrium, the temperature of the arc core can be determined from the electron density measurement [4].

4. Comparison of techniques

The measurements made with the speckle and the interferometric experimental setup allow the two approaches to obtaining spatially resolved information about a circuit breaker arc near current-zero to be compared. It is important to note that both the optical techniques and the test devices used differ. The main difference that stems from the test device is the much more stable arc observed when the gas flow in the arc zone accelerates from rest. In principle, speckle measurements made on such an arc would also show less shot to shot scatter. In other words, this difference is due to the test device and not the measurement technique.

Although the speckle and interferometry techniques are both based on measuring the change in index of refraction between conditions with and without an axially blown arc, there are key differences. The former technique requires a less complex experimental setup. Only a single beam path is required in that case, while the latter approach requires the use of two beam paths, one of which bypasses the test device. The advantage of using a reference beam is that the line-integrated difference in refractive index is measured directly. The speckle technique (similar to schlieren imaging and Hartmann-Schack inteferometry) only allows measurement of the gradient of the refractive index. On the other hand, unlike in the case of speckle imaging, interferometry requires that the coherence length of the laser be sufficient for the reference and probe beams to interfere.

Correct measurement of the distance the diffuser (or speckle plate) is placed out of the focal plane of the camera lens is crucial to accurate speckle measurements and can in some cases limit their accuracy. In the case of interferometry, only the magnification of the imaging optics must be determined accurately, but this is generally straightforward.

The inteferometric technique can also makes better use of the available resolution of the camera, as discussed in [4]. To summarize, the somewhat more complex optical setup required to perform interferometry is justified by its ability to directly determine the index of refraction and by the improved resolution it can provide.

5. Conclusion and outlook

Both speckle and interferometric imaging can be used to obtain information about the density distribution in the arc zone of a circuit breaker during the important current interruption phase. The speckle technique requires a somewhat simpler experimental setup—a reference beam is not needed and the coherence length of the laser does not need to be considered. In addition, the sensitivity of the speckle technique can be tuned by simply shifting the location of the diffuser plate used to generate the speckle pattern. On the other hand, using only a few additional optical elements, the interferometric technique permits direct measurement of the index of refraction (rather than of its gradient) and does not require determination of the diffuser plate defocus distance (the accuracy of which can limit the accuracy of the overall measurement). Thus, while both experimental approaches can be used to study the arc in a circuit breaker near current zero, the interferometric approach is generally to be preferred.

In general, index of refraction based optical diagnostic techniques provide spatially resolved information about the arc and the surrounding gas flow related to the density, arc radius, temperature, conductivity, and electron density that cannot be obtained from conventional measurements of the arc voltage and current. Thus, these methods provide new data that can be used to validate and further develop arc models used in the simulation and design of circuit breakers.

References

[1] Frank Reichert, Jean-Jacques Gonzalez, and Pierre Freton. Modelling and simulation of radiative energy transfer in high-voltage circuit breakers. *Journal of Physics D: Applied Physics*, 45(37):375201, 2012. doi:10.1088/0022-3727/45/37/375201.

[2] Yuki Inada, Shigeyasu Matsuoka, Akiko Kumada, Hisatoshi Ikeda, and Kunihiko Hidaka. Shack-hartmann type laser wavefront sensor for measuring two-dimensional electron density distribution over extinguishing arc discharge. *Journal of Physics D: Applied Physics*, 45(43):435202, 2012. doi:10.1088/0022-3727/45/43/435202.

[3] P C Stoller, E Panousis, J Carstensen, C B Doiron, and R Färber. Speckle measurements of density and temperature profiles in a model gas circuit breaker. *Journal of Physics D: Applied Physics*, 48(1):015501, 2015. doi:10.1088/0022-3727/48/1/015501.

[4] J Carstensen, P Stoller, B Galletti, C B Doiron, and A Sokolov. Measurement of electron and heavy particle density profiles in an axially blown, stable arc. *Physical Review Applied*, Submitted.

[5] M. Stoffels, S. Simon, P. G. Nikolic, P. Stoller, and J. Carstensen. Development of a multiperspective optical measuring system for investigating decaying switching arcs at the nozzle exit of circuit breakers. *Appl. Opt.*, 56(7):2007–2019, Mar 2017. doi:10.1364/AO.56.002007.

doi:10.14311/ppt.2017.1.48
Plasma Physics and Technology **4**(1):48–51, 2017

NUMERICAL ANALYSIS OF LOW VOLTAGE ARC MOTION PROCESS AT VARIOUS FREQUENCIES

J. Yin[a], X. Li[a,*], Q. Wang[b]

[a] *State Key Laboratory of Electrical Insulation and Power Equipment, Xi'an Jiaotong University, Xi'an, Shaanxi, 710049, China*

[b] *School of Sciences, Xi'an University of Technology, Xi'an, Shaanxi, 710048, China.*

[*] xwli@mail.xjtu.edu.cn

Abstract. A three-dimensional (3D) magneto-hydro-dynamic (MHD) model of air arc plasma is built to investigate the frequency effects on the arc motion process with different number of splitter plates. Based on this model, the arc voltage and current density are obtained. The arc motion time is normalized with the frequency and compared at different numbers of splitter plate. The result shows that the normalized time and the arc voltage peak increase with increases of the number of splitter plate.

Keywords: magneto-hydro-dynamic (MHD), eddy current, frequency, motion process.

1. Introduction

In permanent magnet wind power generation systems, the frequency of the output current may vary from several Hz to about 200 Hz. Air circuit breaker (ACB) is a crucial component for controlling and protecting the power distribution network, which requires the ACB should switch off the short-circuit current of various frequencies [1]. Therefore, it is important to study the influence of frequency on the arc motion process.

Many studies have focused on the arc motion of low voltage circuit breaking. Lindmayer et al analyzed the process of arc splitting [2, 3]. McBride et al studied the back commutation phenomena of arc by experiment [4]. Iturregi et al analyzed the influence of the number of splitter plate on arc voltage at direct current (DC) situation [5]. The formulations to calculate eddy currents in electrodes and splitter plate were developed and the eddy current effect on the arc splitting process was simulated [6]. But the influences of frequency and the number of splitter plates at higher frequency on arc motion are not clear.

In this paper, a 3-D MHD model is built to investigate the influence of frequency on air arc motion process with different number of splitter plates. The potential vector approach is adopted to calculate the time-varying electromagnetic field. The eddy current in whole calculated domains is considered. The influences of frequency and the number of splitter plate on the arc motion are analyzed by the arc voltage, the current density distribution sequences and the normalized time.

2. Numerical geometry and method

Fig.1 shows the computational domain for the fluid. A 1/2 symmetric model is adopted. The dimension of the model is $3 \times 11 \times 30$ mm in the x-y-z direction, and the origin of the coordinate is on the symmetry plane ($x = 0$). The far-away domain is built to calculate the magnetic field. The arc current is imposed on the surface "current inlet", and the value is set to $i = 600\sqrt{2}\sin(\omega t + \pi/12)$. The electrical potential is set to zero on the surface "current outlet" as shown in Fig.1. The 0.1 mm sheaths surrounding electrodes and splitter plate are applied to describe the formation of new arc roots and represent the electrode voltage drop. The electrical conductivity of sheath is calculated by nonlinear voltage-current density characteristic [5]. The distance of two electrodes is 8 mm and the materials of electrodes are copper. The splitter plate is ferromagnetic material, and the B-H curve is applied. The transport coefficients and thermodynamic properties of air plasma vary with temperature and pressure [7]. The radiation energy source term is calculated using the net emission coefficient (NEC) method [8].

The electro-magnetic equations are solved by means of ANSYS Emag. And the N-S equations of describe the arc plasma are solved by means of ANSYS Fluent. The date exchange between the two solvers is provided by the coupling server MpCCI [9].

3. Simulation results and analysis

The current density distribution and arc voltage obtained by simulation. The influence of the number of splitter plate and frequency on arc motion will be analyzed.

3.1. Influence of splitter plate number

Fig.2 shows the current density distribution sequences on the symmetry plane of 1SP and 2SP model at f=50 Hz. The negative value of current density indicates the current is along the negative direction of y-axis. We define the four typical position a, b, c, d of arc motion as shown in Fig.2 (a),(b),(c) and (d)

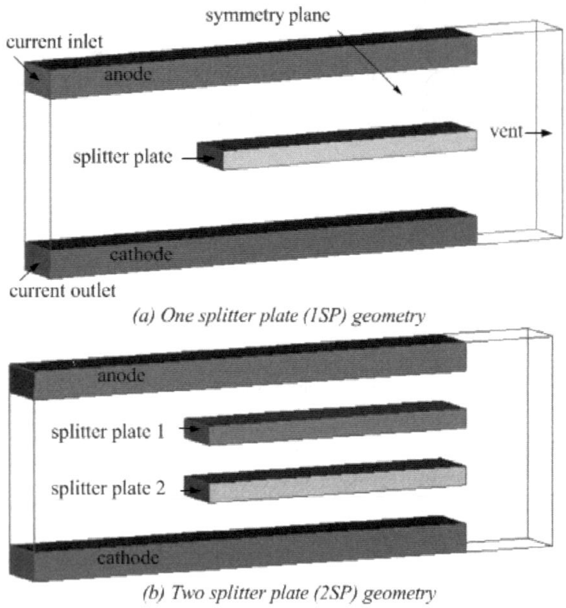

Figure 1. Numerical geometry for fluid domain.

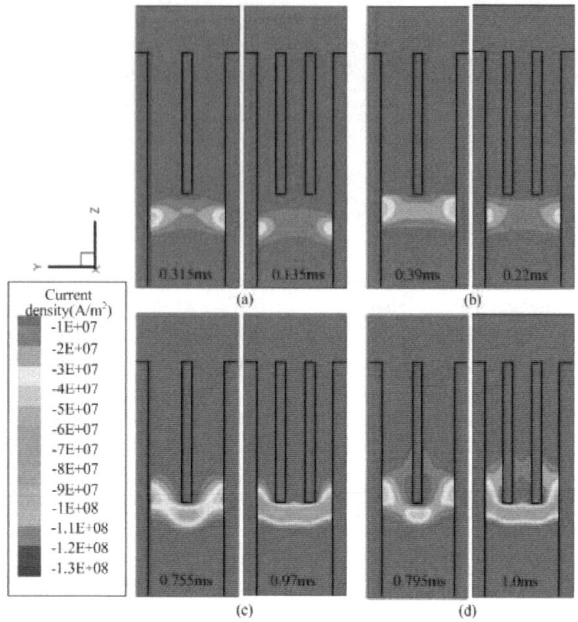

Figure 2. Current density distribution sequences on the symmetry plane at 50 Hz.

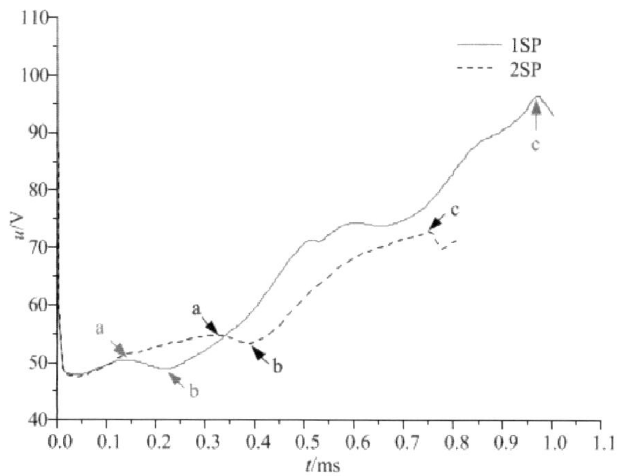

Figure 3. Arc voltage of simulation results at 50 Hz. (a,b,c represent the position a,b,c)

respectively. Fig.3 shows the arc voltages of 1SP and 2SP model. The trend of arc voltage is consistent for 1SP and 2SP model. For 1SP model, after arc ignition, the arc elongated until t=0.315 ms in Fig.2 (a). The arc voltage increases up to 54.7 V. From t=0.315 ms to t=0.39 ms in Fig.2 (b), the arc column is squeezed by splitter plate and the arc voltage decreases to 53.3 V at 0.39 ms. After 0.39 ms, the arc roots move along the electrode surfaces, and the arc column bends and stretches gradually due to the blocking effect of the relative cold splitter plate. The arc voltage achieves the peak value of 72.87 V at t=0.755 ms in Fig.2 (c). At t=0.795 ms as shown in Fig.2 (d), the new arc roots form on splitter plate surface and there are two clearly visible parallel current paths, which leads to the decreases of arc voltage.

For 2SP model, the arc column is squeezed by splitter plate from t=0.135 ms to t=0.22 ms. The arc column bends and the arc voltage increases after t=0.22 ms. At t=0.97 ms, the arc voltage reaches to its peak value of 96.46 V. The new arc roots form on the splitter plate and arc voltage decrease at t=1.0 ms. There is no obvious difference on the arc voltage between 1SP and 2SP model before 0.135 ms.

Fig.4 shows the current density distribution of 1SP and 2SP model at t=0.25 ms when f=50 Hz and f=200 Hz. It is found that the arc roots position of 2SP model are closer to the splitter plate at the same frequency, which because the attraction effect of the ferromagnetic splitter plate is enhanced in 2SP model. The cross-sectional area of arc column is larger at higher frequency due to the arc current increasing with the increase of frequency at the same time.

3.2. Influence of frequency

Fig.5 shows the simulation results of current density under various frequencies (50 Hz, 100 Hz, 200 Hz) for 1SP model. The formation of new arc roots on the splitter plate surface is the fastest at 200 Hz. This is because the Lorentz force of higher frequency is greater at the same time. Fig.6 and Fig.7 show the arc voltage of 1SP and 2SP model under various frequencies. The time of arc voltage reaching its peak will be delayed with the decrease of frequency. The arc voltage peak increase by 0.48 % and 3.6 % for 1SP model and it increase by 1.9 % and 3.8 % for 2SP model, when f increases from 50 Hz to 100 Hz and 200 Hz. The reason is that the effective resistance of arc column increases due to the higher skin effect at higher frequency.

We define the normalized time to analyze the influ-

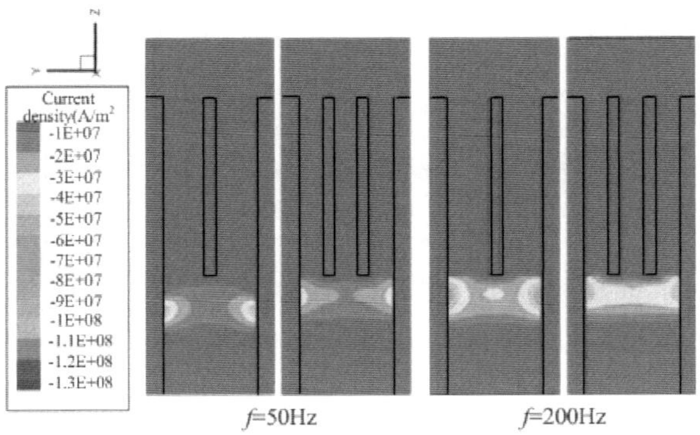

Figure 4. Current density distribution at t=0.25 ms.

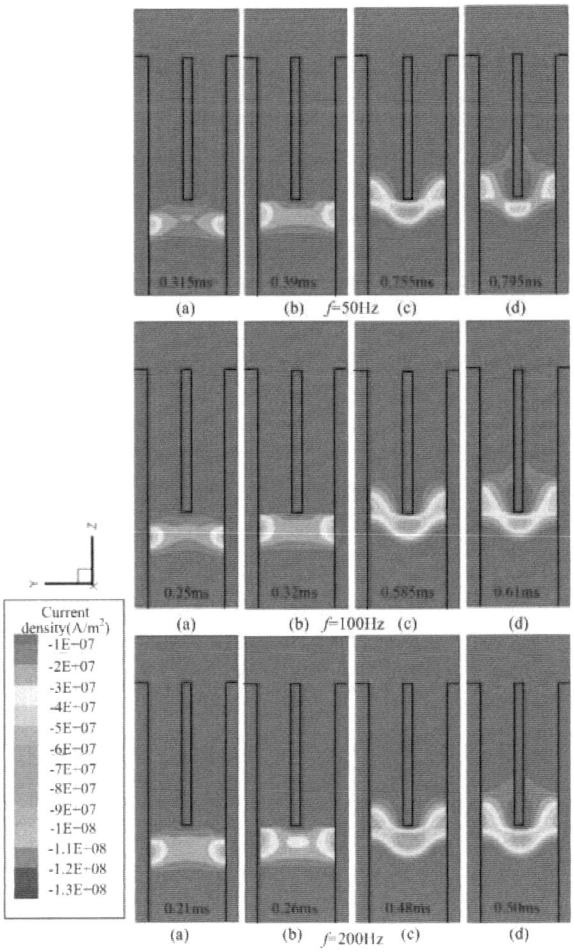

Figure 5. Current density distribution sequences on the symmetry plane under various frequencies.

ence of frequency on arc motion:

$$t_{\mathrm{n}} = \frac{t_{\mathrm{r}}}{T} \quad (1)$$

Where, t_{r} is the time when the arc moves to the same position at various frequencies (50 Hz, 100 Hz and 200 Hz), T is the current cycle of various frequencies. The t_{n} is normalized time of various frequencies in different positions are shown in Fig.8 (1SP) and Fig.9 (2SP). It is found that the normalized time t_{n}

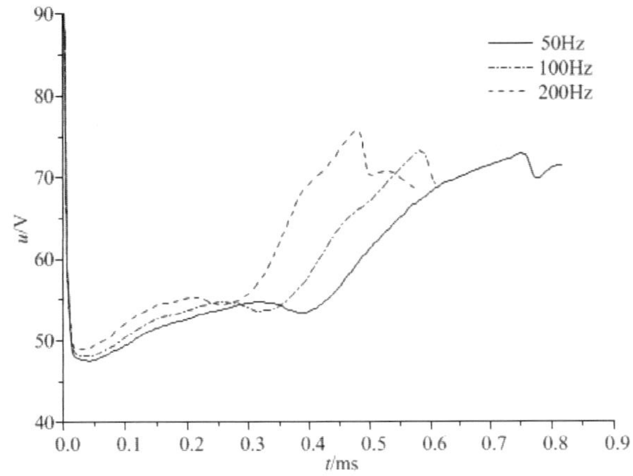

Figure 6. Comparison of the arc voltage under various frequencies 50 Hz, 100 Hz, 200 Hz(1SP).

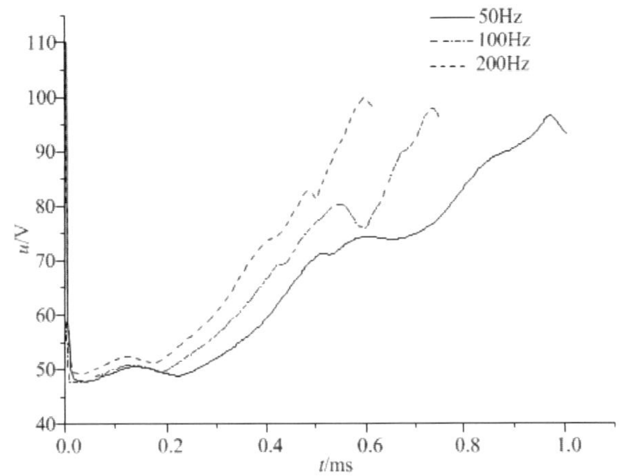

Figure 7. Comparison of the arc voltage under various frequencies 50 Hz, 100 Hz, 200 Hz(2SP).

increases with the increase of frequency because the hindering effect to arc motion by the eddy current in the splitter plate is larger for higher frequency.

In position a, the normalized time increase by 58.7 % and 167 % for 1SP model and it increase by 92.6 % and 278 % for 2SP model, when f increases from 50 Hz to 100 Hz and 200 Hz. In position b, the normalized time increase by 64.1 % and 167 % for 1SP model, and it increase by 72.7 % and 213.6 % for 2SP model, when f increases from 50 Hz to 100 Hz and 200 Hz. Therefore, the hindering effect of eddy current was enhanced with the increase of splitter plate.

4. Conclusions

A 3-D MHD model is built to investigate the influence of frequency on the air arc motion. It can be concluded that:

1. The arc motion velocity of two splitter plates model is faster than the one splitter plate model at the same frequency.

2. The new arc roots formation on splitter plate of

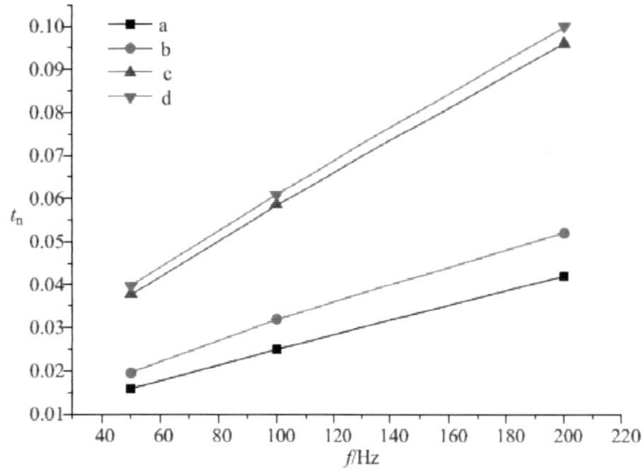

Figure 8. Normalized time of arc motion under various frequencies 50 Hz, 100 Hz, 200 Hz (1SP).

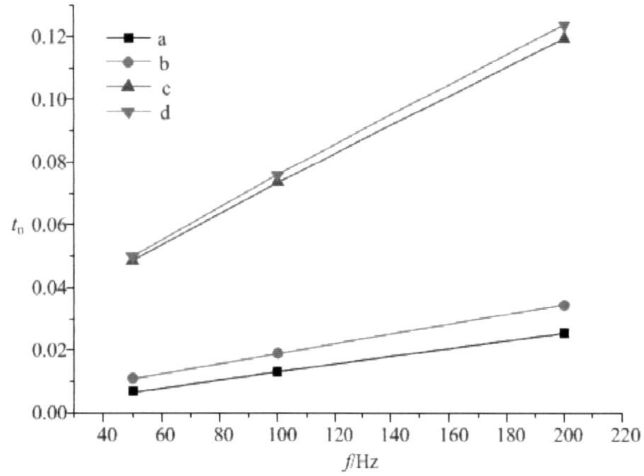

Figure 9. Normalized time of arc motion under various frequencies 50 Hz, 100 Hz, 200 Hz (2SP).

two splitter plates model is slower than one splitter plate model. And the arc voltage peak is higher in two splitter plates model.

3. The normalized time reflects the hindering effect of eddy current on arc motion, and it increases with the increase of frequency. And the growth rate is larger in two splitter plates when f increases from 50 Hz to 100 Hz and 200 Hz.

Acknowledgements

The work was supported by National Natural Science Foundation of China (51507136).

References

[1] J. Qu, Q. Wang, and Z. Liu et al. Influences of closing phase angle and frequency on electro-dynamic stability of air circuit breaker. *IEEE Trans. Compon. Packag. Manuf. Technol.*, 6(2):249–255, 2016. doi:10.1109/TCPMT.2015.2511180.

[2] M. Lindmayer and M. Springstubbe. Three-dimensional-simulation of arc motion between arc runners including the influence of ferromagnetic material. *IEEE Trans. Compon. Packag. Manuf. Technol.*, 25(3):409–414, 2002. doi:10.1109/TCAPT.2002.804604.

[3] M. Lindmayer, E. Marzahn, and A. Mutzke et al. The process of arc splitting between metal plates in low voltage arc chutes. *IEEE Trans. Compon. Packag. Manuf. Technol.*, 29(2):310–317, 2006. doi:10.1109/TCAPT.2006.875902.

[4] J.W. McBride, D. Shin, and T. Bull et al. A study of the motion of high current arcs in splitter plates using an arc imaging system. *28th International Conference on Electric Contacts*, pages 175–180, 2016. URL: http://eprints.soton.ac.uk/id/eprint/402403.

[5] A. Iturregi, B. Barbu, and E. Torres et al. Electric arc in low-voltage circuit breakers: experiments and simulation. *IEEE Trans, Plasma. Sci.*, 45(1):113–120, 2017. doi:10.1109/TPS.2016.2633400.

[6] O. Chadebec, G. Meunier, and V.G. Mazauric et al. Eddy-current effects in circuit breakers during arc displacement phase. *IEEE Trans, Magn.*, 40(2):1358–1361, 2004. doi:10.1109/TMAG.2004.824768.

[7] A.B. Murphy. Transport coefficients of air, argon-air, nitrogen-air, and oxygen-air plasmas. *Plasma Chemistry and Plasma Processing*, 15(2):279–307, 1995. doi:10.1007/BF01459700.

[8] Y. Naghizadeh-Kashani, Y. Cressault, and A. Gleizes. Net emission coefficient of air thermal plasmas. *J.Phys.D: Appl. Phys.*, 35:2925–2934, 2002. doi:10.1088/0022-3727/35/22/306.

[9] Ch. Ruempler and V.R.T. Narayanan. Arc modeling challenges. *Plasma Physics and Technology*, 2(3):261–270, 2015. URL: http://ppt.fel.cvut.cz/articles/2015/ruempler.pdf.

doi:10.14311/ppt.2017.1.52
Plasma Physics and Technology **4**(1):52–55, 2017

INFLUENCING FACTORS ON THE DIELECTRIC BREAKDOWN STRENGTH IN F-GAS FREE HIGH-VOLTAGE SWITCHGEAR

S. Gossmann[a,*], B. Lutz[b], P. G. Nikolic[a]

[a] *Siemens AG, Corporate Technology, Günther-Scharowsky-Str.1, 91058 Erlangen, Germany*
[b] *Siemens AG, Energy Management, Nonnendammallee 104, 13629 Berlin, Germany*
[*] svetlana.gossmann@siemens.com

Abstract. After being in the focus of sciences' and industry's research and development activities for many years, the investigation of possible SF_6 gas-alternatives has been even more intensified after the revision of the European regulation on fluorinated gases in 2014. In this contribution the influencing factors on the dielectric breakdown of clean air are investigated for weak inhomogeneous field and gas pressures up to 10 bar. Modelling approaches and experimental data are compared.

Keywords: F-gas free switchgear, dielectric breakdown, clean air, surface roughness.

1. INTRODUCTION

Nowadays SF_6 (sulphur hexafluoride) is the state-of-the art insulating and arc quenching medium used in gas-insulated switchgear (GIS), enabling safe current interruption and high dielectric strength. At the same time, if released to the environment, SF_6 is a strong greenhouse gas with a global warming potential of 22800 CO_2 (carbon dioxide) mass equivalents [1–3].

After being in the focus of sciences' and industry's research and development activities for many years, the investigation of possible SF_6 gas-alternatives has been even more intensified after the revision of the European regulation on fluorinated gases in 2014. An assessment of the physical properties of gas-alternatives currently under discussion shows, that the design of future switchgear with SF_6 gas-alternatives requires a detailed consideration of the gas and material properties. In this context it is also expected that a significant change of the design criteria, used for SF_6 switchgear for the last decades, could be necessary. Thus a detailed understanding of the underlying physical processes and breakdown mechanisms as well as of influencing factors on these is necessary.

These influencing factors (e.g. surface roughness, gas pressure etc.) are investigated for clean air in this contribution focussing on weak inhomogeneous field and gas pressures up to 10 bar. Based on these investigations modelling approaches for different effects on the dielectric strength (e.g. influence of protrusion shape and dimensions) of clean air (80 % nitrogen, 20 % oxygen) are analyzed and compared. The investigations are supported by experimental results.

2. BASIC THEORY OF STREAMER-DISCHARGE

In this contribution the streamer breakdown mechanism in clean air is investigated. The streamer discharge is driven by impact-ionization and photon emission processes. If electrodes have the same roughness and are made from the same material, the positive streamer is more critical for the dielectric breakdown. If the avalanche reaches its critical size, a significant field increase occurs in front of the avalanche head and at the avalanche tail. Because of this distortion of the electric field by space-charges, photons are emitted from the avalanche head and in the region behind the avalanche tail. New avalanches are initiated, because the electric field there exceeds the critical electric field, at which the ionization coefficient is greater than the attachment coefficient. In Fig. 1 the development of a positive streamer discharge is depicted [4]. An area with active ionization is located round the head of the streamer with positive charge. In the reduced field area photoelectrons induce new electron avalanches, which develop concentrically toward the streamer head. The large number of electrons entering the streamer head neutralizes the positive space-charge at this location. However, during their development a new center with positive space-charge is formed, which virtually shifts the anode towards the cathode.

The ignition condition for the streamer is fulfilled, if the critical number of free moving charge carriers has reached a number of $N_{cr} = 10^6 \ldots 10^8$ [4]:

$$\int_0^{z_{cr}} \alpha_{\text{eff}} \, dz = \ln N_{cr} \qquad (1)$$

where α_{eff} is the effective ionization coefficient as a function of reduced electric field $\frac{E}{p}$ and z_{cr} is the distance ahead the surface, where $\alpha_{\text{eff}} > 0$.

3. ANALYTICAL AND SEMI-EMPIRICAL APPROACHES FOR MODELLING

The quality of surface finish of a metal depends not only upon the material itself but also upon the manufacturing and machining process, respectively. The approach, how to consider the surface roughness and protrusion dimensions, plays an important role for the

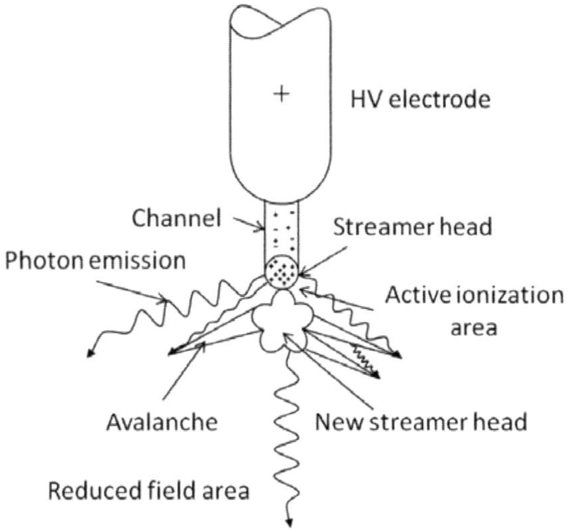

Figure 1. Development of the positive streamer discharge [4]

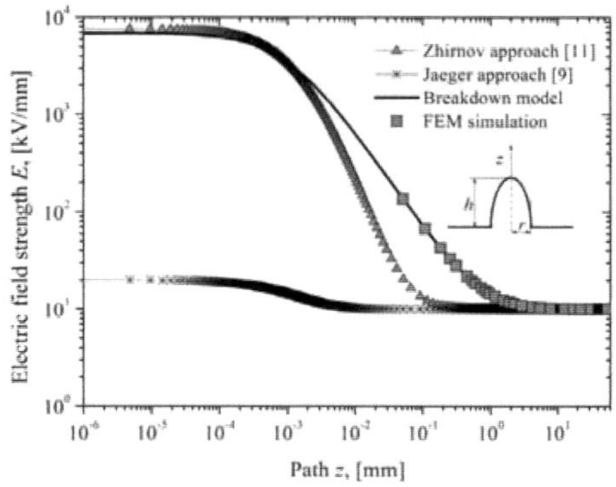

Figure 2. Electric field strength calculated for plane-plane geometry with semi-ellipsoidal pro-trusion with $h = 5\,mm$ and $r = 0.1\,mm$, voltage difference of 600 V and distance between electrodes of 60 mm.

modeling of the streamer formation. There are different methods, how to implement the surface roughness in the modeling, e.g. fractal theory [5], multi-ridge model [6], calculation of enhancement factor [7–9] and the electric field strength along the axis [9–12].

In the breakdown model used in the context of this contribution, the local field enhancement is emulated by a single idealized protrusion in the form of a prolate semi-ellipsoid evolving from a perfectly smooth conductor. The electric field strength $E(z)$ normal to the middle of the tip of the protrusion is described by expression [10]:

$$E(z) = E_0 \left[1 + \frac{2h(\frac{zH}{z^2-H^2}) - \frac{1}{2}\ln(\frac{z+H}{z-H})}{h\,\ln(\frac{h+H}{h-H}) - 2H} \right], \quad (2)$$

where E_0 is the electric field strength for a perfectly smooth conductor; the protrusion in the form of semi-ellipsoid is characterized by its height h and half-width r with $H = \sqrt{h^2 - r^2}$. In order to compare different approaches for emulating the electric field strength along the axis, three approaches, Zhirnov [11] (cf. (3)), Jaeger [9] (cf. (4)) and a breakdown model using (2), with the same electric field enhancement factor β are taken into account for a plane-plane set-up with a semi-ellipsoidal protrusion with tip radius r_c at the high voltage electrode (cf. Fig. 2):

$$E(z) = E_0 \left[\frac{\beta r_c^2}{(r_c + z)^2} + 1 \right], \quad (3)$$

$$E(z) = E_0 \left[1 + \frac{r_c^2}{(z + r_c)^2} \right], \quad (4)$$

For the validation purposes a Finite Element Modeling (FEM) is additionally performed for this plane-plane set-up, considering the same protrusion dimensions as for the analytical approaches.

From the results given in Fig. 2 it is observed, that the analytical expression (2) used in the breakdown model yields good agreement with the simulation data for semi-ellipsoidal protrusion. FEM simulations are also performed for a plane-plane set-up with conical protrusion. Since the model used in the context of this contribution has the prolate semi-ellipsoidal form of protrusion, the same height and tip radius are taken for an additional comparison with FEM simulation for a conical protrusion. Both, simulation and modeling results using (2) are found to be in good agreement. Therefore, for the further breakdown analysis only the proposed form of protrusion, prolate semi-ellipsoid, is used.

4. EXPERIMENTAL APPROACH

Set-ups representing typical GIS arrangements were defined for dielectric breakdown tests. The following requirements had to be fulfilled by the model arrangements:

- ☐ Field utilization factor $\eta = E_{\mathrm{average}}/E_{\mathrm{maximum}}$ comparable to typical GIS arrangements
- ☐ Surface quality comparable to GIS conditions
- ☐ Same electrode materials as in GIS

The distinction between small (i.e. sphere-plane) and big surfaces (i.e. coaxial cylinder) was adopted in order to simulate parts of GIS with smaller and bigger surface area. The model arrangements are shown in Figure 3 where the field utilization factors η of the arrangements are indicated as well.

A variety of different parameters were investigated by high voltage breakdown tests like gas humidity, or the test method [13]. The influence of three parameters, i.e. voltage type, gas pressure and surface roughness, are highlighted in this paper.

According to IEC 62271-1, high dielectric breakdown strength against impulse and AC voltages is required for reliable operation of GIS. Here, 1.2 µs / 50 µs

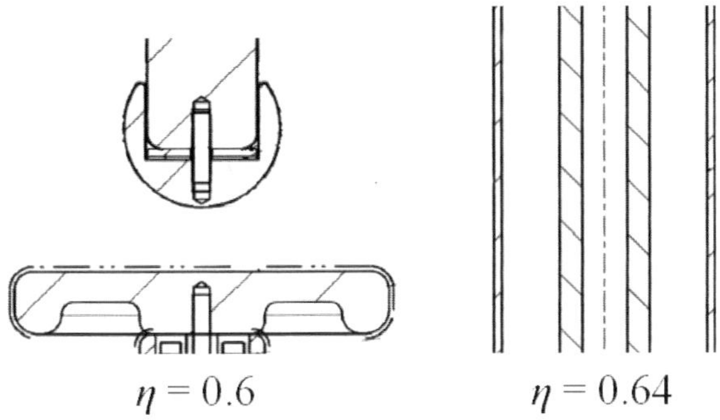

Figure 3. Sphere-plane (left) and coaxial cylinder (right) set-ups for breakdown experiments.

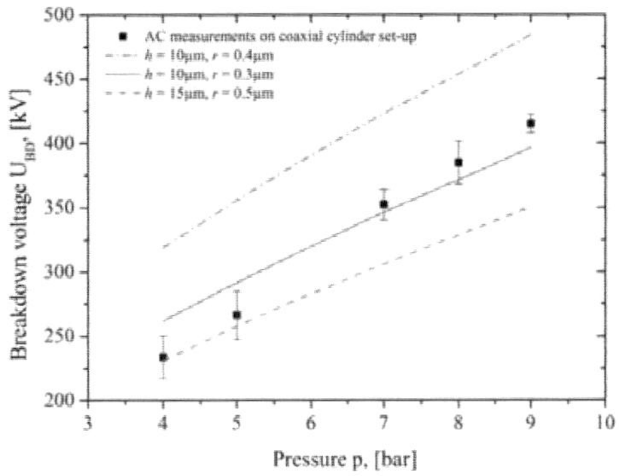

Figure 4. Comparison of experimental and modeling results for critical carrier number of $N_{cr} = 10^8$.

lightning impulse and 50 Hz sinusoidal AC voltage were applied. The up-and-down method [4] was applied to evaluate the breakdown voltage limit values. The choice of gas pressure within a range of 4 bar to 9 bar (absolute value) is important for proper GIS design influencing e.g. size and weight of compartments. Finally, the machining of surfaces influences manufacturing and costs of GIS so that proper surface roughness has to be carefully defined. Here, aluminium electrode surface profiles with mean roughness height of surface profile ranging from 1 µm to 63 µm were evaluated. The experimental results are shown in section 5.

5. DISCUSSION OF MODELING AND EXPERIMENTAL RESULTS

The rough surface has a negative influence on the dielectric strength, especially with increasing gas pressure. These effects were previously investigated experimentally [14, 15] and some empirical models were developed [16]. In such models the influence of the electrode protrusion on the dielectric strength is represented by a roughness factor, which is a function of the gas pressure and the protrusion tip radius. It was found that this roughness factor does not depend on the type of applied test voltage (i.e. alternating current (AC) voltage or lightning impulse (LI) voltage) [17]. For this approach measurements with electrodes with varying surface roughness at different gas pressures are needed. In contrast to this, the streamer model for the breakdown analysis used in the context of this contribution applies the parameters for emulating the same surface roughness for different pressures.

In Fig. 4 the experimental results for the coaxial cylinder set-up with polished surface are compared with the modeling results, obtained for the critical charge carrier number $N_{\mathrm{cr}} = 10^8$ as ignition condition [4]. By this approach with constant critical charge carrier number and varying protrusion dimensions, only rough agreement between experimental and modeling data is observed.

Next to the surface roughness the streamer breakdown mechanism significantly depends on the gas pressure. The streamer head dimension as well as avalanche length decrease with increase of pressure. The space-charge field E_r at the head of the avalanche is also a function of air pressure p. If avalanche has travelled a distance z in a non-uniform field, the space-charge field is given by [18]:

$$E_r = 5.27 * 10^{-7} \frac{\alpha_z \exp(\int_0^z \alpha \, \mathrm{d}z)}{\sqrt{\frac{z}{p}}}, \qquad (5)$$

where α_z is the value of effective ionization coefficient, α corresponding to the external field at the head of the avalanche.

For this reason, in the model used in the context of this contribution, the logarithm of critical number of charge carriers $\ln N_{\mathrm{cr}}$ is taken as a function of pressure, namely $\sqrt{p}$. The breakdown analysis utilizing the proposed model is performed for both, sphere-plate and coaxial cylinder set-ups (cf. Fig. 5). Thereby the sphere-plate is stressed by positive LI voltage 1.2 µs / 50 µs, whereas for the coaxial cylinder geometry AC test voltage is applied. For modeling the LI breakdown voltage a constant ratio between AC and LI breakdown voltages is assumed (cf. [19]). In addition the model is calibrated with the measured breakdown voltage at a pressure of $p = 4$ bar to yield suitable assumptions for the protrusion dimensions in comparison to the given surface roughness of the electrodes. Nevertheless in future investigations also the influence of multiple protrusions with varying dimensions occurring on one electrode surface and their influence on the breakdown process and modeling should be considered. In addition statistical delay times of ignition and spark formation as well as the modeling of the breakdown process in (strongly) non-uniform electric fields will be considered in future development steps of the model and further experimental investigations. Applying the approach described above, the prediction of breakdown voltages for pressure values

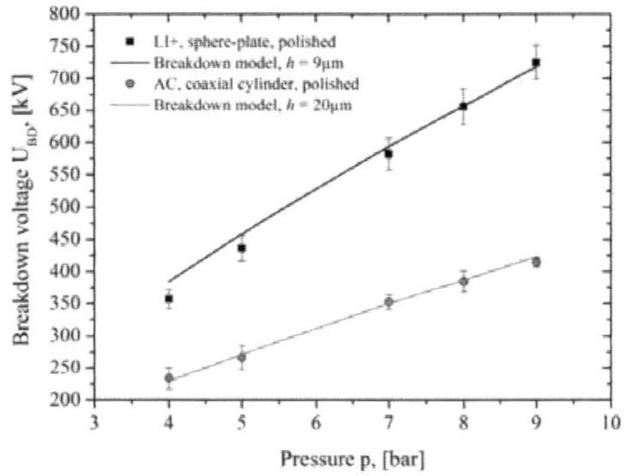

Figure 5. Comparison of experimental and modeling results for ln N_{cr} as a function of pressure $\sqrt{p}$.

of $p > 4$ bar is possible and for both set-ups the experimental and modeling results are in good agreement. It is observed, that the results indicate a slightly higher surface roughness for the coaxial cylinders than for the sphere-plate arrangement. For sphere-plate set-up, protrusion with the height $h = 9\,\mu m$ located at the surface of the sphere needs to be assumed while a protrusion with $h = 20\,\mu m$ is modeled at the surface of the inner cylinder. Both assumptions are in the expected range for the surface roughness of polished surfaces.

6. CONCLUSION

The model described in this contribution is developed for the breakdown analysis in clean air at different pressures taken into account real surface roughness of metal electrodes. The logarithm of the critical number of charge carriers is taken as a function of pressure. For both sphere-plate and coaxial cylinder set-ups, stressed by positive LI and AC voltages, the experimental and modeling results are found to be in good agreement.

Based on the simulation model and experimental results, new and reliable Clean Air GIS design up to 145 kV has been developed [13].

Acknowledgements

The authors would like to thank A. Geisler for his support with the model development as well as W. Hartmann and T. Hammer for the fruitful discussions.

References

[1] UFCCC. *Kyoto Protocol to the united nations framework convention on climate change.* United Nations Framework Convention on Climate Change, 1997.

[2] D. Gautschi and K. Pohlink. Ist SF_6 in Hochspannungsschaltanlagen ersetzbar? - Neue Forschungsarbeiten bringen ermutigende Resultate. *Bulletin Electrosuisse und Verband Schweizerischer Elektrizitätsunternehmen (VSE)*, 105(12):43–46, 2014.

[3] B. Lutz et al. Alternativen zu SF_6 - Aktueller Stand und klimaneutrale Auswege. In *Fachtagung Hochspannungs-Schaltanlagen: Anwendungen, Betrieb und Erfahrungen*, Darmstadt, 2016.

[4] M. Beyer et al. *Hochspannungstechnik.* Springer-Verlag, 1986.

[5] Y. Zhang et al. Study of Microscopic electric Field on contact Surface in Vacuum Interrupters based on Fractal Theory. *ISDEIV*, pages 29–32, 2016. doi:10.1109/DEIV.2016.7748667.

[6] R. Li and Y. Qiu. A new multi-ridge model for electrode surface roughness effect in SF6. In *Proc. 2nd Int. Conf. Properties and Applications of Dielectric Materials*, pages 105–108, 1988. doi:10.1109/ICPADM.1988.38344.

[7] H.G. Kosmahl. Analytic Evaluation of Field Emission Enhancement Factors for Ellipsoidal Cones and Elliptic Cross-Section Wedges. *IEEE Transactions on Electron Devices*, 38(6):1534–1537, 1991. doi:10.1109/16.81650.

[8] C. J. Edgcombe and U. Valdré. Microscopy and computational modelling to elucidate the enhancement factor for field electron emitters. *Journal of Microscopy*, 203(2):188–194, 2001. doi:10.1046/j.1365-2818.2001.00890.x.

[9] D. L. Jaeger et al. Local electrostatic effects of surface structure on field emission. *Journal of Applied Physics*, 93(1):691–697, 2003. doi:10.1063/1.1526934.

[10] V.G. Agapov and M.V. Sokolova. The Influence of Electrode Surface Microwhiskers on a Breakdown Voltage of a Pressurized Gas. In *International High Voltage Symposium*, pages 395–398, Zurich, 1975.

[11] V.V. Zhirnov. On the Cold Emission Mechanism of Diamond Coated Tips. *Journal de Physique*, 6(5):107–112, 1996. doi:10.1051/jp4:1996517.

[12] A. Galdetskiy. Field Enhancement in some Field Emission Structures and Ion Bombardment Suppression by Applied rf Voltage. In *Technical Digest of IVMC*, pages 598–601, 1997. doi:10.1109/IVMC.1997.627658.

[13] N. Presser et al. Advanced insulation and switching concepts for next generation High Voltage Substations. In *paper B-108*, Cigré Paris, 2016.

[14] M. Rossner and A. Hopf. Elektrische Festigkeit von alternativen Isoliergasen bei hohem Druck. *PESS*, page P01.6.

[15] M. Rossner et al. Dielectric Strength of Alternative Insulation Gases at High Pressure in the Inhomogeneous Electric Field. In *EIC*, pages 369–374, 2015. doi:10.1109/ICACACT.2014.7223618.

[16] R. Arora and W. Mosch. *High Voltage and Electrical Insulation Engineering.* Wiley-IEEE Press, 2011.

[17] W. Mosch and W. Hauschild. *Hochspannungsisolierungen mit Schwefelhexafluorid.* VEB Technik Verlag, 1979.

[18] J.M. Meek and J.D. Cragss. *Electrical breakdown of gases.* Oxford University Press, 1953.

[19] N. Koshino et al. Partial discharge and breakdown characteristics of CO_2-based gas mixtures as SF_6 substitutes. In *Gaseous Dielectrics X*, chapter 5, pages 217–222. Springer US, 2004. doi:10.1007/978-1-4419-8979-6_30.

doi:10.14311/ppt.2017.1.56
Plasma Physics and Technology **4**(1):56–59, 2017

ANALYSIS OF RADIATION DISCRETIZATION FOR MODELLING A SPARK GAP FOR SURGE CURRENTS

C. SANDER[a,b,*], J.-E. SCHMUTZ[a], M. KURRAT[b]

[a] *PHOENIX CONTACT GmbH & Co. KG, Flachsmarktstraße 8, D-32825 Blomberg, Germany*
[b] *Institute for High Voltage Technology and Electrical Power Systems, Technical University Braunschweig, Schleinitzstraße 23, 38106 Braunschweig, Germany*
[*] christian.sander1@phoenixcontact.com

Abstract. In this paper we address a method for spectrally resolved radiation modelling in thermal plasmas encountered in surge protective devices based on spark gaps. Compared to most switching applications, power input and plasma pressure are much higher which leads to an optically thick plasma with line broadening and enhanced wall ablation. In this situation it is possible to capture the full effect of spectrally resolved radiation on plasma dynamics by performing line-by-line calculations with downsampled absorption spectra. We show that it is possible to achieve radiation convergence with 1000 lines. Approaches for a further reduction of calculation times using band-averaged models and κ-group models are discussed. The κ-group model is based upon a grouping of the absorption coefficients into subgroups with different ranges of κ before averaging. The spectral calculation results are compared to the approximative methods and significant differences for Rosseland means are observed.

Keywords: spark gap, radiation, spectral, simulation, arc, surge protective device.

1. Introduction

Simulation of surge protective devices (SPDs) [1] has received considerably less attention than other devices employing thermal plasmas such as circuit breakers. Nontheless, the challenges faced here are similar, if not more difficult, due to the higher power density leading to a wider range of temperatures, pressures, and dynamics. Spark-Gap SPDs make use of wall stabilized arcs with small channel diameters which are needed for arc extinction and net follow current suppression. These kinds of arcs are influenced by interaction with the enclosing walls. Ablation occurs mainly because of radiation, so a large emphasis needs to be put on precise modelling of radiative transport. This is a considerable challenge because of the complex spectral dependence of absorption coefficients. Peyrou et al. investigated the precision of methods for solving the radiative transfer equation and the effects of spectral averaging methods[2]. They found reasonably good results using the Rosseland average for a multi-band model of 11 bands in optically thick conditions. Nordborg et al. used a subdivision of the bands according to the value of the absorption coefficient to reduce averaging errors with good results [3]. Randrianandraina et al. use a combination of Planck and natural averages based on frequency of the band and on temperature to obtain better results with a multi-band model[4].

In this paper we'll present time-dependent simulations of an SPD using a downsampled line-by-line method and compare it with band-averaged methods and the κ-group model used by Nordborg et al.[3].

2. Simulation model

The simulation is based on the assumption of local thermodynamic equilibrium and electrical quasineutrality. These assumptions result in the classical Navier-Stokes equations with additional terms for ohmic heating, radiation, electron enthalpy transport and ablation. Magnetic forces are smaller than pressure forces for the conditions discussed here and are neglected for better performance (less than 1 % difference).

$$\frac{\partial \rho}{\partial t} + \vec{\nabla}\cdot(\rho\vec{U}) = S_\rho \tag{1}$$

$$\frac{\partial \rho\vec{U}}{\partial t} + \vec{\nabla}\cdot(\rho\vec{U}\otimes\vec{U}) = \vec{\nabla}\cdot\boldsymbol{\tau} - \vec{\nabla}p \tag{2}$$

$$\frac{\partial \rho(h+k+\frac{p}{\rho})}{\partial t} + \vec{\nabla}\cdot(\rho(h+k)\vec{U}) = S_e + h_{S_\rho}S_\rho \tag{3}$$

In these equations ρ denotes density, $\vec{U}$ velocity, S_ρ the mass source term (see eq. 9), $\boldsymbol{\tau}$ the viscous stress tensor, p pressure, h enthalpy, k kinetic energy and h_{S_ρ} the enthalpy of the ablated mass entering the fluid region. The term S_e contains all sources that increase the internal energy and the kinetic energy:

$$S_e = \frac{\vec{j}^2}{\sigma} + S_R - \vec{\nabla}\cdot(\lambda\vec{\nabla}T) + \frac{5k_B}{2q_e}\vec{j}\cdot\vec{\nabla}T + \vec{\nabla}\cdot(\boldsymbol{\tau}\cdot\vec{U}) \tag{4}$$

Here $\frac{\vec{j}^2}{\sigma}$ is the ohmic heating, S_R the radiative source term, $\vec{\nabla}\cdot(-\lambda\vec{\nabla}T)$ the heat conduction source term, $\frac{5k_B}{2q_e}\vec{j}\cdot\vec{\nabla}T$ the enthalpy transport by electrons with k_B being the Boltzmann constant and q_e the elementary charge and $\vec{\nabla}\cdot(\boldsymbol{\tau}\cdot\vec{U})$ the viscous heat. The electrostatic potential ϕ is calculated as

$$\vec{\nabla}\cdot(\sigma\vec{\nabla}\phi) = 0 \tag{5}$$

which gives the electric current as $\vec{j} = -\sigma\vec{\nabla}\phi$, with σ being the electrical conductivity.

Radiation is considered in the P1-model, which gives the following equations in its spectral form [5]:

$$\vec{\nabla}(\frac{1}{3\kappa_\nu}\vec{\nabla}G_\nu) = \kappa_\nu(G_\nu - 4\pi B_\nu(T)) \tag{6}$$

$$S_R = \vec{\nabla}\cdot(\int \frac{1}{3\kappa_\nu}\vec{\nabla}G_\nu \,\mathrm{d}\nu) \tag{7}$$

To consider the effect of plasma-wall interaction, we model the ablation of the polyoxymethylene (POM) wall by solving a heat balance equation (eq. 8) at the surface, considering heat conduction from plasma into the solid, radiation and the ablation heat flux of the ablated material going into the plasma. Ablation starts once the surface temperature reaches the evaporation temperature of POM which is taken as $T_v = 683\,\mathrm{K}$. We assume that the wall temperature cannot exceed the evaporation temperature, which allows to determine the required ablative heat flux needed for limiting the temperature.

$$q_{c,p} + q_r = q_{c,f} + q_v \tag{8}$$

$$S_\rho = \frac{q_v A}{\Delta h_{v,\mathrm{eff}} V} \tag{9}$$

Here A is the surface area of a boundary cell and V its volume. $\Delta h_{v,\mathrm{eff}} = \Delta h_v + (h(p, T_k) - h(p_0, T_v))$ is the enthalpy change of the gas which is ablated. It is commonly accepted that ablated gas enters the plasma at a much higher temperature than T_v because radiation is absorbed in the intermediate layer close to the wall [6]. We assume a temperature of the ablated gas of $T_k = 3500\,\mathrm{K}$ and an evaporation enthalpy of $\Delta h_v = 1.6\,\mathrm{MJ/kg}$. For electrode erosion we consider a simple model in which the mass source term is proportional to current density [7].

The plasma composition, thermodynamic and transport properties of the plasma are taken from D'Angola et al. [8], hence we assume that the plasma consists of air only. The radiative absorption coefficient $\kappa_\nu(p, T)$ is calculated with SPARTAN [9], which is freely available.

The simulation model was implemented in OpenFOAM [10] using the flow solver published by Kraposhin et al.[11].

In spark gaps an ignition circuit is often used to facilitate plasma ignition at lower voltages than the very high breakdown voltage of the main electrodes (see figure 1). A third electrode is connected by a metal-oxide varistor to the first main electrode and placed near the second main electrode. Ignition then happens by electric breakdown or resistive elements at much lower voltage. The nonlinearity of the varistor characteristic curve is solved by employing the linearity of equation 5, separating it into multiple equations

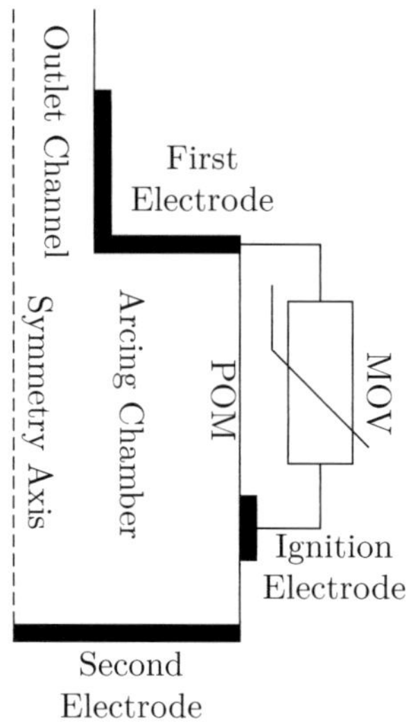

Figure 1. Investigated geometry with ignition circuit.

where only one of the electrodes has a potential $\phi_i \neq 0$ and then using superposition to calculate the full potential. This allows to reduce the iterative procedure of finding the correct potentials to a problem that doesn't require the repeated solving of equation 5.

The simulation is started at ambient conditions. To make the ignition possible, an artificially increased conductivity value is assumed in the area between the ignition electrode and the second electrode for a short amount of time. An 8/20 µs waveform impulse with an amplitude of 15 kA is impressed between the main electrodes. No-slip boundary conditions are used and on the outer side of the solids we apply spatially constant potentials and ambient temperature. The calculations shown here were performed on an Intel i7-4770 using one thread.

3. Averaging methods

We employ the following methods for handling the spectral dependence of radiation:

1. Spectrally resolved absorption coefficients $\kappa_\nu(p, T)$, possibly downsampled for better performance
2. Mean absorption coefficients (MACs) band model for spectrally separated bands $\kappa_i(p, T)$, namely Planck, Rosseland and Natural means
3. κ-group model, based on subdividing each band from 2. into discontinous groups depending on the value of $\kappa_\nu(p, T)$ and averaging these groups resulting in a set of absorption coefficients $\kappa_{i,j}(p, T)$

Spectrally resolved absorption coefficients can return exact results at the price of very high computational cost. Because of this we employ a coarse mesh to be able to handle spectral discretizations of up to 10^5 points. Calculation of MACs are performed using the frequency intervals $[3\cdot 10^{-5}, 0.35, 1.54, 2.65, 2.94, 3.51, 15]\cdot 10^{15}\,\mathrm{s}^{-1}$ ([7] extended for broader spectral

N_ν	U V	$\overline{T}$ K	T_{max} K	$\overline{p}$ bar	m kg	t h
10^5	403.6	28884	44905	936.1	4.22e-9	280
	ΔU %	$\Delta\overline{T}$ %	ΔT_{max} %	$\Delta\overline{p}$ %	Δm %	t %
10^4	0.06	-0.03	-0.07	-0.07	0.10	13
5000	0.04	-0.03	-0.15	-0.13	-0.02	6.5
1000	0.35	-0.20	-0.19	0.02	0.78	1.4
100	n. c.	n. c.	n. c.	n. c.	n. c.	n. c.

Table 1. Absolute values of the most accurate calculation with $N_\nu = 10^5$ frequency sampling points and relative errors with fewer sampling points at $t = 10\,\mu s$

range) and $[0.075, 0.350, 1.540, 2.114, 2.180, 2.430, 2.563, 2.900, 3.260, 3.350, 3.520, 3.750, 4.200, 5.960, 6.561, 7.048, 7.840, 11.00, 15.00] \cdot 10^{15}\,s^{-1}$ (based on $\kappa_\nu(p,T)$ dependency). A method of choosing the subgroups based on the value of κ_ν is required for the κ-group model. Because SPARTAN outputs an irregular frequency axis it was necessary to resample the spectra to a regular axis with $2 \cdot 10^7$ points. Because of the pressure and temperature dependence it is not possible to sort a specific κ_ν value uniquely into one group for all pressures and temperatures. Instead of using global definitions for the boundaries of the groups we define them separately for each pressure and temperature based on the minimum and maximum values of κ_ν encountered in one band. The groups are equidistantly spaced on the logarithmic scale. This method results in a different mapping of κ_ν for each pressure and temperature. We can calculate a mean group for every frequency point and use this as a mapping for all absorption spectra. In the end one of the different absorption means is calculated for the set of frequencies contained in each band/group tuplet resulting in a set of $\kappa_{i,j}(p,T)$ values. If one of these sets is empty it will be dropped.

4. Results

The convergence behaviour of the spectrally resolved calculations is shown in table 1 along with calculation times. The values are evaluated at current maximum ($t = 10\,\mu s$) when a wall stabilized arc is formed. At $N_\nu = 100$ the simulations didn't converge, however for discretizations with $N_\nu \geq 1000$ the error is less than 1 % for all quantities, showing a good convergence for a much lower sampling rate than the complexity of the spectrums suggests. We attribute this to the relatively high optical thickness occuring at large pressures at magnitudes above 100 bar, leading to line broadening which thus requires less points. Moreover the mesh size is larger than the optical thickness for many frequencies which may also have an influence. During the ignition ($t < 4\,\mu s$) there are some deviations in the maximum encountered temperature between 100000 and fewer points (not listed in the table) but this doesn't have a significant influence on the further behaviour. We take the calculation with 100000 sampling points as reference for all further calculations because the difference seen in comparison with fewer sampling points suggests the correctness of this result. Higher discretizations were not performed because of calculation times.

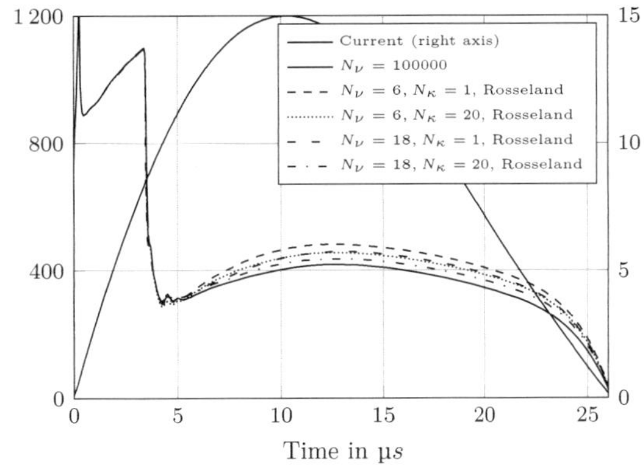

Figure 2. Current impulse and voltages of band models and κ-grouped models compared to the most accurate spectrally resolved calculation. (Left axis: Voltage in V, right axis: Current in kA)

Figure 2 shows exemplary voltage curves of Rosseland-averaged band and κ-group models for the two spectral intervals discussed before. The Rosseland average is displayed because it represents the worst case of the Planck, Natural and Rosseland means. It can be seen that the voltage during the ignition phase and the ignition time are mostly unaffected by the means of radiation approximation. This is largely caused by the low amount of ablation that happens in this phase. Differences start to appear after the ignition when ablation becomes relevant. Increasing the spectral bands and/or number of κ-groups leads to better results in all cases but also increases calculation times (Approximately doubled calculation time for $N_\nu = 6, N_\kappa = 20$ compared to $N_\nu = 6, N_\kappa = 1$).

The average temperature of κ-group models with $N_\nu = 18, N_\kappa = 20$ and different averaging methods is shown in figure 3. All methods show little deviations compared to the exact result, however the Rosseland mean underpredicts the mean temperature the most.

Table 2 shows the percental errors of voltage, average temperature, maximum temperature, average pressure and total ablated mass in the arcing chamber at $t = 10\,\mu s$ and the percentage of calculation time compared to the spectrally resolved calculation with 100000 sampling points. The choice of Rosseland average results in the largest errors of voltage and ablation mass. At the same time the Rosseland average gives a very good fitting to the maximum temperature.

The Rosseland average overpredicts the emission in the arc core, leading to a stronger ablation and a stronger cooling in the boundary regions, thus the arc is more constricted and a larger current density and

N_ν	N_κ	$\overline{\kappa}$	ΔU	$\Delta \overline{T}$	ΔT_{max}	$\Delta \overline{p}$	Δm	Δt
6	1	P	4.98	-0.76	7.05	-8.23	-2.52	0.14
6	1	N	0.89	1.90	5.56	-1.76	1.30	0.12
6	1	R	15.2	-6.98	-0.08	5.97	32.0	0.13
6	20	P	4.17	-0.35	6.76	-3.36	3.00	0.26
6	20	N	2.08	0.03	5.27	-1.66	2.49	0.25
6	20	R	8.86	-4.07	1.49	5.28	21.6	0.27
18	1	P	5.91	-0.49	3.14	-6.35	2.32	0.14
18	1	N	3.03	0.53	6.46	-3.36	1.53	0.14
18	1	R	8.67	-4.10	0.55	1.36	15.9	0.14
18	20	P	4.76	-0.94	4.07	-2.08	5.07	0.48
18	20	N	3.95	-0.49	4.12	-1.50	4.85	0.47
18	20	R	4.41	-1.99	0.25	-0.63	6.03	0.49

Table 2. Relative errors and calculation time (in %) of radiation approximations at $t = 10\,\mu s$.

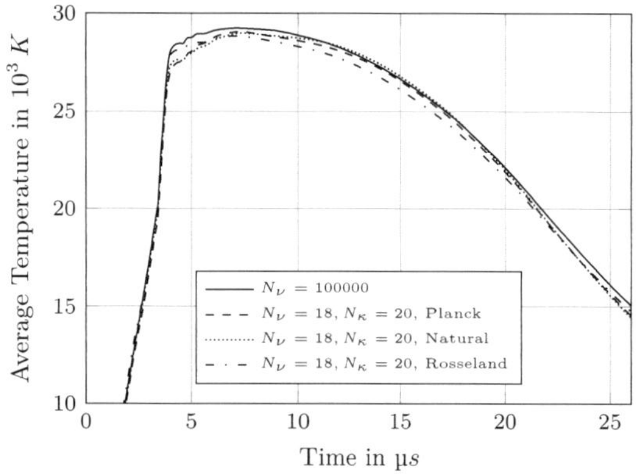

Figure 3. Average temperature of reference calculation compared to different κ-group models radiation approximation methods.

ohmic heating is observed in the core which compensates the stronger emission. The enhanced ablation is also responsible for the lower average temperatures.

The Planck and Natural means both give very good results, with the natural mean being slightly better at most quantities. Using more bands doesn't lead to significantly better results. The reason is most likely an unoptimized choice in band limits and too much variation of the absorption coefficient in the single bands even when more bands are used. Using more κ-groups shows consistent improvements only for the Rosseland mean.

These results suggest to perform further optimizations or to use an averaging method that gives acceptable results at the lowest computation time. In this case we would favor the 6-band natural mean. Further optimization may improve the choice of band limits, employ line limiting methods or switch the averaging method in some ranges of pressures and temperatures.

5. Conclusions

We have investigated the accuracy of various methods of spectral discretizations. The use of downsampled absorption coefficients gave good results, making this a feasible method if the increased calculation time is acceptable. The well known band model gives good results compared to the spectrally resolved result when the Planck or natural means are used. The 6-band natural mean gives errors on the order of approximately 5 % and is more than 10 times faster than the fastest spectrally resolved calculation. However, the Rosseland mean results in a larger error and it cannot be recommended without further improvements. The grouping of the absorption coefficient based on its value gives improved results, however there is still a difference to be observed. It is possible that the accuracy of these approximations can be improved by line limiting or by the selective use of different MACs. It might also be possible to get better results by an optimized choice of band and group limits.

References

[1] Phoenix Contact GmbH & Co. KG. Lightning and surge protection basics. `https://www.phoenixcontact.com/assets/downloads_ed/global/web_dwl_promotion/5131327_TT_Basics_EN.pdf`, 2017.

[2] B. Peyrou et al. Radiative properties and radiative transfer in high pressure thermal air plasmas. *J. Phys. D: Appl. Phys.*, 45(45):455203, 2012. doi:10.1088/0022-3727/45/45/455203.

[3] H. Nordborg and A. A. Iordanidis. Self-consistent radiation based modelling of electric arcs: I. efficient radiation approximations. *J. Phys. D: Appl. Phys.*, 41(13):135205, 2008. doi:10.1088/0022-3727/41/13/135205.

[4] H. Z. Randrianandraina et al. Improvements of radiative transfer calculation for SF_6 thermal plasmas. *J. Phys. D: Appl. Phys.*, 44(19):194012, 2011. doi:10.1088/0022-3727/44/19/194012.

[5] Michael F. Modest. *Radiative Heat Transfer, Third Edition*. Academic Press, 3 edition edition.

[6] D. Godin et al. Modelling and simulation of nozzle ablation in high-voltage circuit-breakers. *J. Phys. D: Appl. Phys.*, 33(20):2583, 2000. doi:10.1088/0022-3727/33/20/310.

[7] C. Rümpler. *Lichtbogensimulation für Niederspannungsschaltgeräte*. Fraunhofer-Verlag, 2009.

[8] A. D'Angola and other. Thermodynamic and transport properties in equilibrium air plasmas in a wide pressure and temperature range. *Eur. Phys. J. D*, 46(1):129–150, 2008. doi:10.1140/epjd/e2007-00305-4.

[9] Lino da Silva, M. An adaptive line-by-line—statistical model for fast and accurate spectral simulations in low-pressure plasmas. *J. Quant. Spectrosc. Radiat. Transfer*, 108(1):106–125, 2007. doi:10.1016/j.jqsrt.2007.03.005.

[10] H. Jasak. OpenFOAM: Open source CFD in research and industry. *Int. J. Nav. Archit. Ocean Eng.*, 1(2):89–94, 2009. doi:10.2478/IJNAOE-2013-0011.

[11] M. Kraposhin et al. Adaptation of kurganov-tadmor numerical scheme for applying in combination with the PISO method in numerical simulation of flows in a wide range of mach numbers. *Procedia Comput. Sci.*, 66:43–52, 2015. doi:10.1016/j.procs.2015.11.007.

doi:10.14311/ppt.2017.1.61
Plasma Physics and Technology **4**(1):61–65, 2017

EXPERIMENTAL STUDY ON ARCING EFFECTS TO CONTACT MATERIALS IN 270V DC

Q.C. Zhu[a], Z.B. Li[a,*], J. Wei[a], Z.X. Liu[a], J. Huang[a], D.D. Zhang[a], M.Hasegawa[b]

[a] *State Key Laboratory of Advanced Electromagnetic Engineering and Technology, Huazhong Univ.of Science and Technology, 430074, Wuhan, China*

[b] *Chitose Institute of Science and Technology, 066-8655, Chitose, Hokkaido, Japan*

[*] zbli@hust.edu.cn

Abstract. As supply DC voltage levels in automobiles, aviation and aerospace are becoming higher, high voltage and large capacity compact DC electromagnetic relays or contactors are becoming more important. In this paper, experiments were conducted at DC 270V/100A in air and nitrogen, respectively, with CuCr40, CuCr45, CuCr50, CuW60, CuW70 and CuW90 as contact materials. Welding resistance (corresponding to electrical contact life), contact erosion and breaking arc durations were measured, and the obtained results were discussed.

Keywords: contact materials, DC voltage, arcing duration, welding, erosion, relay.

1. Introduction

In recent years, solar power plants and energy storage systems have been becoming popular year by year with increasing demands for renewable energy, resulting in increased demands for high voltage and large current switching devices [1–3]. At the same time, electrical load in the fields of electric vehicles and aerospace have been becoming larger, also leading to development of electromagnetic DC relays capable of switching higher voltages and larger currents [4–6]. As reported, higher voltages and larger currents will make it more difficult to extinguish arcs, resulting in more serious contact erosion [7–9]. Thus, it is of great significance to investigate effects of atmospheres and contact materials on contact welding resistance, erosion and breaking arc durations for the purpose of improving reliabilities and electrical lifespans of higher voltage and larger current relays.

In this paper, CuCr40, CuCr45, CuCr50, CuW60, CuW70 and CuW90 contacts were used to conduct switching operations in a resistive DC 270 V/100 A load circuit on a simulation test device in air and nitrogen, respectively, and influences of contact materials and atmospheres on contact welding resistance, erosion and breaking arc durations were analyzed.

2. Experiemntal Method

The experimental schematic circuit diagram used in this study is shown in Figure 1. The main circuit was composed of a DC power source E providing a regulated output voltage of 270 V, a load resistor R of 2.7 Ω, and a pair of electrical contacts K which was set on the simulation test device. The coil control circuit, enclosed with dotted lines, included a DC power source Ec, a rheostat Rc, an electromagnetic coil and a Programmable Logic Controller PLC. The voltage

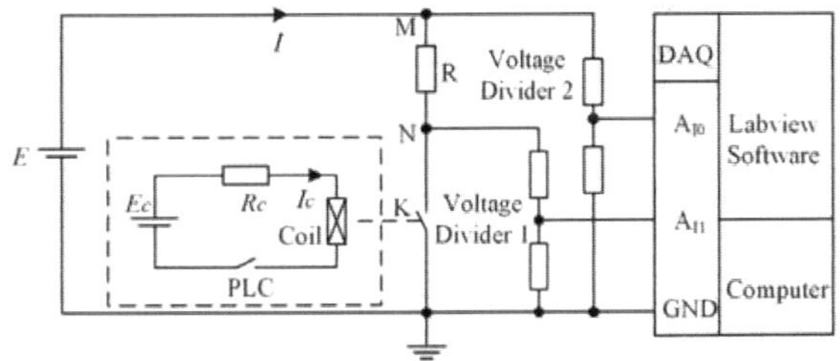

Figure 1. Circuit diagram.

U_N between the pair of contacts K was measured with voltage divider 1, and the voltage U_M was measured with divider 2. The current I flowing through R was calculated as $(U_M$-$U_N)/R$, where U_M might not be equal to the voltage value of the DC power resource E if wire impedances were taken into account. Output voltages of the two voltage dividers were recorded with PC-DAQ data acquisition system which was managed by LabVIEW software (with a sampling frequency of 100 kHz).

The contact materials used in the experiments were CuCr40, CuCr45, CuCr50, CuW60, CuW70, and CuW90. The number represents the weight percentage of Cr or W. For example, CuCr40 means that Cr accounts for 40 %, and Cu accounts for 60 %. Each of these contact materials was of rivet type with the rivet head diameter of 10 mm and the thickness of 3 mm. Before starting each experiment, the contact simulation test device was put into a vacuum chamber that was then filled by air or nitrogen with pressure of $1 \cdot 10^5$ Pa. The gap length for each pair of contacts was 7 mm at maximum, and the contact operation frequency was set as 0.7 s ON and 2.3 s OFF. A pair of permanent magnets was used to obtain transverse magnetic field, and the magnetic flux density between

Contact material	gases	1# N_1	1# N_2	1# N_3	2# N_1	2# N_2	2# N_3	3# N_1	3# N_2	3# N_3	$\overline{N}_1$	$\overline{N}_2$	$\overline{N}_3$	$\overline{N}$
CuCr40	air	48	62	1491	120	2327	497	90	1076	131	86	1365	706	2157
	nitrogen	26	75	99	115	10	71	203	9	66	115	31	79	225
CuCr45	air	82	1033	421	107	1341	30	75	231	587	88	858	346	1302
	nitrogen	240	101	33	42	136	235	38	34	55	107	90	108	305
CuCr50	air	557	82	209	69	85	255	118	1216	86	248	461	183	892
	nitrogen	70	17	11	86	128	55	281	11	67	146	52	44	242

Table 1. Welding results of CuCr in air and nitrogen.

Contact material	gases	1# N_1	2# N_1	3# N_1
CuW60	air	>10k	>10k	>10k
	nitrogen	>10k	4033	>10k
CuW70	air	>10k	>10k	>10k
	nitrogen	>10k	>10k	2663
CuW90	air	>10k	>10k	>10k
	nitrogen	>10k	>10k	>10k

Table 2. Welding results of CuW in air and nitrogen.

the pair of contacts was 80 mT. Three pairs of contact samples (named as 1#, 2#, 3#) were used for each contact material in air and nitrogen, respectively. As the contact surface burn-in treatments, each pair of contacts operated 100 switching (making and breaking) operations in the resistive DC 30 V/10 A circuit before the actual experiment.

CuCr contacts known to have weak welding resistance are likely to weld frequently. Therefore, after their first welding, the contact pair was reopened artificially to resume their operations until they welded again. Such procedures were repeated until the third welding occurred, and the corresponding numbers of operations to the three welding incidents were recorded as N_1, N_2 and N_3, respectively. Thus, the total number of operations corresponding to the three welding incidents was calculated as $N = N_1 + N_2 + N_3$. On the other hand, CuW contacts are of good welding resistance, and the number of operations was set to be 10000, which meant that the experiment would be terminated under the following conditions: when the first welding occurred or the number of operations reached 10000.

3. Results and Analysis

3.1. Contacts' welding resistance

Table 1 and Table 2 respectively show welding results of CuCr and CuW in air and nitrogen. In Table 1, N_1, N_2 and N_3 each represent the numbers of operations corresponding to the three welding incidents for CuCr contacts, while $\overline{N}_1$, $\overline{N}_2$, $\overline{N}_3$ each stand for the average numbers of operations corresponding to the three welding incidents, as calculated according to formula (1). In addition, $\overline{N}$ is the average total number of operations of the material as calculated with formula (2). For CuW contacts, N_1 represents the number of operations at the end of the experiment (N_1>10k indicates that there was still no welding incident when the number of operations reached 10000).

$$\overline{N}_i = \frac{1}{3}\sum_{j=1}^{3} N_i(j\#) \tag{1}$$

$$\overline{N} = \frac{1}{3}\sum_{j=1}^{3}\sum_{i=1}^{3} N_i(j\#) \tag{2}$$

Where $N_i(j\#)$ is the number of operations corresponding to the i-th welding incident of the j-th pair of contacts for CuCr.

The welding results in Table 1 and Table 2 indicate the following points:

1. CuW is of higher welding resistance than CuCr in both air and nitrogen. For example, all of the CuW contacts completed 10000 switching (making and breaking) operations without any welding incidents in air. Moreover, among 9 pairs of CuW contacts in nitrogen, only 2 pairs suffered welding, and the minimum number of operations (2663) was obviously higher than all N_1 of CuCr contacts.

2. Monotonic change of the composition proportion of contact material has no monotonic effects on welding resistance of the material. For example, for CuCr material, when the content of Cr increased monotonically from 40 % to 50 %, the values of $\overline{N}_1$, $\overline{N}_2$, $\overline{N}_3$, $\overline{N}$ in nitrogen did not show monotonic decreases or monotonic increases. In addition, the situation of CuW in nitrogen was similar to CuCr.

3. In general, air realizes better resistance against welding than nitrogen. For instance, all of CuW contacts completed 10000 switching operations without any welding incidents in air, while in nitrogen, there were 2 welding incidents when the numbers of operations reached 4033 and 2663, respectively. The average total numbers of operations $\overline{N}$ of CuCr40, CuCr45, and CuCr50 in air were 2157, 1302 and

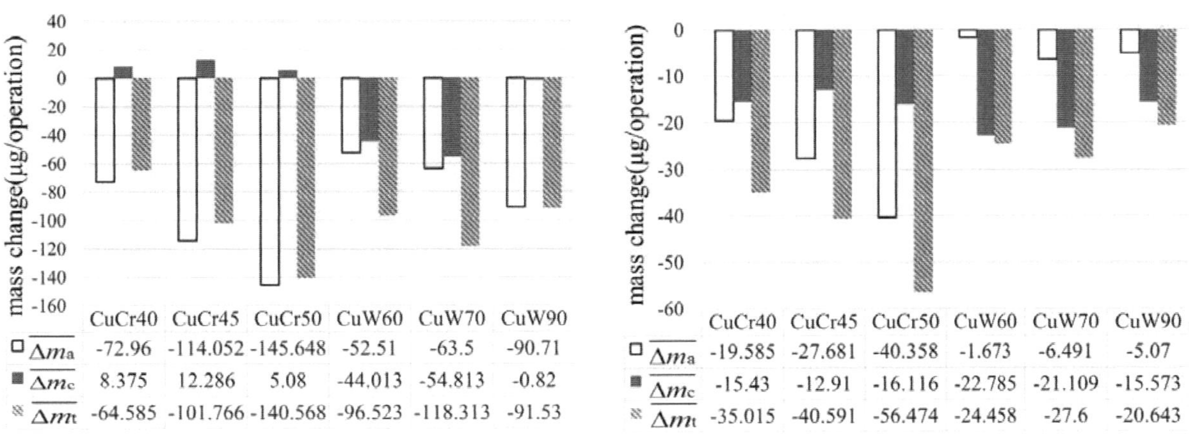

(a) *Results in air* (b) *Results in nitrogen*

Figure 2. Erosion results of CuCr and CuW in air and nitrogen.

892, respectively, which were much higher than the corresponding numbers 225, 305, and 242 in nitrogen.

3.2. Contact erosion

Figure 2 show erosion results of CuCr and CuW in air and nitrogen, where $\overline{\Delta m_a}$ and $\overline{\Delta m_c}$ stand for the average erosion rates of anode and cathode of the material for each switching (making and breaking) operation, respectively.($\overline{\Delta m_a}$=the total anode erosion mass of three pairs of contacts/the total number of operations of three pairs of contacts), and $\overline{\Delta m_c}$ is calculated like $\overline{\Delta m_a}$. $\overline{\Delta m_t}$ represents the average erosion rate of both anode and cathode for each switching operation, and $\overline{\Delta m_t}=\overline{\Delta m_a}+\overline{\Delta m_c}$.

From Figure 2, the following points on erosion results are shown:

1. The erosion mass of CuW is not necessarily lower than that of CuCr in air and nitrogen all the time. The erosion mass is influenced by atmospheres and the electrode polarity. CuW anode contact generally has lower erosion mass than CuCr, while the cathode contact shows the opposite tendencies. For example, comparing the values of $\overline{\Delta m_a}$ in nitrogen, CuW60 to CuW90 were: −1.673, −6.491, −5.07, which were much smaller than the corresponding values: −19.585, −27.681, −40.358 of CuCr40 to CuCr50. In addition, comparing the values of $\overline{\Delta m_c}$ in air, CuW60 to CuW70 were: −44.013, −4.813, which were significantly greater than those of CuCr40 to CuCr50, whose values were: 8.375, 12.286, 5.08 respectively.

2. Monotonic change of the composition proportion of contact material has no monotonic effects on erosion mass of the material. For example, when the content of Cr increased monotonically from 40% to 50% for CuCr material in nitrogen, the values of $\overline{\Delta m_c}$ were: −15.43, −12.91, −16.116 respectively, which decreased at first and then increased. No monotonic increasing or decreasing trends were found there. The $\overline{\Delta m_c}$ values of CuW in air increased at first and then decreased. But for $\overline{\Delta m_a}$ and $\overline{\Delta m_t}$, the values of CuCr and CuW in air and nitrogen generally monotonically increased with the growth of Cr or W content, except for CuW90 in nitrogen.

3. In general, air is more likely to cause high erosion mass than nitrogen. For instance, the values of $\overline{\Delta m_a}$ of CuW60 to CuW90 in air were respectively: −52.51, −63.5, −90.71, while in nitrogen, the values were only: −1.673, −6.491, −5.07. In addition, the values of $\overline{\Delta m_a}$ of CuCr40 to CuCr50 in air were: −72.96, −114.052, −145.648, which were much greater than those (only −19.585, −27.681, −40.358) in nitrogen.

4. Erosion mass is closely related to the electrode polarity. In most cases, the anode erosion mass is greater than the cathode erosion mass, except for CuW in nitrogen, in this study.

3.3. Breaking arc duration

Figure 3 shows the typical voltage waveform during the seperation operation of two electrodes, and the breaking arc duration is defined in the figure.

Table 3 and Table 4 show breaking arc duration results of CuCr and CuW in air and nitrogen, in which $\overline{t}_{b1\#}$, $\overline{t}_{b2\#}$, $\overline{t}_{b3\#}$ respectively represent the average breaking arc durations of the first, second, and third pair of contacts of the material. $\overline{t}_{b1\#}$=the sum of breaking arc duration of all operations of the first pair of contacts/the total number of operations of the first pair of contacts. $\overline{t}_{b2\#}$ and $\overline{t}_{b3\#}$are calculated like $\overline{t}_{b1\#}$. In addition, $\overline{t}_{b}$ is the total average breaking arc duration of three pairs of contacts for the material, and $\overline{t}_{b}$=the sum of breaking arc duration of all operations of three pairs of contacts/the total number of operations of three pairs of contacts.

Contact material	$\bar{t}_{b1\#}$(ms)	$\bar{t}_{b2\#}$(ms)	$\bar{t}_{b3\#}$(ms)	$\bar{t}_{b}$(ms)
CuCr40	13.93	14.36	17.42	15.38
CuCr45	15.11	15.32	14.85	15.13
CuCr50	15.93	14.75	15.93	15.75
CuW60	17.84	17.78	18.25	18.23
CuW70	17.74	15.07	17.88	16.9
CuW90	15.63	17.35	16.66	16.54

Table 3. Breaking arc duration results of CuCr and CuW in air.

Contact material	$\bar{t}_{b1\#}$(ms)	$\bar{t}_{b2\#}$(ms)	$\bar{t}_{b3\#}$(ms)	$\bar{t}_{b}$(ms)
CuCr40	14.63	14.49	13.37	14.07
CuCr45	14.50	13.88	16.95	14.56
CuCr50	15.29	13.71	13.28	13.71
CuW60	12.01	13.72	12.87	12.65
CuW70	11.60	11.79	15.72	12.46
CuW90	12.38	12.63	12.49	12.5

Table 4. Breaking arc duration results of CuCr and CuW in nitrogen.

From Table 3 and Table 4, the following points are shown on the results of breaking arc duration:

1. The breaking arc durations of CuCr and CuW are influenced by the atmospheres. The breaking arc durations of CuW are in general longer than those of CuCr in air, while in nitrogen generally shorter than CuCr, in other words, the breaking arc durations of CuW are influenced more greatly by the atmospheres. For example, in air, the average breaking arc durations of each pair of CuW60 contacts were longer than those of CuCr. Moreover, 1# and 3# pairs of contacts of CuW70 were of similar situation to CuW60. Furthermore, in nitrogen, the average breaking arc durations of each pair of CuW90 contacts were shorter than those of CuCr. At the same time, 1# and 3# pairs of contacts of CuW60, 1# and 2# pairs of contacts of CuW70, were also the same with CuW90.

2. Monotonic change of the composition proportion of contact material has no monotonic effects on breaking arc durations of the material. As for CuCr40 to CuCr50 in air, the total average breaking arc durations $\bar{t}_b$ were: 15.38, 15.13, 15.75, which decreased at first and then increased, and in the case of $\bar{t}_{b1\#}$, $\bar{t}_{b2\#}$ and $\bar{t}_b$ of CuW in nitrogen, the results were similar to the above situations.

3. Comparing the results in air with those in nitrogen, in most cases, the breaking arc durations in nitrogen were likely to be shorter than those in air. For instance, the total average breaking arc durations of CuCr and CuW in nitrogen were smaller than those in air. In addition, among the total 18 numbers of $\bar{t}_{b1\#}$, $\bar{t}_{b2\#}$, $\bar{t}_{b3\#}$ of CuCr and CuW, there were 15 numbers in nitrogen that were less than those in air, just only 3 numbers were contrary ($\bar{t}_{b1\#}$, $\bar{t}_{b2\#}$ of CuCr40 and $\bar{t}_{b3\#}$ of CuCr45, respectively).

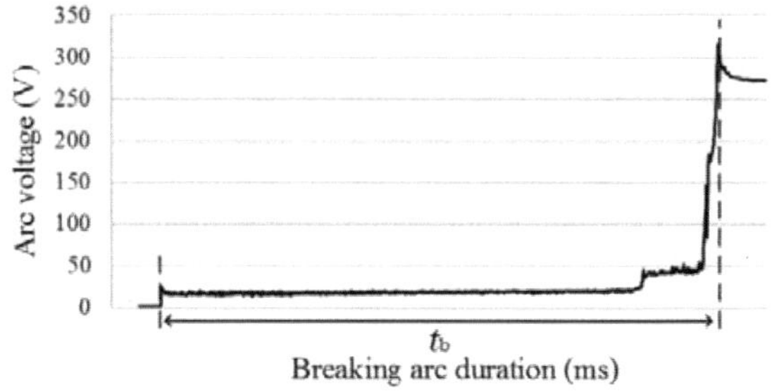

Figure 3. The typical voltage waveform during the seperation operation of two electrodes.

4. Discussions

4.1. Effect of oxide on contact welding resistance

From the welding results of CuCr contacts in Table 1, the values of average number of operations $\overline{N}_1$ corresponding to the first welding incident in air and nitrogen were of the similar order, while the values of $\overline{N}_2$ and $\overline{N}_3$ in air were much larger than those in nitrogen. For instance, the values of $\overline{N}_1$, $\overline{N}_2$, $\overline{N}_3$ of CuCr40 in air were: 86, 1365, 706, while in nitrogen, the values were: 115, 31, 79, and the results of CuCr45 and CuCr50 were similar to CuCr40. The reason for the results may be related to the fact that copper oxides and chromium oxides to be generated in air can reduce welding strengths σ, resulting in enhanced welding resistance[10]. In nitrogen, no oxides are generated and thus such effects will not be expected. At the beginning of the switching operations, the contact surfaces are likely to be slightly oxidized in air, which cannot obviously improve welding resistance. Therefore, the values of $\overline{N}_1$ in air and nitrogen are of the similar order. For $\overline{N}_2$ and $\overline{N}_3$, with the increased oxidation degree, the values in air can be significantly higher than those in nitrogen.

4.2. Differences between anode and cathode erosion

The erosion results show that the anode contact exhibits strong regularities: the mass losses can be observed for all the anode contacts, and with the increased Cr or W content, the erosion mass shows a monotonically increasing trend (except for CuW90 in nitrogen). However, the cathode contact erosion masse shows more influences of materials and atmospheres.

For example, CuCr contacts gained material and mass increased in air, but lost material and mass decreased in nitrogen. The phenomenon can be explained by considering the situation in which the anode contact is primarily bombarded by electrons and then heated, vaporized or lose its mass because of splash, while the cathode contact is likely to be not only heated by arc root but also emit electrons to maintain arc current, resulting in more complex processes of heat input/output and erosion.

5. Conclusions

In this paper, CuCr40, CuCr45, CuCr50, CuW60, CuW70 and CuW90 contacts were used to conduct switching (making and breaking) operations in the resistive DC270 V/100 A load circuit on a simulation test device in air and nitrogen, respectively. The conclusions can be drawn as follows:

1. Comparing CuCr with CuW, CuW shows higher welding resistance in both air and nitrogen;
2. Erosion mass of CuCr and CuW is affected by the electrode polarity and the atmospheres. The CuW anode contact generally has lower erosion mass than CuCr, while the CuW cathode contact has generally higher erosion mass than CuCr;
3. Breaking arc durations of CuCr and CuW are affected by environmental atmospheres. In air, the breaking arc durations of CuW are generally longer than CuCr, while in nitrogen generally shorter than CuCr, in other words, the breaking arc durations of CuW are influenced more greatly by the atmospheres;
4. Monotonic change of the composition proportion of contact material has no monotonic effects on welding resistance, erosion resistance and breaking arc durations.

Acknowledgements

The authors thank for the support from The National Natural Science Foundation of China (contract Number 51377062).

References

[1] D. Volm and F. Winkler. Development of a compact relay for high voltage switching of up to 1000V and 40A. In *Proc. 27th International Conference on Electrical Contacts*, pages 144–148, Dresden, Germany, June 2014.

[2] N. Wakatsuki, T. Kudo, and D. Hara. Development of arc-free DC300V/30A electro-mechanical devices using transient current switching circuits. In *Proc. 27th International Conference on Electrical Contacts*, pages 149–153, Dresden, Germany, June 2014.

[3] K. Yoshida, S. Shimotsuma, K. Sawa, K. Suzuki, and K. Takaya. Various characteristics of electromagnetic contactor when arc discharge are generated only make arc. In *Proc. 62nd IEEE Holm Conference on Electrical Contacts*, pages 215–221, Florida, USA, October 2016.

[4] H. Köepf, E. D. Wilkening, C. Klosinski, and M. Kurrat. Experimental research on direct current switchgear. In *Proc. 60th IEEE Holm Conference on Electrical Contacts*, pages 1–6, New Orleans, USA, October 2014. doi:10.1109/HOLM.2014.7031068.

[5] X. Zhou, X. Cui, M. Chen, and G. Zhai. Experimental study on arc behaviors of a bridge-type contact when opening a resistive load in the range of from 280 VDC to 730 VDC. In *Proc. 60th IEEE Holm Conference on Electrical Contacts*, pages 1–7, New Orleans, USA, October 2014. doi:10.1109/HOLM.2014.7031065.

[6] H. Köepf, E. D. Wilkening, C. Klosinski, and M. Kurrat. Breaking performance of protection devices for automotive dc powertrains with a voltage of 450 V. In *Proc. 27th International Conference on Electrical Contacts*, pages 126–131, Dresden, Germany, June 2014.

[7] Y.Liu, Z.B.Li, K.Wang, and R.H.Wang. Experimental study on arc duration under external transverse magnetic fields in a DC 580 V circuit. In *Proc.26th International Conference on Electrical Contacts*, pages 30–34, Beijing, China, May 2012. doi:10.1049/cp.2012.0617.

[8] T.Atsumi, J.Sekikawa, and T.Kubono. Characteristics of break arcs driven by transverse magnetic field in a DC high-voltage resistive circuit. *IEICE transactions on electronics*, E93-C(9):1393–1398, 2010. doi:10.1587/transele.E93.C.1393.

[9] H.Ono, J. Sekikawa, and T.Kubono. Arc duration of break arcs magnetically blown-out in a DC 450 V resistive circuit. *IEICE transactions on electronics*, E95-C(9):1515–1521, 2012. doi:10.1587/transele.E95.C.1515.

[10] Z.B.Li and G.S.Zhang. The study of the criterion for dynamic welding of electric contact. In *Proc. 16th International Conference on Electrical Contacts*, pages 413–416, Loughborough, UK, September 1992.

doi:10.14311/ppt.2017.1.66
Plasma Physics and Technology 4(1):66–69, 2017

DEVICE FOR PROTECTION AGAINST TRANSIENT AND TEMPORARY OVERVOLTAGE INCLUDING LIMITATION OF THE SPECIFIC ENERGY

A. Ehrhardt*, K. Bühler

DEHN + SÖHNE GmbH + Co.KG., Hans-Dehn-Straße 1, D-92318 Neumarkt, Germany

* arnd.ehrhardt@dehn.de

Abstract. An arrangement is presented which ensures protection against transient and line frequency overvoltage and limits the load on the consumer to a minimum even in case of transient currents. The arrangement consists of a special arrester in the shunt circuit, a fast switch in the series circuit as well as a control and measuring device. The design, mode of operation and parameters of the components of the arrangement are described based on different loads. The protective effect is discussed.

Keywords: overvoltage protection, surge protection, spark gap, semiconductor switch.

1. Introduction

Overcurrent and surge protection are, alongside other protective measures, an inherent part of low-voltage systems and the connected consumers. The requirements on the transient overvoltage protection of low-voltage supply systems are described in IEC 61643-11 [1] and those on line frequency overvoltage protection, for example, home installation, in the EN 50550 [2]. Transient loads occur as a result of, for example, lightning and switching overvoltage. Surge protective devices are installed in the immediate vicinity of the point where the energy and data cables enter buildings or facilities. Beside lightning equipotential bonding these must protect electronic devices against high-energy conducted disturbance variables. Some known faults leading to transient overvoltage are, e.g. body contacts, earth faults, short-circuits and conductor interruption [3]. Line frequency overvoltage results from, among other things, faults in low voltage and medium-voltage systems. In the case of faults in low-voltage systems, the overvoltage level can, depending on the type of fault and the system, reach the line-to-line voltage. Therefore, at a nominal voltage of 230 V with 10 % tolerance, voltages of up to 440 V are possible. Appropriate protective devices are, i.a. coupled with main protective devices like, for example, circuit breakers. In particular, transient load from lighting events results in injected impulse current which can put a heavy load on the consumers connected. The flow of current places especially heavy load on consumers in a low-impedance state until the required sparkover voltage of the protective device, for instance, 1.5 kV is reached. The level of voltage load and also the charge value [As] in connection with the current flow represents a considerable hazard for some components in the consumer devices. The withstand voltage of many electronic components is often considerably lower than 1.5 kV. Components with a thin layer of insulation, like inductances or capacitances can, however, be destroyed or their service life considerably reduced relatively quickly as a result of even quite low overvoltage. Figure 1 shows a simplified view of the timespan of overvoltage which can be tolerated by sensitive electronic components in consumers. More detailed information can be found in, e.g. the ITI (CBEMA) CURVE for AC [4] for under- and overvoltage.

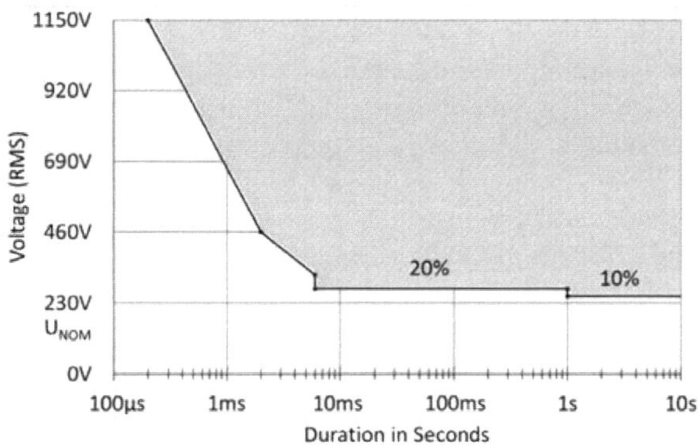

Figure 1. Voltage tolerance curve proper for 230 Volt AC equipment

For instance, the most consumer devices have capacitors in the input circuit, also in DC applications, they can be damaged by impulse current, especially high-level lightning current, before the transient protective devices are triggered. Any existing devices which protect against line frequency overvoltage on the basis of a mechanical switch are ineffective in the face of such loads as a result of the long effective dead time before disconnect the consumer. During this time commonly some 10 ms exists no protection against overvoltages.

The recommended new protective arrangement is capable of limiting the current load of the devices connected, even in the case of injected lightning current and provides uninterrupted surge protection of the transient area until the line frequency protection takes effect.

2. Design and components of the arrangement tested

The arrangement tested is shown in Figure 2. It consists of two basic components: an arrester A which is capable of carrying lightning currents and a semi-conductor based switch, as can be seen in Figure 2. Switch B1 is provided with a control device B3 which is coupled with a monitoring unit B2.

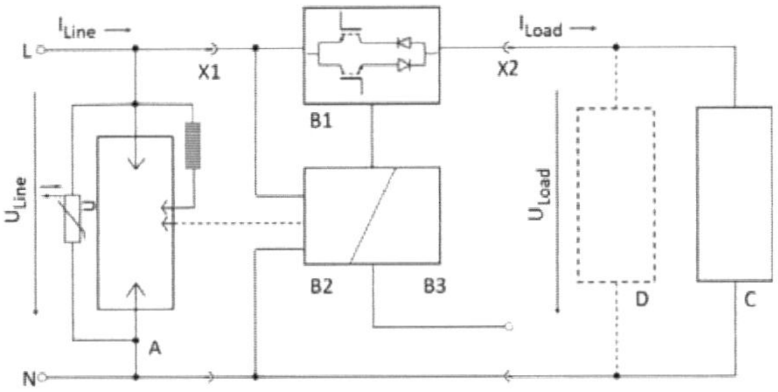

Figure 2. Principle of the recommended protective arrangement A - triggerable spark gap; B1 - switch; B2 - monitoring unit; B3 - control device; C - consumer; D - secondary protection

The voltage, current and current steepness at the switch are monitored. The switch, see also picture within Figure 3, is designed for a nominal current strength of approx. 16 A at 230 V and possesses a peak withstand voltage of 1.2 kV. Passive cooling is sufficient at this capacity. The arrester is executed as spark gap A with a deion chamber and, where necessary, a parallel varistor. The protection level of the spark gap is lower than 1.5 kV. The arrester has an additional control input for activating the spark gap. This input is connected to the control device of switch B2. The switch is put in conductive mode when the voltage level is in the range of the operating voltage. By a nominal current of 10 A the efficiency of the total arrangement is over 99 %. When active, inrush currents of the connected equipment of, for example, 170 A are possible without any functional restrictions. This value can be adjusted within the framework of the performance capacity of the semi-conductor. Likewise, it is possible to realise the functionality of an electronic overcurrent fuse or consider the characteristics of an existing surge protective device D. In the case of deviations in voltage, the current value, the temperature of the switch or the current steepness, switch B1 is put in non-conductive mode. It returns back to conductive mode automatically as soon as the operating conditions are normal again.

3. Mode of operation of line frequency overvoltage

The selected switch can be used at voltages of approx. 440 V AC and, as such, at numerous operating voltages. Corresponding with the selected operating voltage, the power supply of the consumer can be switched off if the threshold value is exceeded by, for example, only 10 % with an accuracy of approx. 2 %. Once the current returns to the permitted range, the consumer C is automatically connected again. The spark gap can remain passive by line frequency overvoltage up to a peak voltage of approx. 800 V. Connection and disconnection takes place via the switch B1 alone. Disconnection takes place after reaching the threshold value of approx. 400 µs. Should the voltage increase to a higher value, the spark gap is set to conductive mode. This can occur repeatedly until the overcurrent protective device is triggered. In addition to the control of the switch through its own monitoring unit, the switch and also the spark gap can be directly and instantaneously controlled by an external signal, e.g. from a consumer to be protected with its own internal monitoring unit. Figure 3 demonstrates the disconnection and connection of a consumer C by the switch at a nominal voltage of 230 V_{eff} and a threshold value of 360 V. When the voltage at the input of switch B1 increases to about the threshold value, the voltage at the output for the end device C is interrupted until the input voltage falls back below the threshold value.

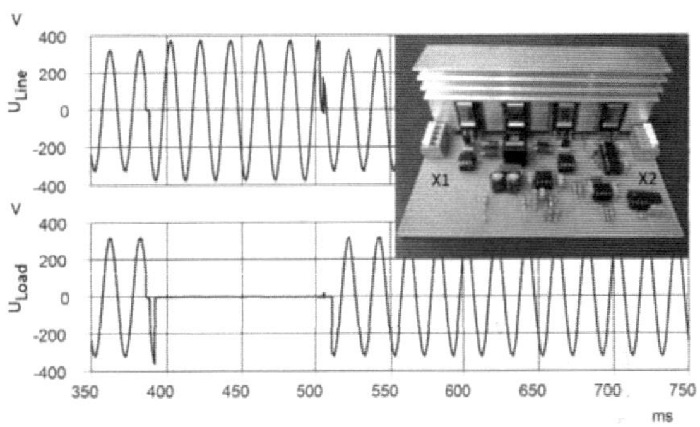

Figure 3. Behaviour of the arrangement in case of line frequency overvoltage top - line voltage; bottom - load voltage, picture of the switch B1 with control and monitoring unit

4. Mode of operation with transient overvoltage

As an example, in consumers with energy storage systems in the input circuit or where the internal surge protection has a low sparkover voltage a certain amount of current from transient interference flows to the end device before sufficient voltage is built up to trigger spark gap A or switch B1. Beside the current intensity, the monitoring unit of the switch also registers the rate of rise in current. Here, the current is measured by a shunt. The rate of rise in current is used to produce an additional voltage drop which has an instant impact on the blocking signal of the switch B1. This makes it possible for the switch to block very quickly, even when the current value is still below the permitted inrush current, for instance

170 A. Blocking the switch leads to a rash build-up of voltage on the spark gap, allowing this protective device to trigger. In addition, the monitoring unit of the switch can control the spark gap via the additional control input which allows it to trigger even at low voltages.

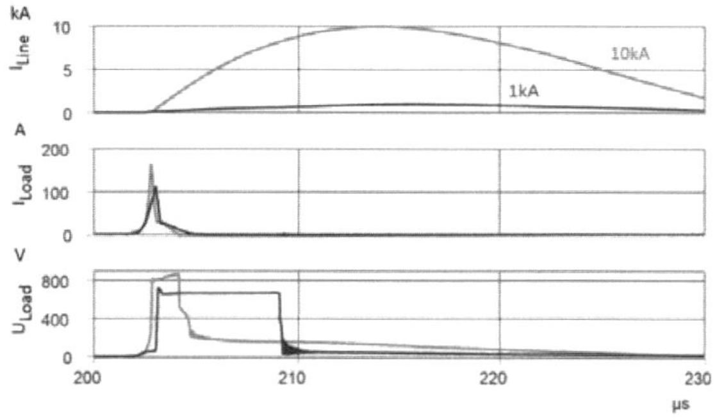

Figure 4. Behaviour of the arrangement in case of transient currents 1 *kA (blue) and* 10 *kA (orange)* 8/20 *µs top - transient impulse line current; middle - transient impulse load current after switch; bottom - load voltage*

This limits the necessary switching capacity of the switch. Very fast ignition can be achieved using a spark gap with diverging electrodes. The extremely low arc voltage after ignition (see Fig. 6) leads to a considerable reduction of the switching capacity of the switch even in highly inductive networks [4, 5]. The ensuing increase in arc voltage in the deion chamber to extinguish follow current remains limited to values under 600 V, whereby falling way below the dielectric strength of the switch. Figure 4 shows the behaviour of the arrangement with transient currents of 1 kA and 10 kA with the pulse form 8/20 µs. The spark gap is not additionally triggered, but rather ignited via the normal trigger switch as a result of the voltage build up from blocking the switch at a current level of under 170 A. Tripping of switch B1 is initiated after just <1 µs by the flow of current in the consumer due to the current steepness.

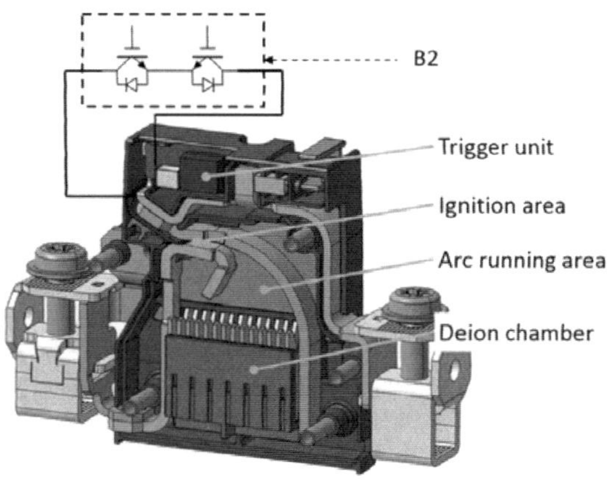

Figure 5. Spark gap with trigger unit and additional actuation by B2

The peak current value is reached at the beginning of the tripping and voltage builds up over switch B1 and thus also over arrester A. After the switch has been triggered and the spark gap ignited, the voltage sinks below 100 V. This considerably relieves the switching capacity of the switch's semi-conductor. The load on the downstream consumer and surge protective device compared with the charge and the current intensity of the transient disturbance is greatly reduced. As well as a voltage-switching element, the triggering has an electronic switch for additional actuation which is actuated by the control unit B3 or the consumer C. The basic construction of the spark gap with additional actuation can be seen in Figure 5. The spark gap with diverging electrodes, trigger circuit and deion chamber is described in detail in [4–6]. The additional actuation on the basis of a semi-conductor is connected with the existing auxiliary electrodes of the spark gap and bridges the transient ignition elements. This arrangement allows the spark gap to trigger by minimal surges. Figure 6 shows such an actuation and subsequent follow current extinction at approx. 230 V mains voltage without pulse stress. After ignition of the spark gap current flows out of the network and is quickly limited when the arc enters the deion chamber.

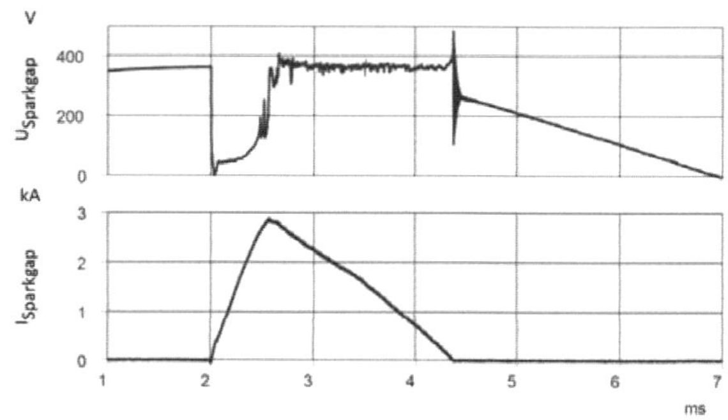

Figure 6. Arc extinction at a mains voltage of 230 *V AC at a prospective short-circuit current of* 25 *kA*

Here, the arc voltage when quenching the follow current remains well under the maximum blocking voltage of the switch so that the latter is not jeopardised by the function of the spark gap. Hence the functionality of the switch and the arrester are coordinated whatever the operating conditions.

5. Protective effect of the arrangement

The explanations show that the characteristics seen in Figure 1 are completely realisable with this arrangement. In contrast, when using normal devices for transient and line frequency surge protection, sensitive devices i.a. between approx. 2 ms – 20 ms are not sufficiently protected. In Figure 7 we can see the load on a commonly used varistors (e.g. S20K275) in consumers C when using a triggerable spark gap at a short distance from the consumer under a load of a lightning current (25 kA 10/350 µs). The current intensity in the varistor can, by all means, increase to readings in the several kilo ampere region before the spark gap is triggered and the varistor unloaded. The thermal load here is in the region of 40 A^2s and the energy load of this varistor lies by approx. 23 J [7].

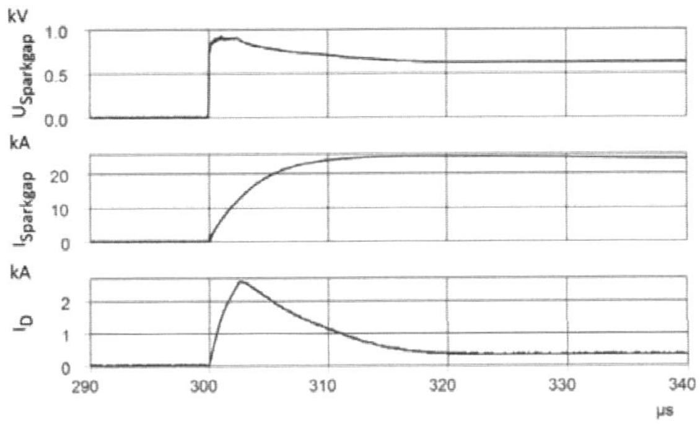

Figure 7. Behaviour of common triggerable spark gap during a lightning current of 25 kA 10/350 µs top - voltage spark gap and consumer, middle - impulse current spark gap, bottom - impulse current load / consumer in case of an varistor (S20K275) within the consumer

With the protective arrangement demonstrated, the current load, see Figure 4, is largely independent of the waveform of the impulse current (8/20 µs and 10/350 µs). The peak value is lower than 250 A, the thermal load lower than 15 mA^2s and the energy load of an SMD varistor (Cu3225K250), even with a low mA point, smaller than 60 mJ.

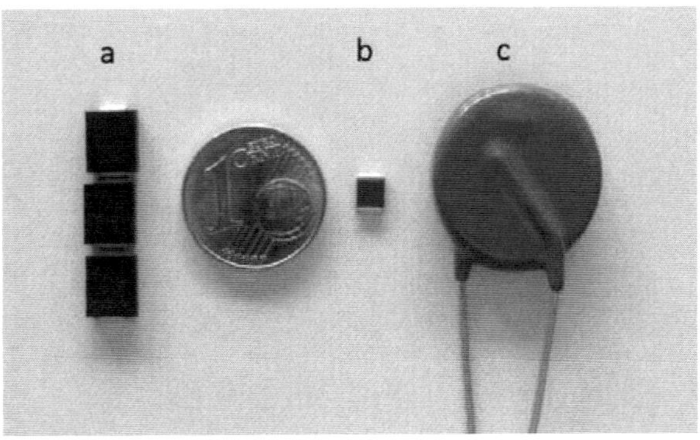

Figure 8. Picture of suitable overvoltage protective components for use as secondary protection D or within consumer C by use of the tested arrangement a - suppressor diodes; b - SMD varistor; c - normal varistor S20K275

Due to the extensive limitation of transient current, components for the secondary protection D (see Fig. 1) with lower performance and very low protection levels, for instance, suppressor diodes or SMD varistors, can be used upstream of or directly in the devices to be protected. The load amount allows the use of SMD components with a considerably lower capacity without additional protective measures, even by exposure to lightning current. Similarly, it is possible to use suppressor diodes with power losses of only 1.5 kW depending on the voltage level. Figure 8 shows some of these components to help us visualise the size and the applications of the protective devices.

6. Conclusions

The protective arrangement presented consisting of a lightning arrester and an electronic switch guarantees complete surge protection for sensitive devices and equipment in case of transient and line frequency disturbances. In addition to surge protection, the intensity and duration of the load on the consumer as a result of transient current can be considerably reduced. The possibility of controlling the switch and the spark gap externally means that the protective device can also be actuated directly by consumers with internal monitoring. The load on the electronic switch when tripping is supported by the almost instantaneous triggering of the spark gap, the low arc voltage of the spark gap after triggering, and the delayed and limited voltage build-up in the spark gap when quenching follow current. The arrangement can be used in AC and, with simplified electronic switches, also in DC power supplies.

References

[1] IEC 61643-11 Ed. 1.0. Low-voltage surge protective devices – Part 11: Surge protective devices connected to low-voltage power systems – Requirements and test methods, 03 2011.

[2] EN 50550:2011+AC:2012+A1:2014. Power frequency protective device for household and similar applications (POP).

[3] T. A. Short. *Electric Power Distribution Handbook*. Second Edition page 571. CRC Press, 2014.

[4] IEC 62305 Ed. 2.0. Protection against lightning – Part 1-4, 12 2010.

[5] L'. Hüttner, L'. Jurcacko, F.Valent, A. Ehrhardt, and S. Schreiter. Fundamental analysis of the encapsulation of a low-voltage spark gap and deion chamber. In *19th International Symposium on Physics of Switching Arc*, pages 229–234, 2011.

[6] A. Ehrhardt, S. Schreiter, U. Strangfeld, and M. Rock. Encapsulated lightning current arrester with spark gap and deion chamber. In *19th International Symposium on Physics of Switching Arc*, pages 165–168, 2011.

[7] B. Leibig and U. Strangfeld. Wirksamkeit von Ableitern für den Blitz- und Ueberspannungsschutz. *etz*, pages 66–69, 08-2010 edition 2010. URL: `http://www.etz.de/844-0-Wirksamkeit+von+Ableitern+fuer+den+Blitz-+und+Ueberspannungsschutz+.html`.

doi:10.14311/ppt.2017.1.70
Plasma Physics and Technology **4**(1):70–73, 2017

OPTIMAL BAND SELECTION FOR THE CALCULATION OF PLANCK MEAN ABSORPTION COEFFICIENTS

M. Gnybida[a,*], Ch. Rümpler[b], V.R.T. Narayanan[a]

[a] *Eaton European Innovation Center, Borivojova 2380, 252 63 Roztoky, Czech Republic*
[b] *Eaton, 1000 Cherrington Pkwy Moon Township, PA 15108, USA*
[*] MykhailoGnybida@Eaton.com

Abstract. Radiative heat transfer is a major heat loss mechanism in thermal plasmas generated during arc flashes/faults in switchgear applications or during high current interruption in low voltage circuit breakers. A common way to calculate the radiation balance is by means of approximate non-gray radiation models like P1 or discrete ordinates (DOM), where the frequency dependent absorption and emission are described in a number of frequency intervals (bands) using a constant absorption coefficient in each band. Current work is focused on finding the optimal number of bands as well as band interval boundaries that provide a reasonable level of accuracy in comparison to a full spectral solution. An optimization procedure has been applied to different SF_6 and copper vapor gas mixtures for an assumed temperature profile. Radiation model results using optimized band averaged absorption coefficients as well as spectral values are provided and discussed for the exemplary temperature profile.

Keywords: radiation, thermal plasma, circuit breaker, SF_6.

1. Introduction

In various applications in the area of low-, medium-, and high-voltage power distribution, thermal plasmas are either used as switching elements for current interruption (e.g. circuit breakers) or arcs can occur as fault conditions (arc flash in switchgear). In order to enhance the design and performance of these devices, modeling approaches that describe the behavior of thermal plasmas have been developed [1] and enhanced in order to include more physical phenomena and cover a broader range of applications [2]. Hereby a set of coupled equations for flow and electrodynamic processes is usually solved (magneto-hydrodynamic approach), in combination with one of various simplified models to consider energy transfer due to radiation.

The simplest approach to model the effect of radiation is the net emission coefficient (NEC) [1], which describes the net radiation from an isothermal sphere or cylinder. But in various applications it is necessary to model not only the energy sink in the arc core due to the net emission, but also the reabsorption in the colder arc fringe and the radiative flux impinging on surfaces needs to be described to achieve higher model fidelity. Radiative heat flux contributes to the ablation of plastic materials, which can substantially influence the arc behavior [3].

Commonly used approximate radiation transfer models are the P1 and the discrete-ordinates model (DOM) [4], where the radiation spectrum is divided into several gray bands. A constant Planck or Rosseland averaged absorption coefficient is derived from the spectral absorption coefficient. Instead of resolving the whole spectrum, the radiation transfer is solved for a small number of bands only, which does reduce the computational effort dramatically while providing reasonable accuracy.

The selection of the number and location of the band limits has an influence on the achievable accuracy and it is not obvious how many bands are sufficient and where to place the bands across the spectrum. To address these questions, numerical optimization procedures have been developed to find an optimal number and distribution of the bands [5, 6]. In this paper, we apply the optimization procedure as introduced in [5, 7] to a specific mixture of SF_6 and copper vapor. Since the amount of metal vapor in the arc chamber or switchgear can vary from 0 to 100% in the real application, it is necessary to quantify the error that is introduced when using the same band limits for various mixture ratios. By using the same band intervals for different mixtures and solving radiation transfer for the full spectrum as well as for the identified bands, the error introduced by the band selection can be identified.

2. Optimization and verification

Planck averaging is used to calculate the mean absorption coefficient $k_{Planck,i}$ for each band i according to:

$$k_{Planck,i} = \frac{\int_{\nu_i}^{\nu_{i+1}} k_\nu B_\nu \mathrm{d}\nu}{\int_{\nu_i}^{\nu_{i+1}} B_\nu \mathrm{d}\nu}, \qquad (1)$$

where k_ν is the spectral absorption coefficient and B_ν represents the black body radiative intensity. The lower and upper interval boundaries (band limits) are denoted by ν_i and ν_{i+1} respectively. We renormalize the line contribution of the spectral absorption coefficients according to [8] with a characteristic plasma absorption length $H = 3 \cdot R_p$, as suggested in [7]. The

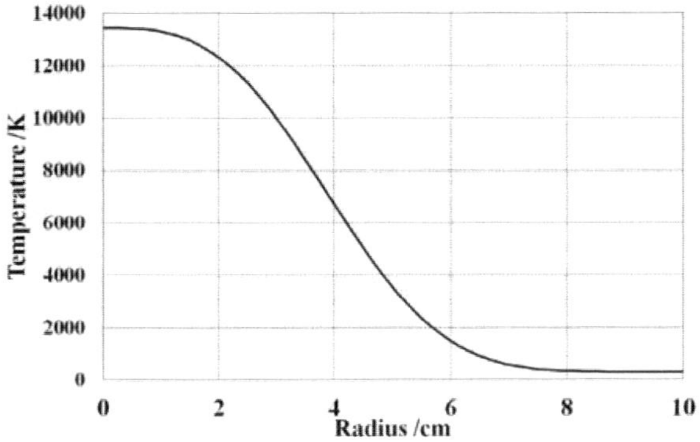

Figure 1. Assumed temperature profile.

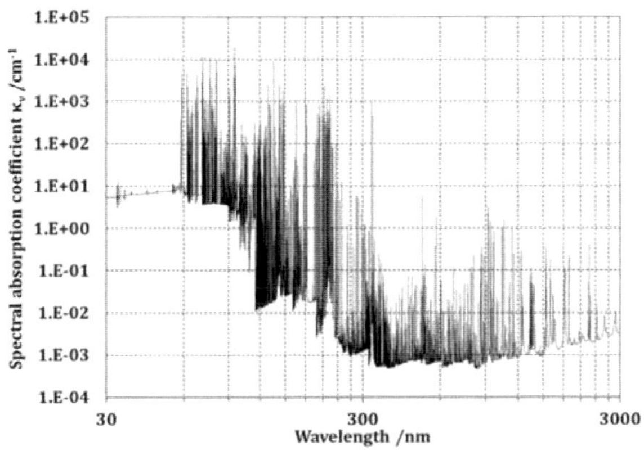

Figure 2. Spectral absorption coefficient for a 90% SF_6 and 10% Cu vapor mixture at 10 000 K and 1 atm.

plasma radius R_p is defined as radius value where the temperature decreased to 50% of the core temperature.

In our optimization procedure, we assume an infinitely long cylindrical calculation domain with 10 cm radius. The domain is evenly filled with a 90% SF_6 and 10% Cu vapor mixture at 1 atm pressure (all gas mixture proportions are noted as mass fractions in this paper). This mixture could be exemplary for the initial phase during an arc flash event in gas insulated medium voltage switchgear. The plasma inside this domain has a temperature distribution along the radius as shown in Fig. 1 with 13 460 K maximum in the arc core and 300 K at the outside boundary. This temperature profile represents a plasma radius of 4 cm according above mentioned definition. The equilibrium species composition for this gas mixture and other mixtures of SF_6 and copper vapor in the temperature range of interest is calculated by minimization of Gibbs free energy. The species composition is the input for the calculation of the spectral absorption coefficients k_ν in a frequency range that covers 10^{12} Hz to 10^{16} Hz with frequency step size of $2 \cdot 10^{10}$ Hz. The calculated spectral absorption coefficient for a 90% SF_6 and 10% Cu vapor mixture at 10 000 K and 1 atm is shown in Fig. 2.

A baseline calculation of the radiation transfer resolving the whole spectrum is the first step in the optimization procedure (spectral solution [5]). This is a computationally expensive step because the radiation transfer needs to be solved for each of the approximately 500 000 points in the spectrum. A spatial discretization of 3.45 mm is used hereby (30 points). In the second step, the radiation transfer is solved for a fixed number of spectral bands using Planck mean absorption coefficients. The optimization algorithm varies the band limits in order to minimize the value of objective function ΔF_ν according to Eq. (2):

$$\Delta F_\nu = \frac{1}{30}\sqrt{\sum_{i=1}^{30} \left(\nabla \cdot F_{\mathrm{exact},i} - \nabla \cdot F_{\mathrm{mean},i}\right)^2}\,, \qquad (2)$$

where $\nabla \cdot F_{\mathrm{exact},i}$ represents the divergence of the radiative flux calculated in the spectral solution for each point i and $\nabla \cdot F_{\mathrm{mean},i}$ represents the divergence of radiative flux calculated using the line limited Planck mean absorption coefficients. This objective function describes the difference between the spectral and the band averaged solution, in particular the energy source/sink term due to radiation transfer. Alternatively, the radiative intensity could be used to define the objective function as well [6].

Based on the variation of ΔF_ν due to varying band limits, the procedure is able to determine an optimal set of band limits. Although the algorithm would be able to find a global minimum, we limit the computational time and abort the calculations if a defined maximum number of iterations is reached. This procedure is repeated for a different number of bands (2 to 6) to analyze the achievable accuracy when using smaller or larger number of bands. Since a larger number of bands increases the computational effort when solving radiation transfer in complex three dimensional CFD models, it is beneficial to use the smallest number of bands that still provides sufficient accuracy.

The optimization procedure delivers a set of band limits for the specified uniform gas mixture, but in realistic applications any mixture of gas and metal vapor could be present. Thus, we need to quantify the accuracy of the radiation transfer model using the band averaged absorption coefficients for various other mixtures. For this evaluation, the difference between spectral solution and band averaged solution is calculated using the 2D radiation transfer code RAT [4], which is enhanced to calculate radiation transfer resolving the whole spectrum or utilizing band averaged coefficients. In this verification step, the same temperature profile as shown in Fig. 1 is used with a spatial resolution of 100×100 points. A relative error norm $\Delta F'_\nu$ is used to quantify the difference between the two approaches [5], where the error is evaluated along the radius only:

$$\Delta F'_\nu = \sqrt{\frac{\sum_i \left(\nabla \cdot F_{\mathrm{exact},i} - \nabla \cdot F_{\mathrm{mean},i}\right)^2}{\sum_i \left(\nabla \cdot F_{\mathrm{exact},i}\right)^2}}\,. \qquad (3)$$

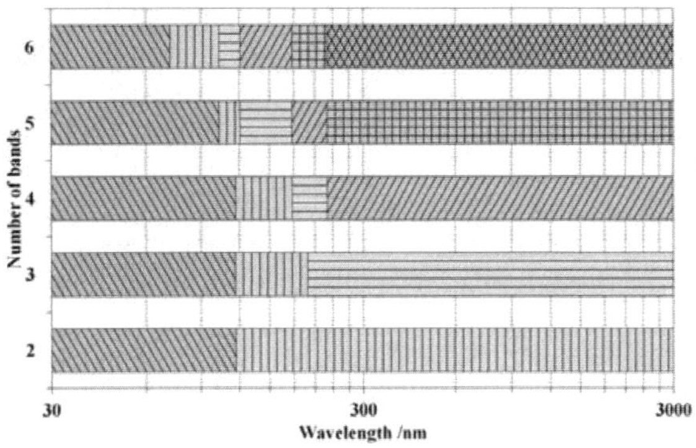

Figure 3. Optimized mean absorption coefficient band boundaries for 90% SF_6 – 10% Cu *vapor mixture.*

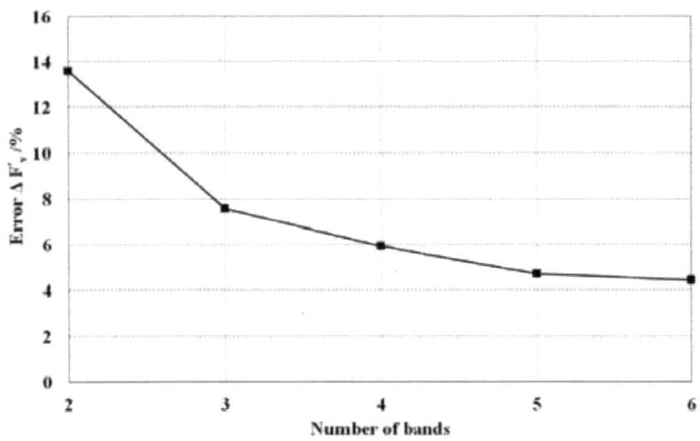

Figure 4. Error norm $\Delta F'_\nu$ *for* 90% SF_6 – 10% Cu *vapor mixture.*

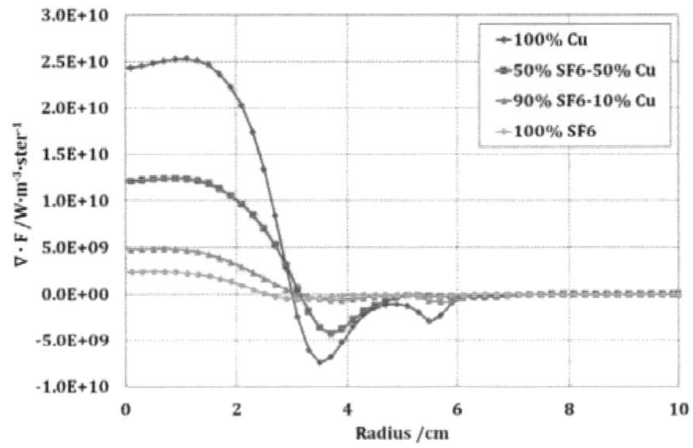

Figure 5. Divergence of radiative flux for various gas mixtures, spectral resolved model.

Band	90% SF_6 10% Cu	50% SF_6 50% Cu	100% SF_6	100% Cu
6 band	4.4%	13.6%	12.0%	14.5%
3 band	8.1%	14.9%	11.9%	16.4%

Table 1. Error norm values for different mixtures and band approximations at 1 atm.

3. Results

The optimization procedure results for 90% SF_6 – 10% Cu vapor mixture at 1 atm are shown in Fig. 3. The derived band boundaries for the different total number of bands are all located in UV region of the spectrum. In case 2 to 4 band numbers, the first interval boundary is located at around 116.5 nm. Also, in case of 5 or 6 bands, the same interval boundary is selected, but more intervals are introduced below and above that wavelength. Comparing the band boundaries with spectral absorption coefficient as shown in Fig. 2, the 116.5 nm interval boundary can be identified as larger change in the continuum contribution. The other interval boundaries cannot be associated with the spectral absorption coefficient that distinctively.

Corresponding error norm values according to Eq. 2 are provided in Fig. 4. The 2–band approximation has an error of 13.6 % which is decreased with increasing number of bands down to 4.4 % for the 6–band approximation. A larger number of bands does improve the accuracy of the radiation model as expected, but the difference between 5–band and 6–band approximation is only about 0.3 %. For this specific gas mixture and temperature profile, solving a 6–band model would not provide a benefit over a 5–band model.

The influence of copper vapor on the radiation transfer can be derived from Fig. 5. The divergence of the radiative flux along the domain radius as result of the spectrally resolved model is shown for these gases/gas mixtures: 100% SF_6, 90% SF_6 – 10% Cu, 50% SF_6 – 50% Cu, and 100% Cu. The divergence of radiative flux in the arc core is increased with larger copper vapor fraction. Compared to pure SF_6, the cooling term is increased by an order of magnitude for pure copper vapor, due to the presence of strong metallic lines in the spectrum.

The influence of different mixtures on the accuracy of the band averaged models is depicted in Fig. 6 for the same four gases/gas mixtures at 1 atm. Spectrally resolved model results as well as the 3–band and 6–band model results are provided, where band averaging is done using band limits according to Fig. 3 (optimized for 90% SF_6 – 10% Cu vapor mixture). In case of pure SF_6, the 3–band model overpredicts the emission by 13.7% ($\nabla \cdot F$ in the arc core), although the reabsorption in the fringe is represented well. This can lead to an overestimation of the radiation losses in an arc model later on. But in case of 50% copper vapor admixture, an underprediction of the reabsorption in the fringe of the temperature profile is noticeable. This can influence the temperature profile as result of an arc model. The error values as provided in Tab. 1 can be used to quantify the accuracy for the whole profile.

The 90% SF_6 – 10% Cu vapor mixture and 6–band approximation has the lowest error value of 4.4%, because this mixture was used to find optimal band boundaries. The error value has a maximum of 16.4% in case of pure copper vapor and 3–band approximation. Reducing the number of bands from 6 to 3

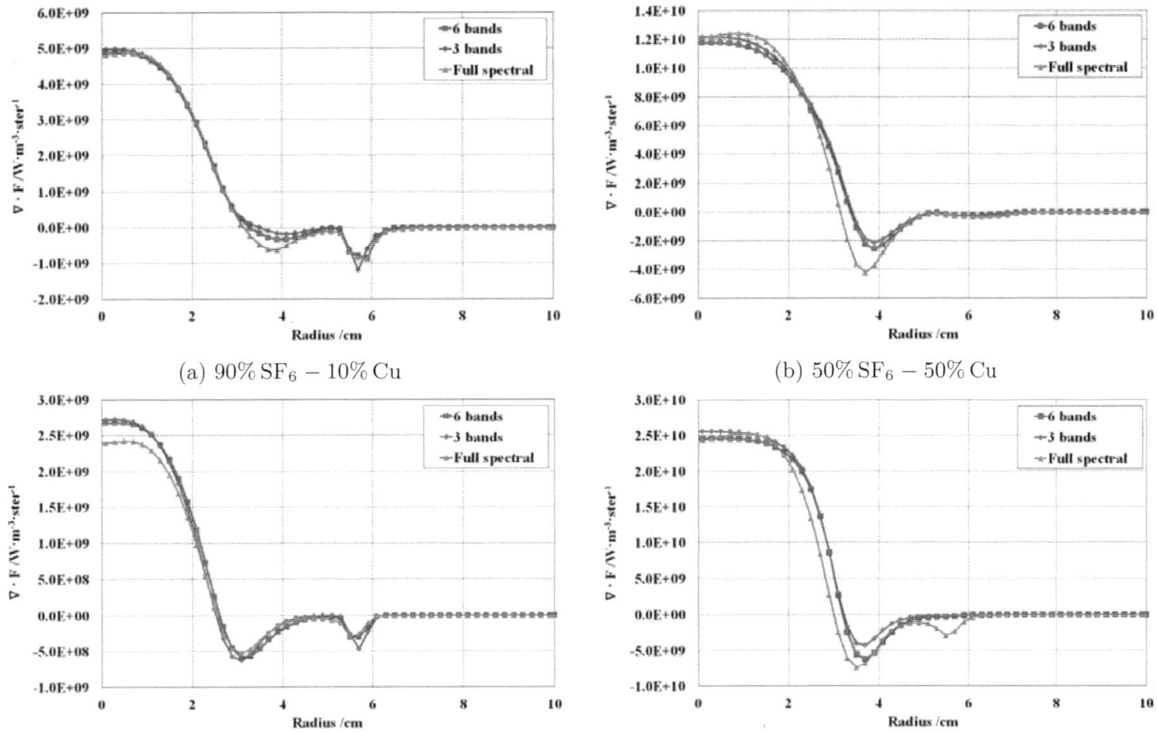

(a) 90% SF_6 – 10% Cu (b) 50% SF_6 – 50% Cu

(c) 100% SF_6 (d) 100% Cu

Figure 6. Divergence of radiative flux for different mixtures at 1 atm.

in case of 90% SF_6 – 10% Cu vapor mixture almost doubles the error from 4.4% to 8.1%, but the error resulting from different gas composition is larger. In this case, a larger number of bands does help to limit the error if gas mixture is varied as well.

4. Conclusions

Approximate radiation transfer models are used in arc modeling and band averaged absorption coefficients are necessary model input. We presented a numerical approach towards an optimal selection of band boundaries for the calculation of Planck mean absorption coefficients. The optimization was performed for a 90% SF_6 – 10% Cu vapor mixture and a defined temperature profile. The comparison between band averaged radiation model and spectral model results allows the quantification of error introduced by band averaging. Based on the results, it is feasible to perform the optimization for a 90% SF_6 – 10% Cu vapor mixture and apply the band limits to various mixtures of SF_6 and copper, at least for the investigated temperature profile. The influence of different temperature profiles is a topic for further investigations.

Acknowledgements

The authors thank P. Kloc and M. Bartlova for providing the radiation database.

References

[1] A. Gleizes, J. J. Gonzalez, and P. Freton. Thermal plasma modelling. *Journal of Physics D: Applied Physics*, 38(9):R153, 2005. doi:10.1088/0022-3727/38/9/R01.

[2] Ch. Rümpler and V.R.T Narayanan. Arc modeling challenges. *Plasma Physics and Technology*, 2(3):261–270, 2015.

[3] Ch. Rümpler, H. Stammberger, and A. Zacharias. Low-voltage arc simulation with out-gassing polymers. In *2011 IEEE 57th Holm Conf. on El. Contacts (Holm)*, pages 1–8, Sept 2011. doi:10.1109/HOLM.2011.6034770.

[4] Michael F. Modest. *Radiative Heat Transfer*. Third edition. Academic Press, 2013.

[5] P. Kloc, V. Aubrecht, M. Bartlova, O. Coufal, and Ch. Rümpler. On the selection of integration intervals for the calculation of mean absorption coefficients. *Plasma Chem Plasma Process*, 35:1097–1110, 2015. doi:10.1007/s11090-015-9648-3.

[6] L. Fagiano and F. Gati. On the order reduction of the radiative heat transfer model for the simulation of plasma arcs in switchgear devices. *J Quant Spec Rad Trans*, 169:58–78, 2016. doi:10.1016/j.jqsrt.2015.10.002.

[7] P. Kloc, V. Aubrecht, and M. Bartlova. Effective plasma radius for planck mean absorption coefficient. In *Proceedings of 21st International Conference on Gas Discharges and their Applications*, pages 89–92, 2016.

[8] H. Nordborg and A. Iordanidis. Self-consistent radiation based modelling of electric arcs: I. efficient radiation approximations. *J Phys D: Appl Phys*, 41(13):135205, 2008. doi:10.1088/0022-3727/41/13/135205.

doi:10.14311/ppt.2017.1.74
Plasma Physics and Technology **4**(1):74–78, 2017

CHEMICAL AND ELECTRON INTERACTION PROPERTIES OF POTENTIAL GASEOUS INSULATION AND ARC-QUENCHING MEDIA

Lili Zhu, Donxian Tanand, Su Zhao, Dengming Xiao*

Department of Electrical Engineering, Shanghai Jiao Tong University, No.1954 Huashan Rd., 200030,Shanghai, People's Republic of China

* dmxiao@sjtu.edu.cn

Abstract. The global warming effects of sulphur hexafluoride (SF_6) lead researchers to look for new environmentally-friendly gas media in high-voltage electrical apparatus. The present paper discusses basic chemical and electron interaction properties that brings high insulation strength and good arc interruption capability for a gas. The conclusion shows that gaseous compounds composed of elements in the upright area of periodic table of elements generally have stronger electronegative ability and good electrical properties. And double bond or triple bond in gas molecule can effectively improve the dielectric strength. For gas mixtures, good cooperation of gases will generate synergetic effect, which makes its dielectric strength higher than the weighted average value of each gas. Some aspects in searching for new arc interruption media are also discussed.

Keywords: chemical property, electron interaction, insulation, arc interruption.

1. Introduction

With increasing concern on the potential effects of global warming, sulphur hexafluoride (SF_6), a common insulation and arc interruption gas medium, has been stipulated to be banned in the year 2020 according to the Kyoto protocol at COP3 (the 3rd Conference of Parties) in 1997 [1]. The WMO (World Meteorological Organization) has recently published Greenhouse Gas Bulletin which shows that the current mole fraction of SF_6 is about twice the level observed in the mid-1990s in the atmosphere [2]. To deal with the issue, studies to search for new environmentally friendly gas to substitute SF_6 in electrical equipment become an urgent problem [3–5]. However, it is too complicated and even impossible to make overall experiments and research for each possible gas. One solution is to refer to general properties of gas with high insulation strength to narrow down the searching scope and make preliminary judgment of insulation property for a certain gas beforehand.

This paper summarized basic chemical and electron interaction properties that will lead to high insulation characteristics for a gas. As the electron attachment ability could reduce the number of free electrons and thus inhibit discharge process, Section 2 lists the values of electronegativity for common elements in period table of elements. The effect of molecule structure on the insulation property is analysed in Section 3, such as the effect of atomic configuration and double bond. For gas mixtures, the cooperation of different gases is introduced in Section 4. Some considerations of SF_6 substitute used for arc interruption are discussed in Section 5.

2. Electronegativity

In chemical molecules, electron pairs are subject to tug competition between adjacent atoms that share it. If one atom has a greater pulling power than the other, the covalent bond owns some ionic characteristics. Then the electron pair tends to be closer to one atom than to the other. The electron pulling force of an atom as part of a bond is called electronegativity.

A simplest analogy is to regard electronegativity as the ability of an atom competing for electrons. It is related to the ionization energy (I) (kJ.mol^{-1}) and electron affinity energy (EA) (kJ.mol^{-1}) of the element. The EA represents the energy released in the formation of negative ion when an atom in ground state obtains an electron. With high ionization energy, electrons are given up reluctantly. While with high electron affinity, no more energy is needed to attach electrons to an atom. Thus, elements with high values of both the two properties are hard to lose electrons and easy to gain them. Halogen elements will develop into a stable electron distribution configuration like inert gases when obtain an extra electron in the outer orbit of electrons, and tend to own strong electron affinity. Additionally, considering the function of covalent bond for atom in molecules, electronegativity is identified as the affinity of atom to the electron in covalent bond in a molecule. By contraries, if the values of ionization energy and the electron affinity are both low, it takes little energy to give up electrons and its inclination to gain other electrons is weak. The electronegativity of such element is low [6].

In the chemical periodic table based on the definite scale derived from Linus Pauling, the ionization energies and electron attachments reach the maximum value at the upper right of the periodic table close to fluorine element. It is obvious that fluorine F, oxygen

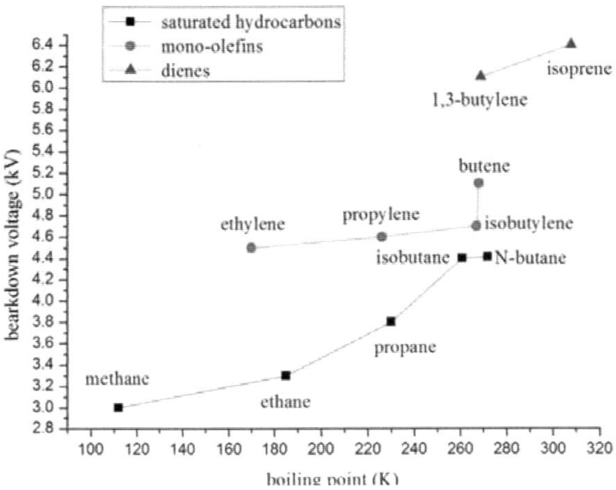

Figure 1. The relationship between the breakdown voltages (pd=50) and the boiling points.

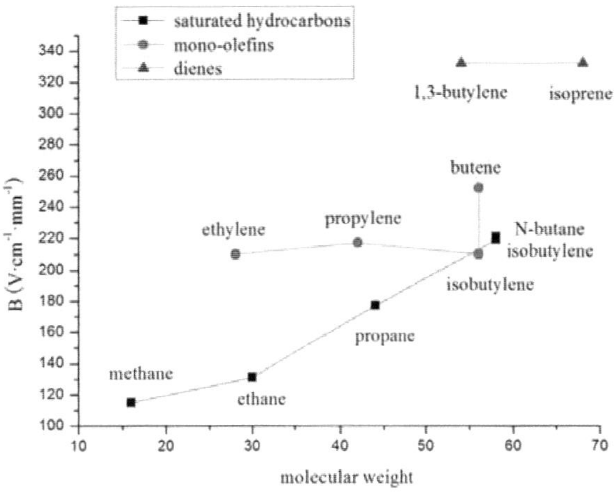

Figure 2. The relationship between the ratio (U/pd) between the breakdown voltages and the pd and the molecular weights.

O, chlorine Cl, nitrogen N, and bromine Br are the elements with the highest electronegativity. Whenever such elements are presented in compound molecules, their atoms are expected to tug strongly on electron pairs shared with their neighbor atom. Electron clouds are still shared with the less electronegative atom, which, however, is unequal. And the electron cloud at the side of the more electronegative element will be denser [7]. As fluorine F has the highest value of electronegativity among all the elements, the expected gases with high insulation strength all contain F element, such as SF_6, CF_3I and c-C_4F_8.

3. The impact of double bond

A.E.D Heylen has compared the electrical insulation strengths of several kinds of hydrocarbons [8, 9]. For alkene, the addition of methyl to mono-olefin will only obtain limited improvement of the dielectric strengths, but raise the boiling points, which limit the utilization as the gaseous media. However, butadiene and isoprene has a much higher dielectric strength than mono-olefins, which show an obvious effect of carbon-carbon double-bond on the improvement of the dielectric strength. This conclusion is also confirmed by the excellent electrical characteristic of butadiene C_4F_6 by comparing to 1-butane through the experiment of J. C. Devins [10, 11].

To further verify the function of the carbon-carbon double bond in the promotion of the electrical property, we compare saturated hydrocarbon, mono-olefin and di-olefin gases. Fig. 1 and 2 show the relationship of the boiling temperature with breakdown voltages under pd=50 (cm·mmHg) and the ratio (U/pd) between the breakdown voltages U and the pd values respectively. And Fig. 3 shows the relationship between the ratio (U/pd) and the molecular weights.

Seen from the above Figures, for a particular gas structure, with larger number of methyl, the boiling point will increase with the molecular weight. For gaseous molecules with similar chemical structure, the

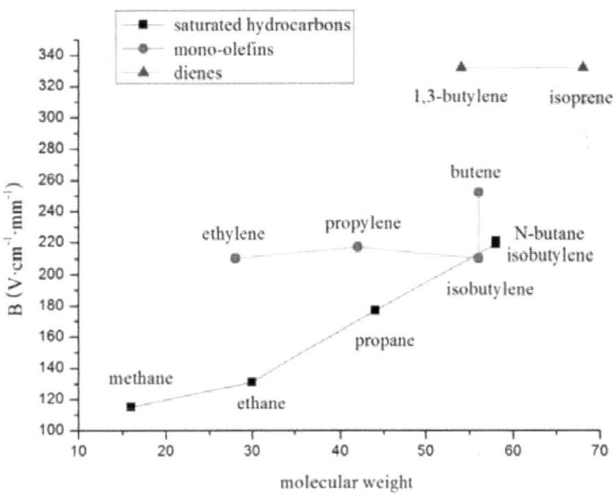

Figure 3. The relationship of the ratio (U/pd) between the breakdown voltages and pd with the boiling points.

containing of more carbon atoms brings better dielectric strength, though with small improving extent, which relates to the growing molecular volumes.

The variation of dielectric strength is significant for gases with similar molecular weights and boiling points but different structures. By the comparisons between ethylene and ethane, as well as butadiene and 1-butene, carbon-carbon double bond substantially improves the electrical performances of a gas. Research shows that the dielectric strength of hydrocarbon compounds largely depends on the integrated cross sections at low electron energy below 4 eV. For hydrocarbon gases, the double bonds or even triple bonds greatly improve the values of electron collision cross sections. The spectroscopy experiment shows that at low electron energy, such as 2 eV, the total collision cross section of ethylene is much higher than that of ethane, both of which reach an extremum at 2 eV. Study of ethylene and acetylene shows that double bond can effectively improve the insulation charac-

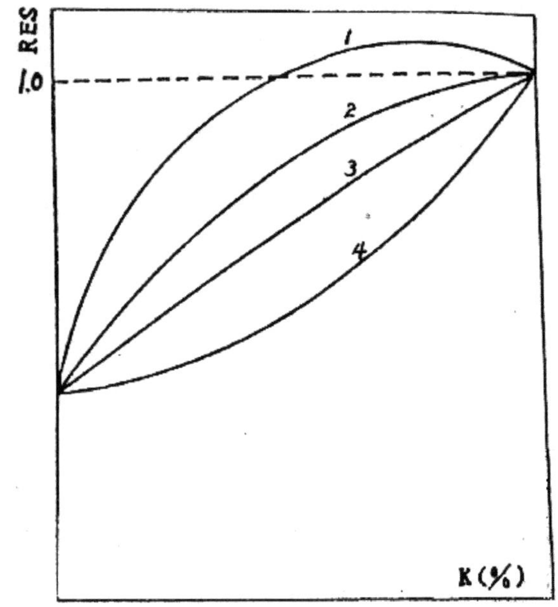

Figure 4. Four kinds of RES for binary gas mixtures.

teristics of the hydrocarbon gases, while the further improvement of dielectric strength brought by triple bond is limited compared to bi-bond hydrocarbon.

4. Synergistic effect

For gas mixtures, the affective factors are more complicated. The dielectric strength is not only related to the electric properties of gas components themselves, but also the synergistic effect between gas components. Fig. 4 represents the relative electric strength (RES) of binary gas mixtures in four different types [11]. Curve 1 refers to a positive synergistic effect, Curve 2 a synergistic effect, Curve 3 a linear effect which can be considered no interaction, and Curve 4 a negative synergistic effect. The mixing ratio k refers to the mole fraction of the gas component with higher dielectric strength in the binary mixtures, and the dielectric strength of this gas is set to be the reference of RES. For example, among SF_6 binary gas mixtures, SF_6/CF_2Cl_2 and SF_6/C_{36} belongs to positive synergistic effect, SF_6/N_2 and SF_6/CO_2 synergistic effect, SF_6/He linear effect, and $SF_6/C_2F_3Cl_3$ negative synergistic effect [10].

Gas mixtures utilized as insulation media in electric apparatus will brings many advantages including gas cost reduction, less environmental impact and wider application range. So taking good use of synergistic effect is significant when searching for new environmentally friendly gas in high voltage engineering. For example, SF_6/N_2 is typical gas mixtures with a synergistic effect, whose electrical strength is much higher than its linear value. When the SF_6 ratio is 50 %, the uniform electrical strength of SF_6/N_2 gas mixtures is over 85 % of that of pure SF_6 gas.

Additionally, the electron-attachment energy of different electronegative gases is best to cover the entire low energy range and thus free electrons with various energies in the range can be captured.

For example, the synergistic effect in electrical insulation characteristics is made good use of in the gas mixtures composed of electronegative gases N_2O, CO_2 and O_2 gases, which have electron affinity at different electron energy levels. The gases N_2O, CO_2 and O_2 have the electron attachment cross section respectively at 0.5–4.0 eV, 4.5–8 eV and 6–10 eV [12]. In other words, such gas mixtures will attach electrons over wide-range of energy level, and effectively suppress the discharging inception and propagation.

5. Interruption capability

Besides dielectric strength, a good arc-extinguishing gas medium should own good thermodynamic properties to satisfy more complicated and serious requirements in a high-pressure and high-temperature environment. The temperature of arcs in SF_6 or air circuit breaker can reach as high as over 30,000 K. Despite of insulation in room temperature, most gases become good conductor of electricity with temperature over 4000 K. The thermal recovery process after current zero largely relies on the thermodynamic and transport properties of high-temperature plasma. Researchers have long been searching for a certain property that has direct relationship with arc quenching capability, but have not reach a universal agreement yet. Recently, J Liu et al. have concluded a decisive influence of the product between density and specific heat ρC_p on the thermal interruption capability of a gas, based on the comparison between ρC_p for air and SF_6. This discovery provides guidance for the search of SF_6 switching gas substitute [13]. Another literature [14] emphases the strong influence from the material properties ρ, ρC_p and ρh of a gas on the interrupting capability of a gas blast arc.

As a widely-used arc quenching medium, SF_6 will also partly decompose after arc. Literature [15] compared the density of fluorine arising from SF_6 and CF_3I which has been broadly proposed as a potential replacement for SF_6. While the fluorine density from SF_6 increased exponentially with the current, that generated from CF3I was small. For CF_3I–CO_2 gas mixtures, the fluorine could not even be detected. This result corresponds to the computational result in literature [16]. But the iodine generated from CF_3I gas brings bad effect on the arc interruption performance, which calls for effective removal of iodine by a gas flow [15]. Through a serious of tests and experiments on a nozzle arc model, disconnecting switches and other arc models, it is concluded that the application of CF_3I–CO_2 mixtures to circuit breaker and switchgear that interrupt large fault currents is difficult, but suitable for small current interruption lower than 4000 A [17], taking into the consideration of its low damaging influence on the atmosphere and global warming. However, its arc interruption mechanism and thus improvement on the switchgear design needs further investigation.

Besides newly gas media, some traditional gases in high voltage engineering including air and CO_2 come to the horizon of the researchers again as they are environmentally friendly. So the substitute of SF_6 in electric and arc extinguishing still has a long way to go.

6. Toxicity and effect on environment

With growing demand and concerning about environmental protection, the development of high voltage, high capacity and compact high-voltage electrical equipment become increasingly urgent, and therefore researchers are seeking for non-flammable, anti-aging excellent insulation. The max temperature using SF_6 is 180 °C, recommended by the International Electrotechnical Commission (IEC). Low fluoride in high temperature converts into acidic substances with water, and the products have an adverse effect on the human body. Then the recombination of F and S atoms (ions) ($S+6F \rightarrow SF_6$) is blocked, which will affect the interruption performance. The decomposition products in the process of SF_6 discharge or arc have highly toxic and corrosive compounds (for example, S_2F_{10}, SOF_2), non-polar pollutants (for example, air, CF_4). SF_6 breakdown voltage are sensitive to water vapor, conductive particles and surface roughness. The non-ideal gas properties of SF_6 appear at low ambient temperatures. For example, in cold climates (about at -50 °C), under normal working pressure (0.4 MPa ~ 0.5 MPa), the portion of the SF_6 gas will be liquefied. Infrared absorption of SF_6 is also effective, due to its chemical inactivity, which is difficult to be removed in the atmosphere. All these undesirable characteristics of SF_6 are called for potential greenhouse gases. One alternative is PFC (perfluorinated hydrocarbons), whose global warming potential GWP is about 1/3-1/4 of SF_6. But their GWP (6000-9200) are still relatively high and their life expectancy in the environment are also long (2600-10000). What is really expected is an environmentally friendly low-GWP value alternative to SF_6 gas. In recent years, researchers have broadly proposed a new environmentally friendly insulating gas, trifluoroiodomethane (CF_32I). GWP values of this gas and CO_2 gas are rather similar, whose lifetime in the environment is only 1-2 days. Since 2007, international researchers continue to publish research reports about CF_3I insulating properties and arc performance on high-level international journals. Theoretical simulation results and experimental data have shown that, the dielectric strength of CF_3I is approximately 1.23 times more than that of SF_6. Considering environmental factors, CF_3I is regarded as an alternative to SF_6 gas in practical applications in the future.

7. Conclusion

Research on SF6 and other potential insulation gas media applied in high voltage electric apparatus shows that

1. Gaseous compounds composed of elements in the upright area of periodic table of elements generally have stronger electronegative ability and good electrical properties. These elements with high values of both the ionization and electron affinity energy are hard to lose electrons and easy to gain them, and thus own good insulation features.

2. Double bond or triple bond in gas molecule can effectively improve the dielectric strength.

3. For gas mixtures, good cooperation of gases will generate synergetic effect, which makes its dielectric strength higher than the weighted average value of each gas.

4. Both decomposed gas and raising rate of recovery voltage affect the arc interruption capability.

5. The toxicity and environmental friendliness of alternative gas, as well as its decomposed products, also should be taken into consideration.

Acknowledgements

Project is supported by the National Natural Science Foundation of China (No. 51337006).

References

[1] Z. S. Zhang, D. M. Xiao, X. L. Liu, W. J. Wang, and Y. A. Wang. Analysis of the insulation characteristics of c-C_4F_8/CO_2 gas mixtures by the monte carlo method. *Journal of Physics D: Applied Physics*, 41(1):015206, 2008. doi:10.1088/0022-3727/41/1/015206.

[2] WMO Greenhouse Gas Bulletin. The state of greenhouse gases in the atmosphere based on global observations through 2014. [2015-11-09].

[3] L. L. Zhu, X. G. Li, and D. M. Xiao. Electron transport coefficients in SF_6 and xenon gas mixtures. *Journal of Physics D: Applied Physics*, 33(23):L145, 2000. doi:10.1088/0022-3727/33/23/102.

[4] L. L. Zhu, Y. Z. Chen, and D. M. Xiao. Electron swarm parameters in and helium gas mixtures. *Journal of Physics D: Applied Physics*, 32(5):L18, 1999. doi:10.1088/0022-3727/32/5/004.

[5] A. R. Ravishankara, R. R. Garcia, S. Solomon, and J. B. Burkholder. Ozone depletion and global warming potentials of CF_3I. *Journal of Geophysical Research*, 99(D10):20929, 1994. doi:10.1029/94JD01833.

[6] W. S. Harwood, R. H. Petrucci, and F. G. Herring. *General Chemistry Principles and Modern Applications*. 8th printing,Eighth Edition. Upper Saddle River, New Jersey, New Jersey: Prentice-Hall Inc., 2002.

[7] P. Atkins and L. Jones. *Chemistry Molecules, Matter, and Change*. 3rd printing. New York: W. H. Freeman and Company, 1997.

[8] A. Heylen. Electric strength, molecular structure, and ultraviolet spectra of hydrocarbon gases. *The Journal of Chemical Physics.*, 29(4):813–819, 1958. doi:10.1063/1.1744595.

[9] A. Heylen and T. Lewis. The electric strength and molecular structure of hydrocarbon gases. *Canadian Journal of Physics*, 36(6):721–739, 1958. doi:10.1139/p58-079.

[10] J. Devins and R. Crowe. Electric strength of saturated hydrocarbon gases. *The Journal of Chemical Physics.*, 25(5):1053–1058, 1956. doi:10.1063/1.1743096.

[11] J. Devins. Replacement gases for sf6. *IEEE Transactions on Electrical Insulation.*, EI-15(2):81–86, 1980. doi:10.1109/TEI.1980.298243.

[12] M. Yoshida, T. Ogawa, H. Kojima, O. Kinoshita, N. Hayakawa, F. Endo, and H. Okubo. Breakdown characteristics of N_2O gas mixtures for quasiuniform electric field under lightning impulse voltage. *IEEE Transactions on Dielectrics and Electrical Insulation.*, 14(6):1492–1497, 2007. doi:10.1109/TDEI.2007.4401233.

[13] J. Zhong, M. Fang, J. Liu, Q. Zhang, and J. Yan. Analysis of the characteristics of dc nozzle arcs in air and guidance for the search of SF_6 replacement gas. *Journal of Physics D: Applied Physics*, 49(43):435201, 2016. doi:10.1088/0022-3727/49/43/435201.

[14] D.T. Tuma. A comparison of the behavior of SF_6 and N_2 blast arcs around current zero. *IEEE Transactions on Power Apparatus and Systems.*, PAS-99(6):2129–2137, 1980. doi:10.1109/TPAS.1980.319791.

[15] H. Katagiri, H. Kasuya, H. Mizoguchi, and S. Yanabu. Investigation of the performance of CF_3I gas as a possible substitute for SF_6. *IEEE Transactions on Dielectrics and Electrical Insulation.*, 15(5):1424–1429, 2008. doi:10.1109/TDEI.2008.4656252.

[16] Y. Cressault, V. Connord, H. Hingana, P. Teulet, and A. Gleizes. Transport properties of CF_3I thermal plasmas mixed with CO_2, air or N_2 as an alternative to SF_6 plasmas in high-voltage circuit breakers. *Journal of Physics D: Applied Physics.*, 44(49):495202, 2011. doi:10.1088/0022-3727/44/49/495202.

[17] H. Kasuya, Y. Kawamura, H. Mizoguchi, Y. Nakamura, S. Yanabu, and N. Nagasaki. Interruption capability and decomposed gas density of CF_3I as a substitute for SF_6 gas. *IEEE Transactions on Dielectrics and Electrical Insulation.*, 17(4):1196–1203, 2010. doi:10.1109/TDEI.2010.5539690.

doi:10.14311/ppt.2017.1.79
Plasma Physics and Technology **4**(1):79–82, 2017

TOWARDS AN EFFICIENT ARC SIMULATION FRAMEWORK

R. Fuchs*, M. Mürmann, H. Nordborg

Institute for Energy Technology, HSR University of Applied Sciences Rapperswil, Oberseestrasse 10, 8640 Rapperswil, Switzerland

* roman.fuchs@hsr.ch

Abstract. Arc simulations require a coupled solution of the flow and electromagnetic equations. Despite of industrial interest, there is no established simulation framework available yet. We assess the usability of STAR-CCM+ for low voltage circuit breaker simulations using a test case of a model arc chamber, since this toolkit allows to define and control the simulation in a single environment. In spite of a partially implemented arc root model, the results agree well with reference data of previous publications.

Keywords: arc simulation, arc root model, splitter plates, low voltage circuit breakers, computational magnetohydrodynamics.

1. Introduction

Electrical arcs occur, for instance, in circuit breakers and other electrical devices that are required for safe operation of a power grid network and its components connected to. Arcs may be created by separating two electrodes while current is flowing (for a detailed description see, e.g., [1]). Hence, a thermal plasma is formed which shows as an intensive source of light and energy at elevated temperature levels. Therefore, circuit breakers must be designed appropriately to withstand the thermal loads and forces due to an explosive gas expansion. The basic principle for arc quenching is to raise its voltage above the grid voltage, which is achieved by arc elongation and efficient cooling of the plasma. Additionally, in low voltage circuit breakers, the preferred method is to drive the arc into a quenching chamber with several ferromagnetic splitter plates. Therein, the arc is split into multiple branches, and because each arc root leads to an additional voltage drop, it becomes extinguished.

Experiments are essential in circuit breaker development, but the short timescales and physical conditions make it difficult to measure relevant quantities in sufficient detail and quality. Therefore, numerical simulations of electrical arcs are an important source of information for industrial manufacturers to optimize their products in size, costs, and extension of operational ranges towards higher currents.

A useful simulation framework must cover the relevant physics including electromagnetism, gas dynamics, thermal radiation, material data at elevated temperature levels, evaporation of materials that are in contact or nearby the arc, as well as their mutually coupled effects. For instance, the electrical conductivity of the plasma is a key material parameter, which depends on the gas temperature, pressure, gas composition, and contamination with metal vapor, and couples the Maxwell and Navier-Stokes equations. Additionally, electrode motion must be included and the framework must allow for mesh deformation, remeshing, and data interpolation in a robust manner.

Moreover, the simulation framework must build on appropriate numerical methods. On the one hand, the gas flow solver should rely on the finite volume method to capture accurately the nonlinear effects of shock waves. On the other hand, the electromagnetic equations are to be solved with edge-based finite elements (FE) if ferromagnetic splitter plates are considered. Consequently, an arc simulation framework consists of two or more separate solvers and, frequently, an additional software level that is responsible for synchronization of the solvers and their data exchange. For example, this methodology is used by [2] and works sufficiently well. However, it requires skilled users in multiple programming languages and touching of low-level operations for implementing advanced physical models. A notable amount of time may be spent on programming and debugging rather than the physical modeling of the circuit breaker.

Our research group started using STAR-CCM+ for arc simulations because of its integrated architecture. This framework consists, among others, of a standard set of flow solvers and a recently added FE magnetic vector potential solver. Surface fluxes and volume sources can be defined via field functions on the GUI level with a comprehensive set of operators and building blocks, and without referencing individual mesh cells. The graphical user interface also allows to define and control moving geometries as well as the required remeshing and data interpolation.[3] Data exchange among the solvers is also included automatically and does not require additional software levels. In total, we expect it to be user-friendly and a promising framework for arc simulations.

The capabilities of STAR-CCM+ are currently being assessed by implementation of previously reported arc simulations. Here, we consider a test case based on the works of T. Rüther [4], A. Mutzke [5], and

J. Rüther [6] who investigated a simplified arc chamber with a single splitter plate. They showed that a nonlinear current-voltage characteristic allows to model the arc splitting process accurately compared to experimental observations.

In this paper, we present results on the selected test case of a simplified arc chamber using STAR-CCM+ as a toolkit for arc simulations. Firstly, we define the model geometry and simulation settings. In the main part, we compare our results to those of [6] and we report on the usability of the software framework. Finally, conclusions are drawn.

2. Simulation model

2.1. Geometry and Mesh

We consider a model arc chamber identical to [6], see figure 1. The parallel arc rails are $1.5 \times 58 \times 4$ mm with 8 mm inner distance, and a splitter plate $1 \times 20 \times 4$ mm located centrally between them. The plasma is computed in a box that starts 16 mm above the lower electrode ends, with an outlet at the upper boundary. The arc is ignited 7 mm below the splitter plate with a radius of 1 mm. An auxiliary cylinder with elliptic cross section (15 and 10 mm semiaxis) contains the plasma and electrode domains so that the magnetic boundary conditions can be specified in the farfield.

The domain is discretized with a polyhedral mesh for all solvers. The cell size is 0.25 mm in the plasma domain, with increasing element size in the elliptic cylinder. Prism layers are specified in the plasma domain with a total thickness of 0.1 mm.

2.2. Governing equations

The compressible Navier-Stokes equations for laminar flow are solved including source terms for the Lorentz force and Ohmic heating. Radiation is included by the discrete ordinate method with mean absorption coefficients given in 6 bands. Absorption coefficients of air and band boundaries are identical to [6].

The electromagnetic equations are solved in the low-frequency and magnetostatic limit. The magnetic vector potential is computed with an iterative solver in finite volumes, because all material data have a relative permeability $\mu_r = 1$. The voltage drop in the arc roots is modeled according to [5], and is specified at the contact interface between the electrodes and the air domain. The characteristic current-voltage curve consists of two exponential functions such that the curve has a peak voltage drop of 19.7 V at a current density of $13\,\mathrm{kA\,m^{-2}}$ and a constant voltage drop of 10 V for large current densities.

In contrast to [6], the energy equation is solved only in the plasma domain with adiabatic boundary conditions. The Ohmic heating due to the voltage drop at the interfaces is neglected. A more complete arc root model is not in the focus of this study and is left to subsequent work that will also include source terms for the mass, momentum, and energy equations due to metal evaporation.

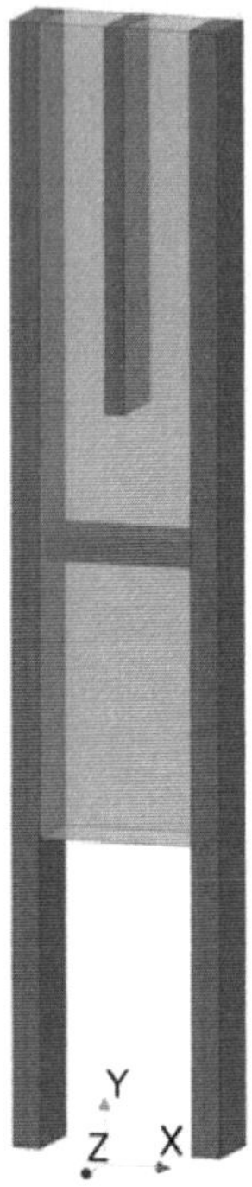

Figure 1. Model arc chamber geometry with the copper parts, the air volume, and the arc column at its initial position.

2.3. Material data

Material data for the copper electrodes and air are provided by STAR-CCM+ using the Equilibrium Air model [7]. The electric conductivity for air is tabulated according to [8], and limited to a minimum value of $\sigma = 10^{-3}\,\mathrm{S\,m^{-1}}$ due to numerical reasons.

2.4. Boundary and initial conditions

A pressure outlet condition is specified at the surfaces located at y = 58 mm; otherwise, wall boundary conditions are used for the gas flow. The gas is initially at rest and at ambient temperature (T = 300 K).

Electric boundary conditions are applied to the lower ends of the parallel rails. The right electrode is at zero potential, while the left electrode is stressed with a sinusoidal total current corresponding to a peak value of 1 kA effective at 50 Hz. A time-delay of 500 µs is included which is in accord with the ignition time in the experiments by [4]; the initial total current is therefore equal to 221 A.

The ignition wire is modeled as an electrically conductive channel with 1 mm radius and conductivity $\sigma = 10^4\,\mathrm{S\,m^{-1}}$ until $t = 15\,\mu s$.

The simulation time step is 0.5 µs uniformly with 40 inner iterations per time step. In the first time step, the solvers are iterated for 100 inner steps to ensure a converged solution. The radiative heat flux is computed only in the first inner iteration per time step to save computing time.

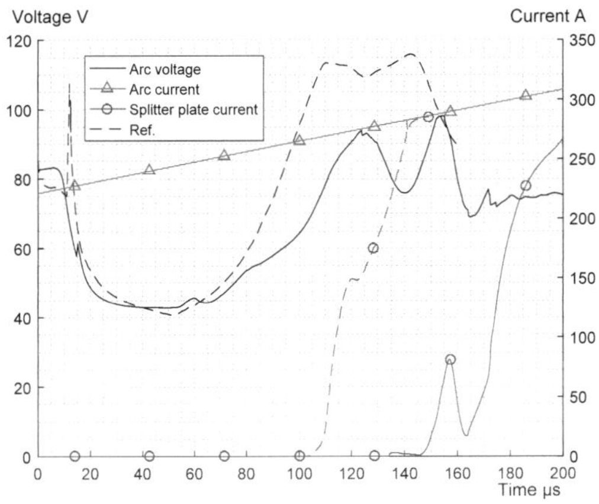

Figure 2. Graphs of arc voltage and currents in comparison to the reference data (figure 3-5 in [6]).

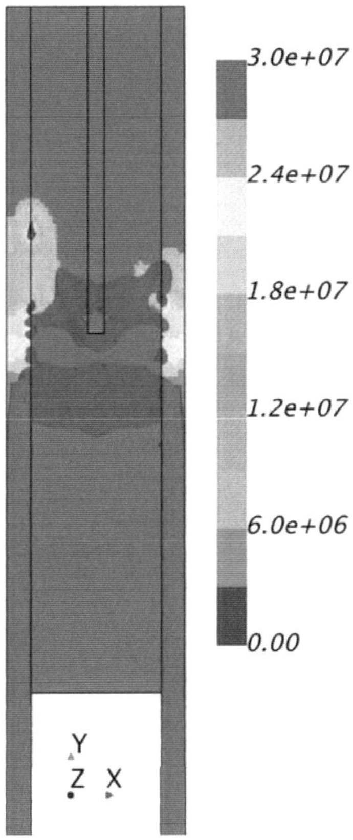

Figure 3. Electric current density at $t = 155\,\mu s$ on central section plane. The isocontour value is at $3 \times 10^7\,\mathrm{A\,m^{-2}}$ and shown in magenta.

3. Results and Discussion

Figure 2 shows the arc voltage, total arc current, and the current through the splitter plate as a function of time from ignition until $t = 200\,\mu s$. Also shown are the reference data from figure 3–5 in [6] which is given until $t = 160\,\mu s$. The total arc current is given as a boundary condition and increases from 221 A to 310 A. Initially, the arc voltage is 83 V and decreases after $t = 6\,\mu s$ towards a minimum value of 43 V at $t = 46\,\mu s$. The jump at $t = 15\,\mu s$ is due to the minimum condition for the plasma electrical conductivity modeling the arc ignition phase. After $t = 55\,\mu s$, the arc voltage increases and shows two maxima of 90 V at $t = 125$ and 95 V at 155 µs, before it tends to a rather constant value of 75 V. The splitter plate current is zero until $t = 135\,\mu s$, when a small current is noted. Then, the current reaches a maximum of 81 A at $t = 155\,\mu s$. After reducing to 18 A in 10 µs, it raises towards the total current. Figure 3 illustrates the situation at the splitter current peak at $t = 155\,\mu s$. We see that a partial current flows through the splitter plate while the major part of the current flows through the plasma below. We also see distinguished arc roots on both rails.

The data is explained as follows with reference to graphical visualizations. Until $t = 55\,\mu s$, the arc remains stationary and heats the nearby plasma resulting in a low arc voltage. Then, the pressure wave reflected from the lower boundary pushes the arc upwards into a cooler domain and results in increasing voltage due to elongated arc length and lower temperature. At $t = 60\,\mu s$, new arc roots are formed symmetrically on the rails reducing the arc voltage slightly. At $t = 80\,\mu s$, the arc reaches the splitter plate. Another pair of arc roots are formed above the lower end of the splitter plate while those at the initial position vanish. Hence, the arc column becomes M-shaped enclosing the splitter plate and while the arc becomes elongated, the arc voltage increases more steeply. Subsequently, the arc stays near the edge of the splitter plate and starts to burn more diffusely below it without being split and moving onto the splitter plate. This is in contrast to [6] and is due to insufficient voltage difference between the arc loop along the interface and the required voltage for arc root formation as modeled by the characteristic current-voltage curve. Additional attempts are required to overcome the arc root formation voltage on the splitter current, explaining the two voltage and current maxima between $t = 120\,\mu s$ and $t = 170\,\mu s$.

Comparison to the reference data shows that our arc voltage is comparable until $t = 70\,\mu s$. Subsequently, different arc splitting behavior is observed which is attributed to the incompletely implemented arc root model and different material data. Other differences exist in the radiation model: here, the discrete ordinate model is used whereas in [6] a combination of net emission coefficients and increased heat conduction due to radiation have been used. Despite these differences, the arc voltage and the timescale for arc splitting are comparable.

The main objective of this study is to evaluate the simulation framework provided by STAR-CCM+ for arc simulations. The results presented above show that it is a viable toolkit. The user-friendly and modern graphical user interface supports the user in setting up the simulation and implementing advanced arc root models. We take it as an advantage that the simulation may be controlled in a single window. Material data can be loaded as multivariate file tables without writing and loading additional user code. Moreover, the user can focus more on plasma modeling rather than programming tasks.

4. Conclusions

We used a model arc chamber referring to a low voltage circuit breaker as a test case to evaluate STAR-CCM+ as a possible step forward to more efficient arc simulations in industry and research. We implemented a macroscopic model for the voltage drop at the electrode- plasma interface as a current-voltage characteristic that is applied at the corresponding geometric interfaces. Further source terms of a complete arc root model will be implemented similarly in future work. Despite the partially implemented model, the results showed that this model allows to capture the main aspects of arc splitting in a quenching chamber. In comparison to [6], differences are attributed to model settings and material data. The simulations ran robustly and are easily handled because the solvers are coupled and managed in a single graphical user interface.

In total, we are satisfied with the capabilities of STAR-CCM+ for arc simulations, and we will continue assessing it in further applications of high- and low-voltage plasma simulations.

Acknowledgements

This work was supported by Siemens PLM software. We want to express our gratitude to Dr Paul Hilscher and Dr Angelo Limone for their support and fruitful discussions.

References

[1] Eduard Vinaricky, editor. *Elektrische Kontakte, Werkstoffe und Anwendungen: Grundlagen, Technologien, Prüfverfahren.* Springer Berlin Heidelberg, Berlin and Heidelberg, 3. edition, 2016.

[2] Christian Rümpler. *Lichtbogensimulation für Niederspannungsschaltgeräte.* Fraunhofer-Verl., Stuttgart, 2009.

[3] Roman Fuchs, Mario Mürmann, Henrik Nordborg, and Paul Hilscher. Electric arc simulations - towards a user-friendly and efficient tool for industry. In *Nafems World Congress*, 2017.

[4] Thomas Rüther. *Experimentelle Untersuchung der Lichtbogenaufteilung an Löschblechen.* Cuvillier, Göttingen, 2008.

[5] Alexandra Miriam Mutzke. *Lichtbogen-Simulation unter besonderer Berücksichtigung der Fusspunkte.* Ingenieurwissenschaften. Verl. Dr. Hut, München, 2009.

[6] Julia Rüther. *Weiterentwicklung und Vereinfachung eines Simulationsmodells für Schaltlichtbögen in Löschblechkammern.* Energietechnik. Verl. Dr. Hut, München, 2014.

[7] R. N. Gupta, K.-P. Lee, R. A. Thompson, and J. M. Yos. Calculations and curve fits of thermodynamic and transport properties for equilibrium air to 30000 K. *NASA STI/Recon Technical Report N*, 92, 1991.

[8] A. D'Angola, G. Colonna, C. Gorse, and M. Capitelli. Thermodynamic and transport properties in equilibrium air plasmas in a wide pressure and temperature range. *The European Physical Journal D*, 46(1):129–150, 2008. doi:10.1140/epjd/e2007-00305-4.

doi:10.14311/ppt.2017.1.83
Plasma Physics and Technology **4**(1):83–86, 2017

DIAGNOSTICS OF NITROGEN-METHANE ATMOSPHERIC GLOW DISCHARGE USED FOR A MIMIC OF PREBIOTIC ATMOSPHERE

V. Mazánková[a,*], L. Töröková[a], D. Trunec[b], F. Krčma[a], S. Matejčík[c], N. J. Mason[d]

[a] *Faculty of Chemistry, Brno University of Technology, Purkyňova 118, 612 00 Brno, Czech Republic*
[b] *Faculty of Science, Masaryk University, Kotlářská 2, 611 37 Brno, Czech Republic*
[c] *Faculty of Mathematics, Physics and Informatics, Comenius University, 842 48 Bratislava, Slovakia*
[d] *Department of Physical Sciences, Open University, Walton Hall, Milton Keynes MK7 6AA, UK*
[*] mazankova@fch.vut.cz

Abstract. The exploration of planetary atmosphere is being advanced by the exciting results of the Cassin-Huygens mission to Titan. The complex chemistry revealed in such atmospheres leading to the synthesis of bigger molecules is providing new insights into our understanding of how life on Earth developed. This work extends our previous investigation of nitrogen-methane (N_2-CH_4) atmospheric glow discharge for simulation chemical processes in prebiotic atmospheres. In presented experiments 2 % of water vapor were addet to nitrogen-methane gas mixture. Exhaust products of discharge in this gas mixture were in-situ analysed by Fourier Transform Infra Red spectroscopy (FTIR). The major products identified in spectra were: hydrogen cyanide, acetylene and acetonitrile.

Keywords: prebiotic atmosphere, Fourier Transform Infra Red spectroscopy, glow discharge.

1. Introduction

More than 60 yers ago, S. Miller and H. Urey [1], in their exciting work, demonstrated the synthesis of several amino acids from a mixture of reducing gases (CH_4, NH_3, H_2O, and H_2) treated with electric discharge. The following explorations showed that a broad array of amino acids could be synthesized, but there was no evidence that all of the fundamental molecules of the RNA genetic code could be produced alongside others in this type of experiment [2, 3]. Recently it was also referred to a positive result on qualitative detection of RNA nucleobases and manifold amino acids from tholines created in a N_2, CH_4, CO mixture [4]. This experiment simulated the atmosphere of Titan by electric discharge. Such experimental results as well as theoretical expectations [5] show that reduced, relatively reactive atmospheres are likely to be more efficient for the synthesis of biomolecules [6]. The gliding arc configuration of atmospheric pressure discharge has been shown to be a good mimic of processes in the prebiotic atmospheres [7] being used to replicate physical and chemical conditions on Titan. It is the only lunar body with significant quantities of methane (CH_4) and nitrogen (N_2) in its atmosphere [8, 9]. The chemical composition of the Titan's atmosphere is considered to be similar to the atmosphere of early Earth and it is favorable for formation of complex molecules containing C, N and H. The solar ultraviolet radiation and magnetospheric electrons are responsible for the generation of primary radicals and other neutral species, which initiate chains of chemical reactions that finally result in the formation of various organic molecules in the Titan atmosphere. This makes Titan as planetary-scale laboratory for the synthesis of complex organic molecules [10].The composition of early Earth atmosphere was deeply discussed in many studies [11–13].

The present work is focused on the experimental study of gaseous products produced in the atmospheric pressure glow discharge fed by N_2, CH_4 gas mixtures with CH_4 concentrations in the range from 1 % to 5 % and admixture of 2 % water vapor (H_2O). The atmospheric DC glow discharge is a source of electrons, ions and neutral radicals. All these species initiate a complex chemical processes under laboratory conditions, which are similar to the processes which may occur in the prebiotic atmosphere. The neutral products generated in the discharge were identified and quantified by the means of the Fourier-Transform-Infra-Red spectroscopy (FTIR). In more detail the influence of the H_2O admixture on production of HCN (the major neutral product) and on the formation of NH_3 has been studied.

2. Experimental setup

The experimental setup schematic drawing is shown in Figure 1. The special high vacuum stainless steel reactor was constructed for our experiments to prevent any oxygen contamination during the experiments. Nitrogen and methane flows were automatically controlled by Bronkhorst controllers. The measurements were carried out at total gas flows 100 sccm at atmospheric pressure and laboratory temperature. The discharge electrode system had the standard configuration of the gliding arc discharge. The discharge was formed in the stable abnormal glow regime, and

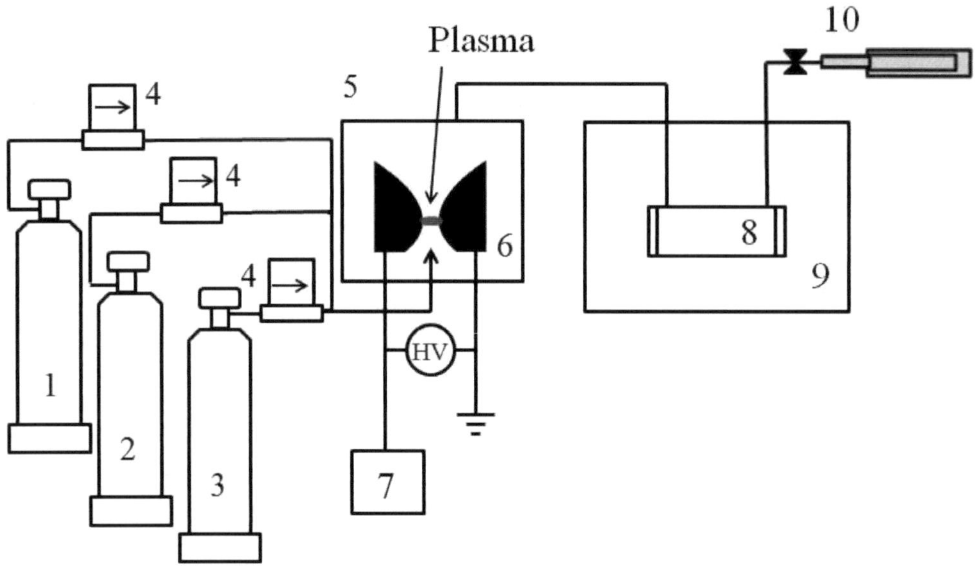

Figure 1. Experimental setup. 1- storage bottle of nitrogen, 2- storage bottle of methane, 3- storage bottle of carbon dioxide, 4- Bronkhorst controllers, 5- reactor body, 6- electrode system, 7- oscilloscope, 8- IR gas cell, 9- FTIR spectrometer, 10- exhaust.

plasma occurred between the electrodes at their shortest distance of 2 mm in the form of a plasma channel of 1 mm in its diameter. The reactor chamber volume was 0.5 l. The discharge was supplied by a DC stabilized HV source. Discharge breakdown voltage was 1500 V, a stable plasma channel was operating at 400 V at current in range 15 mA to 40 mA during all presented experiments. The measurements were performed for different N_2:CH_4 ratios in the range from 1 % to 5 % of methane in nitrogen (both gases having quoted purities of 99.995 %). There was connected bottle gas washing with a high purity water just before the entrance to the reactor. The amount of added water vapour was estimated at 2 % in gas mixture in all performed experiments. The exhaust gas was analyzed in-situ by FTIR spectroscopy using IR cell with total length of 15 cm.

The electrode system is made of stainless steel and it is necessary to clean it after each set of experiments. The deposit created during discharge operation is shown in Figure 2 . Before starting the experiments, the discharge chamber was pumped down to 1 Pa for 1 h and then was filled with the investigated gas mixture up to the pressure of 101 kPa. Atmospheric pressure during the experiments was maintained by a slight pumping through the needle valve.

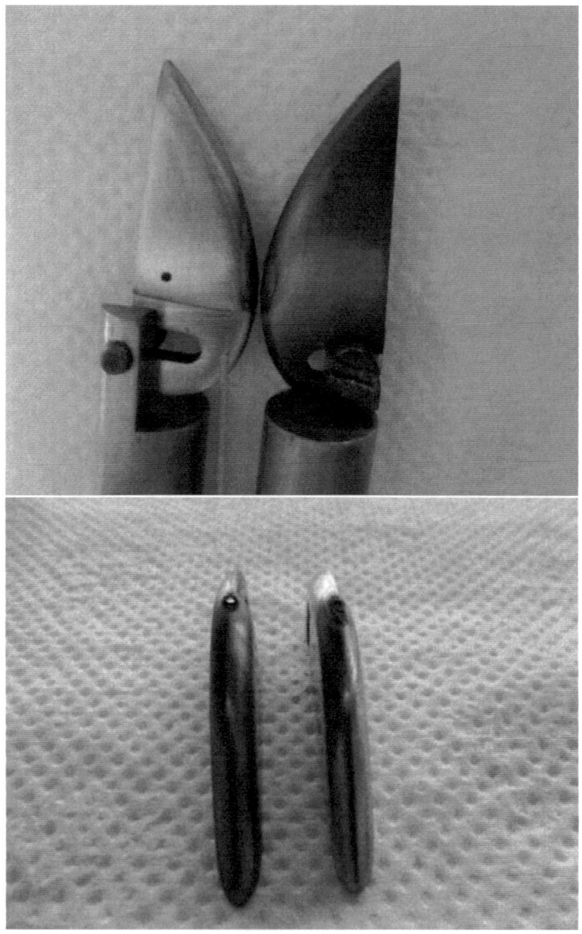

Figure 2. Photos of the electrodes with deposited layers.

3. Results and Discussion

A typical FTIR spectrum showing the products formed in the nitrogen discharge fed by 4 % of CH_4 and 2 % of H_2O is shown in Figure 3. Similar spectra were also observed for other N_2:CH_4:H_2O ratios. Hydrogen cyanide (HCN) was found to be the most abundant product at wavenumbers of 1430 cm^{-1} and of

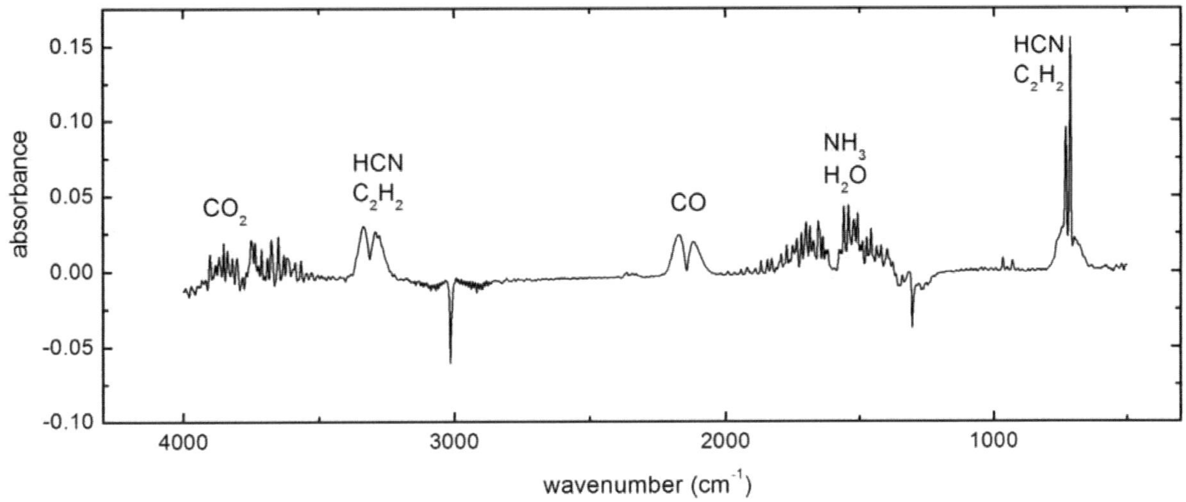

Figure 3. FTIR spectrum recorded in mixture N_2:CH_4 = 96:4 with 2 % H_2O vapor addition at flow rate of 100 sccm and discharge current of 20 mA.

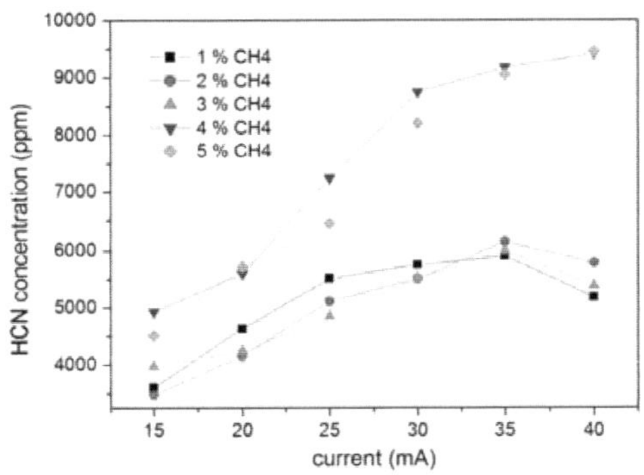

Figure 4. The dependence of hydrogen cyanide concentration on discharge current for different initial concentration of methane.

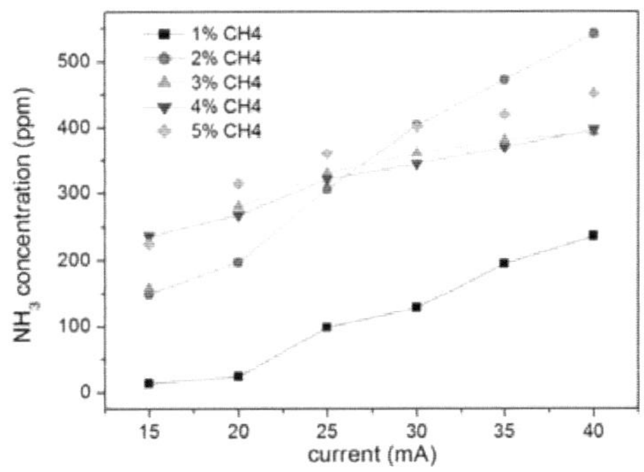

Figure 5. The dependence of ammonia concentration on discharge current for different initial concentration of methane and carbon dioxide.

720 cm^{-1}. This is in agreement with [7]. Ammonia (NH_3), which was identified at 966 cm^{-1}, was also detected in our previous experiments. The other major products were acetylene (C_2H_2) as well as carbon monoxide (CO). These products were recognized in all investigated gas mixtures. The products concentrations depend on the composition of the gas mixtures. Quantitative analysis of selected hydrocarbons were done in dependence on the methane concentration. The absorbance was calculated as an area under the recorded peaks and the concentations of selected products were then calculated. Figures 4 and 5 show dependences of HCN (1430 cm^{-1}) and NH_3 (966 cm^{-1}) concentrations formed under different experimental conditions. The increase of the initial discharge current from 15 mA to 40 mA leads to increase in the product yield of these compounds.

4. Conclusions

The gaseous products formed in the atmospheric glow discharge fed by different mixtures of nitrogen, methane and water were determined by in situ FTIR analysis. The discharge was operated in the flowing regime at different discharge currents at laboratory temperature. An in-situ FTIR technique for the exhaust gas phase sampling was successfully used for chemical analysis to deduce the gas composition in the N_2:CH_4:CO_2 reactive gas mixture mimics of Titan's atmosphere. Various hydrocarbons were observed in all experiments. HCN was identified as the major gas phase product in all of measurements. Others minor products detected were C_2H_2, NH_3, CO_2, CO. These results are consistent with the Titan's atmospheric composition because the same compounds were detected during the Cassini-Huygens space mission. CO could have an effect on the atmospheric reactivity of Titan. The formation of organic molecules incorporating oxygen in gases could occur in the upper atmosphere of Titan where the dissociations of N_2 and CO_2 by VUV photons and magnetospheric electrons are possible. This fact clearly demonstrates that laboratory experiments can be used for prediction of both the presence and possible concentrations of compounds which have not been detected, yet. These simple organics should be tracers of the chemical groups constituting the dusty products.

Acknowledgements

The research was supported by Czech Ministry of Education, Youth and Sports, projects No. COST CZ LD15010 and LD15011 within the collaboration of the COST Actions CM 1401 and TD 1308 and CEEPUS network AT-0063.

References

[1] S. L. Miler. A production of amino acids under possible primitive earth conditions. *Science*, 117(3046):528–529, 1953. doi:10.1126/science.117.3046.528.

[2] T. M. McCollom. Miller-Urey and beyond: What have we learned about prebiotic organic synthesis reactions in the past 60 years?. *Annu. Rev. Earth. Planet. Sci.*, 41:207–229, 2013. doi:10.1146/annurev-earth-040610-133457.

[3] A. Johnson, H. J. Cleaves, J. L. Bada, and A. Lazcano. The diversity of the original pre-biotic soup: Re-analyzing the original Miller-Urey spark discharge experiments. *Orig. Life. Evol. Biosph.*, 39:240–241, 2009.

[4] S. M. Hörst. Formation of amino acids and nucleotide bases in a titan atmosphere simulation experiment. *Astrobiology*, 12:809–817, 2009. doi:10.1089/ast.2011.0623.

[5] C. Chyba and C. Sagan. Endogenous production, exogenous delivery and impact shock synthesis of organic molecules: An inventory for the origins of life. *Nature*, 355:125–132, 1992. doi:10.1038/355125a0.

[6] C. F. Chyba, P. J. Thomas, Brookshaw, and C. L., Sagan. Cometary delivery of organic molecules to the early earth. *Science*, 249:366–373, 1990. doi:10.1126/science.11538074.

[7] L. Torokova, J. Watson, F. Krcma, V. Mazankova, N.J. Mason, G. Horvath, and S. Matejcik. Gas chromatography analysis of discharge products in N_2-CH_4 gas mixture at atmospheric pressure: study of mimic Titan's atmosphere. *Contrib Plasma Phys*, 55:470–480, 2015. doi:10.1002/ctpp.201400052.

[8] A. Coustenis, D.E. Jennings, C.A. Nixon, R.K. Achterberg, P. Lavvas, S. Vinatier, N.A. Teanby, G.L. Bjoraker, R.C. Carlson, L. Piani, G. Bampasidis, F.M. Flasar, and P.N. Romani. Titan trace gaseous composition from CIRS at the end of the Cassini-Huygens prime mission. *Icarus*, 207:461–476, 2010. doi:10.1016/j.icarus.2009.11.027.

[9] S.M. Horst and M.A. Tolbert. The effect of carbon monoxide on planetary haze formation. *Astrophys J*, 781(53), 2014. doi:10.1088/0004-637X/781/1/53.

[10] F. Raulin, C. Brasse, O. Poch, and P. Coll. Prebiotic-like chemistry on Titan. *Chem Soc Rev*, 41:5380–5393, 2012. doi:10.1039/C2CS35014A.

[11] J.F. Kasting and M.T. Howard. Atmospheric composition and climate on the early earth. *Philos Trans R Soc Lond B Biol Sci*, 361:1733–1741, 2006. doi:10.1098/rstb.2006.1902.

[12] J.F. Kasting and S. Ono. Palaeoclimates: the first two billion years. *Philos Trans R Soc Lond B Biol Sci*, 361:917–929, 2006. doi:10.1098/rstb.2006.1839.

[13] S.L. Olson, L.R. Kump, and J.F. Kasting. Quantifying the areal extent and dissolved oxygen concentrations of archean oxygen oases. *Chem Geol*, 362:34–43, 2013. doi:10.1016/j.chemgeo.2013.08.012.

doi:10.14311/ppt.2017.1.87
Plasma Physics and Technology **4**(1):87–90, 2017

IMPLOSIVE THERMAL PLASMA SOURCE FOR ENERGY CONVERSION

J. Sonsky[a,*], V. Tesar[a], J. Gruber[a], A. Maslani[b]

[a] *Institute of Thermomechanics of The Czech Academy of Sciences, Dolejskova 1402/5, 182 00 Prague, Czech Republic*
[b] *Institute of Plasma Physics of The Czech Academy of Sciences, Za Slovankou 1782/3, 182 00 Prague, Czech Republic*
[*] soda@it.cas.cz

Abstract. Laboratory scale thermal plasma source for magnetohydrodynamic or magnetocumulative generator was developed. The thermal plasma is created from combustible stochiometric mixture of hydrogen and oxygen by spherical implosion of convergent detonation wave. Resulting high velocity plasma is observed by capturing emitted light by hi-speed camera to determine plasma velocity and also spectroscopically in order to estimate of plasma temperatures. Construction of the implosion plasma source is discussed.

Keywords: implosion, thermal plasma, detonation wave.

1. Introduction

Combustion of fossil fuels in a process involving production of mechanical work driving a generator with rotating coils in a magnetic field is employed in processes of electricity production. In spite of past developments, the present energy generation processes are still quite complicated and characterised by high losses. In the thermal cycles the losses represent estimated 2/3 of the chemical energy available in the fossil fuel. There is therefore a demand for decreasing these losses preferably by avoiding some of the intermediate conversion steps. Promising possibility is the electricity generation in magnetohydrodynamic generators operating without mechanical moving parts allowing to increase efectivity of the energy conversion by 10 - 20 %. Known for several decades and already tested MHD idea has the weakness of insufficient number of charge carriers in the flow of combustion products through the perpendicularly oriented magnetic field. More ions have to be produced and this can be achieved only at much higher temperatures than obtainable with the currently used deflagration type of combustion. Existing MHD generators built for tests increase the ionisation by combustion in pure high-pressure highly preheated oxygen and seeding the gas with alkali metals. This is expensive and so far relegated the MHD principle to a role of economic failure. Also, the alkali metal approach is expensive due to the need of very special refractory materials necessitated by corrosive character of the additives. A more efficient form of combustion than the usual deflagration is already known. It is detonation, capable of getting nearer to the desirable ionisation goal - but difficult to exploit. Apart from the single-action military uses with explosives, attempts have been made from time to time to apply the detonations in a gas mixture.

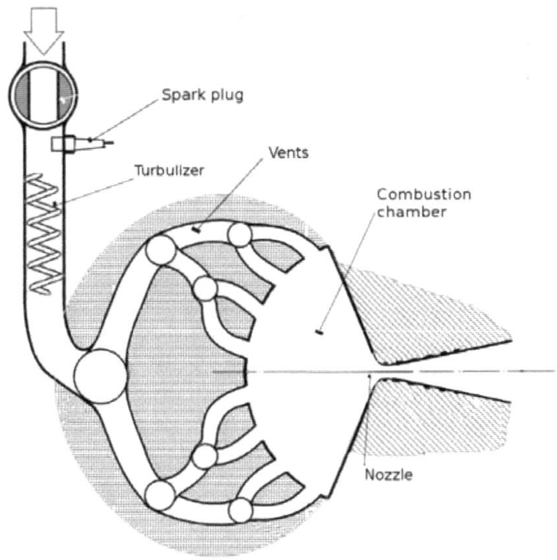

Figure 1. Schematic of the device

2. Experimental model

The idea is to obtain the extremely high temperature combustion by detonation waves focused and comming from mutually opposed directions into the common focal point. In principle, the progress to the centre is obtained from sources distributed over a spherical hollow surface. Attempts to synchronize detonation produced from separate sources are not unsuccessful - the activation energy needed to start the detonation wave directly is relatively high and the need to start large number of waves results in very high energy and power demand. A solution is to create a single detonation wave by means deflagration to detonation transition (DDT) and later bifurcate it to the several identical-length paths to the exits located on the hemisphere surface. Schematic representation of the

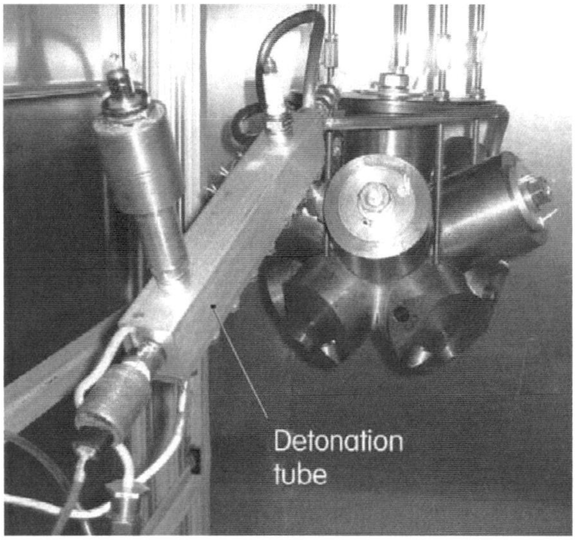

Figure 2. Experimental device

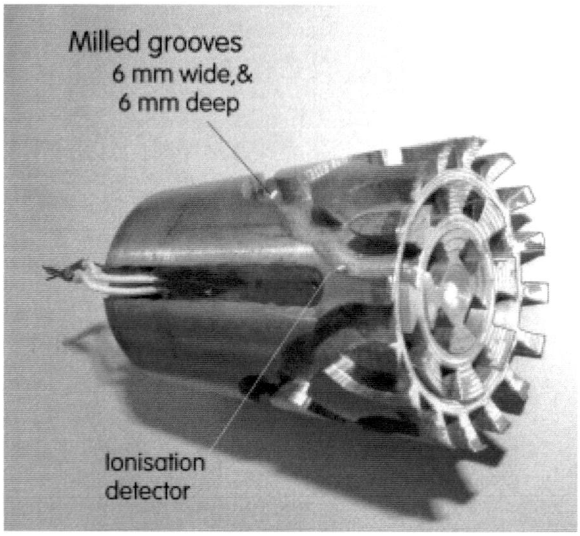

Figure 3. Assembled distributor cones.

principle is presented in Fig. 1.

The main parts shown there are detonation tube, the main body of the combustion chamber, and distributors - a system of progressively branched vents leading the detonation wave from the detonation tube into the chamber internal cavity. The detonation chamber filled with the gas mixture is activated by the spark plug at the end of detonation tube. The mixture starts burning initially in the usual deflagration regime, but generation of turbulence - by turbulator with shape discontinuities at the tube walls - causes a transition into the detonation. The detonation wave then progresses through the bifurcations which divide it and lead to exits positioned on the hemispherical surface of the combustion chamber. There the exiting waves leave the cavities and emerge at the same instant of time (this is secured by exactly the same lengths of the distributor vents) propagating as an implosion towards the common focal point. The instantaneous pressure and temperature values reach extreme magnitudes there, especially the temperature is demonstrably above 10000 K. This leads to the instantaneous ionization - practically a generation of thermal plasma moving at high velocity (commensurable with the speed of propagating detonation waves) through the nozzle. The thermal ionization together with the extreme velocity is essential to generate a high voltage pulse at MHD generator electrodes. For the purposes of this verification study it was sufficient to perform only a single detonation in a simplified model of the device - without the MHD generator. The combustion chamber in the tests was approximately hemispherical, with radius R = 47.55 mm and the apex angle was $\alpha = 180$ deg. There were three main parts of the device: Detonation tube for generation of the primary detonation wave, the distributors vents with sequential bifurcation of the detonation wave into a large number of the entries into the chamber cavity, and the actual combustion chamber

Detonation tube: The body of this detonation tube is of rectangular cross-section, as seen in its external appearance in Fig. 2. It is made of aluminium alloy. The cavity for generation and propagating of detonation wave was made by milling it as square section channel 8×8 mm and covering with top cover plate. Its axial length is 140 mm. Well recognisable in the photograph Fig. 2 is the re-filling valve, electrically operated for admission of the mixture of hydrogen and oxygen in stoichiometric ratio. As a safety measure there are several flame traps inside and in the leading tubing. Activation energy for starting the combustion process is provided by automobile spark plug with added shielding to avoid electrical interference with diagnostics. Transition from the deflagration burning into the detonation is achieved by the presence of turbulizing voids in form of a set of 5 local increases of the channel width. It was made with the same 8 mm milling tool as was used for making the internal 8 mm wide channel. At the exit from the detonation tube the flow was divided (trifurcated) into three metal (cooper) tubes of round cross section, leading to the entrance into distributors.

Distributor vents: The task of all 18 distributors together positioned on the hemispherical wall of the combustion chamber was to divide the incoming detonation wave into 168 inlets leading into the internal chamber cavity. In each distributor, the wave trajectory is progressively bifurcated into the total of 28 inlet exits, all of 6×6 mm square cross-sections. A photograph of typical assembled distributor is presented in Fig. 3, with the exits seen at the right-hand part of the picture. Also visible there are two ionisation probes with their connector leads. The distributors consisted of three cone-shaped components fitting one into another. The channels for delivering the detonation wave were made by milling. From the outer

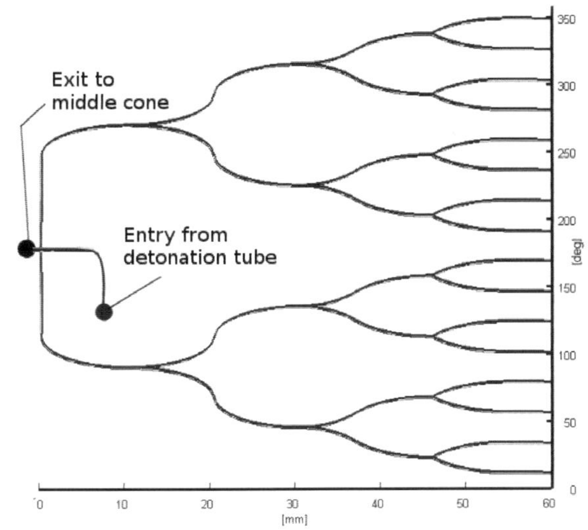

Figure 4. Plot of the enveloped vents.

side(i.e. the side from which the milling tool entered) the grooves were covered - and thus converted into the closed channels - by the conical internal hollow surface of another component fitted from outside. The milling of the channels was performed on milling machine fitted with angular division head. There were two manually input variables. One variable was the axial position distance of the tool axis in 0.1 mm steps, while the other was the angular rotation around the axis of the manufactured component. Plots of mutual values of these two variables - with data points values related to the tool centre - are presented in Fig. 4 for outher cone. Middle and iner cones are similar, they have only eight and four grooves respectively. The basic requirement for proper synchronization of the cumulative implosion is identical lengths of all wave trajectories. The axisymmetric shape of rotatable distributors provided an opportunity for precise mutual synchronization - the effective channel lengths could be varied by small angular rotation of the particular distributor.

Combustion chamber: For ease of manufacturing the hemispherical shape with the large number of distributor exits, it was decided to assemble the chamber from a number of thick-wall steel components. Their inner ends, at which these components were set into mutual contact, were each milled into a shape with five flat (planar) contact surfaces. Put together, in the first approximation the hemisphere of the chamber was assembled as a top half of a dodekahedron covering R = 47.55 mm hemisphere. Opposite to the hemisphere, in the middle of the flat base wall, pentagonal exit hole was left - in a position where in the final setup would the entrance into the MHD generator. This hole serves as a diagnostic access window (in the absence of the generator). The cylindrical outer ends of all six components (top half of dodekahedron planes) of which the chamber was assembled are seen extending out from the body in Fig. 2. They were

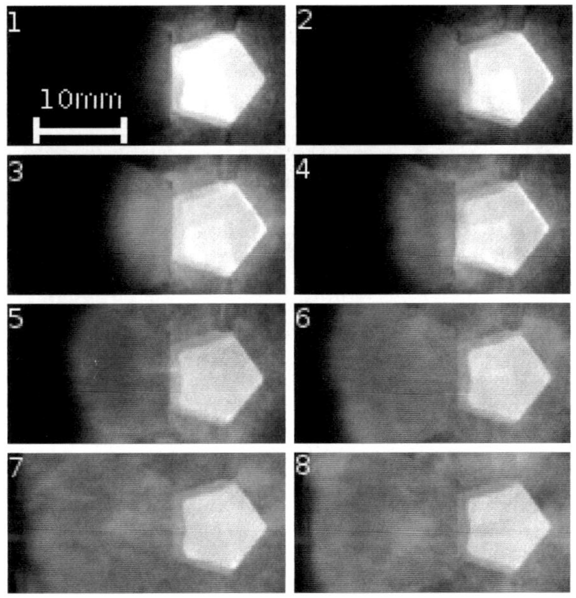

Figure 5. Time assorted camera frames.

provided each with a 6 deg hollow conical cavity of 75 mm length. It was into these conical cavities where the six distributors inserts sets were fitted.

3. Experimental results

Experiments with thermal ionisation by cumulative implosion are still in their early phase. They have, nevertheless, already demonstrated the most important fact: the capability to reach extreme temperature levels in the combustion chamber as well as the correspondingly high ionisation.

Propagation of detonation waves in the distributor cones and formation of the convergent detonation wave were monitored by an array of ionisation probes positioned near the distributor exit openings. The detonation wave itself is weakly conductive so that its passing past the probe was monitored as voltage drop on a charged electrode. The ionisation probes were made from a copper wire of diameter 0.75 mm insulated by PTFE. Active parts of the probes, protruding 2 mm into the distributor channel, had the insulation striped off. The probes were charged via 48 kΩ resistor from an 80 V DC voltage source. Main purpose of these measurements was fine synchronisation of the distributors by means of adjusting their rotary positions. This, depending on the angular adjustment, either prolonged or shortened length of auxiliary vent leading the incoming detonation wave into the first bifurcation inside the distributor. In this way, all 18 distributors could be finely tuned to microsecond range of the detonation waves entrance to hemispherical combustion chamber.

Another already performed set of experiments was measurement of velocity of the plasma ejected from the combustion chamber through its central exit. This was done by analysis of records from high-speed camera

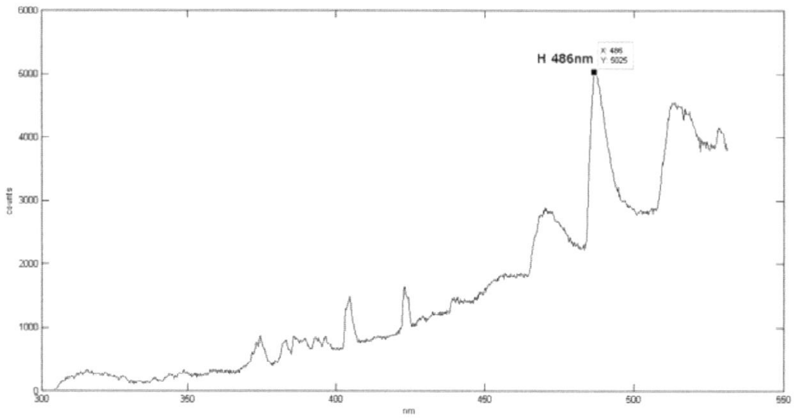

Figure 6. Emission spectra of the thermal plasma.

Phantom v7.3 capable of capturing images at 8.25 μs period, at image resolution 128×64 pixels. The camera field of view was split by an auxiliary mirror to see both the chamber exit and the expanding plasma jet in side view, both in the same frame. The problem of the frame period being nearly the same as the time needed for the plasma to expand through the almost whole camera field of view was solved by using suitable recorded frames from several shots and sorting them in the correct order. The mirror images were calibrated, leading to the length scale shown in the first frame of Fig. 5. From the distances and time scale available from the camera, the plasma front velocity was evaluated to be approximately 2 km/s. The reproducibility of the wavefronts from several shots shows that the expanding plasma is stable.

To check that the resulting plume leaving the combustion chamber exit is really thermal plasma, spectroscopic measurement was performed. The emerging light was collected by closely placed end of fused silica optical fibre 100 μm in diameter. The light was analyzed by IHR370 spectroscope. Results of representative spectroscopic analysis of the plume are in Fig. 6. In the spectra contamination peaks were found, in particular those indicating contamination by alkali metals - potassium and sodium from the electolyser used to generate the combustible gas mixture. Also seen is peak for a trace of lithium, probably from the grease used to seal distributor cones. Much more interesting was H-β line at 486 nm. This line was observable only on "succesfull" shots. When the sharp front of plasma was not observable on camera records (eg. the distributor vents did not fill evenly the chamber by the combustible mixture), the H-β line did not emerge and only alkali metals lines were present. Also the profile of the H-β line shows signs of asymmetric broadening by pressure. Presence of the H-β line was chosen as thermal ionization marker due to low possibility of coincidence with other elements lines.

4. Conclusions

Successful operation of the combustion chamber demonstrated that a promising improvement of efficiency of electricity generation by combustion of gaseous fuels is possible by increasing the degree of ionisation of the combustion product - to the level at which it is possible to employ it in no-moving-parts MHD generator. The new idea, which demonstrated this capability in a laboratory test model, is a combination of detonation combustion and the cumulative effect of waves from many directions. This paper describes design details of the tested model of 47.55 mm radius hemisphere with 168 detonation wave entrances, each of 6 × 6 mm size. While the model is still at an early stage of its development, it was already possible to demonstrate temperatures reaching 10000 K and the corresponding high degree of ionisation. In the overview spectra with low spectral resolution, also indication of lines of atomic iron, aluminium and manganese in the spectral region between 360 nm and 420 nm is registered. Measurement of these lines with high spectral resolution and application of Boltzmann plot will allow better specification of plasma temperature.

Acknowledgements

The work was possible with institutional support RVO: 61388998.

doi:10.14311/ppt.2017.1.91
Plasma Physics and Technology **4**(1):91–94, 2017

STUDY OF THE INSULATING AND ARC-QUENCHING PERFORMANCE OF CO_2 AND ITS MIXTURES: A BRIEF REVIEW

H. Zhao[a,*], W. Gu[b], X. Li[c]

[a] *School of Automation, Northwestern Polytechnical University, Xi'an, Shaanxi, 710072, China*
[b] *Sieyuan Electric Co., Ltd. Shanghai, 201108, China*
[c] *State Key Laboratory of Electrical Insulation and Power Equipment, Xi'an Jiaotong University, Shaanxi, 710049, China*
[*] hzhao@nwpu.edu.cn

Abstract. SF_6, which is widely used as an insulating or arc-quenching medium in electrical apparatus, has a strong global warming potential. Recently, CO_2 and its mixtures attract great attention as a promising alternative gas. This paper presents our calculation and measurement results on the insulating and arc-quenching capabilities of CO_2 and its mixtures, including the discharge parameters, arc characteristics, post-arc breakdown properties, and arc-quenching capabilities. On the basis of our study and other published literatures, a discussion of and perspectives on current research and future research directions are presented as well.

Keywords: insulating, arc-quenching, CO_2.

1. Introduction

SF_6 has been widely used in high voltage apparatus, due to its excellent insulation and arc-quenching performances. However, this gas is a serious greenhouse gas, and was designated a regulated gas at COP3 (3rd Framework Convention on Climate Change) in 1997. As a result, seeking an environmentally friendly gas to replace SF_6 partially or wholly becomes even more urgent and important.

CO_2 have received lots of attentions as a potential candidate for SF_6. However, the dielectric strength of CO_2, 1.93 kV/mm, is significantly lower than that of SF_6, 8.9 kV/mm. Adding a gas with higher dielectric strength into CO_2 may be an effective way to improve its performance. The authors have carried out some investigations on the insulating and arc-quenching performances of CO_2 and its mixtures, and this paper presents our typical results and discussions on the application of CO_2 mixtures for high voltage apparatus.

2. Insulating performance

2.1. Critical reduced electric field strength at room temperature

The critical reduced electric field strength $(E/N)_{cr}$, where E is the electric field strength and N is the number density of particles, was calculated by Boltzmann analysis. The Boltzmann equation was solved by means of the zero-dimensional two-term spherical harmonic approximation. The collision cross section data for the gas molecules were taken from the Plasma Data Exchange Project. The secondary and ionizing electrons after an ionization collision were assumed to have equal energy. In addition, the three-body collisions and Coulomb interactions between charged particles were removed from consideration, and the influence of excited states was also disregarded.

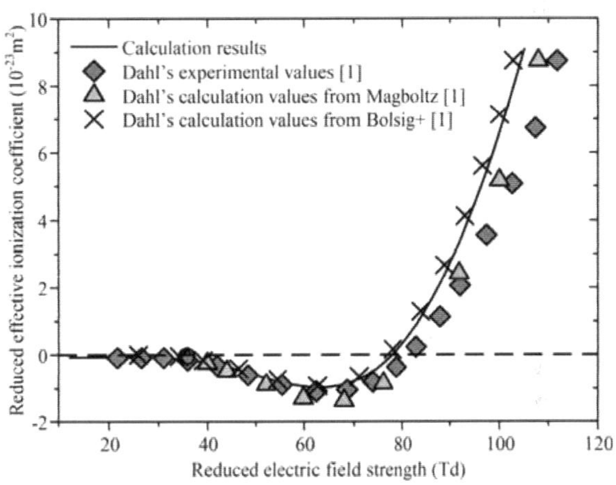

Figure 1. Comparison of $(\alpha - \eta)/N$ in CO_2 with literature data.

Fig.1 compares the calculated values of the reduced effective ionization coefficient $(\alpha - \eta)/N$ of pure CO_2 gas at room temperature with the literature data [1]. A good agreement can be found between the calculated and the literature values [1].

The calculated $(E/N)_{cr}$ of the mixtures of CO_2 with 10 typical gases are plotted in Fig. 2. For comparison, the values of $(E/N)_{cr}$ of the CO_2 mixtures with C_4F_7N and $C_5F_{10}O$ were also derived from the reported experiment value in literatures [2–4].

Among these gas mixtures, the $(E/N)_{cr}$ of CO_2-C_4F_7N possesses the largest values at same mixed ratios, followed by the results of the CO_2 mixtures with $C_5F_{10}O$, CF_3I, c-C_4F_8 and SF_6. However, because of the large molecular weight, the liquefaction

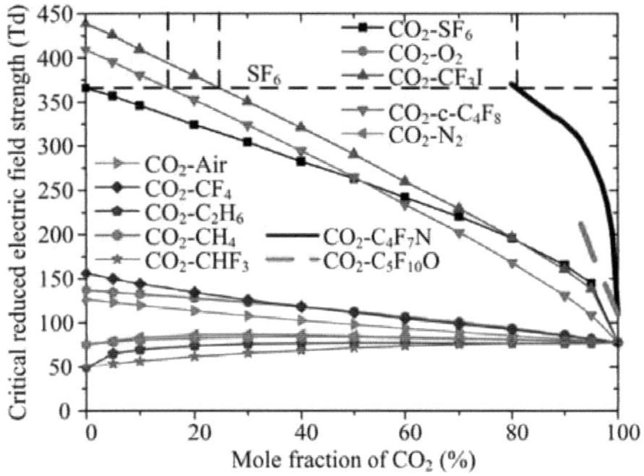

Figure 2. Comparison of $(E/N)_{cr}$ in CO_2 mixtures with different gases.

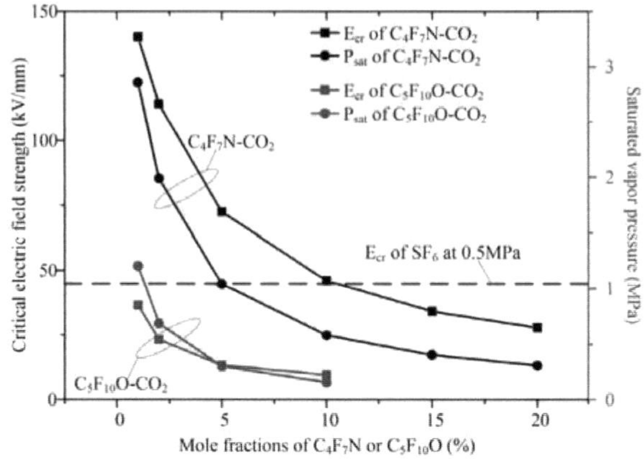

Figure 3. Saturated vapor pressure and E_{cr} characteristics of the CO_2 mixtures with C_4F_7N and $C_5F_{10}O$ at a gas temperature of −15 °C [4]

temperatures of C_4F_7N, $C_5F_{10}O$, CF_3I and c-C_4F_8 are all very high. For example, the boiling points are −4.7 °C, 26.5 °C, −22.5 °C and −8 °C of C_4F_7N, $C_5F_{10}O$, CF_3I and c-C_4F_8 at 0.1 MPa, respectively. Therefore, the analysis of the relations of saturated vapor pressure, mixed ratio and dielectric strength is necessary.

2.2. Saturated vapor pressure and dielectric strength

The saturated vapor pressures for the CO_2-C_4F_7N and CO_2-$C_5F_{10}O$ mixtures were calculated by combining the vapor-liquid equilibrium and Antoine equation in the previous paper [4] with:

$$\log P_1 = A_1 - B_1/(T + C_1) \tag{1}$$

$$\log P_2 = A_2 - B_2/(T + C_2) \tag{2}$$

$$P_1 x = P y \tag{3}$$

$$P_2(1 - x) = P(1 - y) \tag{4}$$

where P_1 and P_2 are the saturated vapor pressures of component 1 and 2, P is the saturated vapor pressure of the gas mixture, T is the liquefaction temperature of the gas mixture, A_1, B_1, C_1 and A_2, B_2, C_2 are the Antoine constants for component 1 and 2, x and y are the mole fractions for liquid phase and vapor phase of component 1 at the moment of vapor-liquid equilibrium.

Fig. 3 shows the saturated vapor pressure and E_{cr} characteristics at saturated vapor pressures of the CO_2 mixtures with C_4F_7N and $C_5F_{10}O$ [4], taking the case of −15 °C as an example. It is seen that the attainable electrical strength of CO_2-C_4F_7N is clear superior to CO_2-$C_5F_{10}O$ at saturated vapor pressures, and the maximal electrical strengths of these two mixtures are both decreased with the reduction of CO_2 concentration. Under the limitation of −15 °C, the maximum allowable operating pressure of 90 % CO_2-10 % C_4F_7N is about 0.58 MPa, and the value of E_{cr} is approximately 45.97 kV/mm, which slightly surpasses the E_{cr} of 44.76 kV/mm in pure SF_6 at 0.5 MPa. However, the dielectric strength of CO_2-$C_5F_{10}O$ is much lower than that in pure SF_6 at 0.5 MPa in the whole range. It indicates that from the view of dielectric strength, the CO_2-C_4F_7N mixture has a greater potential than the CO_2-$C_5F_{10}O$ mixture as an insulating medium.

2.3. Critical reduced electric field strength at elevated temperatures

The critical reduced electric field strengths of the CO_2 mixture with O_2 at elevated temperatures were also studied by Boltzmann analysis, in order to obtain a better understanding of the dielectric property of the hot gas and to provide the fundamental data for evaluating the dielectric recovery performance in CO_2 or its mixtures during the post-arc phase. Fig. 4 presents the $(E/N)_{cr}$ curves for different CO_2-O_2 mixtures at 0.4 MPa and elevated temperatures.

As follows from the figure, at relatively low gas temperatures, raising the mole fraction of O_2 can effectively increase the values of $(E/N)_{cr}$ in the CO_2-O_2 mixture. At gas temperatures above 3500 K, however, the $(E/N)_{cr}$ in the CO_2-O_2 mixture is reduced with the increase of the O_2 concentration. This is mainly because of the electronegativity of O_2. In addition, it is also observed that a slow increase in the $(E/N)_{cr}$ with gas temperature occurs, and this increase appears clearer in the CO_2-O_2 mixture with fewer O_2. This variation is mainly due to the dissociation of CO_2 molecules at high gas temperatures, generating CO and O_2.

The influence of gas pressure on the $(E/N)_{cr}$ in the CO_2-O_2 mixture is shown in Fig. 5. It is clear that the elevation of the gas pressure can greatly delay the variation of the $(E/N)_{cr}$ with the gas temperature to the higher temperature range.

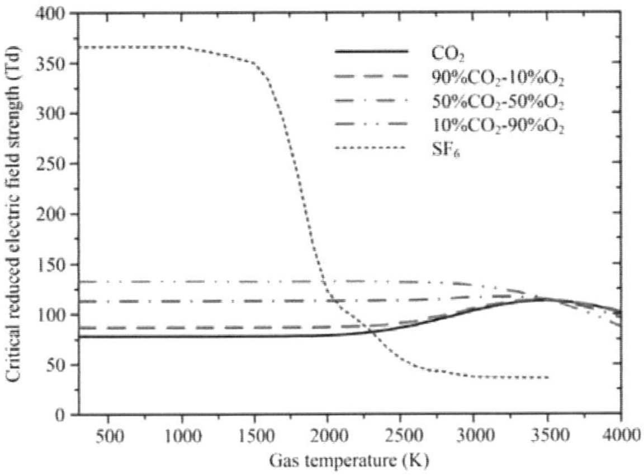

Figure 4. Critical reduced electric field strength in different CO_2-O_2 mixtures at 0.4 MPa and elevated temperatures

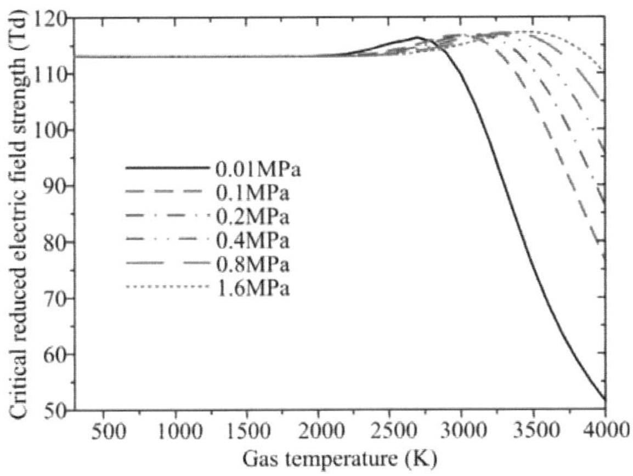

Figure 5. Critical reduced electric field strength in 50 %CO_2-50 %O_2 mixture at different gas pressures and elevated temperatures

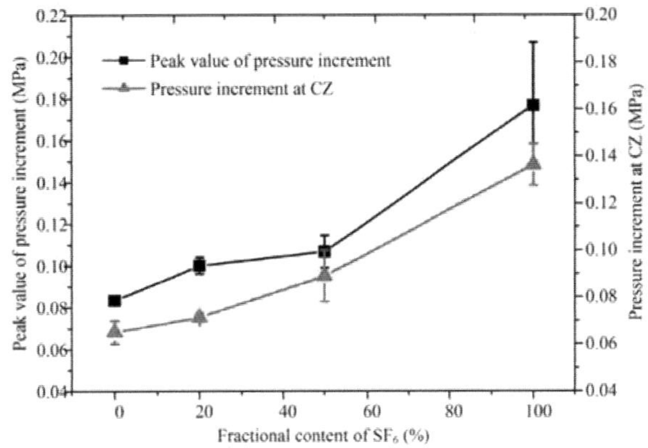

Figure 6. Peak values of the gas pressure increment and gas pressure increments at CZ in different SF_6-CO_2 mixtures at 0.6 MPa and 10 kA [5]

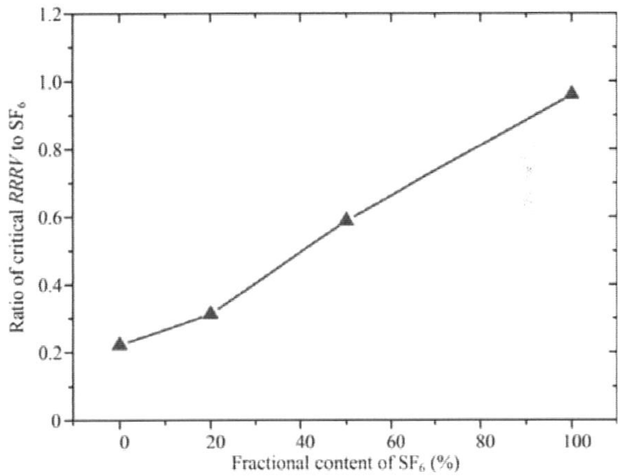

Figure 7. Ratios of critical RRRV to SF_6 in CO_2-SF_6 for different mixed ratios at 0.6 MPa and 10 kA [5]

3. Arc-quenching performance

In order to evaluating the arc-quenching performance of CO_2 and to study its arc characteristics, a series of experiments were carried out based a commercial 126 kV gas-puffer circuit breaker prototype [5].

The transient gas pressures at upstream of the throat of the nozzle were measured, and Fig. 6 gives the peak values of the gas pressure increment and the gas pressure increments at CZ in different SF_6-CO_2 mixtures at the pre-charged gas pressure of 0.6 MPa and the prospective short-circuit current of 10 kA [5]. With increasing SF_6 concentration, both ΔP_{peak} and ΔP_{CZ} increase, and non-linearly variation is also found. ΔP_{peak} and ΔP_{CZ} in CO_2 are about 47.1 % and 47.7 % of that in SF_6, respectively.

The ratios of critical values of RRRV to SF_6 in CO_2-SF_6 for different mixed ratios at the pre-charged gas pressure of 0.6 MPa and the prospective short-circuit current of 10 kA are shown in Fig. 7 [5]. The values of critical RRRV in the CO_2-SF_6 mixture for SF_6 content of 0 %, 20 % and 50 % are about 39 %, 45 % and 70 % of that of SF_6 respectively.

4. Discussions

The values of $(E/N)_{cr}$ in the CO_2 mixtures with $C_5F_{10}O$ and C_4F_7N are clear higher than that in the other mixtures without considering the liquefaction problem. By analyzing the saturated vapor pressure and E_{cr}, it is found that the dielectric strength of the CO_2-C_4F_7N mixture is much superior to the CO_2-$C_5F_{10}O$ mixture. However, the GWP value of the CO_2-$C_5F_{10}O$ mixture is much lower. More detailed and in-depth investigations are necessary. The decomposition when a discharge occurs and its influence on the insulation performance of the gas and harm to human health should be studied. It is also necessary to investigate the breakdown properties of those mixtures under different kinds of electric field, the sensitivity to metallic particles, etc.

The investigation of the interrupting capability of the CO_2-SF_6 mixture shows that large differences exist in the dielectric strength at elevated temperatures and the pressure buildup characteristics in different gases.

The matching characteristic between the gas performance and the GCB structure should be investigated. In addition, the detailed arc-quenching performances of new environment-friendly gases, including C_4F_7N and $C_5F_{10}O$, are also needed to study.

5. Conclusions

Based on the study, the following conclusions could be drawn:

1. The $(E/N)_{cr}$ of the CO_2 mixtures with C_4F_7N, $C_5F_{10}O$, CF_3I, c-C_4F_8 and SF_6 are clear higher than that of other mixtures, and that of CO_2-C_4F_7N possesses the largest values at same mixed ratios.

2. From the view of dielectric strength, the CO_2-C_4F_7N mixture has greater potential than the CO_2-$C_5F_{10}O$ mixture as an insulating medium.

3. Both the interrupting capability and the gas pressure buildup increase with the increase of the SF_6 concentration in CO_2-SF_6 mixtures, and the critical RRRV in the CO_2-SF_6 mixture for SF_6 content of 0 %, 20 % and 50 % are about 39 %, 45 % and 70 % of that of SF_6 respectively.

Acknowledgements

The work was supported by National Natural Science Foundation of China (51607143, 51577143).

References

[1] D. A. Dahl. et al. Obtaining precise electron swarm parameters from a pulsed townsend setup. *Journal of Physics D: Applied Physics*, 45(48):485201, 2012. doi:10.1088/0022-3727/45/48/485201.

[2] S. Silvant. et al. Detailed analysis of live tanks and gis circuit breakers using a new environmental friendly gas. *CIGRE*, pages A3–114, 2016.

[3] P. Simka. et al. Dielectric strength of c5 perfluoroketone. *19th International Symposium on High Voltage Engineering*, pages 23–28, 2015.

[4] X. Li. et al. Insulation performance and application of environment-friendly gases mixtures of c4f7n and c5f10o with co2. *High Voltage Engineering*, 43(3):708–714, 2017.

[5] H. Zhao. et al. Study of the arc interruption performance of sf6-co2 mixtures as a substitute for sf6. *IEEE Transactions on Dielectrics and Electrical Insulation*, 23(5):2657–2667, 2016. doi:10.1109/TDEI.2016.7736824.

doi:10.14311/ppt.2017.1.95
Plasma Physics and Technology 4(1):95–98, 2017

INVESTIGATION ON THE INFLUENCE OF GAS PRESSURE ON CO_2 ARC CHARACTERISTICS IN HIGH-VOLTAGE GAS CIRCUIT BREAKERS

Z. Guo, X. Li*, Y. Zhang, X. Guo, J. Xiong

State Key Laboratory of Electrical Insulation and Power Equipment, Xi'an Jiaotong University, Xi'an 710049, China

* xwli@mail.xjtu.edu.cn

Abstract. CO_2 is identified as a promising alternative gas of SF_6. The magnetohydrodynamics (MHD) arc model is established for a CO_2 circuit breaker. The influence of gas pressure is studied. The simulations are carried out for 0.5 MPa, 0.7 MPa and 0.9 MPa absolute filling pressure, allowing predictions of pressure and temperature distributions. The arc time constant θ and the power loss coefficient Q is extracted. The thermal interruption capability is estimated to grow with increasing filling pressure.

Keywords: CO_2, gas pressure, circuit breaker, Mayr equation.

1. Introduction

SF_6 is currently widely used in high-voltage gas circuit breakers because of its high dielectric strength and excellent arc interruption properties [1]. However, SF_6 is found to have extremely high global warming potential (approximately 23900 times that of CO_2) [2]. Among the alternative gases of SF_6, CO_2 and its mixtures are considered as the promising options and there has been some real products filled with pure CO_2 gas[3, 4].

Although the interruption performance of CO_2 is inferior to SF_6, the gap might become smaller with optimization design of CO_2 circuit breakers based on improved understanding. When CO_2 gas is used as an arc-quenching medium in high-voltage switchgear, the gas pressure may be perhaps several atmospheric pressures. In this paper, the influence of the gas pressure on the interruption performance of the CO_2 circuit breakers is investigated.

2. Simulation model

A 252 kV puffer-type SF_6 circuit breaker with a short-circuit current interrupting capacity of 63 kA is employed in this paper. The dynamic arc behaviors are simulated by adopting a 2D axisymmetric MHD arc model, which takes account of the Joule heating, the electromagnetic effect (the Lorenz force), the radiation loss and the turbulence enhanced mass, momentum and energy transplant. The MHD model is introduced in detail in our previous paper [5].

The plasma medium is assumed to satisfy the conditions of the local temperature equilibrium (LTE) and local chemical equilibrium (LCE). The nozzle ablation process is not considered in this paper. The thermodynamic and transport properties of CO_2 are taken from the study of Tanaka [6]. The net emission coefficient of CO_2 is taken from the study of Billoux [7]. The

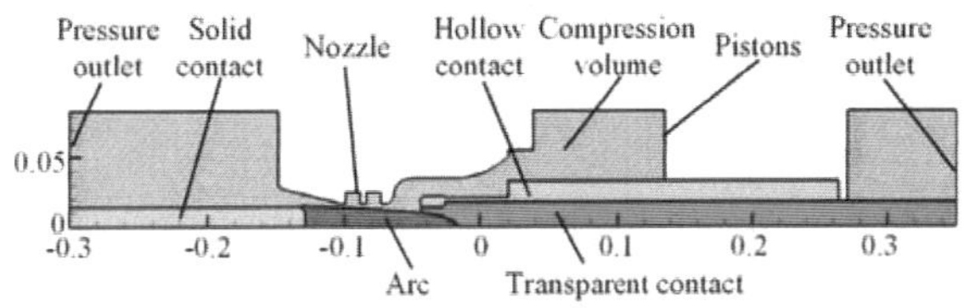

Figure 1. The simplified geometry and computational domain of the 252 kV puffer-type SF_6 circuit breakers

simplified geometry and computational domain of the circuit breakers are presented in Fig1.

The commercial computational software ANSYS FLUENT is used to model the circuit breaker and solve the governing equations. The boundary and initial conditions of the simulation model are listed as follows: (1) The circuit breakers are filled with pure CO_2 at an absolute pressure of 0.5 MPa, 0.7 MPa and 0.9 MPa; the initial temperature is set to 300 K; (2) The movement velocity of the moving contacts is piecewise linearized from the action characteristic experiment result; (3) The electric potential of the solid contact is set at zero; (4) The transparent contact hypothesis is adopted as in some papers; (5) The short-circuit current of 31.5 kA (50 Hz) rms and arcing time of 9.95 ms are adopted.

3. Temperature characteristics

The temperature distribution in the arc extinguish chamber is an important indicator of the interruption process. Fig. 2 shows the temperature distributions in the arc area at 0.5 ms before the current-zero with different filling pressures. The highest temperature is located at the exit of the compression volume. This is because that there is a stagnation zone at the exit of the compression volume where the cold gas from the compression volume gets separated into two parts. The temperature at the arc core increases with the in-

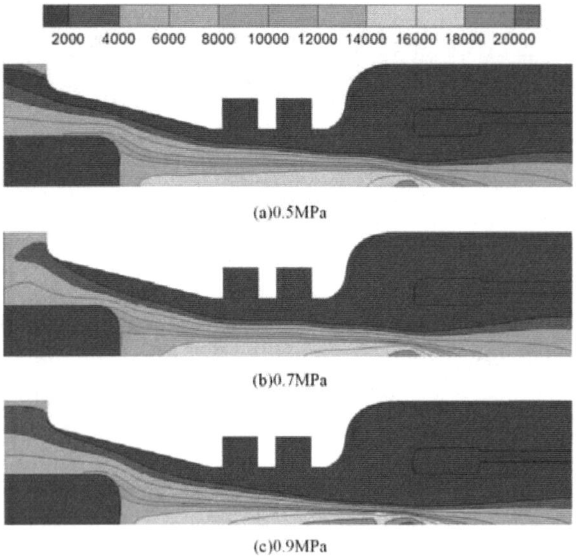

Figure 2. The temperature distributions in the arc area at 0.5 ms before the current-zero with different filling pressures. (Unit: K)

creasing filling pressure. The highest temperatures in the 0.5 MPa, 0.7 MPa and 0.9 MPa cases are 19.454 K, 19.803 K and 20.927 K.

Fig.3 describes the temperature distribution along the radial direction at the nozzle throat and at 0.5 ms before the current-zero with different filling pressures. It is clear that the temperature at the arc core becomes higher when the filling pressure is higher. However, the arc radius, defined as the position of the 4000 K isotherm in this paper as in some other papers, decreases with the increase of the filling pressure. What's more, there is an inflection point on the radial temperature profile around 8000 K in each of the three cases. Fig.4 gives the distribution of C_P along the radial direction at the nozzle throat and at 0.5 ms before the current-zero with different filling pressures. Coincidentally, the inflection points on the temperature profiles corresponds well with the three peaks on the C_P profiles for each cases. The peak value of C_P here is caused by the dissociation of CO and will have detrimental effects on arc's thermal recovery, because this effect will broaden the radiation reabsorption region and make the high temperature arc core less distinctive.

4. Pressure characteristics

Fig.5 gives the pressure distributions in the arc area at 0.5 ms before the current-zero with different filling pressures. There is a distinctive stagnation zone at the exit of the compression volume in each case, with an obvious high pressure region. In the 0.5 MPa case, an arresting shock wave occurs between the solid contact and the nozzle divergence cone. However, in the 0.7 MPa and 0.9 MPa cases, the shock wave at this

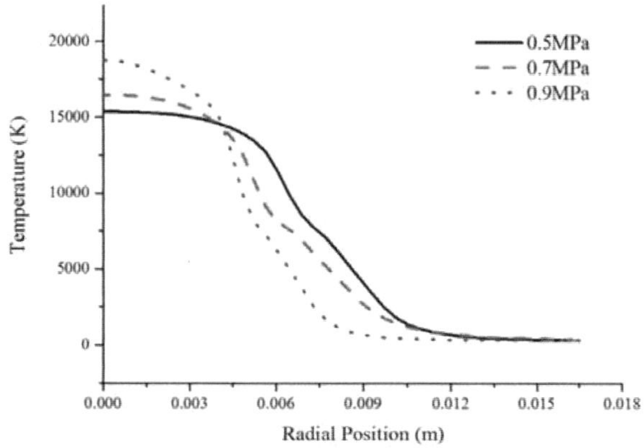

Figure 3. The temperature distribution along the radial direction at the nozzle throat (x=-0.07 m) and at 0.5 ms before the current-zero with different filling pressures.

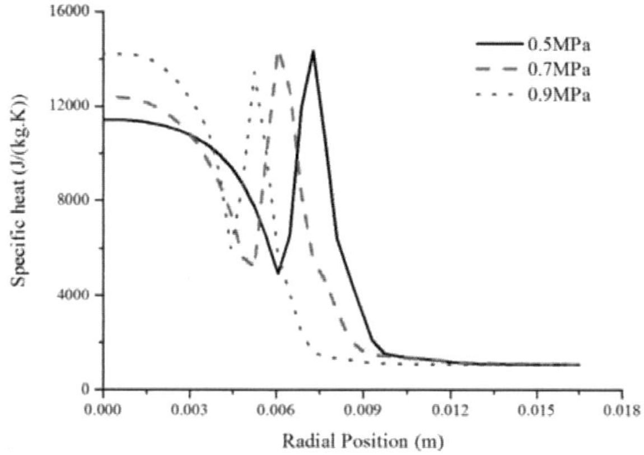

Figure 4. The distribution of the specific heat (C_P) along the radial direction at the nozzle throat (x=-0.07 m) and at 0.5 ms before the current-zero with different filling pressures.

location is much less stronger. Therefore, the filling pressure not only has an impact on the pressure values, but also affects the flow field during the interruption process. Besides, a slight shock wave also shows up inside the hollow contact in each case, which is caused by the contraction-expansion flow field structure.

5. Current zero characteristics

The arc time constant θ and the power loss coefficient Q are commonly used to evaluate the interruption capability of a circuit breaker, and generally give a believable result. Mayr equation is expressed as follow:

$$\frac{1}{g}\frac{\mathrm{d}g}{\mathrm{d}t} = \frac{1}{\theta}\left(\frac{u \cdot i}{Q} - 1\right) \tag{1}$$

Then the time constant θ and the power loss coefficient Q are expressed respectively:

$$\theta = \frac{g_1 i_2^2 - g_2 i_1^2}{i_1^2 \frac{\mathrm{d}g_2}{\mathrm{d}t} - i_2^2 \frac{\mathrm{d}g_1}{\mathrm{d}t}} \tag{2}$$

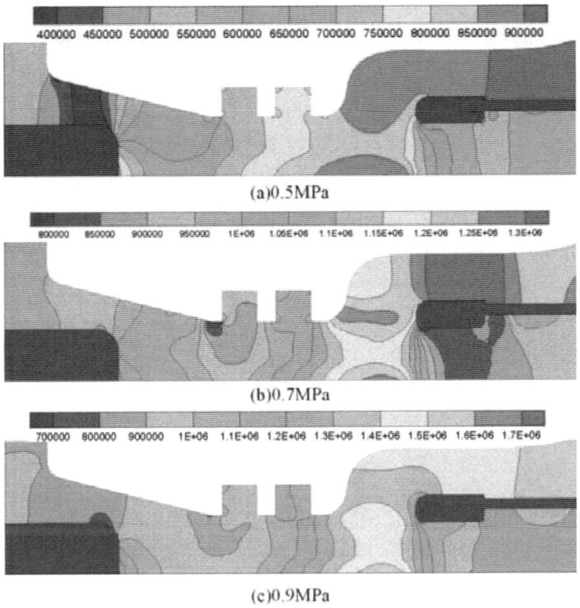

Figure 5. The pressure distributions in the arc area at 0.5 ms before the current-zero with different filling pressures. (Unit: Pa)

$$Q = \frac{i_2^2 \frac{\mathrm{d}g_1}{\mathrm{d}t} - i_1^2 \frac{\mathrm{d}g_2}{\mathrm{d}t}}{g_2 \frac{\mathrm{d}g_1}{\mathrm{d}t} - g_1 \frac{\mathrm{d}g_2}{\mathrm{d}t}} \quad (3)$$

In this paper, we assume that the time constant θ and the power loss coefficient Q keep constant within a simulation time step. Thus, the values of θ and Q at a specific time can be obtained by using two adjacent time steps as the two points required in equation (2) and (3). By this means, the variation of θ and Q over time is available. The θ and Q at the current zero point are obtained and shown in Fig.6. With the increase of the filling pressure, the time constant doesn't show much change with a value around 2.52 µs. However, the power loss coefficient seems to increase monotonically with the increasing filling pressure.

In order to qualitatively evaluate the interruption capability of the circuit breaker, the post-arc current is predicted by using Mayr equation based on the calculated data of θ and Q at the current zero point. The RRRV and relative $\mathrm{d}i/\mathrm{d}t$ are varied to find the interruption limit. The results for the 0.5 MPa case are shown in Fig.7 and Fig.8. In Fig.7, the relative $\mathrm{d}i/\mathrm{d}t$ is fixed as 1. In Fig. 8, the RRRV is set as 2 kV/µs. It is clearly seen that with increasing RRRV or relative $\mathrm{d}i/\mathrm{d}t$, the post-arc current increases effectively, and when RRRV or relative $\mathrm{d}i/\mathrm{d}t$ is beyond the critical value, the current rises sharply and the arc re-ignition happens. The critical RRRV in the 0.5 MPa, 0.7 MPa and 0.9 MPa filling pressure cases is determined to be approximately 0.63 kV/µs, 0.74 kV/µs and 0.81 kV/µs respectively. The critical relative $\mathrm{d}i/\mathrm{d}t$ in the 0.5 MPa, 0.7 MPa and 0.9 MPa filling pressure cases is determined to be approximately 31

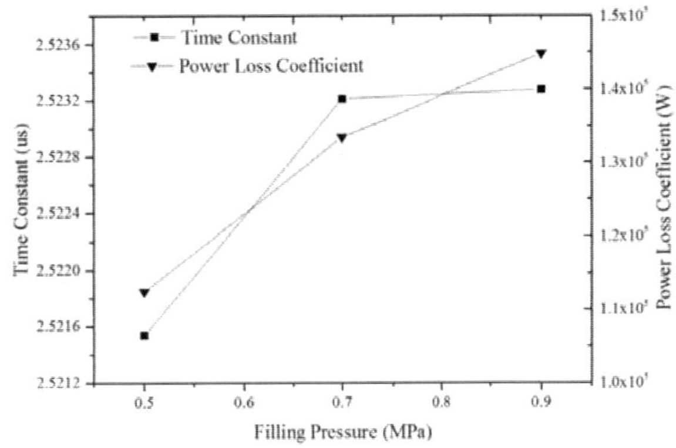

Figure 6. The time constant θ and the power loss coefficient Q at the current zero point with different filling pressures

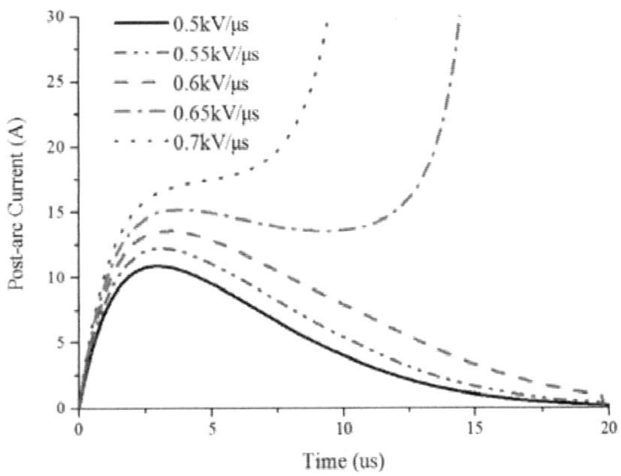

Figure 7. Variation of the post-arc current for different values of RRRV with the filling pressure of 0.7 MPa.

6. Conclusions

The performance of CO_2 gas as an interruption medium in high-voltage circuit breakers is investigated with different filling pressures. It is found that the highest temperature increases with the increasing filling pressure while the arc radius decreases with the increase of filling pressure. Besides, the arc temperature distribution has a close relationship with the specific heat (C_P) of the filling gas. The filling pressure not only has an impact on the pressure values, but also affects the flow field during the interruption process. The post-arc current is predicted by using Mayr equation based on the calculated data of θ and Q at the current zero point. For the 0.5 MPa, 0.7 MPa and 0.9 MPa filling pressure cases, the critical RRRV is determined to be 0.63 kV/µs, 0.74 kV/µs and 0.81 kV/µs and the critical relative $\mathrm{d}i/\mathrm{d}t$ is determined to be 31

Acknowledgements

The work was supported by National Natural Science Foundation of China (51607143, 51577143).

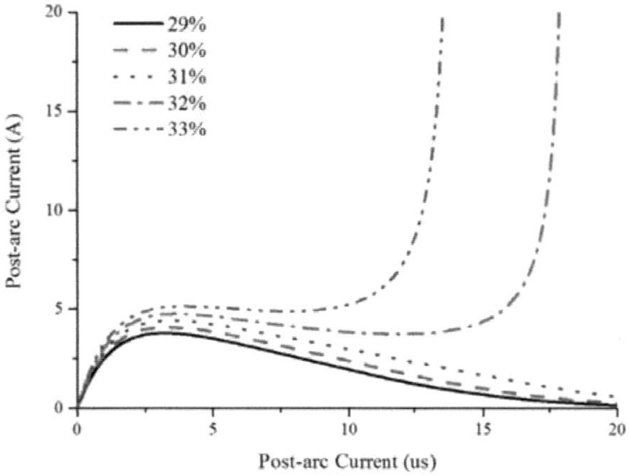

Figure 8. Variation of the post-arc current for different values of relative di/dt *with the filling pressure of 0.5 MPa.*

References

[1] J.O.L. Christophorou and R. Van Brunt. Sulfur hexafluoride and the electric power industry. *IEEE Electrical Insulation Magazine*, 13(20):5, 1997. doi:10.1109/57.620514.

[2] Y. Kieffel and F. Biquez. Sf6 alternative development for high voltage switchgears. *IEEE Electrical Insulation Conference*, pages 379–383, 2015.

[3] T. Uchii, T. Shinkai, and K. Suzuki. Thermal interruption capability of carbon dioxide in a puffer-type circuit breaker utilizing polymer ablation. *Transmission and Distribution Conference and Exhibition: Asia Pacific. IEEE/PES*, pages 1750–1754, 2002. doi:10.1109/TDC.2002.1177719.

[4] M. Seeger. Perspectives on research on high voltage gas circuit breakers. *Plasma Chemistry and Plasma Processing*, 35(3):527–541, 2015. doi:10.1007/s11090-014-9595-4.

[5] X. Jiang, X. Li, H. Zhao, and S. Jia. Analysis of the dielectric breakdown characteristics for a 252 kv gas circuit breaker. *Power Delivery IEEE Transactions on*, 28(3):1592–1599, 2013. doi:10.1109/TPWRD.2013.2254137.

[6] Y. Tanaka, N. Yamachi, S. Matsumoto, and M. Shibuya. Thermodynamic and transport properties of CO_2, CO_2–O_2, and CO_2–H_2 mixtures at temperatures of 300 to 30,000 K and pressures of 0.1 to 10 MPa. *Electrical Engineering in Japan*, 163(4):18–29, 2010. doi:10.1002/eej.20467.

[7] T. Billoux, Y. Cressault, V.F. Boretskij, A.N. Veklich, and A. Gleizes. Net emission coefficient of CO_2–Cu thermal plasmas: role of copper and molecules. *12th High-Tech Plasma Processes Conference*, page 2027, 2012. doi:10.1088/1742-6596/406/1/012027.

doi:10.14311/ppt.2017.1.99
Plasma Physics and Technology **4**(1):99–103, 2017

EVOLUTION OF THE CATHODE SPOT DISTRIBUTION IN AN AXIAL MAGNETIC FIELD CONTROLLED VACUUM ARC AT LONG CONTACT GAP

B. Tezenas du Montcel[a,b], P. Chapelle[a,*], A. Jardy[a], C. Creusot[b]

[a] *Institut Jean Lamour - UMR 7198 CNRS/Université de Lorraine, Laboratoire d'Excellence DAMAS, CS 50840, 54011 Nancy Cedex, France*

[b] *SuperGrid Institute SAS, 130 rue Léon Blum, BP1321, 69611 Villeurbanne, France*

[*] pierre.chapelle@univ-lorraine.fr

Abstract. The distribution of cathode spots in a CuCr25 vacuum arc controlled by an axial magnetic field and ignited on the lateral surface of the cathode is investigated for long gap distances, from the processing of high-speed video images. The processing method includes also estimating the current carried by a single spot and reconstructing the distribution of the current density at the cathode. Various distributions depending partly on the arc current are described.

Keywords: vacuum arc, axial magnetic field, long gap, cathode spot distribution, experimental study.

1. Introduction

Vacuum is one of the most frequently extinguishing medium for medium voltage circuit breakers. At relatively low current or under an axial magnetic field (AMF), a vacuum arc (VA) appears diffuse. The plasma is provided at the cathode by tiny, mobile cathode spots (CS) [1]. CS motion is influenced by various factors, such as surface state, magnetic field and the plasma density near the cathode [2]. The distribution of CSs over the cathode surface has obviously a strong impact on the plasma parameters [3].

Using a numerical processing method of cathode spot images [4], two groups of authors succeeded reconstructing the position of CSs over the cathode surface, from which they have studied the distribution of current density at the cathode [5, 6] and the extension of the area occupied by CSs [2, 7]. These experiments were conducted between static contacts. The arc was triggered thank to an auxiliary electrode inserted at the center of the cathode. Only short contact gaps ranging from 4 to 6 mm in [2, 5, 7] or equal to 11 mm in [6] were considered, with maximal current in 7–25 kA range.

Current efforts to enhance the vacuum circuit breaker voltage range capability impose to employ higher contact gaps (more than 10 mm). However, VA behavior under such conditions is still not fully understood. In this paper, based on a similar method to [4], we investigate the distributions of cathode spots and current density at the cathode in AMF controlled VAs generated between static contacts spaced 20 mm or 30 mm apart, and triggered with an auxiliary electrode situated near the cathode lateral surface. Different evolutions of the CS distribution, mainly depending on the VA current, are described and analyzed.

2. Experimental

2.1. Experimental setup

The experimental setup, illustrated in Figure 1, is composed of a vacuum chamber (10^{-5}–10^{-7} Pa) inside which two fixed 70 mm diameter copper cylindrical contacts surmounted by CuCr25 wafers are mounted. Sixteen turn Helmholtz coils (HC) disposed outside the chamber are used to generate an AMF. As it can be seen in Figure 2, the radial profile of the AMF has a saddle shape with values of 9.6–17.1 mT/kA inside the contact gap region. On the contrary, the time delay between the AMF intensity and the coil current has a bell-shaped radial profile, with values ranging between 3.4 ms and 5.4 ms inside the contact gap region. VA is triggered with the help of an auxiliary electrode situated close to the lateral surface of the cathode and connected to a microsecond High-Voltage (HV) power supply. VA and HC are powered from two independent 50 Hz alternators. A phase shift of 3.4 ms is imposed between the coil current and the arc current to synchronize the AMF and arc current peaks at the edge of the contacts. VAs are filmed with a high-speed camera Photron SA5 with an exposure time of 369 ns and a frame rate in 15000–20000 fps.

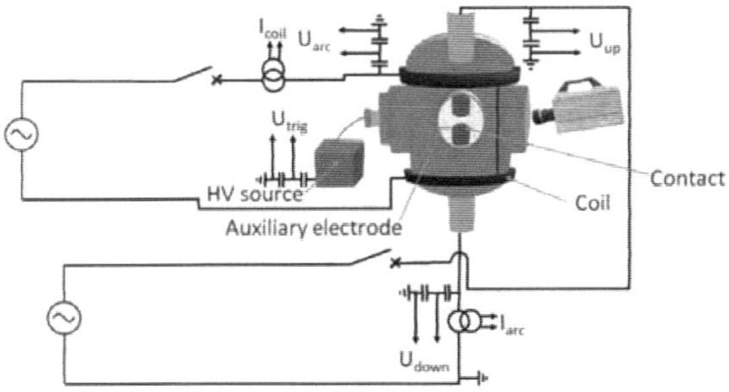

Figure 1. Experimental setup.

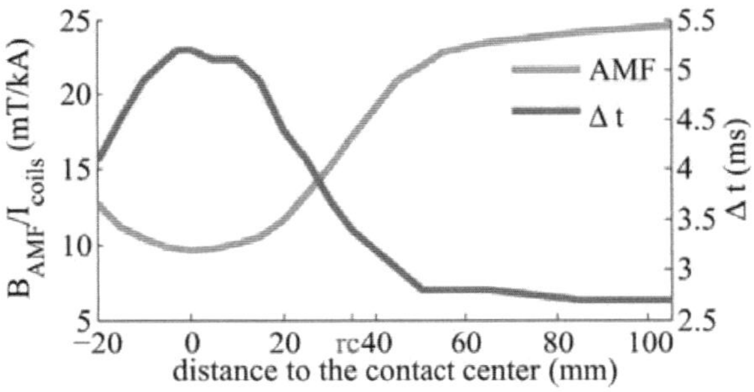

Figure 2. Radial profiles of the axial magnetic field intensity and the time delay (Δt) between B_{AMF} and I_{coil} measured at mid distance between the two contacts.

2.2. Operating parameters

Table 1 shows the operating parameters used during the eleven experiments that are reported in the present paper. Arc current is in 4.9–43.3 kA. All experiments were realized with the AMF oriented towards the cathode, except experiment 7, during which the AMF was oriented towards the anode. The contact gap was either 20 mm or 30 mm.

Experiment	L_{arc} mm	I_{arc} kA	T ms	B_{AMF} mT
1	20	4.9	6.2	−143
2	20	17.2	13.7	−143
3	20	25.7	11.8	−143
4	20	26.7	9.9	−143
5	20	31.1	10.8	−143
6	20	36.3	11.7	−143
7	20	43.3	10.9	269
8	30	8.9	10.9	−145
9	30	11.7	10.9	−144
10	30	18.0	10.9	−143
11	30	24.9	10.9	−141

Table 1. Gap distance, arc current, arc duration and AMF magnitude during each experiment.

2.3. Numerical treatment of cathode spot images

CS distribution was studied with the help of a numerical treatment of cathode spot images developed within the Octave software and inspired by the method proposed by Afanas'ev *et al.* [4]:

1. To detect the CSs, the iso-contours of light intensity on the cathode spot image were calculated with a native function of Octave. Next, these contours were selected according to their intensity, in order to eliminate the darkest contours, as well as according to their size, in order to reject too long and too small contours which cannot surround a single spot. Then, the iso-contour density was estimated by calculating first the centroid position of each contour. When the number of centroids on a pixel reaches a threshold value, this pixel was marked. Finally, spot position was taken as the centroid of connexe area of marked pixels.

2. The image deformation induced by the inclination of the camera was corrected using an OCTAVE native projective transformation function. This function must be supplied with the initial and final positions of four points. For this purpose, we have used the picture of a grid pattern disposed on the cathode surface, which was taken with the same inclination as that during the experiment.

3. From the CS map, the distribution of the current density at the cathode was reconstructed by partitionning the cathode surface into Voronoi cells, each cell enclosing a single spot. The current density in each cell was considered to be inversely proportional to the cell area.

3. Results

3.1. Cathode spots distribution

Because of the lateral trigger system, CSs just after the VA ignition are located on the lateral surface of the cathode. The action of the AMF results then in a very fast retrograde motion of rotation of those spots around the cathode [8]. Eventually, the CSs reach the front surface and spread over it. The temporal evolution of the CS distribution on the cathode surface was investigated using the image processing method described in section 2.3. As shown in Figure 3, various types of evolution of the CS distribution were observed depending partly on the arc current.

1. At low current (up to 11.7 kA for the present conditions, i.e. in Experiments 1, 8 and 9), only a small part of the circumferential edge of the cathode is initially occupied by the cathode spots (Figure 3.a). Subsequently, the occupied area grows in size until the current peak (at around 6 ms). The growth seems to be roughly isotropic as in [2]. The spots form a compact group, and they never manage to fill the whole front surface of the cathode.

2. At moderate current (between 17.2 kA and 31.1 kA for the present conditions, i.e. in Experiments 2 to 5, 10 and 11), the first CSs appearing on the front surface form thin separate strips covering large sections of the cathode circumferential edge (Figures 3.b-c). Next, with the increase of the arc current, the regions occupied by CSs grow along the edge of the cathode, as well as toward the center of the cathode. Eventually, at $t = 4$ ms, the circumference of the cathode is totally covered by the spots, but an empty region is still enclosed within the CS distribution. Then, two types of evolution may be distinguished according to the centering of the empty region with respect to the axis of symmetry of the cathode at $t = 4$ ms:

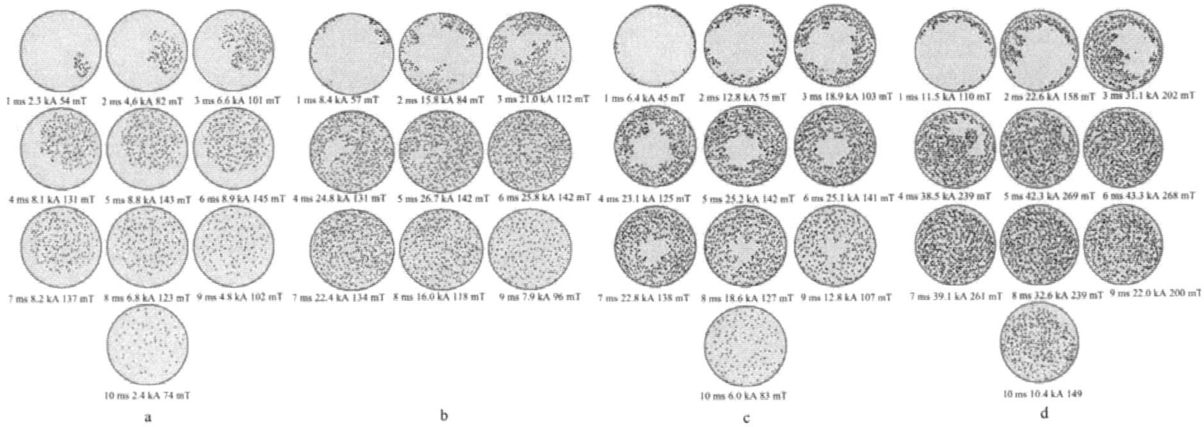

Figure 3. Temporal evolution of the distribution of cathode spots during Experiments (a) 8, (b) 4, (c) 11 and (d) 7.

a. In Experiments 3 and 4, as illustrated in Figure 3.b, the empty region initially relatively decentered and is eventually totally filled by the spots a delay after the current peak.

b. In Experiments 2, 5 10 and 11, as illustrated in Figure 3.c, the empty region is initially relatively centered, and although its size decreases until the current peak; it remains present until the arc extinction.

3. At high current (above 36.3 kA for the present conditions, i.e. in Experiments 6 and 7), the CS distribution evolves similarly to that at moderate current (Figure 3.d). Yet, at the current peak, the top surface of the cathode is always entirely covered by spots. Moreover, the CS distribution seems to reveal an alignment of the cathode spots along spiral arms originating from the center of the cathode.

Note that in all cases, during the decrease of the arc current, the disappearance of CSs seems to happen uniformly over the cathode surface.

3.2. Cathode spot average current

From the data provided by the spot detection procedure, it is possible to determine the average spot current by calculating the ratio of the total arc current to the number of detected spots. We have excluded from our analysis images where a significant proportion of spots could not be detected by our algorithm, because they were located on the lateral surface of the cathode, as a result of either the arc ignition or the ejection of spots from the front surface. Based on arc images taken 1 ms before the arc extinction, an average spot current value of 36.5±2.5 A was obtained for all experiments. This value is similar to the spot current measured by Yu *et al.* [9] for nanocrystaline CuCr25 contact material.

3.3. Radial profile of the cathode current density

The measured spatial distributions of the cathode spots have been used to reconstruct the distributions

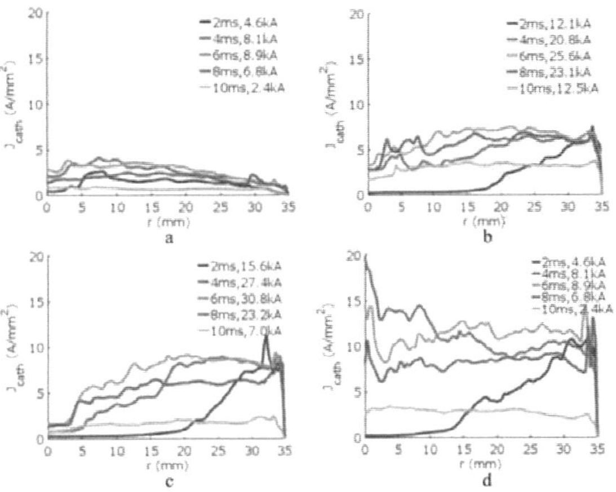

Figure 4. Radial profiles of current density during Experiments (a) 8, (b) 4, (c) 11, (d) 7, each 2 ms.

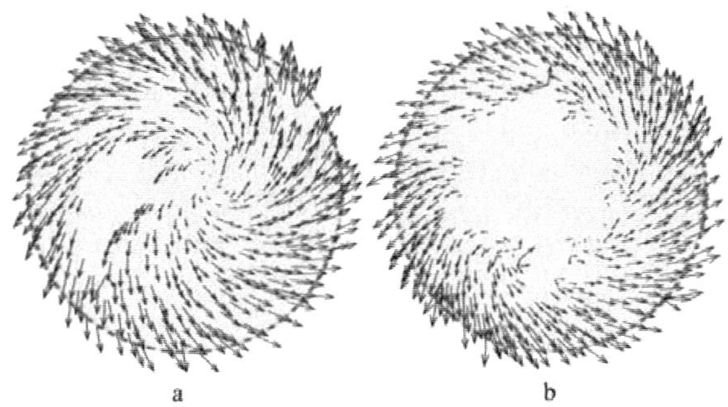

Figure 5. Calculated direction of retrograde motion with Robson drift of cathode spots during Experiments (a) 4 and (b) 11 at 4 ms.

of the current density at the cathode surface, following the method described in Section 2.3. Azimuthally averaged radial profiles of the cathode current density corresponding to the spatial distributions of spots presented in Figure 3 are shown in Figure 4.

1. At low current (Figure 4.a), the current density is approximately uniform up to 20 mm, and then decreases for higher radial positions. This decrease

reflects that the circumference of the cathode is partly non occupied by the cathode spots. After the current peak, the current density decreases approximately in uniform manner, confirming the previous comment made on Figure 3.

2. At higher currents (Figure 4.b-d), the radial profiles of the current density before the current peak are markedly different from that at low current. Due to the initial presence of spots over a large portion of the edge of the cathode, one observes a peak of current density at the cathode edge followed by a sharp fall towards the center of the cathode. As the current rises, the front of current density moves towards the center of the cathode. Nevertheless the current density does not exceed its maximal value reached at $t = 2\,\mathrm{ms}$. At the current peak, when the cathode is fully occupied, the current density is roughly uniform. After the current peak, the current density tends again to decrease uniformly over the surface.

4. Discussion

Figures 3.b-c suggest that the partial or total occupation of the cathode surface by CSs at moderate currents is related to the observed off-centered or centered position of the CS empty region, this characteristic being itself linked to the initial repartition of CSs along the cathode circumference. In Figure 5, based on the measured CS distribution, we have calculated the direction of motion of CSs, following a similar approach to that used in [10]. The plotted vectors represent the retrograde direction of motion associated to the self-generated azimuthal magnetic field, taking into consideration the Robson angular deflection of the motion caused by the AMF [11]. The deflection angle was defined as $\varphi_r = \eta \tan^{-1} \frac{B_{AMF}}{B_t}$ where $\eta \sim 0.8$ on CuCr30 cathode[10]. When the empty region is off-centered (Figure 5.a), the CS motion direction clearly contributes to the filling of the empty region. On the contrary, when the empty region is relatively centered (Figure 5.b), the CS motion direction points only little towards the empty region, which may explain the non-disappearance of this region.

5. Conclusion

The distributions of cathode spots and current density at the cathode in a long gap vacuum arc controlled with an external axial magnetic and ignited on the lateral surface of the cathode have been investigated.

- Various types of CS distribution were observed depending partly on the current. At low currents, CSs form a compact group which does not occupy completely the cathode surface, whereas at high currents the whole cathode surface is covered by CSs. At moderate currents, a CS empty region may subsist at the current peak, which may have its origin in the initial CS distribution along the circumference of the cathode following the arc ignition on the lateral surface of the cathode.
- The contact gap in the range considered in this paper seems to have no impact on the CS distribution.
- A value of the average current emitted by a cathode spot of $36.5 \pm 2.5\,\mathrm{A}$ has been determined.
- In regions occupied by CSs, the current density at the cathode seems roughly uniform.

Acknowledgements

This project was supported by the French Government in the framework of the "Investments for the Future" programme under the project reference (ANE-ITE-002-01).

References

[1] P. Picot. Vacuum switching. Technical Report 198, Schneider Electric, 2000.

[2] X. Song, Z. Shi, C. Liu, S. Jia, and L. Wang. Influence of AMF on the expansion speed of cathode spots in high-current triggered vacuum arc. *IEEE Trans. Plasma Sci.*, 41(8):2061–2067, 2013. doi:10.1109/TPS.2013.2248759.

[3] L. Wang, S. Jia, Z. Shi, and M. Rong. Numerical simulation of vacuum arc under different axial magnetic fields. *J. Phys. D: Appl. Phys.*, 38(7):1034–1041, 2005. doi:10.1088/0022-3727/38/7/011.

[4] V.P. Afanas'ev, A.M. Chaly, A.A. Logatchev, S.M. Shkol'nik, and K.K. Zabello. Computer-aided reconstruction of the cathode images obtained by high speed in high photography of high current vacuum arcs. *IEEE Trans. Plasma Sci.*, 29(5):695–699, 2001. doi:10.1109/27.964456.

[5] A.M. Chaly, A.A. Logatchev, K.K. Zabello, and S.M. Shkol'nik. High-current vacuum arc appearance in nonhomogeneous axial magnetic field. *IEEE Trans. Plasma Sci.*, 31(5):884–889, 2003. doi:10.1109/TPS.2003.818414.

[6] Z. Shi, S. Jia, X. Song, Z. Liu, H. Dong, and L. Wang. The influence of axial magnetic field distribution on high-current vacuum arc. *IEEE Trans. Plasma Sci.*, 37(8):1446–1451, 2009. doi:10.1109/TPS.2009.2019097.

[7] A.M. Chaly, K.K. Logatchev, A.A. Zabello, and S.M. Shkol'nik. High-current vacuum arc in a strong axial magnetic field. *IEEE Trans. Plasma Sci.*, 35(4):939–945, 2007. doi:10.1109/TPS.2007.901975.

[8] B. Jüttner and I. Kleberg. The retrograde motion of arc cathode spot in vacuum. *J. Phys. D: Appl. Phys.*, 33(12):2025–2036, 2000. doi:10.1088/0022-3727/33/16/315.

[9] L. Yu, J. Wang, Y. Geng, G. Kong, and .Z Liu. High-current vacuum arc phenomena of nanocrystalline CuCr25 contact material. *IEEE Trans. Plasma Sci.*, 39(6):1418–1426, 2011. doi:10.1109/DEIV.2010.5625828.

[10] S. Jia, X. Song, Z. Shi, L. Wang, and X. Huo. Investigations on the motion of high-current vacuum-arc cathode spots under a magnetic field. *IEEE Trans. Plasma Sci.*, 39(6):1344–1348, 2011. doi:10.1109/TPS.2011.2127492.

[11] A.E. Robson. The motion of low-pressure vacuum arc in strong magnetic field. *J. Phys. D: Appl. Phys.*, 11(13):1917–1923, 1978. doi:10.1088/0022-3727/11/13/014.

doi:10.14311/ppt.2017.1.104
Plasma Physics and Technology **4**(1):104–107, 2017

FEEDBACKS IN NON-EQUILIBRIUM OXIDATIVE PLASMA REACTING WITH POLYETHYLENE

Kadnikov D. V. *, Ovtsyn A. A., S.A. Shibaev S. A., Smirnov S. A.

Department of technology devices and materials for electronic equipment, Ivanovo State University of Chemistry and Technology, Sheremetevsky Avenue 7, 153000, Ivanovo, Russia

* dkadnikov@rocketmail.com

Abstract. The experimental results on studies of loading effect at the action of a glow DC discharge in oxygen and air on polyethylene film are presented. At the increase in a polymer amount under a treatment the specific rates of etching and CO_2, CO, H_2O, H_2 product formation rates are decreased. The gas temperature is changed slightly. The reduced electric field strength is increased in oxygen plasma and decreased in air plasma.

Keywords: air plasma, oxygen plasma, loading effect, gas temperature, reduced electric field strength, polyethylene.

1. Introduction

It is known that at plasma-chemical etching and modification of polymer materials the rates of plasma-chemical destruction are decreased at the increase in polymer amount. Given effect is usually named with loading effect. For the explanation of such effect it is assumed that the concentration of active species is decreased due to the increase in their consumption rates with the growth of material surface being treated at the fixed formation rates [1, 2]. But this explanation doesn't take into consideration that appearance in a gas phase of etching products has to influence the plasma-physical parameters and, as the result, has to influence the formation rates of active species. As far as we know nobody studied the given problem. For this reason the given article aims to study the effect of products of plasma-chemical destruction of polyethylene film (PE) on the parameters of DC glow discharges in oxygen and air.

2. Experemental

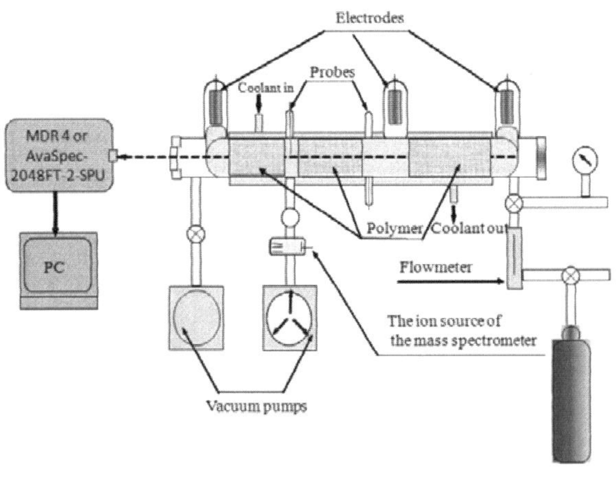

Figure 1. The sketch of experimenal setup.

A DC glow discharge was excited in cylindrical flow reactor from glass. The diameter of inner reactor part was 3 cm. The pressure range was 50-300 Pa and discharge current was 20 - 110 mA (Figure 1). The linear velocity of gas flow was changed in the range of 10 - 50 cm/s at standard conditions. Samples of industrial films of PE of high pressure (RF State Standard 10354-82) of 110 µm thickness were placed in a form of several cylinders on the thermostatically controlled reactor wall (T_W=310 ± 5 K) in a area of positive column. The polymer surface temperature was measured with the vitrified copper-constantan thermocouple. The total length of generatrix of several polymer cylinders was changed in the range of 1.5 - 13.5 cm. The position of extreme sample did not change and the part of inner reactor surface covered with polymer was increased from 4.7 to 42 %. The polymer mass loss was determined with the sample periodic weighting on analytic balance. The rate of oxygen consumption at reactions with polymer and the evolution rates of gaseous products of PE destruction were measured with mass-spectrometers MX-7304 and IPDO - 2A. The surface composition was analyzed by ATR - FTIR method (Avatar - 360). The positive ions flux density on the discharge wall (flat probe), electric field strength, E, concentration of $O(^3P)$ atoms, and $O_2(b^1\Sigma_g^+)$ metastable molecules (emission spectroscopy, MDR - 24 and AvaSpec - 2048FT - 2 - SPU monochromators) was determined by the methods described in [3]. The rotational temperature was determined on distribution of relative intensities for $O_2(b^1\Sigma_g^+, v{=}0 \rightarrow X^3\Sigma_g^-, v{=}0)$ band for O_2 plasma, and for N_2 $(C^3\Pi_u^- \rightarrow B^3\Pi_u^+)$ band for air plasma. It was supposed that rotational temperature is equal to gas one. Using experimentally measured discharge parameters the electron energy function distribution (EEDF) was calculated by the methods described in studies [4–6]. At calculations the electron collisions both with molecules of plasma-forming gases and with gaseous products of etching were taken into consideration. For O_2, CO_2, CO, H_2O, H_2, N_2, NO molecules the cross-section sets of studies [7–13] were used, respectively.

3. Results and discussion

The experiments showed that the stationary rates of films mass loss and evolution of products of plasma-chemical etching to gas phase are established after 5 min of plasma processing. These rates do not change until the 30 - 40% of initial polymer mass was removed [14]. After that time the process rate becomes slower. All data presented below refer to the stationary step of etching. Etching rate was increased with discharge current, linear velocity of gas flow and depended slightly on the gas pressure at other conditions being equal. The film which was situated at gas input had higher etching rate whereas samples situated lower on a gas flow were etched slower by 5 - 15%. The process apparent activation energy of plasma-chemical destruction of PE film determined on the temperature dependence of etching rate was 39±2 and 40 ±3 kJ/mol for oxygen plasma and air plasma, respectively. The activation energy did not depend on the reactor loading degree.

Figure 2 shows the average specific rates of mass loss of PE film (W_{et}) at the treatment in air and oxygen plasmas. These rates depend on the loading degree (ratio of polymer surface being treated to total inner surface of reactor in plasma zone). At any discharge parameters the gaseous products of etching were the same. It was CO_2, CO, H_2O and H_2 molecules. The oxygen only was consumed from gas phase (Figure 3). The given finding is also confirmed that mass loss rates determined on products rate formation are equal to mass loss rates after deduction of oxygen consume. The lack of carbon-containing compounds as a products shows that the destruction of main polymer chain without formation of gaseous products as a primary process, if it take place proceeds with very slowly.

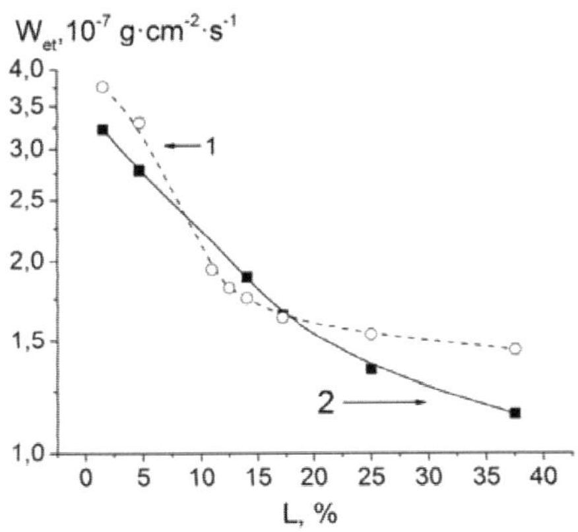

Figure 2. The dependence of etching rate of PE film in oxygen (1) and air (2) plasmas on the degree of reactor loading. The discharge current is 80 mA, the pressure is 100 Pa, the film temperature is 310±5 K.

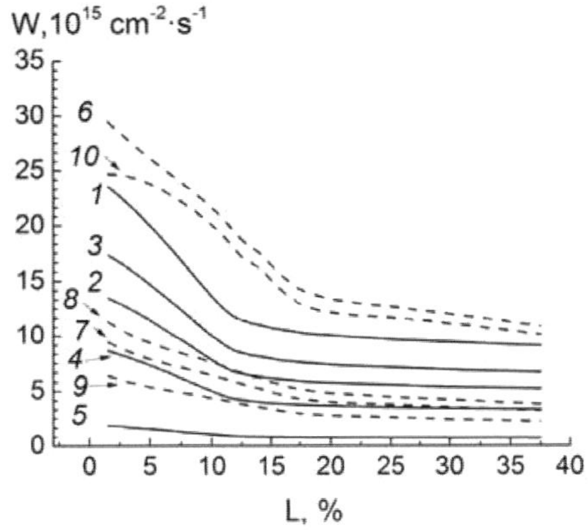

Figure 3. Averaged on sample square the rates of oxygen consumption (1, 2), and evolution rates of CO_2 (2, 7), H_2O (3, 8), H_2 (4, 9) and CO (5, 10) vs reactor loading degree at PE film etching in oxygen (1-5) and air (6-10) plasmas. The pressure is 100 Pa. The discharge current is 80 mA.

That is the destruction of oxidized molecules occurs along polymer chain on constituent their atoms with the different degree of oxidation. The total mole fraction of products in a gas phase reaches of 0.14 for oxygen plasma and 0.12 for plasma of air at the increase in the degree of reactor loading up to 37%. It is important to point out that the gaseous products yield per one consuming oxygen molecule was she same at the increase in amount of polymer treated though the yields are changed at the pressure growth for oxygen plasma. But it doesn't observe for air plasma.

The reduced electric strength measured experimentally for air plasma is decreased slightly at a low discharge current (Figure 4) but it is increased at higher current with PE surface growth. In oxygen plasma the electric field strength growth with reactor

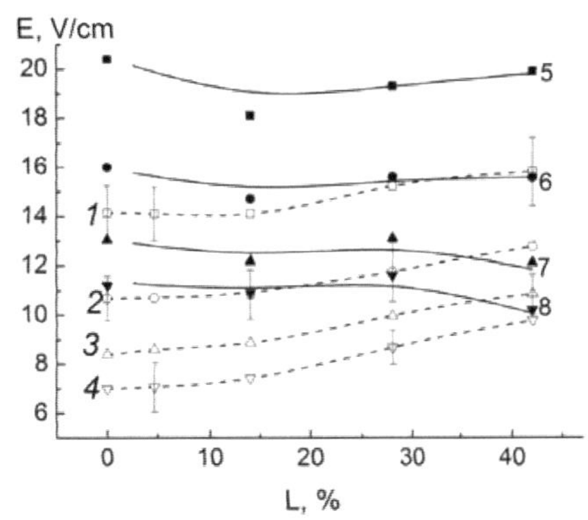

Figure 4. The dependence of electric field strength on reactor loading degree for oxygen plasma (dotted lines) and air (solid lines) at the discharge currents of 20 (1, 5), 50 (2, 6), 80 (3, 7), and 110 mA (4, 8). The gas pressure is 100 Pa.

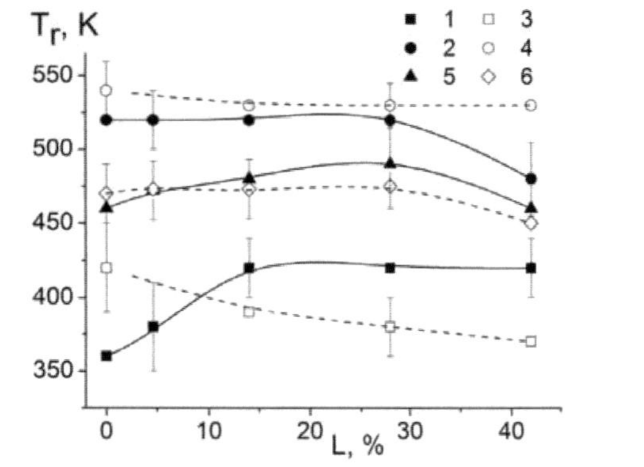

Figure 5. The dependence of rotational temperature of $O_2(b^1\Sigma_g^+)$ in oxygen plasma (1, 2, 5) and rotational temperature of N_2 $(C^3\Pi_u^-)$ in air plasma (3, 4, 6) on loading degree. The discharge currents are 20 (1, 3), 110 (2, 4) and 80 mA (5,6). The pressure is 100 Pa.

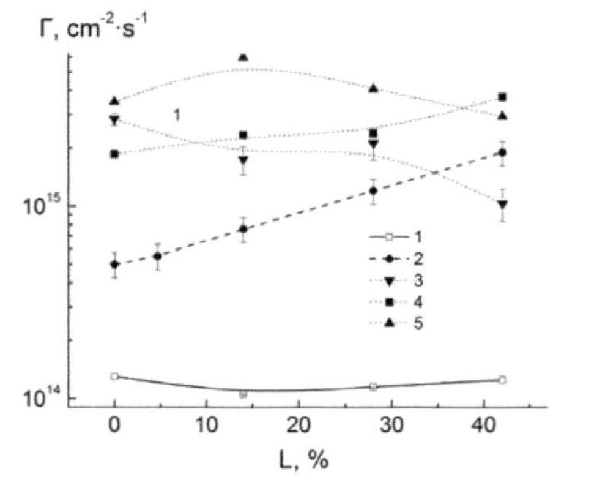

Figure 7. Flux density of positive ions (1, 2), UV - quanta (3), oxygen atoms (4), vibrational excited molecules (5) in air plasma (2-5) and in oxygen plasma (1) at different loading degree. The pressure is 100 Pa. The discharge current is 80 mA.

loading degree with polymer (Figure 4). The similar changes can be explained by changes in balance of formation and loss of charges.

The increase in reactor loading degree with polymer results in a slight changes in rotational temperatures of N_2 $(C^3\Pi_u^-)$ and $O_2(b^1\Sigma_g^+)$ in plasmas of air and oxygen, respectively (Figure 5).

Figure 6 shows the reduced electric field strength, E/N, at the discharges in oxygen and air depending on loading degree. The E/N values are decreased by 15-20% in air plasma. At the same time the E/N growths with the loading degree in oxygen discharge. Using experimental values of E/N and mole fraction of main components of gas phase (N_2, O_2, H_2O, NO, Ar, CO, CO_2, H_2) the EEDF calculation was carried out. Applying these EEDF the composition of plasma species was evaluated by means of common solution of equations of chemical kinetics and vibrational kinetics for ground states of molecules. The last kinetics includes the processes of single-quantum V-V and V-T exchange. It allows estimating the fluxes of active species of oxygen and air plasmas onto polymer surface (Figure 7).

It is necessary to point out the importance taking into consideration the second kind collisions of electrons with the vibrational excited nitrogen molecules and gaseous products of PE destruction. Calculations showed that the second kind collisions influence slightly the kinetic characteristic of electrons determining the EEDF "body". Thus, the difference of electron drift velocity does not exceed of 2%. The rate constants for processes with low threshold energies, for example, excitation of vibrational levels of

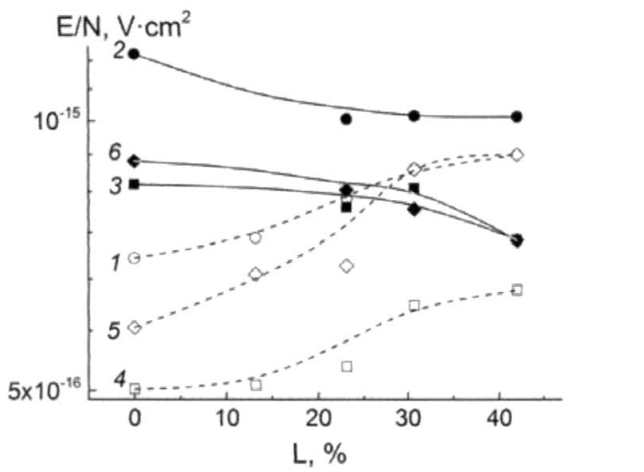

Figure 6. The dependence of reduced electric field strength on loading degree at the discharge in oxygen (1, 4, 5) and air (2, 3, 6). The discharge currents are 20 (1, 2), 110 (3, 4) and 80 mA (5, 6). The gas pressure is 100 Pa.

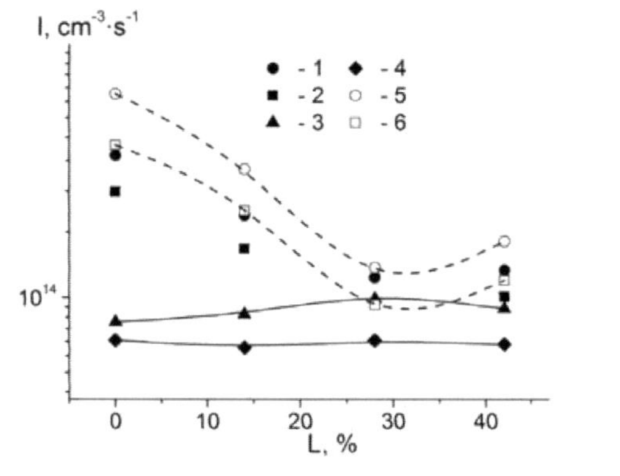

Figure 8. The dependence of emission intensities of lines for transitions $O(3p^3P \rightarrow 3s^3S)$ (1, 5) and $O(3p^5P \rightarrow 3s^5S)$ (2, 6), and bands of N_2 $(C^3\Pi_u^-, V' \rightarrow B^3\Pi_g, V'')$ for transitions $V'=0 \rightarrow V''=3$ (3) and $V'=4 \rightarrow V''=7$ (4) in air plasma (1-4) and oxygen plasma (5,6) on loading degree. Points are the experiment. Lines are calculations. The pressure is 100 Pa, the discharge current is 80 mA..

N_2 ($X^1\Sigma_g^+$) and O_2 ($X^3\Sigma_g^-$) are slightly changed as well. The difference of rate constants of processes of electron impact than higher than E/N is lower and process threshold is higher [15]. For example, the rate constant of $O(3p^3P)$ excitation with electron impact (threshold is 10.99 eV) differ by a factor of 1.5 - 4 depending on the discharge external parameters (E/N and mole fraction of gaseous products).

The adequacy of composition calculations were checked by comparison measured and calculated absolute intensities of atomic lines of oxygen and bands of molecular nitrogen (Figure 8).

Experimental data on kinetic regularities of polymer etching and parameters of non-equilibrium plasma at difference polymer square does not contradict to data of studies[16, 17].

4. Conclusions

Thereby, the loading effect was studied for plasmas of air and oxygen. Gaseous products of polymer etching influence strongly the electrical-physical parameters of discharge and kinetic regularities of plasma-chemical etching. This fact is necessary to take into consideration at further studies and development of new technological processes.

Acknowledgements

This work was supported by the Russian Foundation for Basic Research (RFBR) (Project # 16-32-00404). The authors would like to thank Professor V.V. Rybkin and Smirnov S.A. from Ivanovo State University of Chemistry and Technology for helpful discussions and comments.

References

[1] B. Lamontegne, A.M. Wrobel, G. Jelbert, and M.P. Wertheimer. Large - area microwave plasma etching of polyimide. *J. Phys. D: Appl. Phys.*, 20(7):844–850, 1987.

[2] D. Economau, E.S. Audit, and G. Barno. In situ monitoring of etching uniformity in plasma reactors. *Solid State Technology*, 34(4):107–111, 1991.

[3] V.V. Rybkin and V.A. Titov. Kinetics and mechanisms of oxidative plasma interaction with polymers. In V.Ye. Fortov YU.A. Lebedev, N.A. Plate, editor, *Entsiklopediya nizkotemperaturnoi plazmy*, volume VIII-1. Chemistry of low-temperature plasma. of *B*, page 130. Moscow: Nauka, 2005.

[4] A. M. Diamy, J. C. Legrand, V.V. Rybkin, and S.A. Smirnov. Experimental study and modelling of formation and decay of active species in an oxygen discharge. *Contrib. PlasmaPhys.*, 45(1):5–21, 2005.

[5] S.A. Smirnov, V.V. Rybkin, A.N. Ivanov, and V.A. Titov. The simulation of the processes of formation and decay of neutral particles in dc discharge plasma in an argon-oxygen mixture. *High Temperature*, 45(3):291–297, 2007.

[6] S.A. Smirnov, V.V. Rybkin, and I.V. Kholodkov. Simulation of the processes of formation and dissociation of neutral particles in air plasma: Vibration kinetics of ground states of n2, o2 and no molecules. *High Temperature*, 40(2):161–165, 2002.

[7] Kajita S., Ushiroda S., and Kondo V. Influence of the dissociation process of oxygen on the electron's warm parameters in oxygen. *J. Appl. Phys.*, 67(9):4015, 1990.

[8] Y. Itikawa. Cross section for electron collision with carbon dioxide. *J. Phys. Chem. Ref. Data*, 31(3):749, 2002.

[9] J.E. Land. Electron scattering cross sections for momentum transfer and inelastic excitation in carbon monoxide. *Journal of Applied Physics*, 49:5716, 1978.

[10] Rybkin V.V., Titov V.A., and Holodkov I.V. Kinetic characteristics and cross section for electron collision with water molecules. *J. Izv. Vyssh. Uchebn. Zaved. Khim. Khim. Technol.*, 51(3):617, 2008.

[11] Nishimura H. Yoshino M. Tawara H., Itikawa Y. Cross sections and related data for electron collisions with hydrogen molecules and molecular ions. *J. Phys. Chem. Ref. Data*, 19(3):617, 1990.

[12] Gordiets B.F., Ferreira C.M., Guerra V.L., and et al. Kinetic model of a low-pressure $n_2 - o_2$ flowing glow discharge. *IEEE Transactions on Plasma Science*, 23(4):750, 1995.

[13] Rybkin V.V., Titov V.A., and Holodkov I.V. Kinetic characteristics and cross section for electron collision with no. *J. Izv. Vyssh. Uchebn. Zaved. Khim. Khim. Technol.*, 52(12):3, 2009.

[14] Kuvaldina E.V., Shutov D.A., Rybkin V.V., and Smirnov S.A. Kinetics of gaseous product formation in the surface treatment of polypropylene with nitrogen-oxygen plasmas. *High Energy Chemistry*, 38(3):200, 2004.

[15] Rybkin V.V., Smirnov S.A., and Titov V.A. Populations of lower vibrational levels of n_2 ($x^1\sigma_g^+$) and their effect on some characteristics of electrons in air plasma. *High Energy Chemistry*, 31(5):352, 1997.

[16] Titov V.A., Rybkin V.V., and Smirnov S.A. Physicochemical processes in the nonequilibrium plasma-polymer system. *High Energy Chemistry*, 43(3):172, 2009.

[17] S.A. Smirnov, V.A. Titov, and V.V. Rybkin. Influence of heterogeneous physicochemical processes on the parameters of low-temperature plasma. *Russian Journal of General Chemistry*, 85(5):1260, 2015.

AUTHOR INDEX

AUTHOR INDEX

AUTHOR INDEX

Brno University of Technology
c/o Milada Bartlová - Department of Physics
Technical 2848/8, Zabovresky, 61600
Brno, Czech Republic

CURRAN ASSOCIATES INC.
proceedings.com

ISBN 978-1-5108-9536-2

9 781510 895362